Squaring the Circle

Geometry in Art and Architecture

PAUL A. CALTER

Professor of Mathematics Emeritus
Vermont Technical College

Visiting Scholar
Dartmouth College

WILEY
Publishers Since 1807

To order books or for customer service, please call 1(800)-CALL-WILEY (225-5945).

Printed in the United States of America.

ISBN-13   978-0-470-41212-1

10   9   8   7   6   5   4   3   2

Contents

Preface

Why mention art and mathematics in the same breath, let alone combine them in one book? Let me start by presenting what others have said about the intersection of art and math.

William Ivins, writer and former curator at the Metropolitan Museum of Art, in the Preface to *Art & Geometry*: *"Seen in perspective, art, science, and philosophy are expressions of the same basic intuitions."*

The Greek philosopher Aristotle, in his *Metaphysics*: *"The mathematical sciences in particular exhibit order, symmetry, and limitation; and these are the greatest forms of the beautiful."*

The famous English architect Christopher Wren, in *Parentalia*: *"Geometrical figures are naturally more beautiful than irregular ones; the square, the circle are most beautiful, next the parallelogram and the oval . . ."*

The great English philosopher and logician Bertrand Russell: *"Mathematics possesses not only truth, but supreme beauty—a beauty cold and austere, like that of sculpture, without appeal to any part of our weaker nature . . . sublimely pure, and capable of stern perfection such as only the greatest art can show."*

The English analyst and number theorist Godfrey H. Hardy: *"The mathematician's patterns, like the painter's or poet's, must be beautiful; the ideas, like the colors or the words, must fit together in a harmonious way."*

Wassily Kandinsky, Russian painter: *"The final abstract expression of every art is number."*

Henri Poincaré, French mathematician, physicist, astronomer, and philosopher: *"A scientist worthy of the name, above all a mathematician, experiences in his work the same impressions as an artist; his pleasure is as great and of the same nature."*

Howard Levine, American mathematician, writing in *Leonardo*, 1994: *"Mathematicians and artists are engaged in the ultimate creative activity—creating something out of nothing."*

Jay Kappraff, American mathematician, writing in *Connections*: *"Geometry is the language of the arts and sciences."*

Clearly, putting art and architecture and geometry together is not a new idea, and many books have already been written bridging these topics. As part of the *Mathematics Across the Curriculum* movement, the goal of *Squaring the Circle: Geometry in Art and Architecture* is, in fact, to square the circle between art and architecture and geometry. To square the circle has the geometric meaning of finding a square that has the same perimeter as that of a given circle, but also the symbolic meaning of the reconciliation of opposites—heavenly and earthly, rational and irrational, and so on. Our squaring of the circle will be to show the interconnectedness between two fields of endeavor that at first seem so dissimilar.

There are many possible ways to organize the material in a book such as this, but we have chosen a geometrical organization. Starting with the one-dimensional point, we move through two and three dimensions and end with fractal dimensions. Of the various geometries, we give the most attention to plane and solid Euclidean geometry. However, we also provide a brief treatment of analytic geometry in connection with the conic sections, mention projective geometry when we cover perspective, and give a short treatment of fractal geometry in the late twentieth century.

Topics from art and architecture are grouped with the mathematical topic deemed most suitable, providing examples of the practical application of mathematics. Some chapters have more art than math, and some are the reverse. We did not try to force an artificial balance between the two. In a traditional math book, not every math topic has a direct application; some topics are included to build a foundation. Similarly, we have incorporated art ideas that are not directly traceable to mathematics, which are included for completeness. Within the mathematics framework, we have tried to preserve the chronological order of historical topics as much as possible. Because many of the chapters and even some sections within chapters could be the subjects for entire books, we have had to be economical with space.

With a few exceptions, we focus primarily on Western art and architecture. Existing courses can vary widely in content, as will those just emerging. Our intent is to include the most relevant topics from which instructors can pick and choose in order to design the course of their dreams. Instructors may pick the sections they find most useful, skip others, and fill in where needed with material from our extensive Sources and Bibliography.

Key Features

Our goal was to produce a mathematically accurate and approachable textbook that students, instructors, and general readers alike would enjoy reading and benefit from using. Throughout the text, a variety of key features have been included to help demonstrate the relevance of geometry in art and architecture and the symbiotic relationship between the two areas of study.

Mathematical concepts are explained geometrically, in the language of artists and designers, wherever possible. We aim for an intuitive approach rather than a rigorous approach.

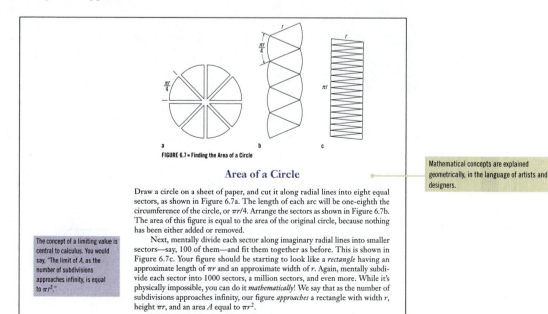

FIGURE 6.7 ▪ Finding the Area of a Circle

Mathematical concepts are explained geometrically, in the language of artists and designers.

Area of a Circle

Draw a circle on a sheet of paper, and cut it along radial lines into eight equal sectors, as shown in Figure 6.7a. The length of each arc will be one-eighth the circumference of the circle, or $\pi r/4$. Arrange the sectors as shown in Figure 6.7b. The area of this figure is equal to the area of the original circle, because nothing has been either added or removed.

Next, mentally divide each sector along imaginary radial lines into smaller sectors—say, 100 of them—and fit them together as before. This is shown in Figure 6.7c. Your figure should be starting to look like a *rectangle* having an approximate length of πr and an approximate width of r. Again, mentally subdivide each sector into 1000 sectors, a million sectors, and even more. While it's physically impossible, you can do it *mathematically*! We say that as the number of subdivisions approaches infinity, our figure *approaches* a rectangle with width r, height πr, and an area A equal to πr^2.

The concept of a limiting value is central to calculus. You would say, "The limit of A, as the number of subdivisions approaches infinity, is equal to πr^2."

Over 300 full-color photographs and fine art images and 800 illustrations are beautifully integrated throughout the text, demonstrating mathematical concepts and their practical applications in art and architecture. The majority of these images, and all of the illustrations, are also available for download from the accompanying Web Resource Center, making it easy for instructors to incorporate them in the classroom, or for students to study them in greater detail on their own.

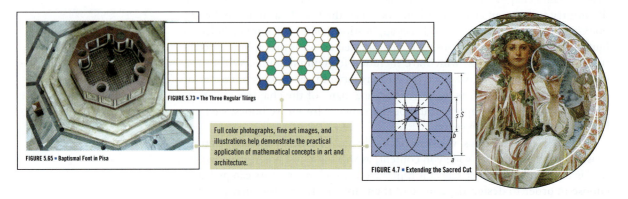

FIGURE 5.65 ■ Baptismal Font in Pisa

FIGURE 5.73 ■ The Three Regular Tilings

Full color photographs, fine art images, and illustrations help demonstrate the practical application of mathematical concepts in art and architecture.

FIGURE 4.7 ■ Extending the Sacred Cut

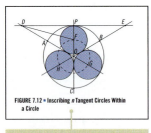

FIGURE 7.12 ■ Inscribing *n* Tangent Circles Within a Circle

The text features a myriad of geometric constructions, which can also be downloaded from the accompanying Web Resource Center for further study.

The text features all the basic geometric constructions including bisecting a line segment, erecting a perpendicular, and so forth, and then gives additional constructions where appropriate. These constructions are also available for download on the accompanying Web Resource Center, so students can study them in greater detail. Constructions that we chose not to include in the text, in order to save space, are suggested as projects. An index to all constructions included in the text is provided in Appendix C.

Each chapter opens with a time line highlighting the historical topics and events students will be studying in the chapter. This feature gives students a brief insight into the relationship between important events in both mathematics and art and architecture in their chronological sequences.

Each chapter opens with a time line highlighting important events in both mathematics and in art and architecture.

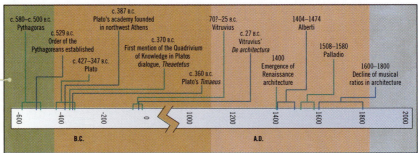

Mathematical equations, formulas, and statements appear after green numbered boxes throughout the text. All of these are collected into the Summary of Facts and Formulas in Appendix F, helping to facilitate reference and review.

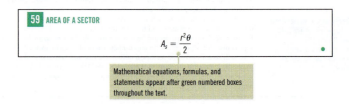

59 AREA OF A SECTOR

$$A_s = \frac{r^2\theta}{2}$$

Mathematical equations, formulas, and statements appear after green numbered boxes throughout the text.

Profiles of the historical figures in mathematics and in art and architecture who played key roles in developing the ideas discussed in this book are presented in boxes at appropriate points in the text. An index to these profiles is located in Appendix B.

To help students develop their problem-solving skills, fully worked examples and accompanying practice exercises are provided in every chapter.

■ **EXAMPLE:** What is the area of a sector that has a central angle of 0.885 rad and a radius of 8.35 cm?

● **SOLUTION:**

$$A_s = \frac{(8.35 \text{ cm})^2 (0.885 \text{ rad})}{2} = 30.9 \text{ cm}^2 \qquad ■$$

Fully worked examples and accompanying practice exercises help develop problem-solving skills.

---

12. Find the area of a sector having a central angle of 0.772 rad in a circle of radius 6.93.
13. Find the area of a sector having a central angle of 55.6° in a circle of radius 77.6.
14. Find the length *AB* of the flying buttress illustrated in Figure 6.15.

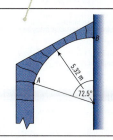

LINES INTERSECTING A CIRCLE

So far we've discovered some interesting facts about the circle itself. However,

---

End-of-chapter Exercises and Projects contain a wide variety of exercises and project ideas that review key concepts for each chapter and encourage students to investigate topics in greater depth. Selected answers to exercises are available in the back of the book, and complete solutions to exercises are available to instructors only on the accompanying Web Resource Center.

Exercises range in difficulty from basic skill-building problems to Mathematical Challenges. Mathematical Challenges aim to extend students' mathematical ability, and although they may be more difficult than other exercises, students with good backgrounds in high school mathematics should be able to work them.

EXERCISES AND PROJECTS

1. Define or describe the following terms:

    mandala   *rota fortuna*   *vanitas*
    tondo   roundel   plaquette

2. Measure the diameters of an assortment of circular items (soup cans, jar lids, etc.). Then, use a strip of paper to measure the circumference of each item. Divide each circumference by the corresponding diameter. Write a

---

Mathematical Challenges

33. *Transformation of Areas.* A problem of interest to the Pythagoreans was how to transform, with compass and straightedge, a polygon into a triangle or square of equal area. This shape is represented in Figure 5.93. Draw an irregular polygon and use the following method to construct, with compass and straightedge, a triangle of equal area:

    ■ Given polygon *ABCDE*, draw line *AC*.
    ■ Draw a parallel to *AC* from *B*, cutting *DC*, extended, at *F*.

    Because triangles *ABC* and *AFC* have the same base *AC* and the same altitude drawn to this base, they are equal in area. That means that quadrilateral *AFDE* has the same area as the original pentagon, but it has one less side. Repeat the process until you find a triangle that has the same area as the original polygon.

34. On page 171 of *The Painter's Manual*, Dürer explains transformation of areas. Compare his method to the one listed in Exercise 33, and write a few paragraphs of explanation.

35. *The Five-Disk Problem.* Five disks, each with radius one unit, are placed as in Figure 5.94. Their points of intersection form the vertices of a regular pentagon, and the circles all pass through *O*, the center of the pentagon. The five disks together cover a circular area of radius *OA*. Find *OA*.

36. *The Pythagorean Theorem Generalized.* The Pythagorean theorem states that the area of a *square* constructed on the hypotenuse of a right triangle equals the sum of the areas of the *squares* on the other two legs. What would happen if we replaced the *squares* in that statement with *regular polygons*?

    a. Use a computer drafting program to construct regular pentagons on the three sides of a right triangle (Figure 5.95), measure their areas, and see if the Pythagorean theorem, generalized, holds.

... w a circle. On-screen, display the ... of the circle, and the ratio of ... g the circle to different sizes,

FIGURE 5.94 ■ The Five-Disk Problem[29]

End-of-chapter Exercises review key concepts and extend students' mathematical ability.

FIGURE 5.95 ■ The Pythagorean Theorem Generalized

Biographical profiles provide more information about key persons in both mathematics and art and architecture.

Some Projects are short, and others are long enough to serve as term projects. They are not marked as such because the same project idea can often be implemented as a short task or expanded into a more substantial piece of work. Students may choose to do one of the given projects, or use them to jog their imaginations into creating a project of particular interest to them. This should be encouraged. The projects can serve as vehicles for individual students to delve deeper into a topic than is possible in the text itself because of its broad coverage.

To encourage students to work together, many projects have been identified as Team Projects, which require collaborative learning. These include outdoor projects, such as laying out geometric figures with ropes and pegs, and indoor projects requiring a large effort best shared by a team of students.

For courses utilizing technology, practice exercises and end-of-chapter Exercises and Projects that can be completed by computer, particularly a CAD program such as The Geometer's Sketchpad, are included in the text. Instructors who see the value in doing at least some of the constructions by hand can direct their students to do so.

Team Projects

33. Use a long knotted rope to make a rope stretcher's triangle. Use it outdoors to lay out a right angle in an open space. Then make three more right angles to form a square. How accurate is your work? Did you come back to the starting point?

34. A ball field, tennis court, or basketball court has many lines that intersect at what appear to be right angles. Using only a tape measure, how can you verify that an angle is indeed 90°?

35. Use an upright or horizontal stick and similar triangles to measure the height or width of a building on campus. Check your results by direct measurement.

36. Construct a measuring device similar to the *radio astronomico* or the *bacolo of Euclid*. Use it to find the width of a building by similar triangles. Verify your results by direct measurement.

37. Check your campus or neighborhood for threefold architectural features: triangular windows, trefoils, triskelions, and so on. Make sketches, take photos, and report your findings.

38. Search books or your neighborhood for examples of the various isometries, as well as scaling. Try to find at least one example of each. Take pictures and make a presentation to your class.

Mathematical Challenges

39. Prove Kepler's statement: *If the sides of a right triangle are in geometric ratio, then the sides are* $1 : \sqrt{\Phi} : \Phi$.

40. *Trigonometric Ratios.* You probably remember the following trigonometric ratios from other mathematics classes. Explain what these ratios are, and give examples of their use.

Projects encourage students to investigate topics in greater depth and provide opportunities for collaborative learning.

4. Using a CAD program, such as The Geometer's Sketchpad, make a sketch to illustrate that the area of a triangle remains unchanged when the top vertex is dragged along a line parallel to the base (Figure 3.75).

5. Demonstrate by construction and measurement that Hero's formula works.

6. Without using the Pythagorean theorem, demonstrate that the 3-4-5 triangle is a right triangle.

7. Show by construction that the incenter of a triangle is the center of the inscribed circle.

8. Show by construction that the circumcenter of a triangle is the center of the circumscribed circle.

9. *Napoleon's Theorem.* Napoleon is known as a brilliant military leader. Few people know that he was also a student of mathematics. In fact, a theorem that he independently solved is named for him.

35a NAPOLEON'S THEOREM

If equilateral triangles are drawn externally or internally on the sides of any triangle, their centers form an equilateral triangle.

a. Create a CAD drawing, such as Figure 3.76, illustrating Napoleon's theorem.

b. Drag a vertex to see if the theorem still holds.

FIGURE 3.75 ■ Dragging One Vertex of a Triangle

FIGURE 3.76 ■ Napoleon's Theorem

Technology exercises and projects build proficiency with CAD programs such as The Geometer's Sketchpad.

28. Write a short paper or a term paper on some topic from this chapter. Come up with your own topic, using these suggestions to jog your imagination:

- Various units of angular measure, degrees, radians, grads, minutes, seconds, etc.
- The Greek mathematician Hero of Alexandria, first century A.D.
- The Greek geometer Pappus of Alexandria, c. A.D. 300
- Various proofs of the Pythagorean theorem
- The use of the "Pythagorean" theorem before the time of Pythagoras
- Various kinds of symmetry
- The rope stretcher's triangle and its use by the Egyptians
- The sacred tetraktys and its importance to the Pythagoreans
- Various shapes of halos and their meaning
- Traditional quilt patterns using triangles, such as *Wild Goose Chase* (pattern # 3537 in Rhemel's *The Quilt I.D. Book*)
- The history of angle trisection, using the outline in Heath, p. 235, as a starting point (see Sources)
- Napoleon as a mathematician
- *Number symbolism* in general
- The symbolism of the number *three* (see Appendix E)

Research paper topics give students the opportunity to develop their writing skills.

To encourage Writing Across the Curriculum and help students develop their writing skills while simultaneously focusing on mathematics in art and architecture, numerous research paper topics are suggested. As with the projects, both long and short topics are available, enabling students to delve deeper into a topic than is possible in the text and further inspiring students to find related and interesting topics of their own.

Sources and Notes at the end of each chapter direct readers to additional information about topics mentioned in the chapter, allowing readers to further investigate topics of personal interest.

The following Appendices are included:

Art–Math Tourist, which provides an index of travel destinations and historical sites referenced in the text

Index to Biographical Profiles, which gives the text location of the various historical profiles featured throughout the text

Listing of Constructions by Chapter, which is a comprehensive index to geometric constructions in the text that can be completed by compass and straightedge, paper folding, or CAD program

Index to Geometric Signs and Symbols, which gives the text location of geometric signs and symbols relevant in both mathematics and in art and architecture

Number Symbolism and Famous Groupings in Art, which contains a wealth of additional interesting information that is related to individual numbers

Summary of Facts and Formulas, which is a comprehensive index to the mathematical equations, formulas, and statements that appear after green numbered boxes throughout the text

Complete *Bibliography*, which is a compilation of the works referred to or consulted in this text to facilitate further research

Selected Answers, which provides numerical answers to computational exercises and projects only. Students can reference these answers to check that they understand basic concepts and are prepared to work more involved or creative applications of the concepts

SOURCES

Aaboe, Asger. *Episodes from the Early History of Mathematics.* New York: Random House, 1964.
Boles, Martha, and Newman, Rochelle. *The Golden Relationship: Art, Math & Nature.* 4 Vols. Bradford, MA: Pythagorean Press.
Campbell, Joseph, with Bill Moyers. *The Power of Myth.* New York: Doubleday, 1988.
Coxeter, H. S. M. *Introduction to Geometry,* 2nd ed. Wiley, 1989.
Dürer, Albrecht. *The Painter's Manual.* Walter Strauss, trans. New York: Abaris, 1977. Original 1525.
Escher, M. C. *The Graphic Works of M. C. Escher.* New York: Ballantine, 1960.
Euclid. *The Thirteen Books of the Elements.* New York: Dover, 1956.
Eves, Howard. *An Introduction to the History of Mathematics.* New York: Holt, 1953.
Fisher, Sally. *The Square Halo.* New York: Abrams, 1995.
Grünbaum, Branko, et al. *Tilings and Patterns: An Introduction.* New York: Freeman, 1989.
Hargittai, István, ed. *Fivefold Symmetry.* New York: World Scientific, 1991.
Jones, Lesley, ed., *Teaching Mathematics and Art.* Cheltenham, UK: Stanley Thornes (Publishers), 1991.
Jung, Carl G., et al. *Man and His Symbols.* New York: Dell, 1964.
Kappraff, Jay. *Connections: The Geometric Bridge between Art and Science.* New York: McGraw-Hill, 1990.
Livio, Mario. *The Golden Ratio.* New York: Broadway Books, 2002.
March, Lionel. *Architectonics of Humanism.* Chichester, UK: Academy, 1998.
Olson, Alton T. *Mathematics Through Paper Folding.* Reston, VA: National Council of Teachers of Mathematics, 1975.
Rehmel, Judy. *The Quilt I.D. Book: 4,000 Illustrated and Indexed Patterns.* New York: Prentice-Hall, 1986.
Runion, Garth E. *The Golden Section.* Palo Alto, CA: Seymour, 1990.
Venters, Diana, et al. *Mathematical Quilts.* Emeryville, CA: Key Press, 1999.

NOTES

1. Hardy, G. H. *A Mathematician's Apology.* London: Cambridge, 1940.
2. From Coxeter's *Introduction to Geometry.*
3. Dürer, p. 147.
4. Kemp, p. 55.
5. Sir Walter Scott describing a wizard in his romantic narrative poem *Marmion* (1808).
6. Faust, Part I, line 1394.
7. Trans. by Alice Raphael, 1930.
8. Dürer, *The Painter's Manual,* p. 145. It is also found in Serlio, Book 1, Chapter 1, Folio 11. This method is originally from "After Ptolemy," *Almagest* Book 1, Chap. 9.
9. Euclid, Vol. 2, pp. 100–104.
10. Euclid, Vol. 2, p. 97.
11. Critchlow, p. 85.
12. Ben-David, Calev, "Ring of the King," *The Jerusalem Report,* Oct. 1995, p. 56.

Sources and Notes direct readers to additional research resources.

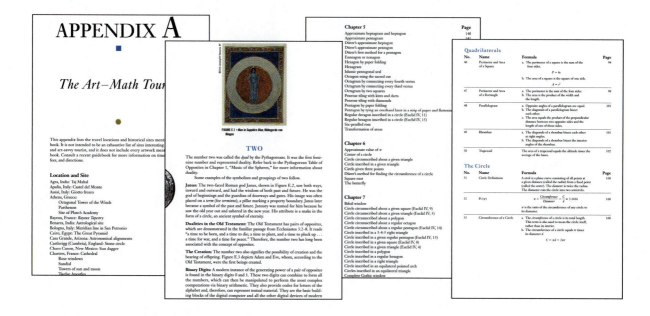

Web Resource Center

A companion Web-based Resource Center is available to students and instructors at www.wiley.com/college/mathstats/online. Key features of the site include:

- Comprehensive collection of full-color fine art and photograph images and illustrations from the text, available for download
- Interactive Geometer's Sketchpad files for selected constructions from the text, available for download
- Links to appropriate journals and organizations

In addition, complete solutions to practice exercises and end-of-chapter Exercises and Projects are available only to instructors to assist in preparing for class quickly and effectively and helping students to succeed in the classroom.

To the Student

Welcome to a fascinating journey through time in which we explore geometry and its relevance and application in the art and architecture of bygone eras through the present day. In this course, you will discover the interconnectedness between these two fields of study, which may at first seem very dissimilar—that is, you will learn to square the circle. You will have the opportunity not only to visualize, but also to experience the intrinsic beauty of geometry through a variety of constructions and projects. In addition, you will uncover the mathematical expression inherent in several well-known pieces of art and architecture, which have been reproduced in this text.

The key to your success in this course is to open yourself up so that you can look at geometry in a totally new way. To be certain, you will learn key geometric concepts and gain a solid foundation in geometry, but you will also discover how these concepts and ideas have been represented in various types of art and architecture for centuries. You will learn things that will enrich not only your mathematical knowledge, but also your creativity in many aspects of your life, from your hobbies to your travels, to perhaps even your future vocation.

Worked-out examples and short exercises are available in each section to help build your proficiency in geometry. Try to solve these problems by yourself; you will be surprised at how quickly you can master the concepts by consistently working the problems. In addition to the short exercises at the end of the sections, try the Exercises and Projects at the end of each chapter. There you will find an array of problems ranging from Mathematical Challenges, which stretch your mathematical abilities, to fantastic projects such as painting, sculpting, and even baking. You will discover that all of these interesting topics are connected to the concepts in geometry that you are learning.

As you can see, geometry isn't confined to working with numerical problems. Just as mathematical expressions can be conveyed through art and architecture, they can also be expertly communicated in essay form. Writing well is a crucial skill in any field, and we encourage you to try the writing projects in the Exercises and Projects at the end of each chapter. Similarly, the significance of technology is apparent in the world around us—at home, in school, and chances are, in your future vocation. We have included a variety of technology exercises in which you can explore The Geometer's Sketchpad (or a similar CAD program) and other programs to re-create some of the constructions you learned in the text and to find new applications of your own.

The text has several other features that you will find helpful for your studies now, and they will remain useful references long after you have finished this course. The Sources and Notes at the end of each chapter are fantastic resources for researching additional information about the topics presented in the chapter. We encourage you to use them to find out more about topics that are personally interesting to you. In addition, the Appendices of this text include a wealth of interesting information, useful indexes, and references to help you study. Take a moment to refer back to the Preface, which gives a brief summary of the content in the Appendices. In particular, be sure to take advantage of Appendix F, the

Summary of Facts and Formulas, which is a helpful compilation of the key geometric concepts and formulas in the text.

For additional resources, visit the Web Resource Center for this text at www.wiley.com/college/mathstats/online. There you will find a comprehensive collection of the photographs and fine art images, illustrations, and interactive Geometer's Sketchpad constructions from the text, available for you to study in greater detail or use to enhance your projects. Simply download the image or construction file you are interested in, enlarge it to the desired size to see the fine details, and print it out for study. In addition, there are several good project ideas that build on the images and The Geometer's Sketchpad constructions discussed in the text.

We hope you find learning about geometry and art and architecture using this text to be an enjoyable experience, and that you take the opportunity to try something new, explore the possibilities, and see things from a different point of view.

To the Instructor

This textbook is an outgrowth of the *Mathematics Across the Curriculum* movement, which has inspired such courses and books as Mathematics and Music, Mathematics and Textile Design, and many others. The material for *Squaring the Circle: Geometry in Art and Architecture* was developed while the author was teaching courses in Art and Geometry under a grant from the National Science Foundation at Dartmouth College. During that time, it became clear that if other similar courses were to follow, a text would be needed to bring many of the required elements together. There had to be a solid core of geometry, of course, as well as extensive examples from the history of art and architecture. Especially important were specific meaningful activities for students to do on their own: constructions, derivations, model making, art projects, suggestions for papers, and so forth. These activities had to be of sufficient quantity for a large class of diverse students, and of high quality to be interesting and motivating to all students.

The emphasis on both mathematics and art and architecture makes the text compatible with a variety of courses, and it can be used in several different ways. Because it includes all the topics needed to give a foundation in geometry, it can be used for a basic mathematics course for liberal arts or art students. No college mathematics prerequisites are required, but a high school mathematics background is assumed. For such a course, a student's interest in art and architecture may serve as an incentive to study related mathematical topics.

Students with some mathematics background, or who are pursuing further study in mathematics, might take a course such as Geometry in Art and Architecture, perhaps offered as a general education elective. In such a case, the student's knowledge of mathematics might be the hook to introduce topics from art and architecture that use mathematical ideas.

An interdisciplinary course combining art, mathematics, and architecture would normally require two instructors: one for art and architecture and one for mathematics. With this text and accompanying Web Resource Center, one instructor could do the job alone. The text and Web-based resources can also serve as classroom supplements in a more traditional mathematics or art history class. A mathematics instructor can use the book to augment a mathematics class with relevant examples from art history, or conversely, an art instructor can use it to augment an art or art history class with relevant examples from mathematics.

In addition to providing students with a solid understanding of mathematical concepts, we encourage instructors to use this text to guide students into new areas of study and experience. The pedagogy is enhanced with an impressive collection of geometric constructions and full-color photograph and fine art images, giving students a visual understanding of key concepts and their relevance in art and architecture throughout the centuries. The majority of images and all of the illustrations and constructions found in the text can also be downloaded from the accompanying Web Resource Center and inserted directly into your PowerPoint presentation, or printed for classroom handouts. Further, the variety of exercises, projects, and writing assignments encourage students to expand their knowledge and express their learning in new and creative ways. For example, students are given the opportunity to work with their hands or with CAD technology to make models, sculptures, and create their own constructions.

Finally, this book will serve as a life-long *resource* for both students and instructors. To this end, it contains Sources and Notes at the end of each chapter and an extensive Bibliography for readers interested in further researching topics of personal interest. The Appendices are also a fantastic resources containing a wealth of interesting and relevant information and quick references back to useful tools in the text.

To assist you in preparing for your course quickly and effectively, be sure visit the accompanying Web Resource Center for this text at www.wiley.com/college/mathstats/online. In addition to a comprehensive collection of the images and constructions found in the text available for download, you will find an array of additional resources, such as the complete solutions to the end-of-section Exercises and the end-of-chapter Exercises and Projects and tips on how to use particular GSP constructions, when appropriate.

We hope that this text and the accompanying Web resources empower you to create just the right course to fit your students' needs and interests, and further enable them to successfully square the circle between geometry and art and architecture.

Acknowledgments

The author and publisher wish to thank the following reviewers for their feedback and contributions.

Pete Anderson, Hermitage High School, Henrico County, VA

Charles R. Dyer, University of Wisconsin–Madison, WI

Diane Favro, University of California at Los Angeles, CA

Rachel Fletcher, New York School of Interior Design, New York, NY

Stephen I. Gendler, Clarion University of Pennsylvania, Clarion, PA

Catherine A. Gorini, Maharishi University of Management, Fairfield, IA

Jay Kappraff, New Jersey Institute of Technology, Newark, NJ

Diane Porter, Troy University, Troy, AL

Sr. Barbara E. Reynolds SDS, Cardinal Stritch University, Milwaukee, WI

Mark A. Reynolds, Academy of Art University, San Francisco, CA

Jim Rose, MFA, Clarion University of Pennsylvania, Clarion, PA

Mark Schlatter, Centenary College of Louisiana, Shreveport, LA

Stephen R. Wassell, Sweet Briar College, Sweet Briar, VA

Carol Martin Watts, Kansas State University, Manhattan, KS

The author would also like to thank Dorothy Wallace of Dartmouth College, Principal Investigator for the *Mathematics Across the Curriculum* project, for her support and encouragement, and the team at John Wiley & Sons: Annie Mac, Allyndreth Devlin, Mae Lum, Martha Emry, Richard Bonacci, and Mike Simpson, for their creativity, thoroughness, and attention to detail, which resulted in this beautiful final result.

About the Author

Paul A. Calter is a Visiting Scholar at Dartmouth College and Professor Emeritus of Mathematics at Vermont Technical College. He is a book review editor of the *Nexus Network Journal* and has interests in both the fields of mathematics and art. He received his B.S. from Cooper Union and his M.S. from Columbia University, both in engineering, and his Masters of Fine Arts Degree from Norwich University. Calter has taught mathematics for over twenty-five years and is the author of ten mathematics textbooks and a mystery novel. He has been an active painter and sculptor since 1968, has had many solo shows and participated in dozens of group art shows, and has permanent outdoor sculptures at a number of locations. Calter developed a course called "Geometry in Art & Architecture," which he has taught at Dartmouth College and Vermont Technical College, and he has given workshops and lectures on the subject. Calter's own art is concerned with astronomical and geometric themes; he searches for a link between the organic and geometric basis of beauty, what has been called the *philosopher's stone of aesthetics*.

Squaring the Circle

"There is geometry in the humming of the strings . . .

. . . there is music in the spacing of the spheres."

PYTHAGORAS

FIGURE 1.1 ■ Pythagoras in Raphael's *School of Athens* (detail)
©Scala/Art Resource, NY

1

Music of the Spheres

We're about to set out on a wonderful journey. We will follow parallel paths through time, from ancient Egypt to the twentieth century; through foreign lands, their art and architecture and their people who have shaped mathematics throughout the centuries; through the geometric figures, from the humble zero-dimensional point to the fractionally dimensioned Sierpinski tetrahedron; and through the geometries, from ancient plane and solid Euclidean geometry to modern fractal geometry. Although the paths are different, we'll try to weave them together into one coherent journey. Along the way, you will be invited to perform simple mathematics exercises, create two- and three-dimensional artworks, write various papers, use a computer, and make a model. At times, you will be encouraged to work in a team or give a presentation to your class. Pick any of these activities that match your own interests.

Our journey begins with Pythagoras and his followers, the Pythagoreans (Figure 1.1), whose ideas dominate much of the material in this book. A great deal of what has been written about Pythagoras and his followers is more myth and legend than historical fact, and we will not try to separate fact from legend

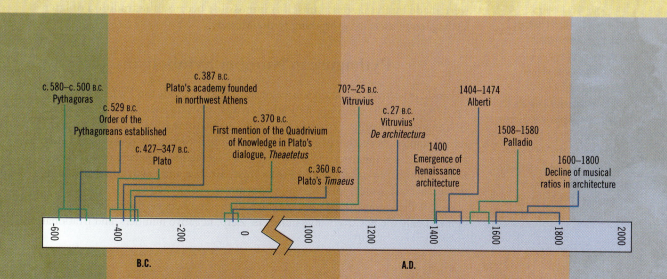

c. 580–c. 500 B.C.
Pythagoras

c. 529 B.C.
Order of the
Pythagoreans established

c. 427–347 B.C.
Plato

c. 387 B.C.
Plato's academy founded
in northwest Athens

c. 370 B.C.
First mention of the Quadrivium
of Knowledge in Plato's
dialogue, *Theaetetus*

c. 360 B.C.
Plato's *Timaeus*

70?–25 B.C.
Vitruvius

c. 27 B.C.
Vitruvius'
De architectura

1400
Emergence of
Renaissance
architecture

1404–1474
Alberti

1508–1580
Palladio

1600–1800
Decline of musical
ratios in architecture

-600 -400 -200 0 1000 1200 1400 1600 1800 2000

B.C. A.D.

here. Instead, we will introduce several discoveries and contributions the Pythagoreans made to present-day mathematics, keeping in mind that we are talking about what *possibly* happened in the fifth century B.C. and what is sometimes called the Pythagorean tradition. We will present a few mathematical ideas from what commentators have described as a much larger philosophical system encompassing ethics, logic, politics, religion, medicine, and so forth.

In this chapter, we will explore the Pythagorean idea of the music of the spheres, which is later used by Plato to describe the creation of the world and appears again in Renaissance architecture. In preparation for the study of the music of the spheres, we will introduce the mathematical concepts of ratio and proportion, which will be essential in demonstrating the origin of the musical ratios, and further, the mathematical construction of the musical scale. We will also explore sequences, means, and some basic geometric constructions that will help us understand how the same mathematical concepts were later used in Renaissance architecture. So let's begin our journey and make some fascinating mathematical discoveries along the way.

PYTHAGORAS AND THE PYTHAGOREANS

According to Pythagorean legend, Pythagoras was one of the greatest philosophers and mathematicians of his time. He is said to have been driven from the island of Samos by his disgust for Polycrates, the tyrant of Samos. In 529 B.C., Pythagoras made the relatively short trip from Samos to Crotone, a Dorian Greek colony in southern Italy. There he gained a large following of students and disciples. He started an academy that gradually formed into a society or brotherhood called the *Order of the Pythagoreans*. The Pythagorean disciplines were said to include silence, music, incenses, physical and moral purification, rigid cleanliness, vegetarianism, pure linen clothes, self-denial, utter loyalty, common possessions, and secrecy, to prevent the Pythagoreans' knowledge from coming into the possession of the profane. These disciplines may have been the roots of disciplines for later monastic orders, such as the Jesuit *Rules of St. Ignatius*.

The works of Pythagoras are known only through the work of his disciples. The Pythagoreans relied on oral teaching, and perhaps their pledge of secrecy accounts for the lack of documents. The oral teachings were eventually committed to writing, but knowing just how many of the "Pythagorean" discoveries were made by Pythagoras himself is impossible because the tradition of later Pythagoreans ascribed everything to the Master.

Pythagorean Number Symbolism

Aristotle is perhaps the main source of information about the Pythagoreans. In his *Metaphysica*, he sums up the Pythagoreans' attitude toward numbers. "The [Pythagoreans were] . . . the first to take up mathematics . . . [and] thought its principles were the principles of all things. Since, of these principles, numbers . . . are the first, . . . in numbers they seemed to see many resemblances to the things that exist . . . more than [just] air, fire and earth and water, [but things such as] justice, soul, reason, opportunity. . . ."

The Pythagoreans did not recognize all the numbers we use today; they recognized only the positive whole numbers. Zero, negative numbers, and irrational

---

PYTHAGORAS
(c. 580–c. 500 B.C.)

Pythagoras was born in Ionia on the island of Samos. One of his teachers was Thales, who is called the father of Greek mathematics, astronomy, and philosophy and who was one of the Seven Sages of Greece.

Following Thales' advice, Pythagoras went to Egypt in his early twenties. According to Iamblichus, "he frequented all the temples . . . , [did] the most studious research . . . [and] he did not neglect any contemporary celebrity, whether a sage renowned for wisdom, or a peculiarly performed mystery. He did not fail to visit any place where he thought he might discover something worthwhile . . . [and] visited all the Egyptian priests, acquiring all the wisdom each possessed. Thus he passed twenty-two years in the sanctuaries of temples, studying astronomy and geometry, and being initiated . . . in all the mysteries of the gods."[1]

He spent another 12 years in Babylonia where he was "associated with the Magi . . . [and where] he studied and completed arithmetic, music, and all the other sciences." He returned to Samos at the age of 56.

numbers didn't exist in their system. Moreover, the Pythagoreans' idea of a number was different from the quantitative one we have today. Now we use a number to indicate a quantity, amount, or magnitude of something, but for the Pythagoreans, each number had its own particular attribute. For example, the number one, or unity, which they called the *monad*, was seen as the source of all numbers. "Unity is the principle of all things and the most dominant of all that is: All things emanate from it and it emanates from nothing. It is indivisible and . . . it is immutable and never departs from its own nature through multiplication ($1 \times 1 = 1$). Everything that is intelligible and not yet created exists in it. . . ."[2]

The number two, the *dyad*, represented duality, subject and object. The Pythagoreans believed the world to be composed of pairs of opposites, as given by Aristotle in this famous table.[3]

The Pythagorean Table of Opposites				
Limit, Unlimited	Odd, Even	One, Plurality	Right, Left	Male, Female
At Rest, Moving	Straight, Crooked	Light, Darkness	Good, Bad	Square, Oblong

With three, the *triad*, that dualism was resolved. The two extremes were united, giving *Harmonia*. This idea of *reconciliation of opposites* will appear again in the discussions of the golden mean and in Chapter 8, "Squaring the Circle." Below is a summary of some of the sometimes fanciful attributes the Pythagoreans gave to numbers:

1. *Monad.* Point. One is the source of all numbers. It is good, desirable, essential, and indivisible.

2. *Dyad.* Line. Two represents diversity, duality, a loss of unity, the number of excess and defect. It is the first feminine number.

3. *Triad.* Plane. By virtue of the triad, unity and diversity of which it is composed are restored to harmony. Three is the first odd, masculine number.

4. *Tetrad.* Solid. This is the first feminine square. It represents justice, and it is steadfast and square. Four is the number of the square, the elements, the seasons, ages of man, lunar phases, and virtues.

5. *Pentad.* This is the masculine marriage number, uniting the first female number and the first male number by addition. It is the number of fingers or toes on each limb and the number of regular solids or polyhedra. It is considered incorruptible because multiples of 5 end in 5.

6. *Hexad.* The first feminine marriage number, uniting 2 and 3 by multiplication. It is the area of a 3-4-5 triangle. It is the first *perfect number*, a number equal to the sum of its exact divisors or factors, except itself. Therefore, $1 + 2 + 3 = 6$.

7. *Heptad.* Seven is referred to as the virgin number, because 7 alone has no factors, and 7 is not a factor of any number within the Decad. Also, a circle cannot be divided into seven parts by any known construction.

8. *Ogdoad.* The first cube.

9. *Ennead.* The first masculine square. Nine is incorruptible, because when it is multiplied by any number it "reproduces" itself. For example, $9 \times 6 = 54$ and $5 + 4 = 9$. Try this again by multiplying 9 by any number, however large.

10. *Decad.* Ten is the number of fingers or toes on a human. It contains all of the numbers; after 10, the numbers merely repeat themselves. It is the sum of the archetypal numbers ($1 + 2 + 3 + 4 = 10$).

Odd numbers were considered masculine; even numbers were feminine. Odds were considered stronger than evens because (a) unlike an odd number, when an even number is halved it has nothing in the center; (b) odd plus even always gives odd; and (c) two evens can never produce an odd, while two odds produce an even. Because the birth of a son was considered more fortunate than the birth of a daughter, odd numbers became associated with good luck. "The gods delight in odd numbers," wrote Virgil in his *Eclogue* viii.

The Renaissance architect Leon Battista Alberti later wrote about odd and even numbers in architecture. "[The Ancients] observed, as to Number, was that it was of two Sorts, even and uneven, and they made use of both . . . for they never made the Ribs of their Structure, that is to say, the Columns, Angles and the like, in uneven Numbers; as you shall not find any Animal that stands or moves about upon an odd Number of Feet. On the contrary, they made their Apertures always in uneven Numbers, as Nature herself has done in some Instances . . . the great Aperture, the Mouth, she has set singly in the Middle."[4] We will have more to say about Alberti later in this chapter.

Figured Numbers

The Pythagoreans represented numbers by using patterns of dots, possibly as a result of arranging pebbles into patterns, as shown in Figure 1.2. There were square numbers, rectangular numbers, and triangular numbers.

Associated with figured numbers is the idea of the gnomon. *Gnomon* means "carpenter's square" in Greek. It is the name given to the upright stick on a sundial. For now, we will use the word *gnomon* as the Pythagoreans did, to refer to an L-shaped border appended to a figured number.

In Figure 1.3a successive L-shaped borders, or gnomons, are added to the monad. To get some idea of Pythagorean thinking, first note that for the monad, the number of points in each gnomon is odd and each successive figure formed is a square. In Figure 1.3b, successive gnomons are added to the dyad. The number of points in each gnomon added to the dyad is even, and each successive figure is a rectangle or oblong. Therefore, the Pythagoreans, in their Table of Opposites, associated odd and even with square and oblong, respectively. Further, each gnomon about the monad forms a square, a stable form whose ratio of width to height never changes. By contrast, each gnomon about the dyad forms a rectangle whose ratio of width to height changes each time. As such, the Pythagoreans associated *limited* with odd and *unlimited* with even.

From these patterns, the Pythagoreans derived relationships between numbers that may have led to the discovery of geometrical theorems. For example, noting that a square number can be subdivided by a diagonal line into two triangular numbers, we can say that a square number is always the sum of two triangular numbers. For example,

$$25 = 10 + 15.$$

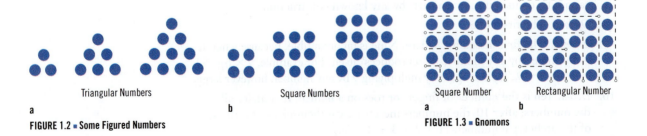

Triangular Numbers   Square Numbers   Square Number   Rectangular Number

a   b   a   b

FIGURE 1.2 ■ Some Figured Numbers   **FIGURE 1.3 ■ Gnomons**

The observation that figured numbers follow certain patterns may have furthered the Pythagorean belief that the study of numbers would lead to the discovery of universal laws.

The Sacred Tetraktys

The triangular number ten, or decad, was especially important to the Pythagoreans and was called the *Sacred Tetraktys*. The prefix *tetra-* means four, and the word *Tetraktys* means a "set of four things." Ten dots form a neat equilateral triangle with four dots on each side, as shown in Figure 1.4.

Ten is important because it is, of course, the number of fingers and the base of the decimal number system. The Pythagoreans also saw significance in ten being the sum of the first four integers:

$$1 + 2 + 3 + 4 = 10.$$

Recall that in Pythagorean number symbolism, the number *one* represented the point, *two* the line, *three* the surface, and *four* the solid. To the Pythagoreans, the Tetraktys represented the continuity linking the dimensionless point with the solid body.

In addition, the Tetraktys, like other triangular numbers, is composed of both odd and even integers (1, 2, 3, and 4). This is in contrast to the square, which is composed of consecutive odd integers only, and the rectangle, which is composed of consecutive even integers only. Because the Pythagoreans felt that the universe was composed of an interweaving of both odd and even, limited and unlimited, they associated the Tetraktys with the cosmos.

The word *Tetraktys* is attributed to Theon of Smyrna (A.D. 100), a Greek mathematician and astronomer. The Sacred Tetraktys was not the only interesting set of four. Here are the ten sets of four given by Theon:

Numbers	1	2	3	4
Geometry	point	line	surface	solid
Elements	fire	air	water	earth
Solids	pyramid	octahedron	icosahedron	cube
Living things	seed	growth in length	growth in breadth	growth in thickness
Societies	man	village	city	nation
Faculties	reason	knowledge	opinion	sensation
Seasons	spring	summer	autumn	winter
Ages of a person	infancy	youth	adulthood	old age
Parts of living things	body	rationality	emotion	willfulness

Whenever we have two or more groups containing the same number of things (say, four), the notion of connections between the groups naturally arises. *Correspondence* refers to the idea that any groups defined by the same number are somehow related. For example, Plato associated the four elements with the four solids (fire with the pyramid, earth with the cube, and so forth). According to Vincent Hopper, the notion of correspondences, such as the relating of the

> "I swear by him who brought us the Tetraktys, which is the source and root of everlasting nature."
> *Pythagorean Oath*

FIGURE 1.4 ■ The Sacred Tetraktys

An important discovery attributed to the Pythagoreans is that the side and diagonal of a square are *incommensurable* (having no common measure). This means that there is no measure that is contained an integral number of times in both the side and the diagonal.

seven planets to the seven days of the week, originated in astrology.[5] That notion was a persistent idea throughout history and will be a recurring theme here. For example, Chapter 4, "Ad Quadratum and the Sacred Cut," will discuss more correspondences between groups of four.

The Quadrivium

Another group of four attributed to the Pythagoreans is the division of mathematics into four groups. This is the famous *Quadrivium of Knowledge*, the four subjects needed for a bachelor's degree in the Middle Ages. The first mention of the quadrivium may have been Plato's dialogue *Theaetetus*, c. 370 B.C.

> "Is Theodorus an expert in geometry?
>   Of course he is, Socrates, very much so.
>   And also in astronomy and arithmetic and music and in all the liberal arts?
>   I am sure he is."

Later, we will discuss the *trivium*, which when combined with the quadrivium comprises the *seven liberal arts*. Each subject is represented in art by an allegorical

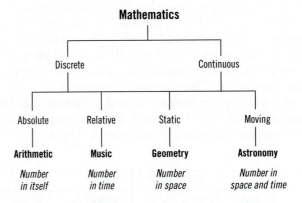

figure, and each is often personified by a particular person. Not surprisingly, arithmetic is personified by Pythagoras and geometry by Euclid.

Our journey will include all the subjects of the quadrivium. Arithmetic, geometry, and music will be included in this chapter; more arithmetic and geometry will be discussed throughout; and astronomy will be included in Chapter 11, "The Sphere and Celestial Themes in Art and Architecture."

RATIO AND PROPORTION

"Remove number from everything, and all will come to nothing."

ISODORE OF SEVILLE (560 – 636)

The geometric concepts of ratio and proportion and their occurrence in art and architecture appear throughout history. The Pythagoreans' musical ratios are discussed at length in this chapter, and the golden ratio, which we treat in detail in Chapter 2, "The Golden Ratio," recurs throughout our studies. In later

SCHOOL OF ATHENS
Raphael, 1510–1511

Although Raphael did not indicate the identities of the figures in this painting (Figure 1.5), scholars agree about whom most of them are. Other identities, based on a new analysis of the painting,[6] are controversial.

Socrates sprawls on the steps, hemlock cup nearby. His student Plato, the idealist, is at center left, pointing upward to divine inspiration. He holds his *Timaeus,* a book that will be discussed in Chapter 2, "The Golden Ratio," and Chapter 10, "The Solids." Beyond him are those philosophers that appealed to intuition and emotions.

Plato's student Aristotle, the man of good sense, is on Plato's left, holding his *Ethics* in one hand and holding out the other hand in a gesture of moderation. Beyond him are representatives of rational activities (logic, grammar, and geometry). As such, Raphael placed the big three of Greek philosophy center stage.

Crito and Apollodorus, Socrates' students, are to his right. They are displaying shock and disbelief at his death. Euclid is shown with a compass, lower right. Beyond Euclid, Ptolemy wears a crown and holds a terrestrial globe, while Zoroaster nearby holds a celestial globe. Diogenes, in black, upper right, stands beneath the statue of Athena, and Pericles, with helmet, stands upper left.

Of course, Raphael did not know what these philosophers looked like. He gave Euclid the face of the architect Bramante, and Plato bears a strong resemblance to Leonardo da Vinci (see Chapter 12, "Brunelleschi's Peepshow and the Origins of Perspective"). Raphael also put himself into the painting, facing the viewer on the right. In front of him is a likeness of Perugino, his friend and teacher.

The figure sitting on the front step, with an elbow on a stone block, resembles Michelangelo. Kenneth Clark points out that he seems to be painted in a different style, that of the Sistine Chapel ceiling. Michelangelo wouldn't let anyone into the Chapel; however, Bramante, who was architect of St. Peters at the time, had the key and he and Raphael entered when Michelangelo was away.[7]

©Scala/Art Resource, NY

FIGURE 1.5 ■ *School of Athens,* Raphael, 1510–1511

chapters, we'll also explore the ratios found in geometric shapes like triangles and rectangles, and we'll document their appearances in art and architecture as well. In this section, we'll provide some mathematical background for calculating ratio and proportion, which will be essential in preparing us for what lies ahead. We'll start our parallel journey through the geometric figures with the simplest, the point and the line.

Points and Lines

A *point* is a geometric element that has position but no dimensions. If the point *P* is thought of as moving to a new position *Q*, its path, called a *locus*, generates a *curve*, as shown in Figure 1.6a. If the point moves without changing direction, it generates a *straight line* (Figure 1.6b). We will simply say *line* when we mean a straight line and use *curve* for other lines.

The straight line has no thickness and extends infinitely far in both directions. A line is considered to have one *dimension*, while a point has zero dimensions. A point on a line divides the line into two half-lines or *rays*, as shown in Figure 1.7a. A *line segment* is the portion of the line between (and including) two points, called *endpoints*. The line segment in Figure 1.7b has endpoints *P* and *Q*. We will refer to this segment as line *PQ* or line *QP*.

> A line subdivided into numbered increments, like a ruler, is called a *number line*.

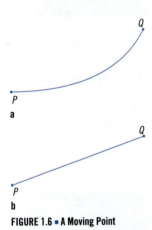

FIGURE 1.6 ■ A Moving Point

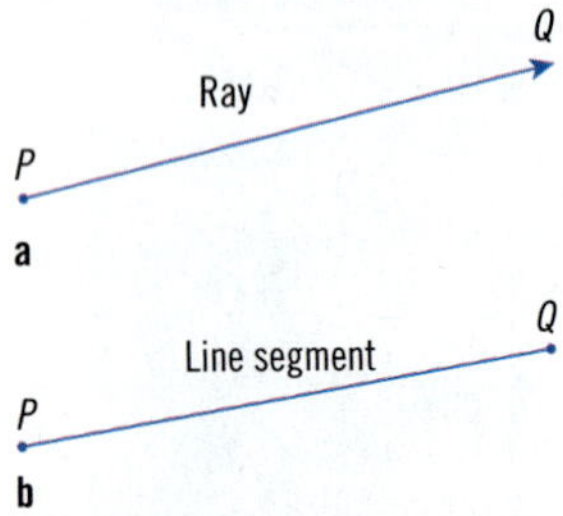

FIGURE 1.7 ■ Ray and Line Segment

FIGURE 1.8 ■ Division of a Line Segment

Ratios

A *ratio* is the quotient of two quantities (that is, one quantity divided by the other). Therefore, the ratio of *P* to *Q* is *P/Q*. A ratio can also be written using the colon (:), as shown here:

$$P : Q.$$

■ **EXAMPLE:** What is the ratio of the smaller segment to larger segment in Figure 1.8?

● **SOLUTION:** The ratio is

$$\frac{ab}{bc}. \qquad ■$$

■ **EXAMPLE:** If a certain floor plan has a width of 7 yd (yards) and a length of 45 ft (feet), what is the ratio of width to length?

● **SOLUTION:**

$$\frac{7 \text{ yd}}{45 \text{ ft}}$$

When writing the ratio of two physical quantities, you can express both quantities using the *same* units, so that they cancel and leave the ratio *dimensionless*. In the previous example, both dimensions can be converted to yards. The units cancel and the result is a simpler dimensionless ratio:

$$\frac{7 \text{ yd}}{15 \text{ yd}} = \frac{7}{15}. \qquad ■$$

> In the following chapter, we will show how to divide a line segment by the golden ratio.

Proportions

When two ratios are set equal to each other, the result is a *proportion*. If the ratio of a to b equals the ratio of c to d, we have the proportion

$$\frac{a}{b} = \frac{c}{d}.$$

This is sometimes written using double colons:

$$a : b :: c : d.$$

Therefore, a proportion relates four quantities, a, b, c, and d. These quantities are called the *terms* of the proportion.

■ **EXAMPLE:** What is the proportion if the ratio of the height of a certain building to its width equals the ratio of 3 to 4?

● **SOLUTION:** The proportion is

$$\frac{\text{height}}{\text{width}} = \frac{3}{4}.$$

■

You solve a proportion just as you would any other equation.

■ **EXAMPLE:** Find x in the proportion

$$\frac{3}{x} = \frac{7}{9}.$$

● **SOLUTION:** Clear fractions by multiplying by the common denominator $9x$:

$$27 = 7x$$

from which,

$$x = \frac{27}{7}.$$

■

The two inside terms of a proportion are called the *means*, and the two outside terms are called the *extremes*.

When the means of a proportion are equal, as in

$$a : b = b : c,$$

the term b is called the *geometric mean* or *mean proportional* between a and c. You can find the mean proportional by solving for b. Multiplying both sides by bc gives

$$b^2 = ac.$$

10 GEOMETRIC MEAN OR MEAN PROPORTIONAL

Geometric mean $b = \pm\sqrt{ac}$ ●

■ **EXAMPLE:** Find the geometric mean between 3 and 12.

● **SOLUTION:** Using Equation 10,

$$b = \pm\sqrt{3(12)} = \pm 6.$$

Therefore, you get *two* answers, $+6$ and -6. You can see that both are correct by placing each into the proportion:

$$\frac{3}{6} = \frac{6}{12}$$

A white numeral in a green box indicates a formula or a statement that is listed in Appendix F of this text for your reference. Although the formulas and statements appear consecutively in Appendix F, they will not usually appear in numerical order throughout the text.

and

$$\frac{3}{-6} = \frac{-6}{12}.$$

Therefore, both 6 and −6 are the geometric means between 3 and 12. ■

Subdividing a Line Segment by a Given Ratio

Later, we will subdivide lines and other geometric figures into various ratios, such as the golden ratio. Here we show how, by example.

■ **EXAMPLE:** Subdivide a line segment of length 148 cm into the ratio of 5 : 9. Work to the nearest tenth of a centimeter.

● **SOLUTION:** Set up the proportion:

$$\frac{\text{shorter segment}}{\text{whole line}} = \frac{5 \text{ parts}}{(5 + 9) \text{ parts}}$$

$$\frac{\text{shorter segment}}{148 \text{ cm}} = \frac{5}{14}$$

$$\text{shorter segment} = \frac{5(148) \text{ cm}}{14} = 52.9 \text{ cm}$$

$$\text{longer segment} = 148 - 52.9 = 95.1 \text{ cm}$$

■

The Rule of Three

Solving proportion problems was an important subject during the Renaissance. This skill was crucial to merchants, who had to deal with problems of pasturage, brokerage, discount, tare allowance, adulteration of commodities, barter, and currency exchange. Not only did every city have its own currency, but each had its own weights and measures. Chapter 10, "The Solids," will discuss how a merchant's ability to gauge volumes of solids helped him to relate to Renaissance paintings.

The *Rule of Three*, also called the *Golden Rule* and the *Merchant's Key*, was the universal mathematical tool of the literate commercial populace during the Renaissance. In his *Del abaco*, Piero della Francesca explained how to solve a proportion: "Multiply the thing one wants to know about by the thing that is dissimilar to it, and divide by the remaining thing. The result is dissimilar to the thing we want to know about."

One *braccio* is about $\frac{1}{3}$ of a person's height, or about 23 in.

■ **EXAMPLE:** Seven bracci of cloth are worth nine lire; how much will five bracci of cloth be worth?

● **SOLUTION:** The thing we want to know about = 5 bracci of cloth.
The thing dissimilar to it = 9 lire.
The remaining thing = 7 bracci of cloth.
Therefore,

$$\frac{5 \times 9}{7} = \frac{45}{7} = 6\frac{3}{7} \text{ lire.}$$

The units are lire, because lire are dissimilar to bracci, the units of the thing about which we wanted to know. ■

SKILLS IN PROPORTION AFFECTED THE WAY PEOPLE VIEWED ART AND ARCHITECTURE

So what does ability to solve proportions have to do with art and architecture? The British historian Michael Baxandall claims that the skills used to solve exchange problems during the Renaissance were the same as those used to make or see pictures. He makes the following points in his text, *Painting and Experience in Fifteenth-Century Italy.*

● Renaissance education placed exceptional value on a few mathematical skills, such as the ability to compute proportions. People did not know more mathematics than we do; however, what they did know, they understood well. Mathematics was a relatively larger part of their intellectual equipment.

● The math skills used by merchants were the same as those used by the artist or architect.

● Due to the status of these math skills in his society, the artist was encouraged to use them in his work.

Because merchants were experienced in manipulating ratios, they were sensitive to pictures and buildings that carried marks of similar processes. The step from computing the proportions of a currency exchange to examining the proportions of a physical body (such as a human head or a building) was a small one.

■ EXERCISES ● RATIO AND PROPORTION

1. Define or describe the following terms:

   straight line   line segment   ray   half-line
   ratio   proportion   dimensionless ratio   means
   extremes   mean proportional   locus

2. Find the value of x in each of the following proportions. Round your answers to the nearest tenth unit, if needed, or leave in fractional form.
   a. $x : 5 = 3 : 10$ b. $x : 4 = 4 : 6$
   c. $3 : x = 4 : 6$ d. $x : 2.75 = 114 : 226$

3. Find the mean proportional between the following quantities.
   a. 2 and 50   b. 4 and 16   c. 6 and 150   d. 5 and 45

4. Subdivide each line segment by the given ratio to the nearest tenth unit.
   a. 200 units long, in a ratio of $3 : 8$ b. 385 units long, in a ratio of $2 : 7$
   c. 285 units long, in a ratio of $2 : 3$ d. 56.8 units long, in a ratio of $3 : 4$

5. If 350 lb (pounds) of marble costs $75, how much will 435 lb of marble cost?

6. If three masons can lay a 15-foot section of brick wall, how many masons are needed to lay 25 feet of a similar wall in the same length of time?

7. If it costs $5843 to paint a particular house having a surface of 3755 sq. ft, how much will it cost to paint a house having a surface of 7325 sq. ft?

8. Prove or demonstrate that the statements listed below are true: In any proportion,

1 The product of means equals the product of extremes.   ●

2 The extremes may be interchanged.   ●

3 The means may be interchanged.   ●

4 The means may be interchanged with the extremes.   ●

FIGURE 1.9 ■ Closeup of Tablet in Rafael's *School of Athens*

PYTHAGORAS AND THE MUSICAL RATIOS

According to legend, the Pythagoreans built an elaborate number lore, but perhaps the numbers that impressed them most were those found in the musical ratios or musical intervals.

The *frequency* of a tone is the rate of vibration (so many vibrations per second) of whatever is producing the tone, such as a vibrating string. A *musical ratio* is the ratio of the frequency of one tone to another, such as between two piano keys. For example, two notes an octave apart have a musical ratio of 2 : 1.

Look again at Pythagoras in the *School of Athens* (Figure 1.1), where we see him explaining the musical ratios to a pupil. A closeup of the tablet he is holding is shown in Figure 1.9 and reveals several elements key to the musical ratios. See if you can identify the following in the closeup:

- The Greek names for the musical ratios (*diatessaron*, *diapente*, and *diapason*)
- The Roman numerals for 6, 8, 9, and 12, which show the ratio of the musical intervals
- The word ΕΠΟΓΛΟΩΝ, the name of the tone, which represents the interval between any two consecutive notes
- The triangular number ten, the Sacred Tetraktys mentioned earlier, inscribed toward the bottom of the tablet

This tablet shows several musical ratios. They are listed in the following chart.

			Greek Term	Latin Term
6 : 12	octave	(1 : 2)	diapason	duplus
6 : 9 or 8 : 12	fifth	(2 : 3)	diapente	sesquialtera
6 : 8 or 9 : 12	fourth	(3 : 4)	diatessaron	sesquitertia
8 : 9	tone	(8 : 9)	tonus	sesquioctavus

Note that all the numbers in the musical intervals, (1 : 2), (2 : 3), and (3 : 4), are contained in the Sacred Tetraktys.

No one knows how the Pythagoreans discovered the musical ratios, but legend says that they found them by experimenting with the *monochord*, a device with a single string and a movable bridge. An illustration of the monochord is shown in Figure 1.10. They moved the bridge to different locations along the string, as shown in Figure 1.11, and noted where the tone produced by the shortened string was harmonious with the tone produced by the string in its original length.

Let's say that the string has a full length of 12 units. (Because we are interested in ratios, the length we choose doesn't matter, and 12 will make the numbers easier.) Placing the bridge so as to shorten the string to 9 units produced a tone harmonious with that of the original 12-unit string. This tone was higher than the original—higher by a ratio of 4 : 3. This musical ratio is called the *fourth*.

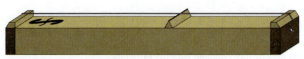

FIGURE 1.10 ■ The Monochord

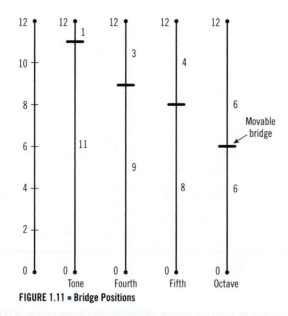

FIGURE 1.11 ■ Bridge Positions

Similarly, shortening the string to 8 units produced a harmonious tone higher than the original by a ratio of 3 : 2. This musical ratio is called the *fifth*. The fourth, the fifth, and the tone can be heard in the familiar song: "Here comes the bride, all dressed in white . . ." The intervals are "Here (fourth) comes the bride, all (fifth) dressed in (tone) white. . . ."

Shortening the string to half its original length, or 6 units, produced a harmonious tone twice as high as the original, or 2 : 1. This musical ratio is called the *octave*. Finally, shortening the string to one-twelfth its original length produced a tone called the *interval*. Therefore, they found that the only pleasant musical intervals could be expressed as the ratio of whole numbers (1 : 2, 2 : 3, and 3 : 4), as shown on the tablet in the *School of Athens* and graphically in Figure 1.12.

Why do some intervals sound pleasant while others sound discordant? To answer this, let's look at the physics of a vibrating string. A string secured at both ends, when plucked or bowed, vibrates as a whole and produces a tone that is called the *fundamental* (Figure 1.13a). Therefore, the D string on a violin will sound a D.

However, the motion of a vibrating string is much more complicated. While vibrating as a whole it will, at the same time, vibrate in halves (Figure 1.13b). Each half-string produces a tone an octave higher than the fundamental. A string will, in addition, vibrate in thirds, fourths, and so on (Figures 1.13c and 1.13d), and produce a whole series of *overtones* or *higher harmonics*. These harmonics are progressively weaker than the fundamental, and they vary from instrument to instrument. They occur naturally whenever a string is plucked or a horn is blown, and they add richness and variety to the tone.

When the fundamental or an overtone of one plucked string exactly matches the fundamental or an overtone of another plucked string, the two strings sounded in unison or in quick succession will sound harmonious. They sound "right." Note, however, that they are all integer ratios of the full string length, and it is these ratios that the Pythagoreans discovered with the monochord.

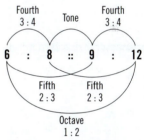

FIGURE 1.12 ■ The Musical Ratios

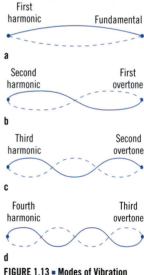

FIGURE 1.13 ■ Modes of Vibration of a String That Is Fastened at Both Ends

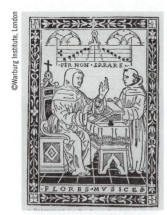

FIGURE 1.14 ■ The Harmonic Scale from *Regule florum musices* (*Rules of Music's Flowers*), Pietro Cannuzio, Florence, 1510

FIGURE 1.15 ■ Pythagoras in *Theorica Musica*, F. Gaffurio, 1492

As such, we have a few pleasant-sounding intervals, the tone, the fourth, the fifth, and the octave, but that is hardly a complete scale. Starting at C, these intervals would give us F, G, and C, an octave higher than where we started. What about the other notes of the scale, such as those depicted in the Italian music books shown in Figures 1.14 and 1.15?

The title page from the music book *Rules of Music's Flowers* (Figure 1.14) dates from nearly the same year as the *School of Athens*. It shows a pattern similar to that on Pythagoras' tablet and also features compasses, which acknowledge a connection between music and geometry. In the upper-left panel of Figure 1.15, we see Iubal (or Jubal) and men hitting an anvil with hammers numbered 4, 6, 8, 9, 12, and 16. (According to the Old Testament, *Iubal* is the "father of all who play the lyre and the pipe."[8]) Other frames show "Pithagoras" or "Pytagora" hitting different sized bells, plucking strings under different tensions, or tapping glasses filled to different depths with water; all of them are marked with those same numbers (4, 6, 8, 9, 12, and 16). In each frame, Pythagoras sounds the ones marked 8 and 16, which is an interval of 1 : 2, the octave. In the lower-right frame, he and Philolaus, another Pythagorean, blow pipes of lengths 8 and 16, again giving the octave. At the same time, Pythagoras holds pipes 9 and 12, giving the ratio 3 : 4, the fourth, while Philolaus holds 4 and 6, giving the ratio 2 : 3, the fifth.

The Greek philosopher Plato later elaborated on the Pythagorean idea of the musical ratios by using mathematical series, sequences, and means to fill the gaps between the known notes. Let's take a look at the mathematics now.

SEQUENCES, SERIES, AND MEANS

In the following section, we will see how Plato used arithmetic, geometric, and harmonic means to develop a musical scale, and later how Alberti and Palladio used these means to determine the proportions of rooms in a building. In the following chapter, we will learn about the Fibonacci sequence, and later the *ad quadratum* and *sacred cut* sequences. Here, we will briefly provide the mathematical background for understanding sequences, series, and means in this chapter, as well as later in our story. Let's start by defining some terms.

A *sequence* is a set of terms arranged in order, such as

$$1, 1, 2, 3, 5, 8, \ldots$$

A *series* is an expression for the sum or difference of the terms of a sequence. The terms are, therefore, connected by plus or minus signs, such as the series

$$1 + 1 + 2 + 3 + 5 + 8 \ldots$$

Arithmetic Progressions

An *arithmetic progression* is a sequence in which each term (after the first) equals the sum of the preceding term and a constant, called the *common difference*. To find the common difference, simply subtract any term from the one following it.

■ **EXAMPLE:** Find the common difference for the arithmetic progression

$$3, 8, 13, 18 \ldots$$

● **SOLUTION:**

Common difference = 13 − 8 = 5.   ■

You can often determine a term of a sequence from those preceding it. The relationship between a term and those preceding it is called a *recursion relationship* or a *recursion formula*. For the arithmetic progression, you can find any term a_n simply by adding the common difference d to the term a_{n-1} immediately preceding it.

12 RECURSION FORMULA FOR AN ARITHMETIC PROGRESSION

$$a_n = a_{n-1} + d$$

■ **EXAMPLE:** Using the recursion formula, find the next term in the following sequence:

$$3, 8, 13, 18 \ldots$$

● **SOLUTION:** We get the common difference by subtracting any term from the one immediately following, so

$$d = 8 - 3 = 5.$$

The next term in the sequence is then $18 + 5$ or 23. ■

> A recursion formula is useful for generating a sequence by using a spreadsheet on a computer. See Problem 11 in the exercise set for this section.

The *arithmetic mean* between two numbers is what we commonly call the *average*, that is, the sum of the numbers divided by two. Therefore, to find the arithmetic mean b between two terms a and c, simply take half the sum of a and c.

9 ARITHMETIC MEAN

$$\text{Arithmetic mean } b = \frac{a + c}{2}$$

We'll see later how Palladio used the arithmetic mean to find room proportions.

■ **EXAMPLE:** Find the arithmetic mean between the numbers 5 and 9.

● **SOLUTION:**

$$\frac{5 + 9}{2} = \frac{14}{2} = 7$$

■

Geometric Progressions

A *geometric progression* is a sequence in which each term (after the first) equals the preceding term times a constant called the *common ratio*. To find the common ratio, divide any term by the one preceding it.

■ **EXAMPLE:** Find the common ratio for the geometric progression

$$5, 20, 80, 320 \ldots$$

● **SOLUTION:**

$$\text{Common ratio} = \frac{80}{20} = 4$$

■

To find any term a_n of a geometric progression, multiply the term a_{n-1} immediately preceding by the common ratio r. Therefore, the recursion formula is as follows.

14 RECURSION FORMULA FOR A GEOMETRIC PROGRESSION

$$a_n = ra_{n-1}$$

■ **EXAMPLE:** Find the next term in the geometric progression of the preceding example.

● **SOLUTION:** The common ratio was 4 and the last term given was 320, so the next term in the progression is

$$4(320) = 1280.$$

To find the geometric mean b between two numbers a and c, take the square root of the product of a and c.

10 GEOMETRIC MEAN OR MEAN PROPORTIONAL

$$\text{Geometric mean } b = \pm\sqrt{ac}$$

■ **EXAMPLE:** The geometric mean b between 3 and 12 is

$$b = \sqrt{3 \times 12} = \sqrt{36} = 6.$$

Harmonic Progressions

A sequence is called a *harmonic progression* if the reciprocals of its terms form an arithmetic progression. The name *harmonic* goes back to the Pythagoreans. Recall that when a stretched string is shortened to $\frac{1}{2}$ its original length, it produces a tone that is "harmonious" with the tone produced by the original length. The same is true when the string is shortened to $\frac{1}{3}$ the original length, and $\frac{1}{4}$, and so forth. The fractions $\frac{1}{2}$, $\frac{1}{3}$, and $\frac{1}{4}$ form a harmonic progression.

The sequence

$$1, \frac{1}{3}, \frac{1}{5}, \frac{1}{7}, \frac{1}{9}, \frac{1}{11}, \cdots$$

is a harmonic progression because the reciprocals of its terms, which are

$$1, 3, 5, 7, 9, 11, \ldots,$$

form an arithmetic progression.

The harmonic mean b between two numbers a and c is equal to twice the product of a and c, divided by their sum.

11 HARMONIC MEAN

$$\text{Harmonic mean } b = \frac{2ac}{a + c}$$

■ **EXAMPLE:** What is the harmonic mean b between 6 and 12?

● **SOLUTION:** By Equation 11,

$$b = \frac{2(6)(12)}{6 + 12} = \frac{144}{18} = 8.$$

The numbers 6, 8, and 12 now form a harmonic progression; their reciprocals form an arithmetic progression. ■

The Three Means by Geometric Construction

Another way to find arithmetic, geometric, and harmonic means is by *geometric construction*. Geometric constructions are an important part of learning geometry, in addition to being fun to do. *Constructions* are created using only a compass and a straightedge. Rulers, squares, and protractors are not allowed. Further, some constructions can be completed with a single setting of the compass; these are called *rusty compass constructions*.

In preparation for finding the three means by geometric construction, let's begin by looking at a simple construction to draw a perpendicular line. Two lines are said to be *perpendicular* if they intersect to form four equal angles. Two lines are said to be *parallel* if they do not intersect, no matter how far they are extended. Many constructions, including that for the three means, require us to draw a perpendicular to a given line, so let's do that construction first. Figure 1.16 illustrates the construction for a perpendicular to a line or to a line segment (the portion of a line between two endpoints).

In Figure 1.16, we are shown the construction of a perpendicular to a line through a given point, which can be either on the line or off the line, as shown. Starting at the given point *P*, swing an arc cutting the given line at *A* and *B*. From *A* and *B*, draw arcs of equal radius that intersect at *C*. Finally, draw line *CP*, the perpendicular to the given line. Further, line *CP* is also the *perpendicular bisector* of line segment *AB*.

Let's proceed now to our constructions of the three means. Palladio explains how to find means in Book 1, Chapter XXIII of his *Four Books of Architecture*, although he does not call them by these names.

To find an arithmetic mean between two lengths, Palladio simply makes a line whose length is the sum of the two lengths, and bisects it. To find the geometric mean or mean proportional, he uses the same method described by Euclid in Book VI, Prop. 13. This method is demonstrated in Figure 1.17, in which we find the mean proportional (geometric mean) between *AB* and *BC*. Lay *AB* and *BC* end to end on the same straight line. Draw a semicircle on *AC*, and erect a perpendicular *BD* at *B*. *BD* is the mean proportional between *AB* and *BC*.

Palladio also gives the construction for finding the harmonic mean, and this is demonstrated in Figure 1.18. To find the harmonic mean between *AB* and *BC* in the rectangular floor plan *ABCD*, first find the arithmetic mean between *AB* and *BC*. This can be done by calculation or by construction. Next, extend *AB* by that amount to *E*. Draw *EC* and extend to where it intersects *AD* extended at *F*. Length *DF* is the harmonic mean proportional between *AB* and *BC*.

We will add to these constructions as we go along. To get an idea of the full range of constructions given in this book, turn to the complete listing in Appendix C.

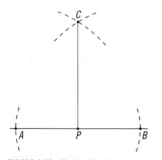

FIGURE 1.16 ■ Perpendicular to a Line through a Given Point Not on the Line

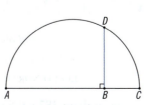

FIGURE 1.17 ■ Finding the Mean Proportional

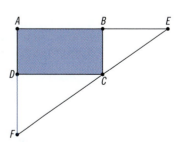

FIGURE 1.18 ■ Finding the Harmonic Mean

■ EXERCISES ● SEQUENCES, SERIES, AND MEANS

1. Define the following terms:

sequence	series	arithmetic progression
geometric progression	harmonic progression	common difference
common ratio	arithmetic mean	geometric mean
harmonic mean	mean proportional	recursion formula
geometric construction	rusty compass	perpendicular
parallel	construction	

2. Find the common difference in these arithmetic progressions:
   a. 125, 131, 137, 144 . . .   b. 54, 46, 38, 30 . . .

3. Find the common ratio in these geometric progressions:
   a. 25, 75, 225, 675 . . .   b. 1.54, 3.85, 9.625 . . .

4. Insert an arithmetic mean between these numbers:
   a. 12 and 18   b. 283 and 495

5. Insert a geometric mean between these numbers:
   a. 5 and 20   b. 4.68 and 8.26

6. Insert a harmonic mean between these numbers:
   a. 1 and 10   b. 3.92 and 8.83

7. Given the numbers 10 and 20, insert an arithmetic mean, a geometric mean, and a harmonic mean.

8. Why is the harmonic mean called "harmonic"?

9. Draw a sequence of squares, starting with side 1 and increasing by .5, to illustrate an arithmetic progression.

$$1, 1.5, 2, 2.5, \ldots$$

10. Draw a sequence of squares on diagonals, that is, where the side of each successive square is equal to the diagonal of the preceding square. Use this to demonstrate a geometric progression with a common ratio equal to $\sqrt{2}$.

$$1, \sqrt{2}, 2, 2\sqrt{2}, 4, \ldots$$

11. Choosing any numbers you want, use the recursion formulas for the arithmetic and geometric progressions to generate progressions on a computer.

12. Using only a compass and a straightedge, repeat the constructions for the three means.

13. Using only the Line and Circle tools of a computer drafting program, repeat the constructions for the three means.

14. Try to devise ways to reproduce the construction of a perpendicular by paper folding. If you get stuck, see Olson's *Mathematics Through Paper Folding*.

PLATO AND THE MUSICAL RATIOS

". . . the safest general characterization of the European philosophical tradition is that it consists of a series of footnotes to Plato."

ALFRED NORTH WHITEHEAD

Plato acquired property about 387 B.C. in the northwestern outskirts of Athens, a site that had been an olive grove, park, and gymnasium sacred to the legendary

PLATO (c. 427–347 b.c.)

Plato was born in Athens, and in his youth he was interested in politics. He eventually became a disciple of Socrates and adopted Socrates' style of debate, in which truth was pursued through questions and answers. Plato saw Socrates die in 399 B.C. at the hands of the Athenian rulers. In Raphael's *School of Athens* (Figure 1.5), Socrates is prone, with a cup nearby. Plato's final years were spent lecturing at his Academy and writing. He died at about the age of 80 in Athens.

Plato shared the Pythagoreans' love of numbers. In his *Epinomis*, Plato wrote, "Numbers are the highest degree of knowledge, and Number is Knowledge itself." Plato loved geometry. When asked "What does God do?" Plato supposedly replied, "God geometrizes . . . [Geometry is] pursued for the sake of the knowledge of what eternally exists, and not of what comes for a moment into existence, and then perishes, . . . [it] must draw the soul towards truth and give the finishing touch to the philosophic spirit."[9]

FIGURE 1.19 ■ Plato and Aristotle in Raphael's *School of Athens*

Plato even defined beauty in geometric terms. "I do not mean by beauty of form such beauty as that of animals or pictures, . . . but . . . understand me to mean straight lines and circles, and planes and solid figures which are formed out of them by turning lathes and rulers and measures of angles; for them I affirm to be not only relatively beautiful, like other things, but eternally beautiful. . . ."[10]

Although Plato loved geometry, he had a low opinion of art. He taught that since our world is a copy or image of reality, then a work of art is a copy of a copy, three steps away from reality. He wrote that "painting and . . . the whole art of imitation, is busy about a work which is far removed from the truth . . . and is its mistress and friend for no wholesome or true purpose . . . it is the worthless mistress of a worthless friend, and the parent of a worthless progeny."[11] Raphael had placed Plato and his student Aristotle in the center of his *School of Athens* (Figure 1.19).

hero Academus. The school he started there, often described as the first university, offered courses in astronomy, biology, mathematics, political theory, and philosophy. Over the doors to his academy were the words

αγεωμετρητοζ μηδειζ εισιτω

which mean, "Let no one destitute of geometry enter my doors." Compare this with what Leonardo later wrote in one of his notebooks, "Let no one read me who is not a mathematician." Plato's Academy was the inspiration for a Platonic academy in Renaissance Florence, started by Lorenzo de Medici and led by the neo-Platonist Marsilio Ficino. It included Pico della Mirandola, Cristoforo Landino, Angelo Poliziano, and sometimes Michelangelo. It was also the inspiration for another academy near Vicenza, started by Count Trissino and attended by Palladio. The ruins of the academy still exist.

Plato left many writings, but his love of geometry is especially evident in his *Timaeus*. Written toward the end of Plato's life, the *Timaeus* describes a conversation among Socrates, Plato's teacher; Critias, Plato's great grandfather; Hermocrates, a Sicilian statesman and soldier; and Timaeus, a Pythagorean, philosopher, scientist, general, contemporary of Plato, inventor of the pulley, and the first to distinguish among the harmonic, arithmetic, and geometric progressions. In the book, Timaeus pays homage to Pythagoras and describes the geometric creation of the universe.

Creation of the World by the Musical Ratios

The *Timaeus* also mentions *Atlantis*, a superior civilization that was submerged in one day and night by floods and earthquakes. Only Egypt, being dry, survived to preserve and hand down the old traditions.[12]

In his *Timaeus*, Plato says that the creator made the *world soul* out of various ingredients and formed it into a long strip.[13] This strip was made of a flexible material, which would later be cut lengthwise and bent into celestial circles. The strip was then marked at intervals as follows:

- First, the creator took one portion from the whole,   (1 unit)
- next, a portion double the first,   (2 units)
- a third portion half again as much as the second,   (3 units)
- the fourth portion double the second,   (4 units)
- the fifth three times the third,   (9 units)
- the sixth eight times the first,   (8 units)
- and the seventh 27 times the first   (27 units)

Therefore, we get the seven integers (1, 2, 3, 4, 8, 9, 27) that contain the Monad, source of all numbers, the first even and first odd, and their squares and cubes.

The seven integers Plato used to describe the creation of the world by musical ratios are often shown arranged into two series called Plato's lambda, after the Greek letter lambda (λ).

After the Monad, the left branch contains the first even number, its square, and its cube; the right branch contains the first odd number, its square, and its cube. These seven numbers contain all the musical consonances. Plato also believed that they "embrace the secret rhythm in the macrocosm and the microcosm alike."[14] Plato's lambda appears in the allegory to arithmetic shown in Figure 1.20.

If you mark off Plato's seven numbers on a musical staff, starting, say, at low C, you get four octaves and a bit more as shown in Figure 1.21. It is the beginning of a musical scale, but it has many gaps.

However, Plato goes on to *fill* each interval with an arithmetic mean and a harmonic mean, as follows. Let's apply Equations 9 and 10 to the first octave,

FIGURE 1.20 ■ Arithmetic Portrayed as a Woman

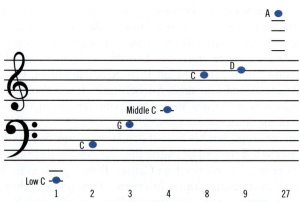

FIGURE 1.21 ■ The Seven Numbers of Plato's Lambda as a Musical Scale

starting from low C in Figure 1.21. The means between 1 and 2 are

$$\text{Arithmetic mean} = \frac{1 + 2}{2} = \frac{3}{2}$$

$$\text{Harmonic mean} = \frac{2(1)(2)}{1 + 2} = \frac{4}{3}$$

So the numbers in the first octave are

$$1, \frac{4}{3}, \frac{3}{2}, \text{ and } 2.$$

The ratio between the first and second numbers (1 and 4/3) is 3 : 4, the fourth. We get the same ratio between the third and fourth numbers (3/2 and 2). The ratio between the first and third numbers (1 and 3/2) is 2 : 3, the fifth. We get the same ratio between the second and fourth numbers (4/3 and 2). *These are the same intervals that the Pythagoreans found pleasing, but Plato found them from arithmetic calculations alone,* and not by experimenting with stretched strings to determine which ratios sounded best. The English philosopher and Plato scholar Francis Cornford wrote, " . . . Plato has constructed a section of the diatonic scale, whose range is fixed by considerations extraneous to music."[15]

If we now add these arithmetic and harmonic means on our original seven-note scale, and similarly insert arithmetic means between higher octaves, we get the scale shown in Figure 1.22.

However, there were still gaps. Plato took the geometric interval between the fourth and the fifth as a full tone. It is found by dividing 3/2 by 4/3.

$$\frac{3}{2} \div \frac{4}{3} = \frac{3}{2} \times \frac{3}{4} = \frac{9}{8}$$

Plato then filled up the scale with intervals of 9/8, as shown in Figure 1.23. Starting, say, at middle C, multiplying by 9/8 takes us to D, and multiplying D by 9/8 gives E. Multiplying E by 9/8 would overshoot F, so Plato stopped at F. This left an interval of

$$\frac{4}{3} \div \frac{81}{64} = \frac{4}{3} \times \frac{64}{81} = \frac{256}{243}$$

between E and F. This ratio is approximately equal to half that of the full tone, and so is called a *semitone*. Note that two semitones approximately equal one full tone.

$$\text{Two semitones: } \frac{256}{243} \times \frac{256}{243} \cong 1.110$$

$$\text{One full tone: } \frac{9}{8} = 1.125$$

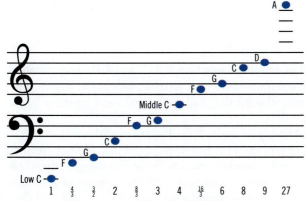

FIGURE 1.22 ■ Musical Scale Showing Fourths and Fifths

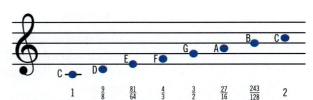

FIGURE 1.23 ■ One Octave of the Musical Scale Showing Tones and Semitones
Only one octave is shown here, but the procedure can be repeated for the remaining octaves.

Ascending from F to B, we proceed again by factors of 9/8. The interval from B to C is once more the semitone 256/243.

Note that while the fourths and fifths were found using arithmetic and harmonic means, the whole tone intervals were found by geometric means. As such, the intervals C to D to E and F to G to A to B form two *geometric progressions*.

This mathematically constructed scale is close to a modern scale, but there are differences. For comparison, Table 1.1 lists the major diatonic scale, starting from middle C. The table also shows the frequency of each note, in hertz (Hz), cycles per second, with A above middle C equal to 440 Hz, by international agreement.

TABLE 1.1 ■ The Major Diatonic Scale

Note		Frequency, Hz, Diatonic Scale in Key of C	Interval between Notes	Interval from C	Frequency, Hz, Equally Tempered Scale
Do	C	264		1:1, unison	261.6
			9/8		
Re	D	297		9:8, major second	293.7
			10/9		
Mi	E	330		5:4, major third	329.6
			16/15		
Fa	F	352		4:3, perfect fourth	349.2
			9/8		
So	G	396		3:2, perfect fifth	392.0
			10/9		
La	A	440		5:3, major sixth	440.0
			9/8		
Ti	B	495		15:8, major seventh	493.9
			16/15		
Do	C	528		2:1, perfect octave	523.3

This too, is not the end of the story. With the diatonic scale, the twelve tones within an octave, including the semitones, are not equally spaced. This would require that a piano, for example, be retuned so that it could play in each key. To avoid this, instruments are now tuned to the *equally tempered scale*, popularized by J. S. Bach. Here, the twelve semitones are equally spaced by an interval of $2^{1/12}$, or the twelfth root of two. Therefore, multiplying $2^{1/12}$ by itself twelve times gives 2, or an octave. The frequency of each note, in the equally tempered scale, is also given in the table. As with the diatonic scale, A = 440 Hz is taken as standard.

Music of the Spheres

By experimenting with plucked strings, the Pythagoreans discovered that the intervals that pleased people's ears were

- Octave   1 : 2
- Fifth   2 : 3
- Fourth   3 : 4

All of them are contained in the simple numbers 1, 2, 3, and 4, the very numbers in their beloved Sacred Tetraktys, which added up to the number of fingers. This staggered the Pythagoreans, who felt they had discovered some basic law of the universe.

Quoting Aristotle's *Metaphysics* again, "[the Pythagoreans] saw that the . . . ratios of musical scales were expressible in numbers . . . and that all things seemed to be modeled on numbers, and numbers seemed to be the first things in the whole of nature, they supposed the elements of number to be the elements of all things, and the whole heaven to be a musical scale and a number."

It seemed clear to the Pythagoreans that the distances between the planets would have the same ratios as the ones that produced harmonious sounds in a plucked string. Their solar system consisted of spheres rotating in circular orbits, each sphere giving off a sound, the way a projectile makes a sound as it swishes through the air. The closer spheres gave lower tones, while the farther spheres moved faster and gave higher-pitched sounds. All combined into a beautiful harmony, *the music of the spheres*. This idea that creation was closely linked to music was continued by many writers over the centuries.

Plato told how the universe was created according to the musical ratios. The Cambridge scholar E. M. W. Tillyard wrote, "But there was the further notion that the created universe was itself in a state of music, that it was one perpetual dance."[16] In his *Republic*, Plato says of the cosmos, "Upon each of its circles stood a siren who was carried round with its movements, uttering the concords of a single scale."

The historian of art and architecture Rudolph Wittkower wrote, "the doctrine of a mathematical universe which . . . was subject to harmonic ratios, was triumphantly reasserted by a number of great thinkers. . . ."[17] Among them, Isidore of Seville (560?–636), a medieval encyclopedist, wrote, "Nothing exists without music; for the universe itself is said to have been framed by a kind of harmony of sounds, and the heaven itself revolves under the tones of that harmony." William Shakespeare (1564–1616), in *The Merchant of Venice*, wrote,

> "There's not the smallest orb which thou behold'st
> But in his motion like an angel sings . . ."[18]

The astronomer Johannes Kepler (1571–1630) wrote that he wished "to erect the magnificent edifice of the harmonic system of the musical scale . . . as God, the Creator Himself, has expressed it in harmonizing the heavenly motions."[19] Later he wrote, "I grant you that no sounds are given forth, but I affirm . . . that the movements of the planets are modulated according to harmonic proportions." Quoting the English poet John Dryden (1631–1700),

> "From harmony, from Heav'nly harmony
> This universal frame began;
> From harmony to harmony
> Thro' all the compass of the notes it ran,
> The diapason closing full in man."[20]

Finally, the English philosopher Anthony Shaftsbury (1671–1713) related human nature to the musical ratios.

> "Virtue has the same fix'd Standard.
> The same Numbers, Harmony, and Proportion will have place in Morals;
> and are discoverable in the Characters and Affections of Mankind."[21]

THE MUSICAL RATIOS IN ARCHITECTURE

"Geometry . . . is of much assistance in architecture . . . it teaches us the use of the rule and compasses . . . and rightly apply the square, the level, and the plummet."[22]

VITRUVIUS

A major figure in our study of geometry in architecture is the Roman architect, Vitruvius. We provide a brief biographical sketch here, and refer to his writings throughout this text.

VITRUVIUS (70–25 B.C.)

Vitruvius, whose full name is Marcus Vitruvius Pollio, was probably born in Formiae (now Formie), Italy. He was an architect and an artillery engineer, probably in the service of the first emperor Augustus. Little is known of Vitruvius' life, but he is best known as the author of the famous treatise *De architectura* (*Ten Books on Architecture*), a handbook for architects and the oldest surviving work on the subject.

De architectura is above all a practical guide to materials, methods, and design. It covers almost every aspect of architecture: city planning, building materials, fortifications, temples, colonnades, private houses, public buildings, baths, basilicas, theaters, gymnasiums, harbors, farmhouses, military engines, public spaces such as the forum, floor, and stucco decoration, hydraulics, clocks, mensuration, and astronomy. Vitruvius was the first to discuss the classical orders of architecture: Doric, Ionic, and Corinthian. Interestingly, he urges the architect to understand music.

Much of the material appears to be based on his own experience, as well as on works by Greek architects such as Hermogenes. Apparently, Vitruvius seeks to preserve the classical Greek tradition in architectural design. Vitruvius' stated wish was that his name be honored long after his death. This was certainly the case, for the *Ten Books on Architecture* became the chief authority on ancient classical architecture, and his writings have been studied ever since the Renaissance.

Systems of Architectural Proportions

"Throughout the history of architecture there has been a quest for a system of proportion that would facilitate the technical and aesthetic requirements of a design."[23]

KAPPRAFF

Architects and builders have always sought systems of proportions to create buildings that were visually pleasing. One of the best-known authorities on the subject was Roman architect Vitruvius' famous treatise *De architectura*, or *Ten Books on Architecture*, in which he compares proportion in architecture to proportion in the greatest work of art, the human body. In *De architectura* or *Ten Books on Architecture*, he wrote, "Symmetry is a proper agreement between the members of the work itself, and relation between the different parts and the whole. . . ."[24] Later he added, "Therefore since nature has proportioned the human body so that its members are duly proportioned to the frame as a whole . . . in perfect buildings the different members must be in exact symmetrical relations to the whole general scheme."[25] In this instance, Vitruvius uses "symmetrical relations" to mean "having the same proportions," rather than some kind of mirror symmetry.

The key elements required in such a system of proportions were later defined by American mathematician Jay Kappraff. He wrote, "Such a system [of proportions] would have to ensure a repetition of a few key ratios throughout the design, have additive properties that enable the whole to equal the sum of its parts, and . . . be adaptable to the architect's technical means. The repetition of ratios enables a design to exhibit a sense of unity and harmony of its parts. Additive properties enable the whole to equal the sum of its parts in a variety of different ways, giving the designer flexibility to choose a design that offers the greatest aesthetic appeal while satisfying the practical considerations of the design. Architects and designers are most comfortable within the realm of integers, so any system based on irrational dimensions or incommensurable proportions should also be expressible in terms of integers to make it computationally

> The repetition of ratios gives a structure a degree of self-similarity, a property of fractals that will be discussed in Chapter 13, "Fractals."

acceptable."[26] In short, such a system of proportions would use a repetition of a few key ratios, have additive properties, and use ratios of whole numbers.

There are three main systems of proportion in architecture, and in this section we'll explore a system based on the Pythagorean musical ratios. This system was primarily used by Renaissance architects like Alberti and Palladio. In Chapter 2, we'll explore a system based on the golden ratio, such as Le Corbusier's Modulor, and in Chapter 4 you will be introduced to a system based on the square, which was apparently used by the Romans.

The Musical Ratios in Renaissance Architecture

According to Rudolph Wittkower, "The conviction that . . . each part of a building . . . has to be integrated into one and the same system of mathematical ratios may be called the basic axiom of Renaissance architects . . . the architect is by no means free to apply . . . a system of ratios of his own choosing . . . and proportions in architecture have to embrace and express the cosmic order. But . . . what are the mathematical ratios that determine the harmony in this macrocosm and microcosm? They had already been revealed by Pythagoras and Plato. . . ."[27]

Wittkower further observes that Renaissance architects were convinced of the Pythagorean vision of the harmonic structure of the universe, and that some inborn sense makes us aware of that harmony. Therefore, if a building contains the same mathematical proportions present in the musical ratios, this "inner sense tells us . . . when the building we are in partakes of the vital force which lies behind all matter and binds the universe together."[28]

As you saw earlier, inserting arithmetic and harmonic means in a geometric progression enabled Plato to mathematically construct a musical scale. These means were also used in Renaissance architecture, but as Kappraff suggests, in ratios of whole numbers. Take a series of numbers in a geometric progression, such as

$$1, 2, 4, 8 \ldots$$

Between each pair of numbers, insert an arithmetic mean and a harmonic mean. First, however, multiply by six to avoid fractional quantities:

$$6, 12, 24, 48 \ldots$$

Inserting arithmetic and harmonic means between each pair of numbers gives:

$$6, 8, 9, 12, 16, 18, 24, 32, 36, 48 \ldots$$

- The geometric progression (6, 12, 24. . .) determines each octave [1 : 2].
- The harmonic mean within each octave, say 6 : 8, determines the fourth [3 : 4].
- The arithmetic mean within each octave, say 6 : 9, determines the fifth [2 : 3].
- The harmonic and arithmetic means within each octave determine the tone [8 : 9].

According to Wittkower, "Whenever one meets ratios of [this series] it is safe to presume that this is not casual but are the result of reflections which depend directly or indirectly on the Pythagoreo-Platonic division of the musical scale."[29] We'll now see how two Renaissance architects, Alberti and Palladio, used the three means and the musical ratios.

Two strings tuned to the same pitch will both vibrate even if only one is plucked, a phenomenon of *resonance*. In the case of strings, the air between the strings transmits the vibrations from one to another. However, some have speculated that a form of resonance might be the reason that we relate so intensely to the archetypal musical ratios, even for visual phenomena.

LEON BATTISTA ALBERTI (1404–1474)

Alberti has been called the prototype of the "Renaissance Man." He was born into a wealthy merchant-banker family, received classical Latin training at a boarding school in Padua, and studied law at the University of Bologna. He obtained a position at the Vatican and eventually took holy orders, although his later interests and activities were entirely secular.

Alberti seems to have collaborated with the Florentine cosmographer Paolo Toscanelli, who provided Columbus with the map that guided him on his first voyage, which led to Alberti's treatise on geography. After this, he wrote the first Italian grammar book, hoping to promote the Tuscan vernacular, and wrote a work on cryptography.

His study of ancient architectural practices resulted in his *De re aedificatoria* (*Ten Books of Architecture*) of 1452. It became a bible of Renaissance architecture and won him his reputation as the "Florentine Vitruvius." His architectural works include the façades of S. Maria Novella, home of Masaccio's *Trinity*, and the Palazzo Rucellai, both in Florence; Tempio Malatestiano in Rimini; and in Mantua, the Churches San Sebastiano and Sant' Andrea.

Alberti and His *Ten Books of Architecture*

The idea that the same ratios that are pleasing to the ear would also be pleasing to the eye appears in the writings of Plato, Plotinus, St. Augustine, and St. Aquinas, but the most direct statement comes from Alberti. In his *Ten Books of Architecture*, he wrote, "indeed I am every day more and more convinced of the truth of Pythagoras' saying, that Nature is sure to act consistently . . . I conclude that the same numbers by means of which the agreement of sounds affect our ears with delight are the very same which please our eyes and our mind. We shall therefore borrow all our rules for finding our proportions from the musicians."[30] In other words, what sounds good also looks good.

In Book IX, Chapter V of his *Ten Books of Architecture*, Alberti gives definitions of the octave, the fifth, the fourth, and the tone, as found by Pythagoras. In Chapter VI, he defines the three means: arithmetical, geometrical, and *musical*, what we call the harmonic mean. In this same chapter, he recommends proportions for floor plans based on the musical ratios. Here Alberti distinguishes between small, medium, and large floor plans. For each, he recommends particular musical ratios for length to width.

- For the smallest, he recommends the square (1 : 1), the sesquitertia (6 : 8), or the sesquialtera (6 : 9).

- For medium plans, he recommends either (a) the double square or octave (1 : 2) or (b) the sesquialtera doubled. In other words, the ratio 6 : 9 applied twice. For a plan whose width is 4 units, applying the ratio 6 : 9 gives 6, and then applying that ratio to 6 gives 9. As such, the final plan would be 4×9 units. This method is called *generation of ratios*. It is a means by which Alberti arrives at a ratio that is not one of the simple ratios (1 : 2, 2 : 3, 3 : 4) but is generated from those simple ratios.

- For the longest floor plans, he suggests (a) a triple proportion, two octaves; (b) a quadruple proportion, three octaves; or (c) a ratio of 3 to 8 (1 : 2 followed by 3 : 4).

Wittkower has written, "Proportions recommended by Alberti are the simple relations of one to one, one to two, one to three, two to three, three to four, etc., which are the elements of musical harmony. . . ."[31] He goes on to give as an example the façade of S. Maria Novella in Florence (Figures 1.25 and 1.26). In Figure 1.26, the façade is shown divided into rectangles with simple musical ratios. The whole façade is inscribed in a square of, say, 2 units. The lower story is comprised of two squares, and the upper square fits into one square. Therefore, the main parts are in the ratio of 1 : 2, the octave. The central bay of the upper story is a square of $\frac{1}{2}$ unit. Two squares of $\frac{1}{2}$ unit enclose the pediment and entablature. The entrance bay is $\frac{1}{2}$ unit by $\frac{3}{4}$ unit, for a ratio of 2 : 3.

Although Alberti's system of proportions met two of the three requirements mentioned earlier, the repetition of a few key ratios and the use of whole number ratios, it lacked the third. According to Kappraff, "Although the Renaissance system of Alberti succeeded in creating harmonic relationships in which key proportions were repeated in a design, it did not have the additive properties necessary for a successful system. It is fascinating that a system of proportions used by the Romans and the system . . . developed by Le Corbusier, known as the Modulor, both conform to the relationships inherent in the system of musical proportions . . . with the advantage of having additive properties."[32]

FIGURE 1.25 ■ Façade of S. Maria Novella, Florence

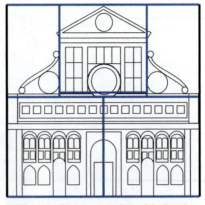

FIGURE 1.26 ■ Wittkower's Analysis of the Façade of S. Maria Novella

Palladio and the *Four Books of Architecture*

" . . . in all fabrics is it requisite that their parts should correspond together, and have such proportions, that there may be none whereby the whole cannot be measured, and likewise all the other parts." [33]

PALLADIO

At the end of 20 years of intensive building, Palladio published *I quattro libri dell' architettura* in 1570. This work was a summary of his studies of classical architecture and, as examples, he included a number of his own designs. In the preface, Palladio cites Vitruvius as his "master and guide," acknowledges his debt to Alberti, and calls his patron, Trissino, "the Splendor of our times." The title page depicts allegories of Geometry and Architecture holding tools of the trade. Some of these same tools appear on the base of the statue of Palladio in Vicenza.

The first book contains studies of materials, the classical orders, and decorative ornaments. The second contains designs for domestic structures like town and country houses. The third has designs for public buildings, bridges, and town planning. The fourth book deals with temples.

Although Palladio does not mention the musical ratios by name, he does advocate them in several places. For example, in Book 1, Chapter XXII, he recommends seven shapes for the floor plans of rooms:[34]

- Round (rarely)
- Square
- Length equal to the diagonal of a square   (the root-2 rectangle)
- Square and a third   (3 : 4 or the fourth)
- Square and a half   (2 : 3 or the fifth)
- Square and two-thirds   (3 : 5 or the square plus a fifth)
- Two squares   (1 : 2 or the octave)

All the rectangular rooms (except the root-2 rectangle, which we will discuss in Chapter 4) are given simple whole-number ratios that are the musical ratios or derived from them.

ANDREA PALLADIO (1508–1580)

Born Andrea Di Pietro Della Gondola, Andrea was given the name Palladio by a patron (Count Gian Giorgio Trissino, a humanist poet and scholar) as an allusion to Pallas Athena of Greek mythology.

Palladio is regarded as the greatest Italian architect of the sixteenth century. He is noted for his treatise, *The Four Books of Architecture* (1570), and for his many palaces and villas still standing in northeastern Italy. His name appears today in the popular *Palladian window* (one composed of a large central section surmounted by a semicircular window, flanked by two narrower, rectangular windows).

Palladio was born in Padua, was apprenticed to a sculptor in his youth, and was later a stone mason. His projects included the villa of Count Trissino, at Cricoli near Vicenza, intended as an academy for Trissino's pupils. This *Academia Trissiniana* was an echo of the Medici Platonic academy in Florence and a distant echo of the Pythagorean and Platonic academies. Trissino took

FIGURE 1.27 ■ Statue of Palladio in Vicenza

an interest in Palladio and strove to provide him with a humanist education.

At the Villa Trissino, Palladio made contacts and may have met the architect Sebastiano Serlio, whose books on architecture were to be an inspiration to him. On three visits to Rome with Trissino, he saw works by the architects Bramante, Peruzzi, and Raphael, and he took measurements of ancient Roman structures. One result of these trips was the publication of his small book *The Antiquities of Rome* (1554), which remained a popular tourist guide for 200 years.

In about 1540, Palladio designed his first villa and his first palace, to be followed by a long and intensive career of building design. Trissino, a great advocate of Vitruvius, had introduced the writings of that architect to Palladio, who later made the plates for Daniele Barbaro's edition of Vitruvius' treatise *On Architecture*, published in 1556. A statue of Palladio (Figure 1.27) stands in Vicenza, the home of many of Palladio's buildings.

Palladio goes on to recommend proportions for room heights in Book I, Chapter XXIII. If a room is 6 × 12 ft, its height should be 9 ft (the arithmetic mean between 6 and 12), or its height should be 8 ft (the harmonic mean). If a room is 4 × 9 ft, its height should be 6 ft (the geometric mean between 4 and 9). Also in Chapter XXIII are the geometric constructions for finding the three means that were presented earlier.

An example of Palladio's use of means to find the height of a room occurs in his Villa Rotonda, pictured in Figure 1.28. Each large corner room measures 26 × 15 ft. The arithmetic mean of these two dimensions gives the height of the room.

$$\frac{26 + 15}{2} = 20.5 \text{ ft}$$

Decline of the Musical Ratios in Architecture

The use of the musical ratios in art and architecture did not end with Palladio. Wittkower mentions their advocacy by Francesco Giorgi, Inigo Jones, Henry Wotton, Joshua Reynolds, Francois Blondel, and Bernardo Antonio Vittone. However, the end of their use was in sight. Wittkower goes on to chart the decline.

© Luke Daniek

FIGURE 1.28 ■ Villa Rotonda in Vicenza

In 1663, the French architect and physician Claude Perrault rejected the notion that certain ratios were in themselves beautiful, but only because people were used to them. He wrote that musical ratios cannot be translated into visual ratios. In his *Vita di Andrea Palladio* (1762), the Italian architect Tomasso Temanza wrote that proportion in music is completely different from proportion in architecture. He maintained that (a) the eye is not capable of judging simultaneously the various ratios in a room, length to width, length to height, and width to height; and (b) the proportions perceived will vary with the angle of vision under which a building is viewed. He also wrote that the use of harmonic proportions would lead to sterility. The Italian architect Guarino Guarini (c. 1670) wrote, "To please the eye one must take away from or add to the proportions, because one object is placed at eye level, another at great height, another in an enclosed space, and yet another in the open air."

In his *Analysis of Beauty* (1753), the English painter and engraver William Hogarth rejected any connection between mathematics and beauty. He refers to "the strange notion that if certain divisions of a string produce harmony to the ear . . . similar distances in . . . form . . . would delight the eye." The Scottish philosopher and historian David Hume, in his *Of the Standard of Taste* (1757), even rejected the notion that beauty is inherent in object, such as a building. He said, "beauty and deformity . . . are not qualities in objects, but belong entirely to the sentiment. . . ." In other words, beauty is in the eye of the beholder. The British statesman and orator Edmund Burke, in *On the Sublime and Beautiful* of 1757, denied that beauty had "anything to do with calculation and geometry." The English philosopher Lord Kames (Henry Home), in his *Elements of Criticism* (1761), noted that as we move about a room, the proportions of length to breadth vary continuously.

The Scottish clergyman and author Archibald Alison wrote *Essays on the Nature and Principles of Taste* (1790). For him, the trains of thought and the associations produced by artworks make them beautiful, and abstract or ideal standards destroy their function. Richard Payne Knight, in his *Analytical Inquiry into the Principles of Taste* (1805), wrote that the same spatial proportions that make one animal beautiful make another absolutely ugly, but the same ratios that produce harmony in a fiddle also produce harmony in a flute or a harp. Therefore, musical harmony and spatial proportions can have nothing in common. Finally, the English art critic John Ruskin, in *The Seven Lamps of Architecture* (1849), declared that possible proportions are as infinite as possible airs in music, and inventing beautiful proportions must be left to the inspiration of the artist.

SUMMARY

Now that we have completed the first leg of our journey, let us, like the two-faced god Janus who could see both past and future, look back at what we've covered and peek ahead to what is to come.

In this chapter, we explored the mathematical concepts of ratio and proportion, followed by sequences, series, and means. The three means were later used to describe how Plato constructed the musical scales, and how Renaissance architects computed building proportions. Sequences and series will be covered again in our discussions of the Fibonacci sequence, Chapter 2, and again in Roman architecture, Chapter 4. We also studied the first boxed and numbered formulas. More of them will be added in the coming chapters. For a look at the entire list of formulas, turn to Appendix F.

We also introduced Pythagoras, some Pythagorean number lore and musical discoveries, and the quadrivium. We touched on three subjects of the quadrivium (music, geometry, and arithmetic). We will discuss astronomy in Chapter 11, "The Spheres and Celestial Themes in Art and Architecture," where we expand on the idea of the music of spheres and celestial themes in art. Geometry will, of course, be the backbone of every chapter. In this chapter, we discussed plane Euclidean geometry, showing geometric figures of one and two dimensions, and we will stay with two-dimensional figures through Chapter 9. We will cover solid Euclidean geometry and three-dimensional figures in Chapters 10 and 11, touch on projective geometry in Chapter 12, and introduce fractal geometry and fractal dimensions in Chapter 13.

We'll add to Pythagorean number symbolism by talking about the symbolism of particular numbers as we get to them in the course of the book, and provide a complete summary in Appendix E. We will note geometric symbols, such as the Tetraktys and pentagram of the Pythagoreans. More of these symbols are listed in Appendix D. For recurring ideas and art motifs, we have discussed correspondences and reconciliation of opposites. We will add the trivium to the quadrivium, which together comprise the recurring motif of the seven liberal arts.

We performed some basic geometric constructions, followed by constructions for the three means. We will have a great many more constructions in later chapters. You can see a listing of them in Appendix C.

The geographical locations given here were ancient Crotone and Renaissance Italy. The other places we will visit are listed in Appendix A. In this chapter, Pythagoras, Plato, Vitruvius, Alberti, and Palladio, all of whom we will encounter again, were introduced. Our entire cast of characters can be found in Appendix B.

The main examples of art and architecture used in this chapter were the proportions of buildings. We discussed the need for systems of proportions in architecture, and cited some examples from Alberti and Palladio. We saw that the Renaissance analogy between audible and visible proportions echoed, for them, the harmonic structure of all creation. Music had a particular attraction; it was one of the quadrivium, while art and architecture were not. Its study and use helped to raise the status of art, which was considered a manual occupation. We will see in Chapter 12 that this was one of the reasons why artists studied perspective.

We then traced the decline in the use of the musical ratios in architecture, one of the main objections to it being that these proportions are not even visible to the viewer, given different angles of vision, limited fields of view, perspective, foreshortening, obstacles blocking vision, and so forth. It is hard to imagine that Renaissance architects such as Alberti, who developed the science of perspective, did not realize this at the time. Their point, however, was that church proportions should reflect the perfection of heaven even if they couldn't be seen. Returning to Alberti's statement, "I conclude that the same numbers by means of which the agreement of sounds affect our ears with delight are the very same which please our eyes *and our mind*." (Emphasis mine.) Wittkower adds, "It follows that perfect proportions must be applied to churches, whether or not the exact relationships are manifest to the 'outward' eye."[35]

The musical ratios were the basis for only one system of proportions. We will describe another system (the modulor system) based on the golden ratio in Chapter 2, and the Roman system of proportions (ad quadratum) based on the square in Chapter 4. All of that is ahead of us. For now, we'll proceed to the second leg of our journey, an investigation of the fascinating golden ratio.

EXERCISES AND PROJECTS

1. Define or describe the following terms:

figured numbers	gnomon	Sacred Tetraktys	correspondences
quadrivium	monochord	Plato's lambda	musical ratio

2. What are the three main systems of architectural proportions?

3. Name three requirements of an ideal system of proportions.

4. Given any two numbers an octave apart, say *a* and *2a*, prove algebraically that their arithmetic mean will give an interval of a fifth, that the harmonic mean will give an interval of a fourth, and that the interval between the two means gives the tone.

5. Reconstruct the harmonic scale using Plato's method. You can find his method in *Timaeus* para. 35–37. You may choose to crunch the numbers using a spreadsheet.

6. Make a painting, graphic design, poster, screensaver, or quilt in which the dimensions or locations of most of the elements form an arithmetic, geometric, or harmonic series. Write a few paragraphs explaining your design.

7. Make a design, as in the preceding project, based on the musical ratios.

8. Design a small building in which the façade and floor plan are based on the musical ratios.

9. Make a working model of Pythagoras' monochord. Devise and conduct some experiments with it, and summarize your results in a short paper.

10. Instead of making a monochord, do the previous project using one string of a violin or guitar. Use a pencil underneath the string as the movable bridge. Make sure the string does not touch the frets of the guitar during your experiments.

11. A *sonometer* is a physics laboratory apparatus that can be used as a monochord. If you can find one at your school, use it to do the previous project.

The next three projects will add to our growing list of geometric constructions. See Appendix C for the complete listing.

12. *Drawing a Parallel.* Reproduce the construction given in Figure 1.29, which shows one way to construct a parallel line through a given point. Starting at the given point *P*, draw a line that intersects the given line *L* at some point *A*. Using a compass setting *AP*, draw an arc from point *P* to locate point *C*. Draw a line from *C* to some point *B* on *L*. Bisect *BC*, locating midpoint *D*. Finally, draw line *PD*, the parallel to *L*.

13. *Subdividing a Line Segment into Equal Parts.* Do the construction for subdividing a line segment *AB* into three equal parts shown in Figure 1.30. First draw any line *L* through *A*. Using a compass, lay off three equally spaced points *C*, *D*, and *E* on *L*. Draw *BE*, and then draw lines through *C* and *D* parallel to *BE*. Points *F* and *G* will subdivide *AB* into three equal parts. This construction can be easily modified to subdivide *AB* into any number of equal parts.

14. *Proportional Division of a Line Segment.* Do the construction shown in Figure 1.31, which demonstrates the proportional division of a line segment into a ratio of 2 : 3. To begin, draw any line *L* through *A*. Using a compass, lay off five equally spaced points on *L*. Draw *BD*, and from the 2/5 point *C*, draw *CE* parallel to *BD*. Point *E* will subdivide *AB* in a ratio of 2 : 3. Other ratios can be constructed in a similar manner.

15. Write a report about any book in the list of sources at the end of this chapter.

16. Write a short research paper or an in-depth term paper. Feel free to come up with your own topic, using the following suggestions to jog your imagination.

   ■ How proportions were solved during the Renaissance using the so-called *rule of three*. Research the rule of three, describe it in a paper, and give examples of its use.
   ■ The Pythagorean legend
   ■ Pythagorean arithmetic, outlined in Heath, Chapter III (see Sources)
   ■ Pythagorean number lore
   ■ Pythagorean geometry, as in Heath, Chapter V
   ■ The quadrivium and the trivium, the seven liberal arts
   ■ The history of the gnomon, outlined in Heath, p. 79
   ■ The physics of the vibrating string
   ■ The various kinds of means, including the three given in this chapter, and seven more, as described in Heath, p. 87
   ■ Raphael's *School of Athens*

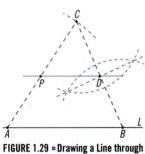

FIGURE 1.29 ■ Drawing a Line through a Given Point *P* Parallel to a Given Line *L*

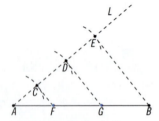

FIGURE 1.30 ■ Division of a Line Segment into Equal Parts

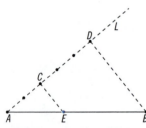

FIGURE 1.31 ■ Proportional Division of a Line Segment

- ■ Plato, as outlined in Heath, p. 285
- ■ Plato's Academy
- ■ Plato's *Timaeus*
- ■ The historical development of musical scales
- ■ Vitruvius' *Ten Books on Architecture* and the impact it has had on the field
- ■ The three classic orders of architecture
- ■ Palladio's *Fugal System of Proportions*, described in Wittkower, p. 126 (see Sources), showing how Palladio coordinated the proportions of each room with *other* rooms in a building
- ■ Leonardo's advocacy of the musical ratios, illustrated by his statement that "music is the sister of painting." Both convey harmonies, music by chords and painting by proportions.
- ■ Leonardo's musical abilities
- ■ How musical intervals and linear perspective are subject to the same numerical ratios, for objects of equal size receding at equal intervals diminish in harmonic progression

- ■ Literary references to the *Music of the Spheres*
- ■ Guido Monaco and the invention of musical notation
- ■ Music as one of the subjects of the quadrivium (Figure 1.32)

17. Make an oral presentation to your class for any of the previous projects or papers.

18. If you play an instrument or sing, make a short musical presentation to your class in which you demonstrate the musical ratios mentioned in this chapter.

FIGURE 1.32 ■ Relief Carving of *Musica* at Pisa Duomo

Mathematical Challenges

The following problems require more mathematics than is presented in the text. Try some of them if you have the mathematical background.

19. Given an arithmetic mean A, a geometric mean G, and a harmonic mean H between the positive numbers a and b, where $a \neq b$, show that

$$A \geq G \geq H.$$

20. A driver goes a certain distance at a speed of 50 mi/h and returns to the starting point at 60 mi/h.
    a. Show that the arithmetic mean $(50 + 60)/2$ does *not* give the average speed for the entire round trip.
    b. Show that the harmonic mean *does* give the correct average speed for the round trip.

21. Prove that Euclid's construction in *Elements*, Book VI, Prop. 13, does indeed give the mean proportional.

22. The construction for the mean proportional can be used to geometrically find the square root of a number. Use it to find the square root of 7.64. Check your answer by calculator.

23. Prove that Palladio's construction does give the harmonic mean.

SOURCES

Aaboe, Asger. *Episodes from the Early History of Mathematics*. New York: Random House, 1964.

Alberti, Leon Battista. *The Ten Books of Architecture*. New York: Dover, 1986. A 1775 edition of the work written in 1452.

Cornford, Francis. *Plato's Cosmology*. New York: Harcourt, 1937.

De Vogel, C. J. *Pythagoras and Early Pythagoreanism*. Assen, the Netherlands: Van Gorcum, 1966.

Euclid. *The Thirteen Books of the Elements*. New York: Dover, 1956.

Grabow, Stephen. "Frozen Music: The Bridge Between Art and Science," article in *Companion to Contemporary Architectural Thought*, Farmer et al., eds. New York: Routledge, 1993, pp. 438–443.

Guthrie, Kenneth S. *The Pythagorean Sourcebook and Library*. Grand Rapids, MI: Phanes,1987.

Heath, Sir Thomas. *A History of Greek Mathematics*. New York: Dover, 1981. First published in 1921.

Hersey, George L. *Architecture and Geometry in the Age of the Baroque*. Chicago: University of Chicago, 2000.

Hersey, John. *Possible Palladian Villas*. Cambridge, MA: MIT Press, 1922.

Hersey, John. *Pythagorean Palaces*. Ithaca, New York: Cornell University Press, 1976.

Kappraff, Jay. *Connections: The Geometric Bridge Between Art and Science*. New York: McGraw-Hill, 1990.

March, Lionel. *Architectonics of Humanism*. Chichester, UK: Academy, 1998.

Newman, James R., ed. *The World of Mathematics*. New York: Simon and Shuster, 1956.

Palladio, Andrea. *The Four Books of Architecture*. New York: Dover, 1965.

Papadopoulos, Athanase. "Mathematics and Music Theory: From Pythagoras to Rameau." *The Mathematical Intelligencer,* Winter 2002, pp. 65–73.

Philip, J. A. *Pythagoras and Early Pythagoreanism*. Toronto: University of Toronto, 1966.

Plato. *Timaeus*. Edited and translated by John Warrington. London: Dent, 1965. Original c. 360 B.C.

Rowe, Colin. *The Mathematics of the Ideal Villa and Other Essays*. Cambridge, MA: MIT Press, 1976.

Vitruvius. *The Ten Books on Architecture*. New York: Dover, 1960.

Wittkower, Rudolf. *Architectural Principles in the Age of Humanism*. New York: Random House, 1965.

NOTES

1. Guthrie, p. 61.
2. Theon of Smyrna, *Mathematics Useful for Understanding Plato*, p. 66.
3. Aristotle, *Metaphysics* (I. 5 986 a 23).
4. Alberti, Book IX, Chapter V.
5. Hopper, p. 90.
6. Daniel Orth Bell. "New Identifications in Raphael's School of Athens." *Art Bulletin,* Dec. 1995, p. 639.
7. Clark, *Civilization*, p. 132.
8. Genesis 4:21.
9. *Republic*, p. 527.
10. Philebos, p. 51.

11. *Republic,* p. 603.
12. *Timaeus,* pp. 37–38.
13. *Timeaus,* p. 35 B, C.
14. Wittkower, p. 104.
15. Cornford, p. 72.
16. Tillyard, p. 101.
17. Wittkower, p. 142.
18. *Merchant of Venice,* V, I, 57.
19. *Harmonice Munde* (1619).
20. *A Song for St. Cecilia's Day.* 1687.
21. In "Advice to an Author," *Characteristicks,* 1737, I. p. 353.
22. Vitruvius I, I, 4.
23. *Proceedings of Nexus '96,* p. 115.
24. Vitruvius I, II, 4.
25. Vitruvius III, I, 4.
26. Kappraff in *Proceedings of Nexus '96,* p. 115.
27. Wittkower, p. 101.
28. Wittkower, p. 27.
29. Wittkower, p. 112.
30. Alberti, Book IX, Chapter V.
31. Wittkower, p. 45.
32. Kappraff in *Proceedings of Nexus '96.*
33. Palladio, Book IV, Chap. V.
34. See March, pp. 277–278, for an extensive list of ratios used by Palladio.
35. Wittkower, p. 27.

"Geometry has two great treasures: one is the theorem of Pythagoras; the other, the division of a line into extreme and mean ratio. The first we may compare to a measure of gold; the second we may name a precious jewel."[1]

JOHANNES KEPLER

FIGURE 2.1 ■ Euclid in Raphael's *School of Athens* (detail)

2

■

The Golden Ratio

●

In the previous chapter, we introduced ratios, sequences, and series and their importance in determining the musical ratios. This chapter will explore another ratio, the golden ratio, which has been called *the world's most astonishing number*.[2] We will begin with an investigation of the golden ratio as defined by Euclid (Figure 2.1) and discuss its relationship to the golden sequence and the Fibonacci sequence. The Fibonacci numbers, which are amazing in themselves, are closely tied to the golden ratio and to the musical ratios of the preceding chapter.

In addition, we will explore whether or not the golden ratio was used in architecture by looking at three specific examples: the Great Pyramid, the Parthenon, and the architecture of Le Corbusier. The question of whether or not the golden ratio can be found in paintings and sculpture will be discussed in later chapters.

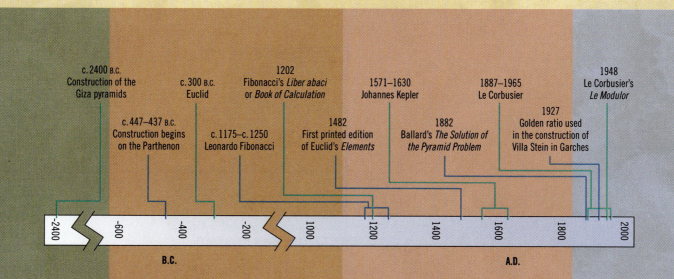

c. 2400 B.C.
Construction of the
Giza pyramids

c. 447–437 B.C.
Construction begins
on the Parthenon

c. 300 B.C.
Euclid

c. 1175–c. 1250
Leonardo Fibonacci

1202
Fibonacci's *Liber abaci*
or *Book of Calculation*

1482
First printed edition
of Euclid's *Elements*

1571–1630
Johannes Kepler

1882
Ballard's *The Solution of
the Pyramid Problem*

1887–1965
Le Corbusier

1927
Golden ratio used
in the construction of
Villa Stein in Garches

1948
Le Corbusier's
Le Modulor

-2400   -600   -400   -200   1000   1200   1400   1600   1800   2000

B.C.   A.D.

THE GOLDEN RATIO

The first known definition of the golden ratio is given by Euclid in his famous *Elements*. He does not use the more modern term *golden ratio* but refers to it as the *extreme and mean ratio*. In Book IV, Definition 3, he states,

> "A straight line is said to have been cut in extreme and mean ratio when, as the whole line is to the greater segment, so is the greater to the less."

Let's rephrase Euclid's definition of the extreme and mean ratio and refer to it by its modern name, *the golden ratio*.

6a THE GOLDEN RATIO

The golden ratio divides a whole so that the smaller part is to the larger as the larger is to the whole.

Figure 2.2 shows a line divided into two parts: a smaller part, which we call one unit, and a larger part. Thus the ratio of the larger part to the smaller is, by definition, the *golden ratio*.

In Euclid's Book II, Proposition 11, the goal is to subdivide a line segment *AB* at *H* so that the rectangle of length *AB* and width *BH* has the same area as the square of side *AH*, as shown in Figure 2.3. We can see that *H* subdivides *AB* into the golden ratio by equating the two areas:

$$(AB)(BH) = (AH)(AH)$$

or

$$\frac{AB}{AH} = \frac{AH}{BH}.$$

In other words, the whole *AB* is to the larger segment *AH* as the larger segment *AH* is to the smaller segment *BH*, precisely our definition of the golden ratio.

The golden ratio, or extreme and mean ratio, has also been called the *golden section*, *golden mean*, *golden number*, and *divine proportion*. It is often denoted by the Greek letter *Phi* (Φ) after the Greek sculptor Phidias, who supposedly used the golden ratio in his sculpture, and we will follow that practice.

Algebraic Derivation of the Golden Ratio

Recall that in the preceding chapter, you learned how to subdivide a line so that the parts of the line were in any given ratio. Figure 2.2 illustrates the subdivision of a line into a specific ratio, the important golden ratio. In the figure, a line is

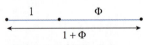

FIGURE 2.2 ■ The Golden Ratio

Propositions regarding the extreme and mean ratio are scattered throughout the *Elements*. We first come across it in disguised form in Book II, Proposition 11, where *extreme and mean ratio* is not even mentioned, and the construction for cutting a line in extreme and mean ratio is given in Book VI, Proposition 30. We show these in the exercises following this section. Construction of golden triangles and the use of the golden ratio in the pentagon are located elsewhere in the *Elements,* and we will cover these as we get to those geometric figures.

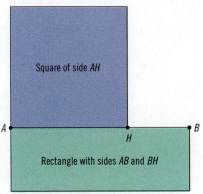

FIGURE 2.3 ■ Euclid Book II, Proposition 11

divided into two parts: a smaller part, labeled 1 or the unit, and a larger part, labeled Φ. According to the definition of the golden ratio, the ratio of the smaller part of the line (1 unit) to the larger part (Φ units) must equal the ratio of the larger part (Φ) to the whole line (1 + Φ), giving us

$$1 : \Phi = \Phi : (1 + \Phi)$$

or

$$\frac{1}{\Phi} = \frac{\Phi}{1 + \Phi}.$$

In other words, Φ is the *mean proportional* between the smaller part and the whole. Note the simplicity of this proportion, which contains just two quantities (1 and Φ) instead of the usual three or four quantities found in a proportion.

We can determine the value of the golden ratios, and hence investigate some of the amazing properties of the golden ratios, by solving for Φ using the quadratic formula. You might remember how to solve quadratic equations from other mathematics classes. If not, here's a quick reminder. A *quadratic equation* is an equation of the form $ax^2 + bx + c = 0$. It contains the variable (x in this case) squared, and often the same variable, not squared, and a constant term. When the terms are arranged in decreasing order of the exponents, as they are here, the equation is said to be in *general form*.

■ **EXAMPLE:** Write the equation $3 + 5x = 2x^2$ in general form, and identify the coefficients *a*, *b*, and *c*.

● **SOLUTION:** Rearranging the terms gives $2x^2 - 5x - 3 = 0$. Here, $a = 2$, $b = -5$, and $c = -3$.   ■

One way to solve a quadratic equation is by the *quadratic formula*.

5   QUADRATIC FORMULA

The solution to the quadratic equation $ax^2 + bx + c = 0$ is

$$x = \frac{-b \pm \sqrt{b^2 - 4ac}}{2a}.$$ ●

■ **EXAMPLE:** Solve the quadratic equation from the preceding example.

● **SOLUTION:** Substituting into the quadratic formula with the values found above, $a = 2$, $b = -5$, and $c = -3$, gives us

$$x = \frac{-(-5) \pm \sqrt{(-5)^2 - 4(2)(-3)}}{2(2)}$$

$$= \frac{5 \pm \sqrt{25 + 24}}{4}$$

$$= \frac{5 \pm \sqrt{49}}{4} = \frac{5 \pm 7}{4}$$

$$= 3 \quad \text{and} \quad -\frac{1}{2}.$$ ■

Similarly, to solve for Φ, we begin with the equation we derived from the definition of the golden ratio:

$$\frac{1}{\Phi} = \frac{\Phi}{1 + \Phi}.$$

Next, multiplying both sides by Φ and rearranging the terms into general form gives us

$$\Phi^2 = 1^2 + \Phi$$

or

$$\Phi^2 - \Phi - 1 = 0.$$

Then, by the quadratic formula,

$$\Phi = \frac{-(-1) \pm \sqrt{(-1)^2 - 4(1)(-1)}}{2(1)}.$$

Taking the positive value of Φ gives us

6b THE GOLDEN RATIO

$$\Phi = \frac{1 + \sqrt{5}}{2} \cong 1.61803\ldots \qquad \bullet$$

The approximate value 1.61803 is a nonterminating, nonrepeating decimal that we rounded to five decimal places. Taking the negative value, we get

6c THE GOLDEN RATIO

$$\Phi = \frac{1 - \sqrt{5}}{2} \cong -0.61803\ldots \qquad \bullet$$

Sometimes, we want the ratio of the smaller part (1) to the larger part (Φ). If $\Phi/1 \approx 1.61803$, then $1/\Phi$ is

7a RECIPROCAL OF THE GOLDEN RATIO

$$\frac{1}{\Phi} = \frac{2}{1 + \sqrt{5}} \cong 0.61803 \qquad \bullet$$

Notice that this value is of the same absolute magnitude as the value obtained in Equation 6c. You can compute another relationship for $1/\Phi$ by dividing each term in the equation

$$\Phi^2 - \Phi - 1 = 0$$

by Φ, and rearranging them.

7b RECIPROCAL OF THE GOLDEN RATIO

$$\frac{1}{\Phi} = \Phi - 1 \qquad \bullet$$

Finally, let's get another relationship by adding Φ and 1 to both sides of the equation:

$$\Phi^2 - \Phi - 1 = 0.$$

We obtain

$$\Phi^2 - \Phi - 1 + \Phi + 1 = 0 + \Phi + 1,$$

or the square of the golden ratio.

8 SQUARE OF THE GOLDEN RATIO

$$\Phi^2 = \Phi + 1 \qquad \bullet$$

In other words, you can compute the square of Φ simply by adding 1 to Φ. Some unusual properties for this amazing number are quickly becoming evident.

Incommensurability of the Golden Ratio

"The golden mean is quite absurd;
It's not your ordinary surd.
If you invert it (this is fun!),
You'll get itself reduced by one;
But if increased by unity,
This yields its square, take it from me."[4]

PAUL S. BRUCKMAN

Two line segments are said to be *commensurable* if each is a whole-number multiple of some unit of length. For example, a line segment having a length of 3.5 feet (ft) is commensurable with a line segment having a length of 2.5 yards (yd). Both have a common unit of measure, the inch (in.). Therefore, 3.5 feet contain exactly 42 inches, and 2.5 yards contain exactly 90 inches. Similarly, a line segment having a length of 3.7 feet is commensurable with a line segment having a length of 2.9 yards. Both have a common unit of measure, the tenth of an inch (0.1 in.). Therefore, 3.7 feet contain exactly 444 tenth-inches, and 2.9 yards contain exactly 1044 tenth-inches.

Another example of commensurability is that a centimeter is commensurable with an inch; 1 inch equals 2.54 centimeters, exactly. If you use a line segment of 0.01 inch long as the unit of measure, then 1 inch contains exactly 100 such units, while 1 centimeter contains exactly 254 such units. One last example of commensurability is that the line segments of length $\frac{5}{2}$ and $\frac{3}{5}$ are also commensurable. If you take $\frac{1}{10}$ as the unit of measure, the first line segment contains exactly 25 units, and the second contains exactly 6 units.

Two line segments that are *not* commensurable are called *incommensurable*. Another interesting property of the golden ratio is that it is an incommensurable quantity. When two parts of a line segment are divided by the golden ratio, they have no common measure. Also, the length of the original segment is incommensurable with the length of either segment.

The diameter and the circumference of the same circle comprise another example of incommensurable quantities. It is impossible to find any unit of measure, no matter how small, that is exactly contained by both. That is, there is no unit for which both the diameter and the circumference are exact multiples of that unit. The ratio of circumference to radius is the irrational number π.

The side and the diagonal of the same square form one last example of incommensurable quantities. As with the diameter and circumference of a circle, it is impossible to find any unit of measure that is exactly contained by both. The ratio of the diagonal to the side is the irrational number $\sqrt{2}$.

The discovery of incommensurable quantities, presumably by the Pythagoreans, made it necessary to extend the number system to what are called *irrational numbers*. An irrational number is a real number that cannot be expressed either as an integer or as a ratio of two integers. Therefore, $\sqrt{2}$ (the square root of two) is the irrational number giving the length of the diagonal of a square of side 1, and $\sqrt{5}$ is the irrational number that arises from an algebraic derivation of the golden ratio. Written in decimal form, an irrational number is always infinite and nonrepeating. It cannot be expressed exactly with a decimal number of finite length.

Geometric Construction of the Golden Ratio

Finding Φ by geometric construction is easy; just follow these steps, which are illustrated in Figure 2.4. Start with a square *ABCD*, 1 unit on a side, and bisect it

Proving that a number is irrational is not easy. See the exercises following this section for a proof that $\sqrt{2}$ is irrational.

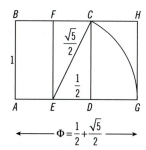

FIGURE 2.4 ■ Geometric Construction of the Golden Ratio

Can you prove that the construction in Figure 2.4 gives the golden ratio? See the exercises following this section.

vertically with line *EF*. Then swing an arc of radius *EC* to the extension of *AED*. *AG* will equal Φ. Rectangle *ABHG* is called a *golden rectangle*, which will be discussed in Chapter 4, "Ad Quadratum and the Sacred Cut."

The Golden Sequence

A *golden sequence* is a geometric progression with a first term of 1 and a common ratio of Φ.

16 GOLDEN SEQUENCE

$$1 \quad \Phi \quad \Phi^2 \quad \Phi^3 \quad \Phi^4 \quad \Phi^5 \ldots$$

●

We have already shown in Equation 8 that

$$\Phi^2 = 1 + \Phi.$$

Then multiplying by Φ gives

$$\Phi\Phi^2 = \Phi (1 + \Phi)$$
$$\Phi^3 = \Phi + \Phi^2.$$

Multiplying by Φ again gives

$$\Phi^4 = \Phi\Phi^3 = \Phi (\Phi + \Phi^2)$$
$$= \Phi^2 + \Phi^3.$$

In other words, *each term in the golden sequence equals the sum of the two preceding terms.* Such a sequence is sometimes called an *additive* sequence. As such, the golden sequence is both an additive sequence and a geometric progression.

Geometric Construction of the Golden Sequence

The golden sequence is easily constructed by compass once the lengths 1 and Φ have been determined. Just follow the steps illustrated in Figure 2.5. To begin, lay off the given lengths 1 and Φ. With radius $1 + \Phi$, swing an arc from *A* and locate point *B*. The distance *AB* is equal to Φ^2. With radius $\Phi + \Phi^2$, swing an arc from *B* and locate point *C*. Then distance *BC* is equal to Φ^3. Continue for the other terms in the sequence.

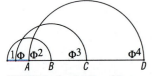

FIGURE 2.5 ■ Construction of the Golden Sequence

■ EXERCISES ● THE GOLDEN RATIO

1. Define or describe the following terms:

golden ratio	extreme and mean ratio	irrational number
divine proportion	incommensurable	golden number
commensurable	golden mean	quadratic formula
golden section	quadratic equation	golden sequence

2. In your own words, explain what the golden ratio is.

3. Repeat the algebraic derivation of the golden ratio.

4. Construct the golden ratio using the method in Figure 2.4.

5. Prove that the construction in Figure 2.4 does, in fact, give the golden ratio. You will need to use the Pythagorean theorem (Equation 39).

6. Figure 2.6 illustrates another construction of the golden ratio. Perform this construction. (This construction is attributed to Hero [or Heron] of Alexandria, a Greek mathematician and physicist of the first century A.D.) Construct a right triangle *ABC*, where *BC* is half *AC*. From *B*, swing an arc

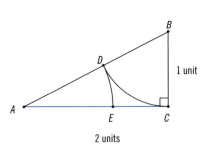

1 unit

2 units

FIGURE 2.6 ■ Hero's Construction of the Golden Ratio

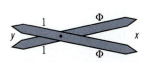

a

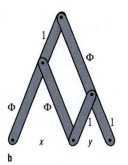

b

FIGURE 2.7 ■ Dividers for the Golden Ratio

from *C*, locating *D*. From *A*, swing an arc from *D*, locating *E*. Then *AE* divides *AC* by the golden ratio.

7. Demonstrate that the construction in Figure 2.6 gives the golden ratio.

8. Make one of the golden ratio dividers shown in Figure 2.7. Use it to quickly lay out the golden ratio. In (a), this is simply two strips of cardboard, wood, or metal, hinged at a point that subdivides each leg in the golden ratio. The ratio of the openings, *x/y*, is then Φ. In (b), each of the two long legs has a hinge pin that subdivides it in the golden ratio. The two shorter legs are proportioned as shown. The ratio of the openings, *x/y*, is then Φ.

9. Prove that the devices in the preceding exercise give the golden ratio.

10. Given a line segment 15.8 cm in length, calculate the point that will subdivide the line in the golden ratio. Check your calculation using your golden ratio dividers.

11. Pick any two numbers (for example, 15.3 and 18.5) and add them (getting 33.8 in this case).

    15.3   18.5   33.8

    Then add the last number to the one before it.

    15.3   18.5   33.8   52.3

    Again add the last number to the one before it.

    15.3   18.5   33.8   52.3   86.1

    Using a spreadsheet or by hand, repeat this sequence for as long as you want. Then divide each number by the one before it to obtain another sequence of numbers. Does the sequence of quotients appear to be approaching a particular value? Can you determine what that value is?

12. Enter Φ into your calculator and square it. What do you see? Then enter Φ into your calculator and take its reciprocal. What do you see? Can you explain these results?

13. Repeat the construction to subdivide a line in a Φ series.

14. *Euclid's Proposition 11 in Book II, by CAD.* Construct the square and rectangle shown in Figure 2.3 with a computer drafting program. Have the program measure the area of each and verify that they are equal.

15. *Euclid's Division of a Line in Extreme and Mean Ratio, by CAD.* Perform the construction in Figure 2.8. Let *AB* be the line that is to be divided using the golden ratio. Construct a square *ABCD* on *AB*. Using a computer drafting program, construct rectangle *DHFG* equal in area to square *ABCD*, so that *AEFG* is a square. Then *E* cuts *AB* in the golden ratio. Demonstrate by measurement that this is so.

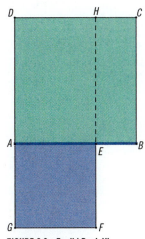

FIGURE 2.8 ■ Euclid Book VI, Proposition 30

16. Manually repeat the construction in Euclid Book II, Proposition 11 (see Figure 2.3).

   a. Subdivide a line segment AB at H so that the rectangle of length AB and width BH has the same area as the square with side AH.

   b. Show that H subdivides AB in the golden ratio by equating the two areas.

17. *Proof that the Square Root of 2 is Irrational.* Try to follow this proof. Consider presenting it to your class. We assume that $\sqrt{2}$ is *rational*, so that it can be expressed as the ratio of two integers a and b.

$$\sqrt{2} = \frac{a}{b}$$

where, without loss of generality, we can require that the fraction a/b is reduced to lowest terms. That is, a and b have no common factors (other than 1). Removing fractions, we get

$$a = \sqrt{2}b.$$

Squaring both sides gives

$$a^2 = 2b^2.$$

From this, we can conclude that a^2 is even, and also that a is even. Can you explain why?

   Now, let's substitute $2c$ for a, getting

$$4c^2 = 2b^2$$

or

$$2c^2 = b^2.$$

Now, we can conclude that b^2 and hence b are even, so b must have a factor of 2. However, this is impossible because we required that a and b have no common factors (we had shown in the preceding step that a is even and so must have a factor of 2). Therefore, our starting assumption that $\sqrt{2}$ is rational must be false.

THE FIBONACCI NUMBERS AND THE GOLDEN RATIO

Now that you've been introduced to the golden ratio, let's look at a famous sequence of numbers that is closely intertwined with that ratio, the *Fibonacci sequence*. This sequence is related to the previously described musical ratios that found their way into architectural proportions. Following this section, we will investigate whether or not the golden ratio was also the basis of architectural proportions.

The Fibonacci Sequence

The *Liber abaci*, or *Book of Calculation*, was the first work by a European on Indian and Arabic mathematics. Despite its name, the book is about how to calculate *without* using an abacus. (At that time in Italy, the word *abaci* generally meant "practical calculation.") Written in 1202, it successfully promoted Hindu–Arabic numerals over the cumbersome Latin numerals in use until then. The publication of this book brought Fibonacci wide recognition. However, much of his fame today rests on a single problem tucked among many others in Chapter 12 of his book. "A certain man had one pair of rabbits in a certain enclosed place, and one wishes to know how many are created from the pair in one year when it is the nature of them in a single month to bear another pair, and in the second month those born to bear also."[5]

Let's make a table showing pairs of rabbits each month. Let *I* stand for one pair of infant (nonproductive) rabbits, and *A* stand for one pair of adult (productive) rabbits. Let's begin with a pair of infants in January who are then adults in February. They produce an infant in March and in every following month. Their March infant becomes an adult in April and produces an infant in May and in every following month, and so on. Here are the results for seven months. Note that, in each month, every *A* is replaced by *AI*, while every *I* is replaced by *A*.

		Total Pairs
January	*I*	1
February	*A*	1
March	*A I*	2
April	*A I A*	3
May	*A I A A I*	5
June	*A I A A I A I A*	8
July	*A I A A I A I A A I A A I*	13

The total number of pairs per month, both infant and adult, is shown in the column to the right. They form the Fibonacci sequence.

17 FIBONACCI SEQUENCE

$$1 \quad 1 \quad 2 \quad 3 \quad 5 \quad 8 \quad 13 \quad 21\ldots$$

Note that each term (except the first two) is found by adding the two terms immediately preceding it. Recall from the preceding section that the golden sequence had the same property and was called an additive sequence.

In general, the Fibonacci sequence is described by the following formula.

18 RECURSION FORMULA FOR THE FIBONACCI SEQUENCE

$$F_n = F_{n-1} + F_{n-2}$$

This sequence has dozens of interesting properties, some of which are given in the exercises at the end of this section. There is even a journal, the *Fibonacci Quarterly*, devoted to the study of Fibonacci numbers. The property of most interest to us is how the sequence relates to the golden ratio. Let's divide each term of the Fibonacci sequence by the term immediately preceding it. These quotients, each rounded to three decimal places, produce the numbers shown below in the second row:

1	1	2	3	5	8	13	21	34...
1.000	2.000	1.500	1.333	1.600	1.625	1.615	1.619...	

Notice that the ratio of each term to the one immediately preceding it appears to oscillate about the golden ratio (approximately 1.618), first smaller than Φ, then larger. This ratio gets closer to Φ as we move farther out in the sequence. We say that the ratio of the terms *approaches* Φ. There is a compact way of writing this idea, using *limit notation*. If F_n is any term of a Fibonacci sequence and F_{n-1} is the preceding term, we have

19 RATIO OF TWO CONSECUTIVE TERMS IN THE FIBONACCI SEQUENCE

$$\Phi = \lim_{n \to \infty} \frac{F_n}{F_{n-1}}$$

In words: Φ is the limiting value, as *n* approaches infinity, of the quotient F_n over F_{n-1}. ●

FIGURE 2.9 ■ Statue of Fibonacci in the Campo Santo, Pisa

You can use a formula to find any term of a Fibonacci sequence. This formula is sometimes named for the French mathematician Jacques Phillipe Marie Binet (1786–1856), although it was probably known earlier.

20a BINET'S FORMULA

$$F_n = \frac{1}{\sqrt{5}}\left[\left(\frac{1+\sqrt{5}}{2}\right)^n - \left(\frac{1-\sqrt{5}}{2}\right)^n\right]$$

You might recognize the quantity in the first set of parentheses as Φ and that within the second set as $1/\Phi$. Substituting these gives a simpler form for Binet's formula.

20b BINET'S FORMULA

$$F_n = \frac{1}{\sqrt{5}}\left[\Phi^n - \left(-\frac{1}{\Phi}\right)^n\right]$$

■ **EXAMPLE:** Find the tenth term of the Fibonacci sequence.

● **SOLUTION:** Substituting into Equation 20b with $n = 10$,

$$F_{10} = \frac{1}{\sqrt{5}}\left[\Phi^{10} - \left(-\frac{1}{\Phi}\right)^{10}\right]$$

$$= \frac{122.99 - (-0.00813)}{\sqrt{5}} = 55.$$

We are using irrational numbers, which in decimal form are nonterminating, so it is not possible to work to their full accuracy. However, we know that Fibonacci numbers are *whole* numbers, so we can round the result to the nearest integer. ■

Glance back at Table 1.1 in Chapter 1, which shows the diatonic musical scale. You can see that the note A is at an interval of 5/3 from C, a major sixth. *Both 5 and 3 are Fibonacci numbers*. Further, you can obtain a minor sixth by taking the interval from E (330 Hz) to high C (528 Hz). This interval, 528/330, reduces to 8/5; these numbers are *also Fibonacci numbers*. As you have seen, each of these quotients gives an approximate value for Φ.

Golden number advocates claim that the major sixth and the minor sixth are especially pleasing intervals to the ear, and they credit the approximation to Φ as a reason for this.

■ EXERCISES • THE FIBONACCI NUMBERS AND THE GOLDEN RATIO

1. Define or describe the following terms:

   Fibonacci number   recursion formula
   Fibonacci sequence   Binet's formula

2. Use a spreadsheet to generate the first 50 Fibonacci numbers.
   a. In a separate column, compute the ratio of each number to the one before it. What do you see?
   b. Demonstrate that the sum of the first n terms $F_0, F_1, F_2, \ldots F_n$, increased by 1, equals F_{n+2}.
   c. Demonstrate that the square of any term is either 1 unit more or 1 unit less than the product of the two terms adjacent to it, or

   $$(x_{n+1})(x_{n-1}) = x_n^2 + (-1)^n.$$

d. Demonstrate that the difference of the squares of two Fibonacci numbers with subscripts that differ by 2 is also a Fibonacci number or

$$(F_{n+1})^2 - (F_{n-1})^2 = F_{2n}.$$

3. Start with the number 1. In the next row, replace the number 1 with the number 10, getting 10. Then in subsequent rows, replace each 1 with 10 and each 0 with 1. You should get

```
          1
         1 0
        1 0 1
       1 0 1 1 0
      1 0 1 1 0 1 0 1
```

Now, count the number of 1s in each row. What do you see?[6]

THE GOLDEN RATIO IN ARCHITECTURE

As you saw in Chapter 1, "Music of the Spheres," many architects were mindful of the proportions they used in their works throughout much of art history. In Chapter 1, we also explored the use of the musical ratios in building design, and now we will explore the use of the golden ratio for the same purpose.

Let's start with a structure where the golden ratio seems to be present but where its use may not have been intentional, follow with a temple widely claimed to contain the golden ratio but where its presence is not convincing, and end with a villa where the golden section was expressly used by the architect.

The Great Pyramid

The three pyramids of the ancient Giza Necropolis are located in what is today Cairo, Egypt. Each pyramid was built during the lifetime of a single king and was intended to help that king ascend to the gods. Most were made in the Fourth Dynasty of the Old Kingdom, about 2800 B.C. Herodotus, known as *the father of history*, saw the pyramids around 440 B.C. and, in his *History*, wrote that the pyramids, already ancient, were covered with a mantle of highly polished limestone with joints so fine they could hardly be seen.[7]

The pyramids are supposed to have many "secrets," and many claims have been made about them, including the following:

- They are models of the earth.
- They form part of an enormous star chart.
- Their shafts are aligned with certain stars.
- They are part of a navigational system to help travelers in the desert find their way.

Although these and other claims have been made, we will examine only the two that have direct bearing on our exploration of the links among art, architecture, and geometry. One claim is that the Great Pyramid, the oldest of the pyramids and the last remaining of the Seven Wonders of the World, contains the squaring of the circle. This claim will be investigated in Chapter 8, "Squaring the Circle." Another claim is that the Great Pyramid contains the golden ratio; in particular, that the ratio of the slant height *s* to half the base *b* is the golden ratio. Let's examine this claim in greater detail now.

If you take a cross section through the Great Pyramid, you'll get a triangle, often called the *Egyptian triangle*, which supposedly contains the golden ratio. Figure 2.10 illustrates the approximate dimensions of the Great Pyramid and the Egyptian triangle.[8] Because the base dimension is not exactly the same for each side, this illustration uses their average, for simplicity. The slant height is calculated from the pyramid's height and base dimension. Using these dimensions, we can determine whether or not the Great Pyramid contains the golden ratio.

Dividing slant height *s* by the half-base *b* gives

$$186.6 \div 115.25 = 1.619,$$

which agrees with our value for the golden ratio ($\Phi = 1.618$) at least to two decimal places. Therefore, within the accuracy of the original dimensions, *the ratio of the slant height to half the base is equal to the golden ratio.*

Any pyramid with a square base can be described using only two numbers (for example, the base dimension and the height). However, the *proportions* of such a pyramid can be fully described using only *one* number: the ratio of any two of the pyramid's dimensions.

Let's describe the Great Pyramid in terms of ratios. If we let half the base of the Great Pyramid be one unit, then the slant height of each face is equal to Φ. The pyramid height *h* can be found by the Pythagorean theorem (Equation 39):

$$\Phi^2 = h^2 + 1^2.$$

We've already shown (Equation 8) that

$$\Phi^2 = \Phi + 1.$$

Equating Φ^2 in each of the previous equations, we get

$$\Phi + 1 = h^2 + 1,$$

which means that h^2 must equal Φ. Therefore,

$$h = \sqrt{\Phi}.$$

Therefore, the sides of the Egyptian triangle are in the ratio

$$1 : \sqrt{\Phi} : \Phi,$$

as shown in Figure 2.11.

The Egyptian triangle is also known as the *Kepler triangle*, after the German astronomer Johannes Kepler (1571–1630). In a letter to a former professor, Kepler stated a theorem, which can be paraphrased as follows: *If the sides of any right triangle are in geometric ratio, then the sides must be* $1 : \sqrt{\Phi} : \Phi$.

The American writer Peter Tompkins writes, "The measurements made . . . definitely prove the occurrence of the Golden Section throughout the architecture

> The Pythagorean theorem states that in a right triangle, the square of the hypotenuse (the longest side) equals the sum of the squares of the other two sides. We will cover this theorem in detail in Chapter 3, "The Triangle."

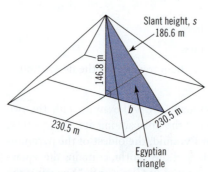

FIGURE 2.10 ■ Approximate Dimensions of the Great Pyramid

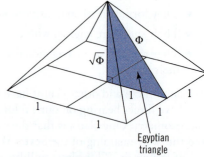

FIGURE 2.11 ■ Great Pyramid Dimensions in Terms of Φ

THE STAR CHEOPS

On his way to Australia to become chief engineer of the Australian railways, a British railway engineer, Robert Ballard, saw the pyramids. From a moving train, he observed how the relative appearance of the three pyramids on the Giza plateau changed. He concluded that they were used as sighting devices, and in 1882 he wrote a book with the grand title *The Solution of the Pyramid Problem*.[9]

He also noted that the cross section of the Great Pyramid is made up of two Egyptian triangles. He then constructed what he called a *Star Cheops,* which he said "is the geometric emblem of extreme and mean ratio and the symbol of the Egyptian Pyramid Cheops." For us, the Star Cheops provides an easy way to construct a model of the Great Pyramid.

Figure 2.12 demonstrates the construction of a Star Cheops, using the following steps. First, draw line segment *OB* of any convenient length. Using one of the preceding constructions for Φ, subdivide *OB* at *A* by the golden ratio. Next, draw circles of radii *OA* and *OB*. Finally, circumscribe a square about the smaller circle and draw four triangles, each with its base on the square and a

vertex on the outer circle. Triangle *ABC* is an Egyptian triangle, so triangle *BCD* has the proportions of one face of the Great Pyramid. Therefore, the triangles, when folded up and joined at their edges, will form a pyramid having the same proportions as the Great Pyramid.

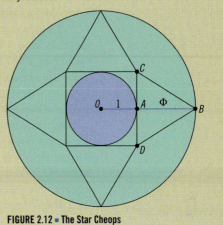

FIGURE 2.12 ■ The Star Cheops

of the Old Kingdom."[10] In fact, Φ can be found in the Great Pyramid. The question is, Was its use intentional? Scholars who think it was usually base their position on Herodotus' statement, "the pyramid itself . . . is square, each face is eight plethra, and the height is the same. . . ."

What Herodotus' statement really means is a source of controversy. From our survey of the dimensions, it is clear that the height of the pyramid is not the same as the side. However, noting that the *plethron* is the name of both a unit of length and a unit of area, Herodotus' statement has been interpreted to mean the following: *The area of one face equals the square of the height of the pyramid,* as shown in Figure 2.13. Let's verify this:

$$\text{Area of a face} = \tfrac{1}{2}(230.5 \text{ m})(186.6 \text{ m}) = 21{,}506 \text{ m}^2$$

$$\text{Square of the height} = (146.8 \text{ m})^2 = 21{,}550 \text{ m}^2$$

The two areas agree to within 0.2%, a good indication that this interpretation of Herodotus' remark is correct. However, other scholars question Herodotus' statement itself, charging that it is a gross misinterpretation or mistranslation of the original.

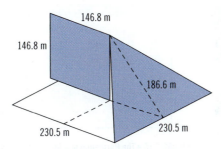

FIGURE 2.13 ■ Area of the Face of the Great Pyramid

In contrast to those who find the golden ratio wherever they look, others, such as American mathematician George Markowsky, find it almost nowhere. In an article entitled "Misconceptions about the Golden Ratio," Markowsky writes, "A variety of people have looked for Φ in the dimensions of the Great Pyramid of Khufu (Cheops). . . . This idea sounds like something dreamt up to justify a coincidence rather than a realistic description of how the dimensions of the Great Pyramid were chosen. It does not appear that the Egyptians even knew of the existence of Φ much less incorporated it in their buildings." We'll refer to this article again when Markowsky attempts to debunk other claims about the golden ratio.

Looking to explain the shape of the Great Pyramid without using Φ, investigators have found at least two possible explanations: the *slope theory* and the *measuring wheel theory*. The slope or *seked* theory maintains that builders simply maintained a constant slope while adding stones; that is, they kept the same number of vertical units for each horizontal unit. The choice of a particular slope gave the pyramid its proportions, with no consideration of the golden ratio. The measuring wheel theory suggests that the Egyptians laid out the pyramid using a simpler version of the modern measuring wheel, a small bicycle wheel geared to a counter that is often used to measure distances along roads and driveways. For example, take a wheel of any diameter, and make the base of the pyramid a square, one wheel-revolution on a side. Then make the pyramid height equal to two diameters of the wheel. The result is a pyramid that has the shape of the Great Pyramid and that contains the golden ratio.

To make the actual Great Pyramid, you could use a wheel 0.7337 meter in diameter. As such, 100 revolutions would equal one side of the base of the Great Pyramid, and 200 diameters would equal the height of the Great Pyramid.

> Can you prove the measuring wheel assertion? See the exercises at the end of the chapter.

The Parthenon

The Parthenon, pictured in Figure 2.14, is the major temple dedicated to the Greek goddess Athena. It is located on the Acropolis, a hill overlooking Athens, Greece. The name *Parthenon* comes from the cult of Athena Parthenos (Athena the Virgin). It was built in the middle of the fifth century B.C. and is generally considered to be the finest example of the Doric order, one of the three classical Greek architectural orders. Many claims have been made that the Parthenon was designed using the golden ratio in the façade as well as throughout the structure. Drawings, such as Figure 2.15, are frequently used to show the façade or some other feature of the Parthenon enclosed in a golden rectangle. In this drawing,

FIGURE 2.14 ■ The Parthenon

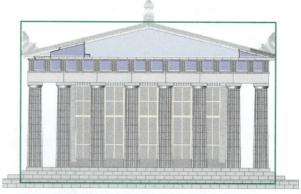

FIGURE 2.15 ■ The Parthenon in a Golden Rectangle[11]

the golden rectangle has the second row of steps as its width and the distance from the peak, now gone, to the first row of stones from the ground as its height.

It may be true that the Greeks built the golden ratio into the Parthenon. However, when you look at Figure 2.15, you may feel that the sides of the rectangle were selected somewhat arbitrarily. For instance, why choose one row of stones as the bottom of the rectangle and not another? Figures 2.16 through 2.22 illustrate the results when we try some alternative choices for the sides of the rectangle.

In Figure 2.16, we fit our façade into an 8 : 5 rectangle, or the musical ratio earlier identified as the minor sixth, by using the same width as that given in Figure 2.15 and redrawing the height. In Figure 2.17, we can fit our façade into a different 8 : 5 rectangle by using the width of the entablature as the width of the rectangle and the distance from the peak to the second step from the top as the height of the rectangle.

We can also fit the façade into a 3 : 2 rectangle, or the musical ratio called the fifth, by using the width given in Figure 2.17 but taking the height from the peak to the lowest step, as shown in Figure 2.18. In Figure 2.19, we fit our façade into a 5 : 3 rectangle by using the same width as that given in Figure 2.18, but measuring to a higher step. This is our musical major sixth.

Figure 2.20 shows the columns fitting nicely into two root-2 rectangles, and in Figure 2.21 we include the horizontal bands above the columns to get a root-5 rectangle. Finally, in Figure 2.22, we show the façade fitting into two squares.

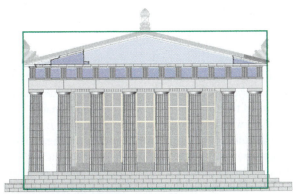

FIGURE 2.16 ■ The Parthenon in an 8 : 5 Rectangle

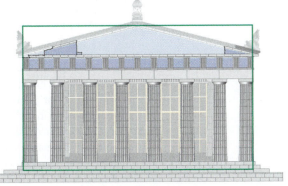

FIGURE 2.17 ■ The Parthenon in Another 8 : 5 Rectangle

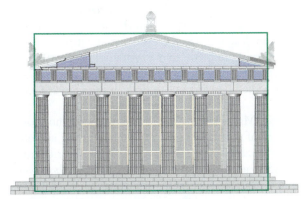

FIGURE 2.18 ■ The Parthenon in a 3 : 2 Rectangle

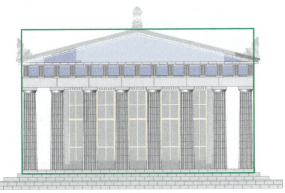

FIGURE 2.19 ■ The Parthenon in a 5 : 3 Rectangle

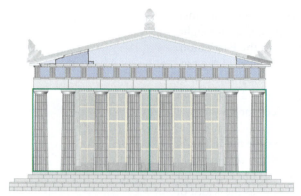

FIGURE 2.20 ■ The Parthenon in Two Root-2 Rectangles

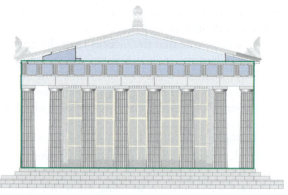

FIGURE 2.21 ■ The Parthenon in a Root-5 Rectangle

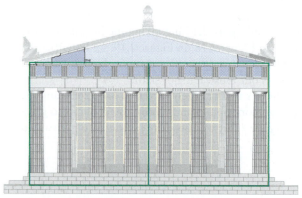

FIGURE 2.22 ■ The Parthenon in a Double-Square Rectangle

Fitting the façade of the Parthenon into a single rectangle is not the whole story. Researchers point out that *subdivisions* of the façade show the golden ratio, as do the floor plan, subdivisions of the floor plan, placement of colonnades and altars, and so forth. However, even with the simple example used here, it is clear that different results can be obtained depending on the choice of measuring points. There are many different drawings of the Parthenon with conflicting dimensions. Markowsky writes, "The dimensions of the Parthenon vary from source to source probably because different authors are measuring between different points. With so many numbers available, a golden ratio enthusiast could choose whatever numbers gave the best result."[12]

The sacred cut will be discussed in Chapter 4, "Ad Quadratum and the Sacred Cut."

Also, the numbers can yield to whatever axe the author has to grind. For instance, Tons Brunés, an enthusiast for what he called the *sacred cut*, not surprisingly finds that the Parthenon design is based on the sacred cut. Wittkower stated it well: "It is true, that in trying to prove that a system of proportion has been deliberately applied by a painter, a sculptor, or an architect, one is easily misled into finding those ratios which one sets out to find. In the scholar's hand dividers do not revolt."[13]

The American astrophysicist Mario Livio wrote, "The two main shortcomings of claims about the presence of the golden ratio in architecture or in works of art, on the basis of dimensions alone [are], (1) they involve numerical juggling, and (2) they overlook inaccuracies in measurements."[14] The numbers were juggled in the previous examples. In reference to Livio's second point, note, for example, that the ratio 8 : 5 equals 1.600 and $\Phi = 1.618$, approximately. Their difference is about 0.018, or 1.1%. Therefore, the measurements required to distinguish between a ratio of Φ and that of 8 : 5 would have to be accurate to about 1% or better. Their initial construction may not have been made to that accuracy, and now the structures are old, weathered, and perhaps damaged by war or shifted by earth movements. By comparing once more Figures 2.15 and 2.16, you can see that replacing the golden rectangle with an 8 : 5 rectangle would make no apparent difference.

Le Corbusier's *Modulor*

In what is probably his most important writing, *Le Modulor*, Le Corbusier described a system of architectural proportions based on the human body and on the golden ratio. He started with a single number: the distance from the ground to the navel of a man 183 cm (approximately 6 ft) tall. He assumed that the man's navel subdivides his height by the golden ratio, placing it 113 cm from the ground. This number, 113, became the source of a sequence of numbers, all in the Φ sequence. The numbers above 113 are obtained by successive multiplication by Φ, and those below 113 are computed by successive division by Φ.

… 10   16   27   43   70   **113**   183   296   479   775 …

Le Corbusier called this sequence of numbers his *Red Series*.

Next, Le Corbusier took his single source number, 113, and doubled it to get 226 cm. Again, he formed a Φ sequence around 226 and got the rounded values:

… 33   53   86   140   **226**   366   592   958 …

He called this sequence of numbers his *Blue Series*. The number 226 is significant because it is the approximate distance, in centimeters, from the ground to the tips of the fingers when our 6-foot person stretched his arm up overhead. The height from the ground to his wrist is 86 cm when his arm is hanging down by his side (as in Figure 2.23).

To obtain a set of rectangles on which to base architectural designs, Le Corbusier made a grid, with each series represented on both the horizontal and vertical axes. Figure 2.24 shows this grid in which he combined both the

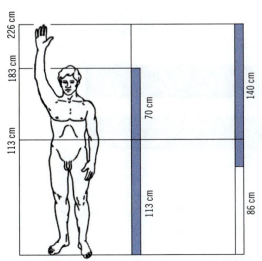

FIGURE 2.23 ■ Artist's Rendering of Le Corbusier's Modulor System of Proportions

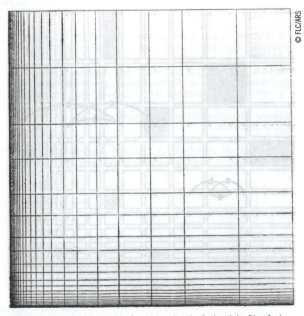

© FLC/ARS

FIGURE 2.24 ■ A Grid Formed by Superimposing the Red and the Blue Series

Red and the Blue Series, which was called the Modulor. He said, "It is a flawless fabric formed of stitches of every dimension, from the smallest to the largest, a texture of perfect homogeneity."[15] To illustrate the use of these rectangles in dividing up an area, Le Corbusier created what he called The Panel Exercise, which is shown in Figure 2.25. "You take a square, say, and divert yourself by dividing it up in accordance with the measures of the Modulor. This game can be played indefinitely."[16] Le Corbusier first used the golden ratio in 1927 at the Villa à Garches. An illustration of the ground plan and elevation (shown in Figure 2.26) shows the golden ratio. In fact, the elevation contains the notation $A : B \cdot B : (A + B)$, a clear expression of the golden ratio.

According to the German art historian Marcus Frings, the Modulor has some deficiencies.[17] Although Le Corbusier meant for it to be used for all dimensions, vertical and horizontal, he bases it solely on the vertical dimension.

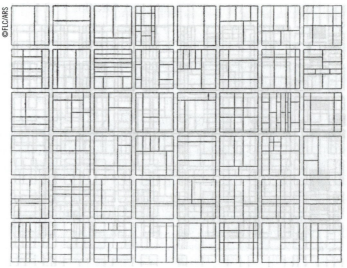

FIGURE 2.25 ■ The Panel Exercise

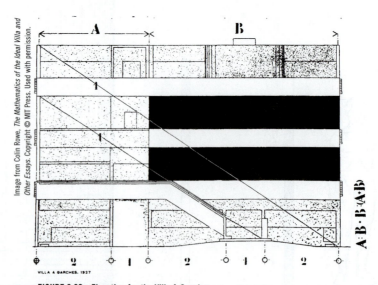

FIGURE 2.26 ■ Elevation for the Villa à Garches

Further, because the Blue and the Red Series can be combined, the system becomes so elastic that the golden ratio is hard to detect.

As you can see, there is no general agreement about the use of the golden ratio in architecture, and the subject is still being debated. We'll later see that the same is true about the use of the golden ratio in painting and sculpture.

SUMMARY

In this chapter, we continued our mathematical development of ratio and proportion, sequences and series, and solutions of quadratic equations. The chapter defined the golden ratio, derived a formula for it, and gave its geometric construction. You were introduced to the Fibonacci sequence and how it was related to the golden ratio and the musical ratios. You were also introduced to Euclid, Fibonacci, and Le Corbusier.

You saw that the Great Pyramid does, within the accuracy of the measured dimensions, contain Φ, whether or not its use was intentional. On the other hand, it was clear that the Parthenon façade would fit into a number of different rectangles as well as the golden rectangle. Only with Le Corbusier's Modulor was there no doubt about the use of Φ, as it was the express intention of the architect.

The use of the golden ratio in art and architecture is a controversial subject, with articles still being published on both sides of the debate. Some skepticism is needed regarding claims about the use of the golden ratio, or, for that matter, any other system of proportions. You will encounter this problem again when we examine the presence of certain proportions in other works of art. As stated earlier, this chapter is only an introduction to the golden ratio. We have not yet run out of gold. Yet to come are the following:

- Golden rectangle, golden rhombus, golden gnomons, and the golden ratio in art
- Golden ratio in the pentagon and pentagram
- Use of golden triangles in Penrose tilings
- Golden spiral and the golden ratio in nature
- Golden parallelepiped, golden cuboid, and the golden ratio in the dodecahedron
- Golden fractal tree

However, before we get to them, we'll discuss the *golden triangle*, as well as many other triangles, in the next chapter.

EXERCISES AND PROJECTS

1. Define or describe the following terms:

Kepler triangle	Egyptian triangle
Star Cheops	Modulor

2. Find the answer to Fibonacci's rabbit problem. How many pairs of rabbits will there be after one year?

3. Le Corbusier assumed that a person's navel divided one's height by a ratio of Φ. Try to verify this by direct measurement of a number of classmates' heights. Summarize your findings in a short report. Look up *mean* and

standard deviation in an elementary statistics textbook, and provide these numbers with your report.

4. *Squaring Rectangles.* If you multiply the first and second Fibonacci numbers (1 and 1), then the second and third (1 and 2), and the third and fourth (2 and 3), the sum of these three products equals the square of the last number used (3^2):

$$(1)(1) + (1)(2) + (2)(3) = 3^2.$$

   a. Use a spreadsheet to show that this property holds true, for an odd number of products, for more than just the first four numbers in the sequence (but always starting with 1).

   b. This property can be illustrated by rectangles that completely fill a square (Figure 2.27).[18] Construct such a figure depicting Fibonacci numbers other than those shown.

   c. Use your construction as the basis for a graphic design or painting.

5. Demonstrate that the sum of any ten consecutive Fibonacci numbers is always exactly divisible by 11.

6. Use a spreadsheet to compute the first 50 Fibonacci numbers using Binet's formula (Equation 20) and using the recursion formula (Equation 18). Compare the values given by each.

7. Play Le Corbusier's Panel Exercise to create a pleasing design. Add color. Use your design to make a painting, graphic, wallpaper pattern, wrapping paper, greeting card, screensaver, tiling, mosaic, or a fabric design such as a pillow, tee shirt, or quilt.

8. Make a graphic design that incorporates the golden ratio or the Fibonacci numbers in a significant way. Write a few paragraphs explaining your design.

9. Design your own floor plan or façade (preferably both) with proportions that feature the golden ratio. Build a model of your design.

10. Draw the Star Cheops. Use it to make a model of the Great Pyramid.

11. Make a cutaway model of the Great Pyramid that clearly demonstrates the proportions mentioned in this chapter.

12. Prove that the measuring wheel mentioned earlier, with a diameter of 0.7337 m, will give the dimensions of the Great Pyramid when used as described.

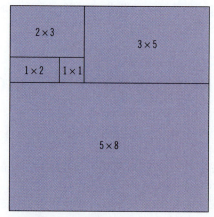

FIGURE 2.27 ■ Squaring Rectangles

13. Make a small measuring wheel with which you can lay out the dimensions of a pyramid with the same proportions as the Great Pyramid. Demonstrate it to your class.

14. Write a report on any book in the list of sources at the end of this chapter.

15. Get a copy of the *Fibonacci Quarterly* from your college library. Write a short paper on one of the articles in the journal.

16. Read Markowsky's refutation of the occurrence of the golden ratio in the Great Pyramid and elsewhere. Write a paper stating whether you agree or disagree and give your reasons.

17. Write either a short research paper or a term paper. Come up with your own topic, using these suggestions to jog your imagination. Possible topics include the following:

   ■ The golden mean in the pyramids of Egypt
   ■ The measurement of the pyramids, described in Heath, p. 129 (see Sources)
   ■ The golden mean in Egyptian temples
   ■ The golden ratio in the sculpture of ancient Greece
   ■ The golden ratio in the pottery of ancient Greece
   ■ The *geometric style* in Greek pottery
   ■ The golden ratio in temples of ancient Greece, especially the Parthenon
   ■ Alexandria at the time of Euclid
   ■ Leonardo Fibonacci and the Fibonacci sequence
   ■ The golden ratio in the growth of living things
   ■ The *Da Divina Proportione* (Luca Pacioli's book on the golden section)
   ■ Johannes Kepler (1571–1630), the German astronomer who believed that God used the golden ratio as a tool for creating the universe
   ■ Adolph Zeising (1810–1876), who introduced the term *Goldener Schnitt* (golden cut)
   ■ The architect Ernst Neufert (1900–1986), who advocated the golden section in his *Bauordnungslehre* of 1943
   ■ Gustav Theodor Fechner (1801–1887), the German physicist, philosopher, and experimental psychologist who experimented on the supposed appeal of the golden section
   ■ Jay Hambidge and the Dynamic Symmetry Movement
   ■ Le Corbusier's Modulor System
   ■ The origin of the word *golden* in the golden ratio, discussed in Herz-Fischler, Appendix I (see Sources)

18. Aristotle, in his *Nicomachean Ethics*, used the idea of the golden mean, or *middle path*, between extremes. For example, being friendly is the mean between being surly and being obsequious. Similarly, courage is the mean between cowardice and foolhardiness. The same idea is expressed as the Buddha's middle path between self-indulgence and self-renunciation. Research and write a paper on this theme.

19. Read Colin Rowe's *The Mathematics of the Ideal Villa and Other Essays*, and compare the use of mathematics between Palladio and Le Corbusier.

20. Find and read Millay's entire poem "Euclid Alone Has Looked on Beauty Bare." Write a few paragraphs of commentary.

21. Some composers, such as the Hungarian Béla Bartók (1881–1945), have supposedly used the golden ratio in their works. Investigate some such claims and write a paper about your findings.

22. Make an oral presentation to your class about any of the above projects or papers.

23. Make a "Great Pyramid" cake and share it with your classmates.

Mathematical Challenges

24. The quadratic equation can be solved by the quadratic formula (Equation 5). Another way to solve a quadratic is by *completing the square*. If you are familiar with this method, use it to derive Equation 6b and compare your results with those obtained by quadratic formula.

25. Derive the quadratic formula by completing the square, given the equation $ax^2 + bx + c = 0$.

26. The golden ratio is given by the following continued fraction:

$$\Phi = 1 + \cfrac{1}{1 + \cfrac{1}{1 + \cfrac{1}{1 + \cdots}}}$$

Demonstrate by calculation, by hand, or with a spreadsheet, that this is true.

27. Prove that the golden ratio is given by the continued fraction in the preceding problem.

28. The golden ratio is given by the following radical expression:

$$\Phi = \sqrt{1 + \sqrt{1 + \sqrt{1 + \sqrt{1 + \cdots}}}}$$

Demonstrate by calculation, by hand, or with a spreadsheet, that this is true.

29. Derive the radical equation used in the preceding problem. *Hint:* Start by taking the square root of both sides of Equation 8.

SOURCES

Fibonacci, Leonardo. *Fibonacci's Liber Abaci, a Translation into Modern English of Leonardo Pisano's Book of Calculation.* Lawrence Sigler, Trans. New York: Springer, 2002.

Frings, Marcus. "The Golden Section in Architectural Theory." *Nexus Network Journal:* http://www.nexusjournal.com/Frings.html.

Herz-Fischler, Roger. *A Mathematical History of Division in Extreme and Mean Ratio.* Ontario: Laurier, 1987.

Huntley, H. E. *The Divine Proportion.* New York: Dover, 1970.

Kappraff, Jay. *Connections: The Geometric Bridge between Art and Science.* New York: McGraw-Hill, 1990.

Le Corbusier. *The Modulor.* Cambridge, MA: Harvard University Press, 1966.

Linn, Charles. *The Golden Mean.* New York: Doubleday, 1974.

Livio, Mario. *The Golden Ratio.* New York: Broadway Books, 2002.

Markowsky, George. "Misconceptions about the Golden Ratio." *College Mathematics Journal,* Jan. 1992.

Pedoe, Dan. *Geometry and the Visual Arts.* New York: Dover, 1976.

Rowe, Colin. *The Mathematics of the Ideal Villa and Other Essays.* Cambridge, MA: MIT Press, 1976.

Runion, Garth E. *The Golden Section*. Palo Alto, CA: Seymour, 1990.

Thompson, Darcy. *On Growth and Form*. New York: Dover, 1992. First published in 1942.

Tompkins, Peter. *Secrets of the Great Pyramid*. New York: Harper & Row, 1971.

Wittkower, Rudolf. *Architectural Principles in the Age of Humanism*. New York: Random House, 1965.

NOTES

1. Kepler, "Mysterium cosmographicum (Cosmic Mystery)," Chapter 12 of the 1621 edition. Roger Herz-Fiscler, "A Mathematical History of the Golden Number," p. 175.
2. This phrase is in the title of Livio's book, which has served as a main source for this chapter.
3. Coxeter, *Introduction to Geometry*, p. 175.
4. This is the first verse of a poem called "Constantly Mean" by Paul S. Bruckman, published in the *Fibonacci Quarterly*, 1977.
5. This translation is from Sigler, p. 404.
6. Livio, p. 212.
7. *Herodotus Book II*, para. 124.
8. Pyramid dimensions are from Tompkins, p. 209 and p. 364, based on a survey by J. H. Cole, a professional surveyor. Obsolete units like the plethron and cubit are here given in meters. The dimensions are rounded to the nearest tenth of a meter. The pyramid is ancient and crumbling, which makes measuring it imprecise. The surveys show some variation in the length from side to side, and also from researcher to researcher.
9. New York: Wiley, 1882.
10. Tompkins, p. 195.
11. The façade drawing is from *Abbildungen zur algemeinen Bauzeitungen*, Vienna 1838, in Brunés, I, 302.
12. Markowsky, p. 2.
13. Wittkower, p. 126.
14. Livio, p. 46.
15. Le Corbusier, p. 84.
16. Le Corbusier, p. 92.
17. Frings, *The Golden Section in Architectural Theory*.
18. Livio, p. 104.

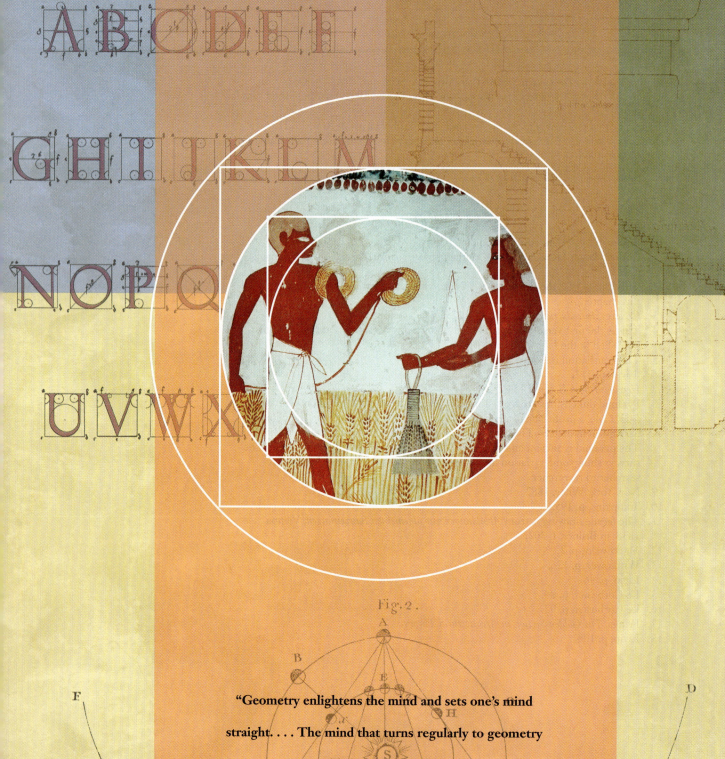

"Geometry enlightens the mind and sets one's mind

straight. . . . The mind that turns regularly to geometry

is unlikely to fall into error."

THE *MUQUADDIMAH* OF IBN KHALDÙN

FIGURE 3.1 ■ The Harpedonaptai; Rope Stretchers or Engineers (detail)
©Werner Forman/CORBIS

3

The Triangle

In this chapter, we'll continue our exploration of geometry in *two dimensions*, also known as the *plane*. We'll start by demonstrating how an angle is formed with two intersecting lines and go on to consider the angular relations among several intersecting lines. Using this knowledge of angles, we'll begin our study of the triangle. We'll discover relationships among the angles of a triangle, both internal and external, and find its perimeter and area. We'll define various lines that can be drawn inside a triangle and how they give us several kinds of centers for a triangle. We'll examine the different types of triangles, particularly the right triangle and the important Pythagorean theorem.

Then we'll illustrate how one triangle can be related to another by congruency or similarity and how to change one triangle into the other. We'll also discuss symmetry, a topic that will reappear many times in later chapters.

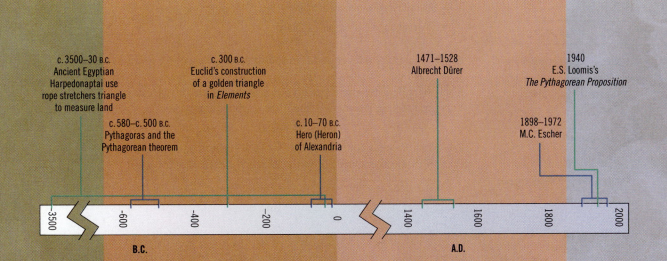

c. 3500–30 B.C.
Ancient Egyptian
Harpedonaptai use
rope stretchers triangle
to measure land

c. 580–c. 500 B.C.
Pythagoras and the
Pythagorean theorem

c. 300 B.C.
Euclid's construction
of a golden triangle
in *Elements*

c. 10–70 B.C.
Hero (Heron)
of Alexandria

1471–1528
Albrecht Dürer

1898–1972
M.C. Escher

1940
E.S. Loomis's
The Pythagorean Proposition

-3500   -600   -400   -200   0   1400   1600   1800   2000

B.C.   A.D.

ANGLES BETWEEN LINES

Additional information on the geometry of triangles can be found in *Technical Mathematics*.

As discussed in Chapter 1, "Music of the Spheres," a ray is a half-line. An *angle* is formed when two rays intersect at their endpoints. The angle shown in Figure 3.2 can be designated in any of the following ways: angle *ABC*, angle *CBA*, angle *B*, angle *θ*, or ∠*B*, where the symbol ∠ means angle. The point of intersection is called the *vertex* of the angle, and the two rays are called the *sides* of the angle. We often think of an angle as being generated by a ray turning from some *initial position* to a *terminal position*, as shown in Figure 3.3. One *revolution* is the amount a ray would turn to return to its initial position.

The *degree* and the *radian* are the commonly used units of angular measure. We'll use only decimal degrees here, and we'll cover radians in Chapter 6, "The Circle." A degree (represented by the symbol °) is equal to $\frac{1}{360}$ of a revolution. Therefore, there are 360 degrees in one revolution. A fractional part of a degree may be expressed as a common fraction ($35\frac{1}{4}°$), as a decimal fraction (35.25°), or as *minutes and seconds* (35°15'00"). A *minute of arc* (symbolized by ') is equal to $\frac{1}{60}$ of a degree, and a *second of arc* (symbolized by ") is equal to $\frac{1}{60}$ of a minute.

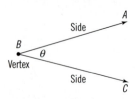

FIGURE 3.2 ■ An Angle

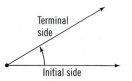

FIGURE 3.3 ■ An Angle Formed by Rotation

21 UNITS OF ANGULAR MEASURE

$$1 \text{ revolution} = 2\pi \text{ radians} = 360°$$
$$1° = 60 \text{ minutes}$$
$$1 \text{ minute} = 60 \text{ seconds}$$

Types of Angles

Figure 3.4 shows four different angles that can be formed using varying revolutions: (a) a *right angle* ($\frac{1}{4}$ revolution), usually marked with a small square at the vertex; (b) an *acute* angle (less than $\frac{1}{4}$ revolution); (c) an *obtuse* angle (greater than $\frac{1}{4}$ revolution but less than $\frac{1}{2}$ revolution); and (d) a *straight* angle ($\frac{1}{2}$ revolution). Two angles are said to be *complementary* if their sum is a right angle and *supplementary* if their sum is a straight angle, that is, 180 degrees.

Angles *A* and *B* in Figure 3.5 are called *opposite* or *vertical* angles. Further, angles *A* and *C* are called *adjacent* angles. They have a common side and a common vertex. It should be clear from the figure that:

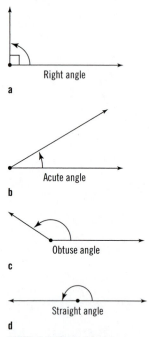

FIGURE 3.4 ■ Types of Angles

22 OPPOSITE AND ADJACENT ANGLES

For two intersecting lines
- Opposite angles are equal to each other.
- Adjacent angles are supplementary.

■ **EXAMPLE:** Find angles *A* and *B* in the structure shown in Figure 3.6.

● **SOLUTION:** Because angle *A* is opposite to the 34° angle, we can say that *A* = 34°. Further, angle *B* and the 34° angle are supplementary, so
$$B = 180° - 34° = 146°.$$
For brevity, we'll simply say "*B* = 146°" rather than "angle *B* = 146°" or the even more cumbersome but more correct "the measure of angle *B* = 146°." ■

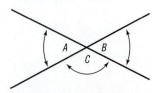

FIGURE 3.5 ■ Opposite and Adjacent Angles

Transversals

A *family of lines* is a group of lines that are related to each other in some way. Figure 3.7a shows a family of parallel lines. Recall that lines that do not intersect are called *parallel lines*, and lines at right angles to one another are said to be *perpendicular*. Figure 3.7b shows a family of lines that all pass through the same point.

A *transversal* is a line that intersects a family of lines, such as line *T* in Figures 3.8a and 3.8b. In Figure 3.8a, two parallel lines are cut by a transversal *T*. Angles *A*, *B*, *G*, and *H* are called *exterior angles*, and *C*, *D*, *E*, and *F* are called *interior* angles. Angles *A* and *E* are called *corresponding* angles. Other corresponding angles in that figure are *C* and *G*, *B* and *F*, and *D* and *H*. Angles *C* and *F* are called *alternate interior angles*, as are *D* and *E*. These angles are related in the following way:

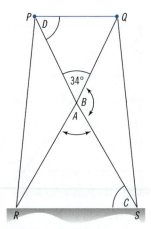

FIGURE 3.6 ■ A Structure

23 PARALLELS CUT BY A TRANSVERSAL

If two parallel lines are cut by a transversal,

- Corresponding angles are equal.
- Alternate interior angles are equal.

Therefore, in Figure 3.8a,

$$A = E, C = G, B = F \quad \text{and} \quad D = H.$$

Also,

$$D = E \quad \text{and} \quad C = F.$$

a

b

FIGURE 3.7 ■ Families of Lines

■ **EXAMPLE:** The top girder *PQ* in the structure shown in Figure 3.6 is parallel to the ground, and angle C is 73°. Find angle *D*.

● **SOLUTION:** Two parallel lines, *PQ* and *RS*, are cut by transversal *PS*, so

$$D = C = 73°.$$ ■

Cutting parallels by two transversals is the basis for one of our basic constructions: proportional division of a line segment (refer back to Figure 1.31). When two or more parallel lines are cut by *two* transversals, as shown in Figure 3.9, the portions of the transversals lying between the same parallels are called *corresponding segments*. The following theorem applies.

a

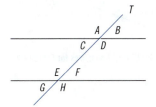

b

FIGURE 3.8 ■ Transversals

24 PARALLELS CUT BY TWO TRANSVERSALS

When a number of parallel lines are cut by two transversals, the ratios of corresponding segments of the transversals are equal.

Therefore, in Figure 3.9,

$$\frac{a}{b} = \frac{c}{d} = \frac{e}{f}.$$

■ **EXAMPLE:** Figure 3.10 depicts a portion of an archeological survey map. Find the distances *PQ* and *QR*.

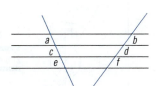

FIGURE 3.9 ■ Parallels Cut by Two Transversals

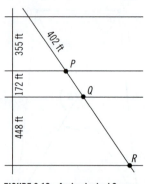

FIGURE 3.10 ■ Archeological Survey Map

● **SOLUTION:** From Statement 24,

$$\frac{PQ}{172} = \frac{402}{355}$$

$$PQ = \frac{402}{355}(172) = 195 \text{ ft.}$$

Similarly,

$$\frac{QR}{448} = \frac{402}{355}$$

$$QR = \frac{402}{355}(448) = 507 \text{ ft.}$$

■

■ EXERCISES ● ANGLES BETWEEN LINES

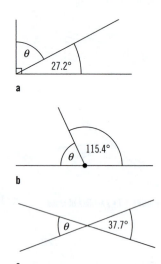

FIGURE 3.11 ■ Figures for Exercise 2

1. Define the following terms:

parallel lines	perpendicular lines	family of lines
transversal	initial position	terminal position
minute of arc	second of arc	right angle
acute angle	obtuse angle	straight angle
opposite angles	vertical angles	adjacent angles
complementary angles	supplementary angles	alternate interior angles

2. Find angle θ in Figures 3.11a, b, and c.

3. Find angles A, B, C, D, E, F, and G in Figure 3.12.

4. Find distance x in the street layout of Figure 3.13.

5. Find angles A and B in Figure 3.14.

6. Three parallel steel cables hang from a girder to the deck of a bridge (Figure 3.15). Find the distance x.

7. *Angle Bisection.* The method used to bisect an angle is similar to the construction for bisecting a line segment. Work it out on your own, and write instructions for doing this construction.

8. *Angle Bisection by Paper Folding.* Try the exercise depicted in Figure 3.16, and explain why it works. Given angle *ABC*, fold along a crease through *B* so that line *AB* lies along line *BC*. The crease then bisects the given angle.

9. *Angle-Trisection Device.* The Danish mathematician and astronomer Asger Aaboe describes a simple angle-trisector device (Figure 3.17) that he calls a *tomahawk*.[1] The cut-out shape is inserted into an angle α so that it touches the angle's sides at *P* and *Q*.

   a. Demonstrate by measurement or by a CAD construction that $\beta = \alpha/3$.

   b. Make and use the device. Measure with a protractor to demonstrate whether the device works.

   c. Prove that it works.

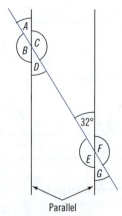

FIGURE 3.12 ■ Two Parallel Streets Cut by a Third

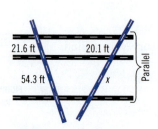

FIGURE 3.13 ■ A Street Layout

FIGURE 3.14 ■ Intersecting Girders

FIGURE 3.15 ■ Steel Bridge Cables

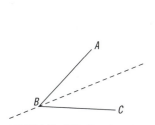

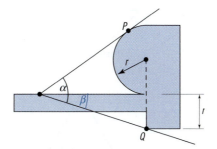

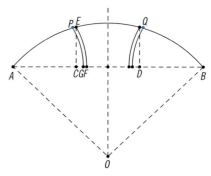

FIGURE 3.16 ■ Angle Bisection by Paper Folding

FIGURE 3.17 ■ Tomahawk Angle-Trisector

FIGURE 3.18 ■ Dürer's Approximate Trisection of an Arc and an Angle

10. *Dürer's Approximate Trisection of an Arc and an Angle.* Do the construction of Figure 3.18, and follow these steps to trisect an angle *AOB*. Locate points *A* and *B* equidistant from *O*. Draw line segment *AB*. Trisect *AB* (we learned how to trisect a line in Chapter 1), getting points *C* and *D*. Draw arc *AB* from *O*, and erect a perpendicular from *C* to the arc at *E*. With center *A*, swing an arc from *E* to *F*. Trisect line segment *CF*. With center *A*, draw an arc from trisection point *G* to arc *AB*, getting *P*. Lay off distance *AP* from *B* to obtain point *Q*. Points *P* and *Q* approximately trisect arc *AB*, and lines *OP* and *OQ* (not shown) approximately trisect angle *AOB*.

Determine, by measurement, how accurate the construction is, given your chosen angle.

> We will discuss the German artist Albrech Dürer throughout this book. A biographical sketch of him is included in Chapter 5, "Polygons, Tilings, and Sacred Geometry."

TRIANGLES

"Philosophy is written in this grand book the universe . . . but the book cannot be understood unless one first learns to comprehend the language and to read the alphabet in which it is composed. It is written in the language of mathematics, and its characters are triangles, circles, and other geometric figures . . . without these, one wanders about in a dark labyrinth."

GALILEO, IN *IL SAGGIATORE*

A *polygon* is a plane figure bounded by straight line segments, which are called the *sides* of the polygon. The word comes from the Greek *poly*, which means "many," and *gon*, which means "angle." The sides intersect at points called the *vertices*. The angle between two sides is called an *interior angle* or *vertex angle*. A polygon that has three such angles is called a *triangle*. Triangles may be defined in terms of their sides or in terms of their angles.

> A three-sided polygon could be called a *trigon*, but it is customary to use the word *triangle*.

Figures 3.19a through 3.19f depict several classifications of triangles:

a. An *acute triangle* is one in which all interior angles are acute.

b. An *obtuse triangle* is one in which one interior angle is obtuse.

c. A *scalene triangle* is one in which no two sides are equal.

d. An *isosceles triangle* is one in which at least two sides are equal.

e. An *equilateral triangle* is one in which three sides are equal.

f. A *right triangle* is one in which one interior angle is a right angle.

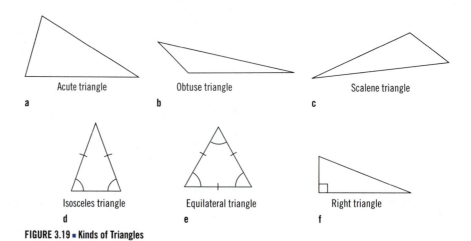

FIGURE 3.19 ■ Kinds of Triangles

An *oblique triangle* contains no right angle. These definitions are not all mutually exclusive. For example, an equilateral triangle is also an acute triangle, an isosceles triangle, and an oblique triangle. Acute, obtuse, scalene, and equilateral triangles are all oblique.

In this section we'll discuss the various properties of triangles, from their angles to their centers.

Sum of the Angles

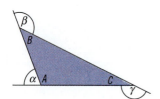

FIGURE 3.20 ■ Exterior Angles

We have already seen that an interior angle is the angle between two sides of the triangle, but what about exterior angles? Imagine a point moving from C to A in triangle ABC, which is shown in Figure 3.20. At A, the point has to change direction by an angle α to head toward B. Angle α is called an *exterior angle* of the triangle. It is the angle between a side of a triangle and an extension of the adjacent side.

At B, the point has to change direction by an angle β to head toward C. At C, the point has to change direction by angle γ to return to the original direction. The three changes in direction (that is, the three angles) have to add up to a complete revolution, so that

$$\alpha + \beta + \gamma = 360°.$$

Using this information about the exterior angles of a triangle, we can also determine the relationship between the interior angles of a triangle. Notice that α and A are supplementary angles, so $\alpha = 180° - A$. Similarly, $\beta = 180° - B$, and $\gamma = 180° - C$. This gives us

$$(180° - A) + (180° - B) + (180° - C) = 360°.$$

This relationship can be simplified into the following:

31 SUM OF THE INTERIOR ANGLES OF ANY TRIANGLE

The sum of the three interior angles A, B, and C of any triangle is 180°.

$$A + B + C = 180°$$ ●

■ **EXAMPLE:** Find the interior angle A of a triangle if the other two interior angles are 38° and 121°.

● **SOLUTION:** By Equation 31,

$$A = 180° - 121° - 38° = 21°.$$ ■

Further, we can determine how an exterior angle is related to the interior angles using the same formula. If you apply Equation 31 to Figure 3.21,

$$A + B + (180 - \theta) = 180.$$

From this, you can deduce the following:

FIGURE 3.21 ■ An Exterior Angle

32 EXTERIOR ANGLE OF A TRIANGLE

An exterior angle equals the sum of the two opposite interior angles.

$$\theta = A + B$$

●

Perimeter and Area of a Triangle

The *perimeter* of any plane figure is simply the distance around it, or the sum of the sides. To find the perimeter of a triangle, just add the lengths of the three sides.

The *area* of a plane figure is a measure of the region within the boundaries of that figure. We can find the area of a triangle if we know the measurements of its altitude and base. See Figure 3.22. The *altitude* of a triangle is the perpendicular distance from a vertex to the opposite side, which is called the *base*.

To demonstrate finding the area of a triangle, let's draw triangle *ABC*, as shown in Figure 3.23a. If side *AB* is the *base* of the triangle, then *CE* is the *altitude* of the triangle. Draw lines parallel to *AB* and *BC*, as shown in Figure 3.23b, forming what is called a *parallelogram* or a four-sided figure with opposite sides that are equal and parallel. The figure clearly shows that the area of the parallelogram equals the base *AB* times the altitude *CE*. (If you were to cut off the triangular portion *BEC* and move it to *AFD*, it would just complete the rectangle *DCEF*; see Figure 3.23c.) The area of triangle *ABC* is just half that, or half the base times the altitude.

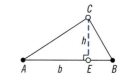

FIGURE 3.22 ■ Altitude and Base

29 AREA OF A TRIANGLE

The area of a triangle equals one-half the product of the base *b* and the altitude *h*.

$$A = \frac{bh}{2}$$

●

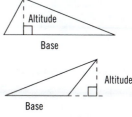

a

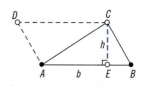

b

■ **EXAMPLE:** What is the area of a triangle having a base of 52 cm and an altitude of 48 cm?

● **SOLUTION:**

$$\text{Area} = \frac{52(48)}{2} = 1248 \text{ cm}^2$$

■

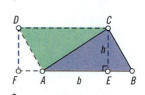

c

FIGURE 3.23 ■ Area of a Triangle

■ **EXAMPLE:** The triangular pediment of a building façade shown in Figure 3.24 has a base of 30 m and an altitude of 4 m. Find its area.

FIGURE 3.24 ■ Example of a Pediment on a Building Façade

● **SOLUTION:** The base of the triangle is 30 m and the altitude is 4 m, so

$$\text{Area} = \tfrac{1}{2}(30)(4) = 60 \text{ m}^2.$$ ■

If the altitude of a triangle is not known, but the three sides *a*, *b*, and *c* are known, you can use *Hero's formula* to compute the area. Hero is the same mathematician whose construction for the golden ratio was used in Chapter 2, "The Golden Ratio."

30 **HERO'S FORMULA**

$$\text{Area of triangle} = \sqrt{s(s - a)(s - b)(s - c)},$$

where *s* is half the perimeter,

$$s = \frac{a + b + c}{2}.$$ ●

■ **EXAMPLE:** Find the area of a triangle with sides 3.25, 2.16, and 5.09.

● **SOLUTION:** First, find the half-perimeter *s*:

$$s = \frac{3.25 + 2.16 + 5.09}{2} = 5.25.$$

The area is then

$$\text{Area} = \sqrt{5.25(5.25 - 3.25)(5.25 - 2.16)(5.25 - 5.09)} = 2.28.$$ ■

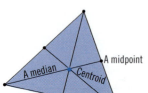

FIGURE 3.25 ■ Medians and Centroid

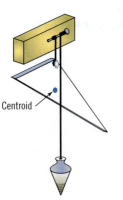

FIGURE 3.26 ■ Finding the Centroid by Experiment

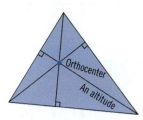

FIGURE 3.27 ■ Orthocenter

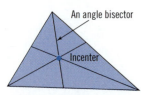

FIGURE 3.28 ■ Incenter

The Centers of a Triangle

A triangle can have four kinds of centers: centroid, incenter, circumcenter, and orthocenter. Centroids of plane areas are important in many calculations, such as measuring forces on walls, dams, and wind vanes, and measuring stresses in architectural members. The architectural importance of the incenter, circumcenter, and orthocenter will become clearer in Chapter 7, "Circular Designs in Architecture," when we do constructions involving both triangles and circles. Until then, consider these relationships among the four centers.

A *median* of a triangle is a line from a vertex to the midpoint of the opposite side. The *centroid*, also called the *median point*, of a triangle is the point of intersection of its three medians, as shown in Figure 3.25. It is located at $\tfrac{2}{3}$ the distance along a median from a vertex to the opposite side.

If a triangle (or any other shape) were cut from sheet material of uniform thickness, say cardboard, the *centroid* would be the "center of gravity" of the sheet. If suspended from any point, as in Figure 3.26, the sheet would swing so that the centroid would be directly below the point of suspension.

The *orthocenter*, as shown in Figure 3.27, is located at the intersection of the three altitudes of a triangle.

The *incenter*, which is the center of the inscribed circle, is located at the intersection of the three angle bisectors of the triangle, as shown in Figure 3.28. Recall that an *angle bisector* is a line that subdivides an angle into two equal parts.

The *circumcenter*, as shown in Figure 3.29, is located at the intersection of the three perpendicular bisectors of the sides. Recall from Chapter 1 that a *perpendicular bisector* is a line drawn at right angles to a given line at its midpoint.

The centers of a triangle can be summarized as follows:

33 CENTERS OF A TRIANGLE

a. The *centroid* is the point of intersection of the medians.

b. The *orthocenter* is the point of intersection of the altitudes.

c. The *incenter* is the point of intersection of the angle bisectors.

d. The *circumcenter* is the point of intersection of the perpendicular bisectors of the sides.

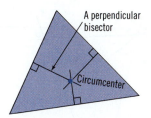

FIGURE 3.29 ■ Circumcenter

A circle is said to be *inscribed in* a triangle if all three sides of the triangle are tangent to the circle, as shown in Figure 3.30a. Conversely, the triangle is said to be *circumscribed about* that circle. A circle is said to be *circumscribed about* a triangle if it passes through all three vertices of the triangle, as shown in Figure 3.30b. Conversely, the triangle is said to be *inscribed in* the circle.

In Chapter 7, you'll study the incenter of a triangle, which is the center of the inscribed circle, and the circumcenter of a triangle, which is the center of the circumscribed circle.

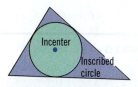

a

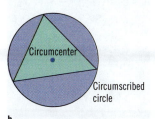

b

FIGURE 3.30 ■ Inscribed and Circumscribed Circles and Triangles

The Golden Triangle

Now, let's define a *golden triangle*, also called the *sublime triangle*. An isosceles triangle, as you've seen, is a triangle with two equal sides, called the *legs*, and a third side, called the *base*. A golden triangle is an isosceles triangle in which the ratio between the legs and the base is the golden ratio (Φ). There are two types of golden triangles, *acute* and *obtuse*, depending on whether we take the ratio of leg to base or the ratio of base to leg, as shown in Figure 3.31. Euclid explains how to construct a golden triangle, although he doesn't refer to it by that name. In Book IV, Proposition 10, he states that the object is "To construct an isosceles triangle having each of the angles at the base double [the measure] of the remaining one."[2]

The following construction of the golden triangle given in Figure 3.32 is somewhat similar to Euclid's. On a line segment AO, locate point D that subdivides line OA by the golden ratio, letting s be the larger part. Swing arcs from A and D with radius s. Label their intersection as B. Finally, draw AB, OB, and DB.

Because AO was subdivided by the golden ratio,

$$\frac{AO}{s} = \frac{AO}{AB} = \Phi.$$

Therefore, in isosceles triangle AOB, the ratio of side to base is Φ. It is called an *acute golden triangle*. Further, triangle OBD is also isosceles. The ratio of its base OB to a side OD is Φ. It is called an *obtuse golden triangle*. Triangle ABD is also an acute golden triangle.

> We'll use golden triangles to construct penrose tilings in Chapter 5.

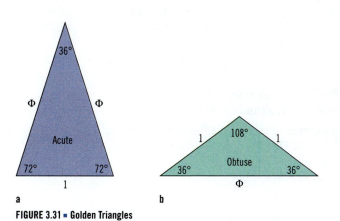

a　　　　　　　　　　b

FIGURE 3.31 ■ Golden Triangles

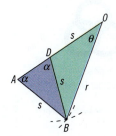

FIGURE 3.32 ■ Construction of the Golden Triangle

FIGURE 3.33 ■ Golden Gnomons

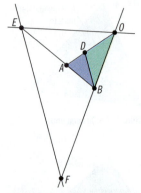

FIGURE 3.34 ■ *Creation of Adam* (detail), Della Quercia, c. 1430

Trefoil

Triquerta

Tricepts

Borromean rings

FIGURE 3.35 ■ Other Threefold Designs

From Ernst Lehner, *Symbols, Signs, and Signets* (Dover Publications, Inc.)

To find the magnitude of the angles, note that triangle *OBD* is isosceles, so angle *OBD* equals θ. By Equation 32, angle α, an exterior angle to triangle *OBD*, equals the sum of the two opposite interior angles, or 2θ. Therefore, in triangle *AOB*, the base angles are each 2θ and the angle at *O* is θ, for a sum of 5θ. These angles must also add up to 180°, so

$$5\theta = 180°$$
$$\theta = 36°.$$

Also,

$$\alpha = 2\theta = 72°.$$

There is another way to look at Figure 3.32. Start with acute golden triangle *ABD*, and append the obtuse golden triangle *OBD* to it. This will create the acute golden triangle *OAB*. Therefore, the obtuse golden triangle can be thought of as a *gnomon*, a figure added to another to get a larger figure having the same shape. You could then append obtuse golden triangles *OAE*, *BEF*, and so forth to get ever larger acute golden triangles, as demonstrated in Figure 3.33. This figure is sometimes called *whirling golden triangles*.

Triangular Halos and Triangular Designs

Triangles and triangular designs can be found throughout the history of art and architecture. One common usage of triangles is to represent the Trinity in Christianity: God the Father, the Son, and the Holy Spirit. For example, a halo, or *nimbus*, is a zone of light often depicted behind the head or body of a sacred figure in religious paintings and sculptures. Most halos are circular, but they come in other shapes as well. A triangular halo, such as the one shown in Figure 3.34, is used only when representing God the Father. Triangular windows are common in churches, perhaps also representing the Trinity.

Other common three-branched designs that resemble triangles are frequently used in churches. One such design is the three-lobed trefoil window. Similar three-branched or three-cornered designs are shown in Figure 3.35. They include the *triquerta*, the *tricepts*, and the *Borromean rings*. A related design (not shown) is the three-legged Greek *triskelion*. Although these designs resemble triangles, they are not triangles.

■ EXERCISES ● TRIANGLES

1. Define or describe the following terms, using an illustration if needed:

triangle	vertex	interior angle
exterior angle	acute triangle	obtuse triangle
scalene triangle	isosceles triangle	equilateral triangle
oblique triangle	right triangle	altitude of a triangle
base of a triangle	perimeter	

2. Find the area of each of the following triangles:
   a. Base = 33, altitude = 48
   b. Base = 2.43, altitude = 3.81
   c. Base = 213, altitude = 418

3. Find the area of each of the following triangles:
   a. Sides 44, 37, 51
   b. Sides 6.82, 4.28, 5.33
   c. Sides 837, 684, 774

4. What is the cost, at $1125 per acre, of a triangular piece of land with a base of 828 ft and an altitude of 412 ft?

5. At $3.50 per sq. yd, find the cost of paving a triangular courtyard with a 105-ft base and 21-yd altitude.

6. In Figure 3.36, a highway cuts a corner from a parcel of land. Find the acreage of the triangular lot *ABC* (1 acre = 43,560 sq. ft).

7. A surveyor starts at *A* (Figure 3.37) and lays out lines *AB*, *BC*, and *CA*. Find the three internal angles of the triangle formed.

8. *Structures.* Three supports pinned at their ends to form a triangle will not distort the structure the way a four-sided framework would. Because of their rigidity, triangles are frequently used by engineers and architects. In this exercise, a beam *AB* is supported by two crossed beams as shown in Figure 3.38. Find distance *x*.

9. Write a computer program or create spreadsheet formulas that will accept the three sides of any triangle as input. Then compute the area using Hero's formula.

10. *Constructing an Equilateral Triangle.* Devise a procedure for constructing an equilateral triangle using a compass and straightedge. Write up your procedure and add it to your growing list of geometrical constructions.

11. *Euclid's Construction of an Equilateral Triangle.* Using Figure 3.39 as a guide, write a method for constructing an equilateral triangle.

12. *Folding an Equilateral Triangle.*[3] Starting with any rectangle *ABCD*, as shown in Figure 3.40 (an $8\frac{1}{2} \times 11$ in. sheet works fine), perform the following steps. Fold midline *EF*. Unfold. Fold *A* onto *EF* so that the crease passes through *B*. Mark points *G* and *J*. Unfold. Fold along *BJ* to bring *G* onto *BD*. Mark *H*. Fold along *GJH* to complete the triangle *BGH*. Explain why triangle *BGH* is equilateral.

13. *Sum of the Angles of a Triangle by Computer Drawing Program.* Draw a triangle, and display the measure of all the internal angles and their sum. What sum do you get? Drag the triangle to different shapes while watching the sum of the angles. What do you conclude?

14. *Exterior Angle.* In the drawing from Exercise 13, extend one side to form an external angle. Display its measure, as well as the sum of the two opposite interior angles. Drag the triangle to different shapes while watching those figures. What do you find?

15. *Sum of the Angles of a Triangle by Paper Folding.*[4] Starting with any triangle *ABC*, as shown in Figure 3.41, perform these steps. Fold an altitude *BD*. Fold vertices *A*, *B*, and *C* onto *D*. Does this make clear how *A*, *B*, and *C* are related? What do you conclude?

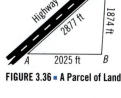

FIGURE 3.36 ■ A Parcel of Land

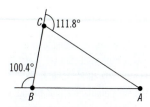

FIGURE 3.37 ■ A Surveying Layout

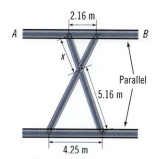

FIGURE 3.38 ■ A Structure

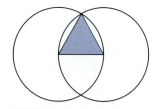

FIGURE 3.39 ■ Construction from Euclid, Book IV, Proposition 1

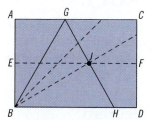

FIGURE 3.40 ■ Folding an Equilateral Triangle

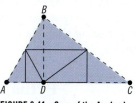

FIGURE 3.41 ■ Sum of the Angles by Paper Folding

16. *Area of a Triangle by Paper Folding.* Use the folding from Exercise 15. What are the final shape and dimensions of the figure after folding? How is the figure's area related to the area of the original triangle?

17. Define the following terms:

centroid   median   orthocenter
incenter   circumcenter   inscribe
circumscribed

18. Given any triangle, do the construction for the centroid, the incenter, the circumcenter, and the orthocenter.

19. Repeat the constructions in Exercise 18 for an equilateral triangle. What do you find?

20. *Euler Line.* Repeat the construction in Exercise 18 using a computer drawing program. Draw a line between each pair of centers (that is, centroid to incenter, centroid to circumcenter, and so forth). Then drag the triangle to change its shape. Do you see any relationship among the various centers? Based on your observations, what do you suppose an *Euler line* is? Use a mathematical dictionary to check your supposition.

21. Show by construction that the orthocenter of an obtuse triangle lies outside the triangle.

22. Where is the orthocenter of a right triangle?

23. Demonstrate by construction that any triangle having two equal altitudes is isosceles.

24. Show by construction that any triangle having two equal medians is isosceles.

25. *Centroids by Paper Folding.* Cut a triangle from a sheet of paper. Fold each side to find its midpoint, and then fold from each midpoint to the opposite vertex to locate the centroid.

26. *Orthocenter by Paper Folding.* Cut a triangle from a sheet of paper. Then fold each altitude to locate the orthocenter.

27. *Incenter by Paper Folding.* Cut a triangle from a sheet of paper. Then fold each angle bisector to locate the incenter.

28. *Circumcenter by Paper Folding.* Cut a triangle from a sheet of paper. Then fold the perpendicular bisector of each side to locate the circumcenter.

29. *Orthic Triangle.* The orthic triangle is the triangle formed by connecting the feet of the altitudes of any triangle. Draw an orthic triangle. The circumcenter of the orthic triangle is called the *nine-point center.* Find it.

> The Euler line is named for the Swiss mathematician Leonhard Euler (1707–1783), said to be the most prolific mathematician in history.

> In Chapter 7, we'll use the nine-point center to draw a *nine-point circle.*

RIGHT TRIANGLES

A *right triangle,* as mentioned in the preceding section, is a triangle in which one interior angle is a right (90°) angle. This section will explain how to find the length of a missing side of a right triangle, and it will introduce some special right triangles.

Pythagorean Theorem

The Pythagorean theorem describes the relationship among the three sides of a right triangle, and it is one of the most important theorems in mathematics.

This theorem has been proved numerous times since its discovery during the time of Pythagoras. In his book *The Pythagorean Proposition*, E. S. Loomis presented 350 demonstrations of the theorem.

To illustrate the theorem, let's draw a right triangle *ABC*, as shown in Figure 3.42a. Put the right angle at *C*, and draw legs *a* and *b* and hypotenuse *c*. To make things easier, we'll always draw a right triangle the same way—with *C* always the right angle (opposite *c*, the hypotenuse), with leg *a* opposite angle *A*, and leg *b* opposite angle *B*. Next, draw a square of side *c* and in it place four right triangles identical to the one you just drew, as shown in Figure 3.42b. The four triangles will fit perfectly, without gaps, because the two triangles meeting at any corner of the outer square form a right angle at that corner. The inner square formed by this construction has a side equal to *b − a*.

The area of the outer square is c^2, which is equal to the sum of the area of the inner square plus the areas of the four right triangles. Equating these and simplifying gives us the following:

$$c^2 = (b - a)^2 + 4\left(\frac{ab}{2}\right)$$

$$= b^2 - 2ab + a^2 + 2ab.$$

From this, we get the Pythagorean theorem.

39 PYTHAGOREAN THEOREM

The square of the hypotenuse of a right triangle is equal to the sum of the squares of the two legs.

$$a^2 + b^2 = c^2$$

●

The converse is also true: If the square of the hypotenuse equals the sum of the squares of the legs, the triangle is a right triangle. The terms of Formula 39 can, of course, be given in any order, but they are easier to memorize when given in alphabetical order, as shown here. Remember that the Pythagorean theorem applies *only* to right triangles.

■ **EXAMPLE:** What is the length of the hypotenuse if the sides of a right triangle have lengths of 3 and 4?

● **SOLUTION:**

$$3^2 + 4^2 = c^2$$
$$c^2 = 9 + 16 = 25$$
$$c = 5$$

The 3-4-5 right triangle shown in Figure 3.43 provides a good illustration of the theorem. The squares are built on each leg of the triangle and have areas of 9 and 16. The square built on the hypotenuse has an area of 25. Therefore, 9 + 16 = 25.

■

> This proof of the Pythagorean theorem is an example of a *dissection proof*, where one geometric figure is dissected and the pieces reassembled in another way so as to prove something.

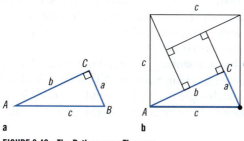

a

FIGURE 3.42 ■ The Pythagorean Theorem

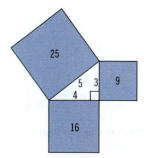

FIGURE 3.43 ■ Illustration of the Pythagorean Theorem

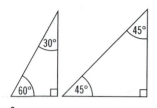

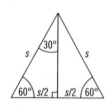

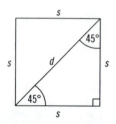

FIGURE 3.44 ■ Right Triangles

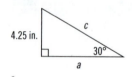

a

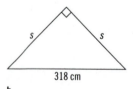

b

FIGURE 3.45 ■ Example Diagrams

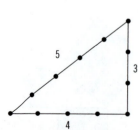

FIGURE 3.46 ■ The Rope Stretcher's Triangle

Two Special Right Triangles

The 30-60-90 right triangle and the 45° right triangle shown in Figure 3.44a are the familiar drafting triangles that you can buy in a bookstore. You can obtain a 30-60-90 right triangle from an equilateral triangle by drawing an altitude from any vertex to the opposite side bisecting that side, as shown in Figure 3.44b. Note the following:

30-60-90 RIGHT TRIANGLE

In a 30-60-90 right triangle, the short leg is half the hypotenuse. ●

You can obtain the 45° right triangle from a square by drawing its diagonal as shown in Figure 3.44c. By the Pythagorean theorem,

$$d^2 = s^2 + s^2$$
$$= 2s^2$$
$$d = \sqrt{2}s.$$

45° RIGHT TRIANGLE

In a 45° right triangle, the hypotenuse is $\sqrt{2}$ times a leg. ●

■ **EXAMPLE:** Find c in Figure 3.45a.

● **SOLUTION:**

$$c = 2(4.25) = 8.50 \text{ in.}$$ ■

■ **EXAMPLE:** Find s in Figure 3.45b.

● **SOLUTION:**

$$s = 318 \div \sqrt{2} = 225 \text{ cm}$$ ■

The Rope Stretcher's Triangle

A practical application of any triangle is to use it for *triangulation*. Triangulation is a way to locate a point by sighting it from two or more directions. One such instance of triangulation using right triangles takes us back to the origins of geometry in ancient Egypt. *Geometry* means "earth measure" (*geo + metry*).

According to Herodotus, the Nile flooded its banks every year, obliterating the markings for fields. The engineers of the time, known as Harpedonaptai, or rope stretchers, used knotted ropes to redraw the boundaries of fields after the annual flooding, as shown earlier in Figure 3.1. In order to measure the land, the engineers used the rope stretcher's triangle, which is illustrated in Figure 3.46. Notice that the rope stretcher's triangle is a 3-4-5 right triangle.

To the Egyptians, flooding symbolized the cyclic return of the primal watery chaos. Geometry, which was used to reestablish the boundaries, was seen as restoring law and order on earth. We'll see this notion again: that geometry is sacred because it represents order.

On the other hand, restoring order did not always have a sacred purpose. Herodotus says of a certain king Sesostris, "This king divided the land . . . so as to give each one a quadrangle of equal size and . . . on each imposing a tax. But everyone from whose part the river tore anything away . . . he sent overseers to measure out how much the land had become smaller, in order that the owner might pay on what was left. . . . In this way, it appears to me, geometry originated. . . . " Even then, taxes were important.

The rope stretcher's triangle, or the 3-4-5 right triangle, is also called the *sacred triangle*, the *rope knotter's triangle*, the *Egyptian triangle*, and the *Pythagorean triangle*. Vitruvius also used the 3-4-5 triangle in his work, noting that "Pythagoras has shown that a right angle can be formed without the contrivances of the artisan . . . the result which carpenters reach very laboriously, but scarcely to exactness, with their squares. . . . If we take three rules, one three feet, the second four feet, and the third five feet in length, and join these rules together with their tips touching to form a triangular figure, they will form a right angle."

■ EXERCISES ● RIGHT TRIANGLES

1. Define the following terms:

    right triangle   Pythagorean theorem
    hypotenuse   leg

2. Find the missing part in right triangle *ABC*.
    a. If $a = 283$ and $b = 462$, find c.
    b. If $a = 4.83$ and $c = 6.02$, find b.
    c. If $b = 33.6$ and $c = 49.7$, find a.

    For practice, try doing these exercises using only geometry, even if you know some trigonometry.

3. Use the Pythagorean theorem to find the missing leg or hypotenuse in the right triangles in Figure 3.47.

4. The house rafters shown in Figure 3.48 are 24 ft wide. The ridge is 9 ft higher than the side walls, and the rafters project 1.5 ft beyond the sides of the house. How long are the rafters?

5. A house (Figure 3.49) that is 50 ft long and 40 ft wide has a pyramidal roof with a height of 15 ft and with the peak *P* being directly over the center of the rectangular floor. Find the length *PQ* of a hip rafter (one that reaches from a corner to the peak of the roof).

6. What is the length of the handrail on a flight of stairs that has 16 steps, as shown in Figure 3.50? On this staircase, each step is 30.5 cm wide and 22.3 cm high.

7. A vertical pole 145 ft high is supported by three guy wires attached to the top and reaching the ground at distances of 60 ft, 108 ft, and 200 ft from the foot of the pole. What is the length of each wire?

8. A 39-ft ladder just reaches the top of a building when its foot stands 15 ft from the building. How high is the building?

9. A ladder 32 ft long stands flat and vertical against the side of a building. How many feet must the bottom end be pulled away from the wall to cause the top end to lower by 4 ft?

10. What is the side of a square with a diagonal that is 50 m?

11. A room is 20 ft long, 16 ft wide, and 12 ft high. What is the diagonal distance from one of the lower corners to the opposite upper corner?

12. Two streets, one 16.2 m wide and the other 31.5 m wide, meet at right angles. What is the distance between opposite corners of the intersection?

The rope stretcher's triangle is numerologically significant because $3 + 4 + 5 = 12$, the number of the zodiac.

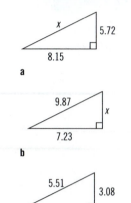

FIGURE 3.47 ■ Some Right Triangles

FIGURE 3.48 ■ House Rafters

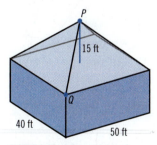

FIGURE 3.49 ■ A Pyramidal Roof

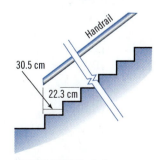

FIGURE 3.50 ■ A Handrail

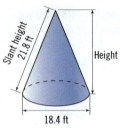

FIGURE 3.51 ▪ A Conical Roof

13. A rectangular park 156 m long and 233 m wide has a straight path running through it from opposite corners. What is the length of the path?

14. The slant height of a conical roof to a round tower is 21.8 ft and the base diameter is 18.4 ft, as shown in Figure 3.51. How high is the cone?

15. Using a computer drawing program, draw any right triangle. Then construct a square on each side. Have the computer display the area of the square on the hypotenuse and the sum of the areas of the squares on each leg. How are they related? Drag a corner of the triangle (making sure it remains a right triangle) to different positions. What can you say about the areas?

16. Write a computer program that will accept as input the two sides of any right triangle. Then compute and display the length of the missing side.

CONGRUENCE, SYMMETRY, ISOMETRY, AND SIMILARITY

"Tyger! Tyger! burning bright

In the forests of the night,

What immortal hand or eye

Could frame thy fearful symmetry?"

WILLIAM BLAKE (1757–1827), "THE TYGER"

Congruent Triangles

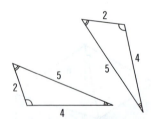

FIGURE 3.52 ▪ Congruent Triangles

Two triangles that are identical except for location and orientation are said to be *congruent*. The angles and the sides of one triangle are equal to the angles and the sides of the other, as shown in Figure 3.52. One of the triangles can be made to exactly overlay the other by sliding, turning, or flipping over, or by any combination of these motions.

Suppose you have a cardboard cut-out of a triangle, and trace its outline on a sheet of paper. Then slide, turn, and flip the triangle, tracing the outline in each new position. All the triangles so traced are congruent with each other.

Try to draw a triangle when you are given two angles and a side. Can you draw it in more than one way? If so, are the triangles you draw congruent with each other? Draw some triangles when you are given two sides and the included angle. Are the triangles congruent? Next, draw triangles given three sides. Are they congruent? You may have concluded that you get congruent triangles with each of the three sets of given information.

43 CONGRUENT TRIANGLES

Two triangles are congruent if any one of the following is true.

a. Two angles and a side of one are equal to two angles and a side of the other (ASA), (AAS).

b. Two sides and the included angle of one are equal, respectively, to two sides and the included angle of the other (SAS).

c. Three sides of one are equal to the three sides of the other (SSS).

Mirror and Rotational Symmetry

When half a figure is the mirror image of the other half, the figure has *mirror symmetry* or *bilateral symmetry*, as shown in Figure 3.53. The line that divides the figure into symmetrical halves is called the *axis of symmetry* or *mirror line*. Mirror symmetry is also called *axial symmetry*, or *symmetry with respect to a line*, the line being the axis of symmetry. A perpendicular plane through the mirror line is called a *mirror plane*.

In Figure 3.54a, the two points *P* and *Q* are symmetric about an axis *M* if *M* is the perpendicular bisector of line *PQ*. In Figure 3.54b, the two triangles are symmetric about an axis if each vertex of one is symmetric with the corresponding vertex of the other. Two symmetrical triangles are always congruent. Lastly, in Figure 3.54c, a single triangle is said to be symmetric with respect to an axis if that axis divides the triangle into two congruent triangles. An isosceles triangle has one axis of symmetry, while an equilateral triangle has three. We also say that a figure on one side of the axis is a *reflection* of the corresponding figure on the other side of the axis.

A figure has *rotational symmetry* about a point *O* if its appearance is unchanged by a rotation about *O* of less than one revolution. Imagine a cardboard triangle pinned to a board. If in a rotation of less than one revolution, the appearance of the triangle is unchanged, then it has rotational symmetry about the pin location. If it takes *n* rotations to return the figure to its original orientation, the rotational symmetry is *n-fold*. For example, the equilateral triangle in Figure 3.55 has *threefold* rotational symmetry about a point *O* because it takes three rotations (of one-third revolution each) to return the triangle to its original position.

Rotational symmetry is also called *central symmetry*, *kaleidoscope symmetry*, and *symmetry about a point*. If a plane figure has rotational symmetry about a point *O*, then the perpendicular through *O* is called the *axis of rotation*.

Transformations and Isometries

For the purposes of this discussion, we'll use the term *transformation* to mean a motion that carries each point *P* in the plane into some other point *P'* in the plane, with a one-to-one correspondence between the two points. We'll discuss two kinds of transformations: *congruent transformations* (or *isometries*), which preserve lengths, and *similarity transformations*, which do not.

An *isometry* (from *iso* + *metry*, or *equal in measure*) is a transformation that preserves length. It is also called a *congruent transformation*. Congruent triangles are also said to be *isometric*. Recall the cardboard triangle we used to demonstrate rotational symmetry. The motions we used to move it from place to place are called *isometries* because they did not change the size or shape of the triangle. The triangle *after* any isometry or combination of isometries was congruent with the triangle *before* the isometries.

The various isometries shown in Figure 3.56 are described in Statement 97.

FIGURE 3.53 ■ Mirror Symmetry

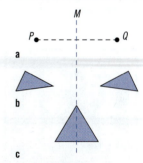

FIGURE 3.54 ■ Symmetry about an Axis

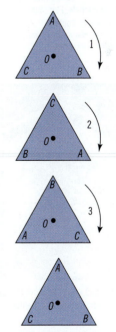

FIGURE 3.55 ■ Rotational Symmetry

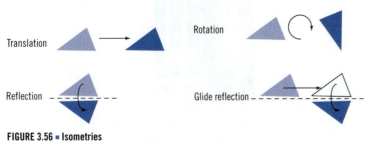

FIGURE 3.56 ■ Isometries

FIGURE 3.57 ■ Fretwork
From F. M. Hessemer, *Historic Designs and Patterns in Color* (Dover Publications, Inc.)

FIGURE 3.58 ■ *Sky and Water* (detail), M. C. Escher, 1938
©The M. C. Escher Company-Holland. All rights reserved; www.mcescher.com

FIGURE 3.59 ■ Mosaic Floor Pattern, Orsanmichele, Florence, Mid-Fourteenth Century
From F. M. Hessemer, *Historic Designs and Patterns in Color* (Dover Publications, Inc.)

FIGURE 3.60 ■ Rose Window, St. Paul's Cathedral, San Diego, California

97 ISOMETRIES

Translation: A shift in position of a given distance

Rotation: A turning through an angle

Reflection: A mirror image about some *mirror line*

To these isometries, we often add the following:

Glide reflection: A combination of translation and reflection

We say that a figure is *symmetrical* if we can apply certain isometries or symmetry operations that leave the figure unchanged. The various kinds of isometries can be found in many instances of art and architecture. A few examples are shown in Figures 3.57 through 3.64.

Repeated patterns, such as the fretwork in Figure 3.57, are examples of *translation.* This classical pattern is also called a labyrinth fret, key pattern, Greek key, or meander. The woodcut by M. C. Escher, which is shown in Figure 3.58, is another example of translation. Note the isometries, which were regular features of Escher's work.

Figure 3.59 is an example of fourfold *rotational symmetry,* and the rose window in Figure 3.60 exhibits twelvefold rotational symmetry.

The Celtic *Tree of Life* in Figure 3.61 is a delightful but inexact example of *mirror symmetry.* Windows and doorways, such as the window at St. Paul's shown in Figure 3.62, provide good examples of mirror symmetry.

The edge of the Celtic carving *St. Vigean's Stone* shown in Figure 3.63 shows *glide reflection,* a combination of translation and reflection. In Figure 3.64, we see *glide reflection* in the floral motif and translation in the row of diamonds.

FIGURE 3.61 ■ Celtic *Tree of Life* (detail), from the *Book of Kells,* c. 800?
From George Bain, *Celtic Art* (Dover Publications, Inc.)

FIGURE 3.62 ■ Gothic Window, St. Paul's Cathedral, San Diego, California

FIGURE 3.63 ■ Edge of *St. Vigean's Stone*, Twelfth Century?
From George Bain, *Celtic Art* (Dover Publications, Inc.)

FIGURE 3.64 ■ Mosaic Band in the Floor of the Baptistery, Florence, Eleventh Century
From F. M. Hessemer, *Historic Designs and Patterns in Color* (Dover Publications, Inc.)

Similar Triangles

Two triangles are said to be *similar* if they have the *same shape*, even if one is larger than the other or is rotated, translated, or reflected (has mirror symmetry) about some axis. That means that the angles of one of the triangles must equal the angles of the other, as in Figure 3.65.

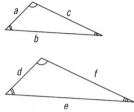

FIGURE 3.65 ■ Similar Triangles

44a SIMILAR TRIANGLES

Two triangles are similar if the angles of one triangle equal the angles of the other. ●

Pairs of equal angles are called *corresponding angles*. Sides that are between the same pair of equal angles, such as sides *a* and *d* in Figure 3.65, are called *corresponding sides*. Sides *b* and *e* are also corresponding sides, as are sides *c* and *f*.

In two similar triangles, if a side of one is, say, twice as long as the corresponding side in the other, then any other side in the larger triangle must be twice as long as the corresponding side in the smaller triangle. If this were not so, the two triangles would not have the same shape, and, hence, would not be similar.

44b CORRESPONDING SIDES OF SIMILAR TRIANGLES

If two triangles are similar, their corresponding sides are in proportion. The ratio between corresponding sides is called the *scale factor*. ●

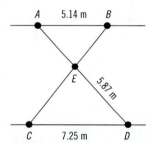

FIGURE 3.66 ■ A Framework

■ **EXAMPLE:** The two beams *AB* and *CD* in a framework shown in Figure 3.66 are parallel. Find distance *AE*.

● **SOLUTION:** Angles *AEB* and *CED* are vertical angles and, therefore, are equal. Further, angles *ABE* and *ECD* are alternate interior angles between parallel lines, and so are equal. Therefore, the upper and lower triangles are similar. Because *AE* and *DE* are corresponding sides, we write

$$\frac{AE}{5.87} = \frac{5.14}{7.25}$$

$$AE = 5.87\left(\frac{5.14}{7.25}\right) = 4.16 \text{ m.} \quad ■$$

Measuring inaccessible distances by means of similar triangles is not new. Figure 3.67, taken from a book printed in 1545, shows an instrument being used to measure a façade. Leonardo sketched a similar device, which he called the *bacolo of Euclid*. The surveyor's instrument called the *cross-staff* is a similar device.

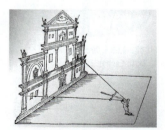

FIGURE 3.67 ■ Instrument Used to Measure the Width of a Façade

Areas of Similar Triangles

We will expand on the idea of similarity with similar polygons in Chapter 5 and similar solids in Chapter 10.

The areas of similar triangles can be determined using the scale factor. Starting with two similar triangles, draw an altitude to the corresponding sides of each. Each altitude divides each triangle into two right triangles. One such right triangle in the larger triangle must be similar to the corresponding right triangle in the smaller triangle. The altitudes of the two similar triangles are in the same proportion as the sides, and they have the same scale factor. Therefore, if the scale factor is, for example, two, then the altitude and the base of the larger triangle are twice the altitude and the base of the smaller. Because the area is the product of base and altitude, the *area* of the larger triangle is *four times* the area of the smaller triangle.

44c AREAS OF SIMILAR TRIANGLES

Areas of similar triangles are proportional to the *square* of the scale factor. ●

■ **EXAMPLE:** If the sides of one triangle are three times the length of corresponding sides of a similar triangle (a scale factor of three), how many times larger is the area of the larger triangle than the smaller?

● **SOLUTION:** The area of the triangles is proportional to the square of the scale factor, so the area of the larger triangle is 3^2 or *nine times* the area of the smaller. ■

Similarity Transformations

FIGURE 3.68 ■ Scaling

The operations that transform one triangle into another, similar triangle are called *similarity transformations*. As with isometric transformations, they include translation, rotation, and reflection, but the concept of *scaling* is added. *Scaling* means that a figure is enlarged or reduced by some scale factor, such as in Figure 3.68. Escher's fish, shown in Figure 3.69, are a good example of scaling.

98 SIMILARITY TRANSFORMATIONS

The similarity transformations are translation, rotation, reflection, glide reflection, and *scaling*. ●

FIGURE 3.69 ■ *Circle Limit II,*
M. C. Escher, 1959

■ EXERCISES • CONGRUENCE, SYMMETRY, ISOMETRY, AND SIMILARITY

1. Define the following terms:

congruent triangles	similar triangles	axis of symmetry
mirror line	mirror symmetry	bilateral symmetry
axial symmetry	rotational symmetry	central symmetry
isometry	corresponding angles	corresponding sides
scale factor		

2. The isosceles triangular window shown in Figure 3.70 has sides parallel to the edges of the roof. Find dimensions *a* and *b*.

3. Identify the kinds of symmetry in each of the shapes in Figure 3.71.

4. Examine an alphabet drawn with simple sans serif block letters and strokes of equal width. Identify which letters have symmetry and name the kind of symmetry.

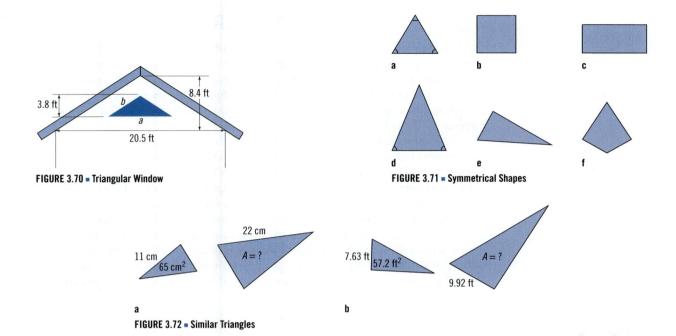

FIGURE 3.70 ■ Triangular Window

FIGURE 3.71 ■ Symmetrical Shapes

FIGURE 3.72 ■ Similar Triangles

5. Figure 3.72 shows two pairs of similar triangles. For each pair, find the scale factor and the area of the larger triangle.

6. *Line Parallel to Triangle Side, Euclid Book VI, Proposition 2.* In Chapter 1, we used this proposition to construct a line through a given point parallel to a given line. Prove Statement 34, which is illustrated in Figure 3.73.

34 LINE PARALLEL TO A SIDE

If a straight line is drawn parallel to one side of a triangle, it will cut the other sides proportionally.

Conversely, if two sides of a triangle are subdivided proportionally, the line joining the subdividing points will be parallel to the third side of the triangle.

7. *Measuring Inaccessible Distances.* In his *Ten Books of Architecture*, the artist and architect Leon Battista Alberti described how to use similar triangles to measure, for example, the height of a tower on level ground. Figure 3.74 shows that the distance from one's eye to the stick is 5 ft, the line of sight cuts the stick 4 ft from the ground, and the tower is 55 ft from the stick. Find the height *h* of the tower.

8. In *The Musgrave Ritual*, Sherlock Holmes calculated the length of the shadow of an elm tree that is no longer standing. He knew that the elm was 64 ft high and that the shadow was cast at the instant the sun was grazing the top of a particular oak tree. Holmes held a 6-ft-long fishing rod vertically and measured its shadow at that instant. He found it to be 9 ft long. How long would the shadow of the now-vanished elm tree be at that instant?

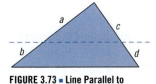

FIGURE 3.73 ■ Line Parallel to Triangle Side

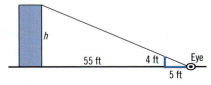

FIGURE 3.74 ■ Measuring the Height of a Tower

SUMMARY

In this chapter, we continued our exploration of the two-dimensional world with the angle and the triangle. To create a good base on which to build later, most of this chapter discussed plane geometry. However, it also gave some examples of the triangle in art and architecture. This was followed with a section on congruence, symmetry, isometry, and similarity, laying the foundation for our exploration of tilings in Chapter 5.

As a contrast with strict geometry and its rigorous proofs, we looked at the misty world of geometric symbolism, such as with the equilateral triangle representing the Trinity. Throughout this book, we'll try to bridge these two worlds. The geometry that permeated the Middle Ages could not help but affect the most prominent architecture of the time: the Gothic cathedrals. Such symbolism and mathematics are ideas buried in the roots of modern civilization. Many such ideas have come down to us over the centuries, in literature and as common art motifs. Consult Appendix E for some of the associations for the number *three*, as well as for famous groupings of three in art. Moving on from the three-sided polygon, in the next chapter we'll add sides and cover the square and rectangle.

EXERCISES AND PROJECTS

1. Define the following terms:

   rope stretcher's triangle   golden triangle   golden gnomon
   triskelion   trefoil

2. How are the angles of a golden triangle related to each other? Use this property to compute them for both the acute and the obtuse golden triangle.

3. Prove that the area of any triangle equals half its base times its altitude.

4. Using a CAD program, such as The Geometer's Sketchpad, make a sketch to illustrate that the area of a triangle remains unchanged when the top vertex is dragged along a line parallel to the base (Figure 3.75).

5. Demonstrate by construction and measurement that Hero's formula works.

6. Without using the Pythagorean theorem, demonstrate that the 3-4-5 triangle is a right triangle.

7. Show by construction that the incenter of a triangle is the center of the inscribed circle.

8. Show by construction that the circumcenter of a triangle is the center of the circumscribed circle.

9. *Napoleon's Theorem.* Napoleon is known as a brilliant military leader. Few people know that he was also a student of mathematics. In fact, a theorem that he independently solved is named for him.

FIGURE 3.75 ■ Dragging One Vertex of a Triangle

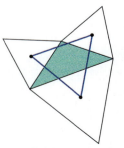

FIGURE 3.76 ■ Napoleon's Theorem

35a **NAPOLEON'S THEOREM**

If equilateral triangles are drawn externally or internally on the sides of any triangle, their centers form an equilateral triangle.   ●

    a. Create a CAD drawing, such as Figure 3.76, illustrating Napoleon's theorem.

    b. Drag a vertex to see if the theorem still holds.

10. *Inner Napoleon Triangle.*

35b INNER NAPOLEON TRIANGLE

The inner Napoleon triangle has the same center as the external triangle.   ●

a. Demonstrate that Napoleon's theorem holds if the equilateral triangles are drawn *internally*, rather than externally.
b. Demonstrate that the inner Napoleon triangle has the same center as the external triangle.

11. *Morley's Theorem.* Demonstrate Morley's theorem, illustrated in Figure 3.77.

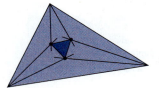

FIGURE 3.77 ■ Morley's Theorem

36 MORLEY'S THEOREM

The lines of trisection of the vertices of a triangle intersect to form an equilateral triangle.   ●

Morley's theorem was discovered in 1899 by Frank Morley, whose son Christopher is the well-known fiction author.

12. *Pappus' Theorem.* Demonstrate *Pappus'* theorem, illustrated in Figure 3.78.

37 PAPPUS' THEOREM

If points *A*, *C*, and *E* lie on one straight line, and *B*, *D*, and *F* lie on another line, then the points of intersection of *AB* with *EF*, *BC* with *DE*, and *CF* with *AD* lie on a third straight line.   ●

13. *Desargues' Theorem.* Desargues' theorem concerns two triangles that are *in perspective*, as shown in Figure 3.79.

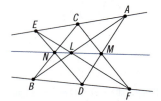

FIGURE 3.78 ■ Pappus' Theorem

45a TRIANGLES IN PERSPECTIVE

Two triangles are said to be *perspective from a point* if the three lines joining pairs of corresponding points meet in a single point.   ●

Desargues' theorem (Figure 3.80) goes on to say that:

45b DESARGUES' THEOREM

If two triangles are perspective from a point, then the three points of intersection of corresponding sides, extended, lie on a straight line.   ●

Demonstrate this theorem with a CAD drawing, such as the one shown in Figure 3.80, dragging various points to be sure that the theorem holds. (Chapter 12 will cover perspective in more detail.)

14. *Square-Root Spiral.* Construct the square-root spiral as shown in Figure 3.81. Start with an isosceles right triangle with legs equal to 1. On its hypotenuse, construct another right triangle, with the other leg equal to 1.

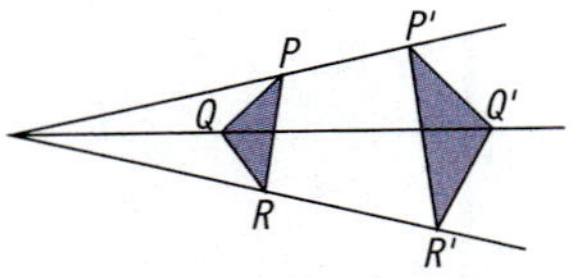

FIGURE 3.79 ■ Triangles Perspective from a Point

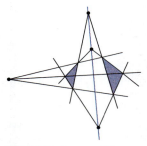

FIGURE 3.80 ■ Desargues' Theorem

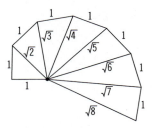

FIGURE 3.81 ■ Square-Root Spiral

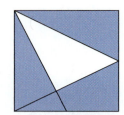

FIGURE 3.82 ■ Triangle within a Square

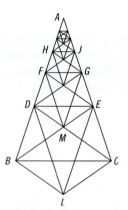

FIGURE 3.83 ■ The Lute of Pythagoras

A *porism* or a *corollary* is an additional inference drawn from the proposition that immediately precedes it.

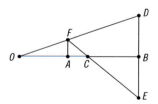

FIGURE 3.84 ■ Pappus' Construction for the Harmonic Mean

Continue adding right triangles as desired. Show that the lengths of the segments radiating from the center to the spiral are equal to the square roots of 1, 2, 3, and so on.

15. Do the following construction from Lawlor of a triangle within a square, as shown in Figure 3.82. Bisect two adjacent sides of the square and connect each to the opposite vertex of the square.
    a. If the side of the square is one unit, what is the length of each diagonal?
    b. How is the length of each diagonal related to the golden ratio?
    c. Identify the kind of triangle formed.

16. Construct the *Lute of Pythagoras*, as shown in Figure 3.83. Given golden triangle *ABC*, perform the following. From *B*, draw an arc with radius *BC* to locate *E*. Similarly locate *D*. From *D* and *E*, draw arcs of radius *DE* to locate *F* and *G*. Repeat the construction upward a few more times.

    Connect *D* and *E* to get the first "rung." Connect *B* to *E* and *C* to *D* to form diagonals that meet at *M*. This forms pentagon *DMEGF*. Repeat upward, getting more rungs, diagonals, and pentagons.

    Connect *F* and *G* to *M*, getting a golden triangle *FGM*. Repeat for the other rungs. Locate *L* by swinging arcs of radius *BD* from *B* and *C*.

17. Euclid's Proposition 8 in Book VI states that if an altitude is drawn to the hypotenuse of a right triangle, the triangles adjoining the altitude are similar to each other and also similar to the original triangle (Equation 41). Can you explain why?

18. *Euclid's Construction of the Mean Proportional.* In the porism following Book VI, Proposition 8, Euclid shows that the altitude drawn to the hypotenuse of a right triangle is the *mean proportional* between the segments of the hypotenuse. Prove or demonstrate that this is true.

19. *Pappus' Construction for the Harmonic Mean.* Prove that the construction in Figure 3.84 gives the required harmonic mean. To find the harmonic mean between two line segments *OA* and *OB*, follow these steps. Erect a perpendicular to *OB* at *B*. Mark points *D* and *E* at equal distances from *B*. Draw *OD*. Erect a perpendicular to *OB* at *A*, which cuts *OD* at *F*. Draw *FE*, cutting *OB* at *C*. *OC* is then the harmonic mean between *OA* and *OB*.

20. *Pappus' Construction for the Three Means.* Demonstrate that the construction in Figure 3.85 gives the three means. To find the three means between *AB* and *BC*, follow these steps. Lay *AB* and *BC* end to end. Draw a semicircle on *AC*. Erect a perpendicular to *AC* at *B* to intersect the semicircle *E*. Draw *DE*. Erect a perpendicular to *DE* through *B*, cutting *DE* at *F*.

    Then for segments *AB* and *BC*, the following are true:

■ *DE* is the arithmetic mean.

■ *BE* is the geometric mean.

■ *EF* is the harmonic mean.

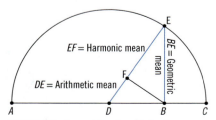

FIGURE 3.85 ■ Pappus' Construction for the Three Means

21. Most computer drafting programs have a *mirror* or *reflect* feature that enables you to place a mirror line in your drawing to reflect figures to the other side of the line. *Translate, rotate,* and *dilate* tools are also common features. Experiment with these tools, and write a paragraph or two on your findings, relating them to the ideas of symmetry presented in this chapter.

22. Create a graphic design using triangles as the main element. Use your design for wrapping paper, a greeting card, a screensaver, or a fabric design for a pillow, tee shirt, quilt, or dress.

23. Repeat the preceding project, using the golden triangle as the main design element.

24. Make a kite in the shape of the Lute of Pythagoras and decorate it appropriately. Fly it on campus.

25. Paste a picture on one side of a piece of thin wood or cardboard. Then cut the sheet apart into shapes representing as many kinds of triangles that you can. Label each piece on the back to make a jigsaw puzzle for children.

26. Make a cardboard cut-out of a triangle. Draw the three medians to locate the centroid at their point of intersection. Hang the cut-out from a push pin so that it can swing freely, and hang a weight on a string from the same point. When the cut-out and the weight have stopped swinging, mark the string's position on the cardboard. Change the suspension point and repeat. Where do the string positions intersect? Can you explain this?

27. Write a report on any book in the list of sources for this chapter or some other source related to this chapter.

28. Write a short paper or a term paper on some topic from this chapter. Come up with your own topic, using these suggestions to jog your imagination:

    ■ Various units of angular measure, degrees, radians, grads, minutes, seconds, etc.

    ■ The Greek mathematician Hero of Alexandria, first century A.D.

    ■ The Greek geometer Pappus of Alexandria, c. A.D. 300

    ■ Various proofs of the Pythagorean theorem

    ■ The use of the "Pythagorean" theorem before the time of Pythagoras

    ■ Various kinds of symmetry

    ■ The rope stretcher's triangle and its use by the Egyptians

    ■ The sacred tetraktys and its importance to the Pythagoreans

    ■ Various shapes of halos and their meaning

    ■ Traditional quilt patterns using triangles, such as *Wild Goose Chase* (pattern # 3537 in Rhemel's *The Quilt I.D. Book*)

    ■ The history of angle trisection, using the outline in Heath, p. 235, as a starting point (see Sources)

    ■ Napoleon as a mathematician

    ■ *Number symbolism* in general

    ■ The symbolism of the number *three* (see Appendix E)

29. William Blake wrote "The Tyger" using the word *symmetry*. Can you find other famous works that examine symmetry? Check *The Divine Comedy* and *Alice in Wonderland* for examples.

30. Make an oral presentation to your class for any of the previous projects.

31. Make a batch of cookies with the shapes of the kinds of triangles described in this chapter. In class, a student must identify the triangle to get the cookie.

32. Make a cake in a square pan. Use it for one of the dissection proofs of the Pythagorean theorem.

Team Projects

33. Use a long knotted rope to make a rope stretcher's triangle. Use it outdoors to lay out a right angle in an open space. Then make three more right angles to form a square. How accurate is your work? Did you come back to the starting point?

34. A ball field, tennis court, or basketball court has many lines that intersect at what appear to be right angles. Using only a tape measure, how can you verify that an angle is indeed 90°?

35. Use an upright or horizontal stick and similar triangles to measure the height or width of a building on campus. Check your results by direct measurement.

36. Construct a measuring device similar to the *radio astronomico* or the *bacolo of Euclid*. Use it to find the width of a building by similar triangles. Verify your results by direct measurement.

37. Check your campus or neighborhood for threefold architectural features: triangular windows, trefoils, triskelions, and so on. Make sketches, take photos, and report your findings.

38. Search books or your neighborhood for examples of the various isometries, as well as scaling. Try to find at least one example of each. Take pictures and make a presentation to your class.

Mathematical Challenges

39. Prove Kepler's statement: *If the sides of a right triangle are in geometric ratio, then the sides are* $1 : \sqrt{\Phi} : \Phi$.

40. *Trigonometric Ratios.* You probably remember the following trigonometric ratios from other mathematics classes. Explain what these ratios are, and give examples of their use.

40 TRIGONOMETRIC RATIOS

a. Sine:

$$\sin \theta = \frac{\text{opposite side}}{\text{hypotenuse}}$$

b. Cosine:

$$\cos \theta = \frac{\text{adjacent side}}{\text{hypotenuse}}$$

c. Tangent:

$$\tan \theta = \frac{\text{opposite side}}{\text{adjacent side}}$$

41. *Archimedes' Method for Angle Bisection.* To bisect angle α, as shown in Figure 3.86, perform these steps. Draw α as the central angle in a circle, intercepting arc PQ. Extend OP to S, and draw an inscribed angle PSQ, intercepting arc PQ. Then angle PSQ is half-angle α. Explain why this construction works.

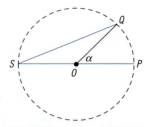

FIGURE 3.86 ■ Archimedes' Method for Angle Bisection

42. *Archimedes' Trisection of an Angle.* Figure 3.87 shows a construction for angle-trisection attributed to Archimedes. It is not a strict compass-and-ruler construction because it allows you to mark the ruler. Reproduce it, and verify by measurement that it works.

    To trisect angle α, choose a point A on one side of the angle, and draw a circle of radius OA. Mark point B. Find a point E so that DE is equal to the radius OB of the semicircle, and so that DE, extended, passes through B.

    You can do this by first marking distance OB on our straightedge. Then position the straightedge so that it passes through B at the same time that the two marks fall on OA, extended, and on the circle. Angle AEB is then one-third angle α.

    Note that this construction is different from others because it required you to *mark the straightedge*. Such methods were known to the ancient Greeks, who called them *neusis* constructions.

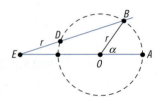

FIGURE 3.87 ■ Archimedes' Trisection of an Angle

43. *Pappus' Trisection of an Angle.* Figure 3.88 illustrates another construction for angle-trisection attributed to Pappus (c. A.D. 300). It also allows the ruler to be marked. Reproduce and explain this construction.

    Given angle α, choose a point A on one side of the angle. Through A, draw lines parallel and perpendicular to the other side of the angle. Fit a line CD, with length that is twice OA, between AB and AD, so that its extension passes through O. Note that this is another *neusis* construction. Angle BOD then trisects the given angle α.

44. *Hero's Formula.* If you are familiar with the trigonometry of oblique triangles, especially the law of cosines, you might enjoy trying to follow the derivation of Hero's formula (Equation 30). See Coxeter's *Introduction to Geometry*, p. 12.

45. A hexagonal floor tile, as shown in Figure 3.89, measures 50 cm across the flats. Find the distance x across the corners. *Hint:* We haven't studied the hexagon yet, but you can solve this exercise by first dividing the hexagon into right triangles.

46. *Dissection Proofs.* The Pythagorean theorem is often proved by means of a dissection proof, such as the one given earlier in Figure 3.42. Figure 3.90 (see next page) illustrates another proof. Finish this proof. Given right triangle *abc*, draw two squares, each having a side $a + b$. Cut one square into five pieces: a square of side c and four copies of the given right triangle. Cut the other square into six pieces: squares of sides a and b and four

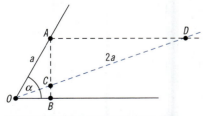

FIGURE 3.88 ■ Pappus' Trisection of an Angle

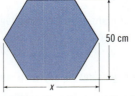

FIGURE 3.89 ■ A Hexagonal Floor Tile

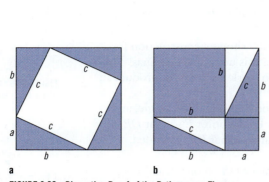

FIGURE 3.90 ■ Dissection Proof of the Pythagorean Theorem

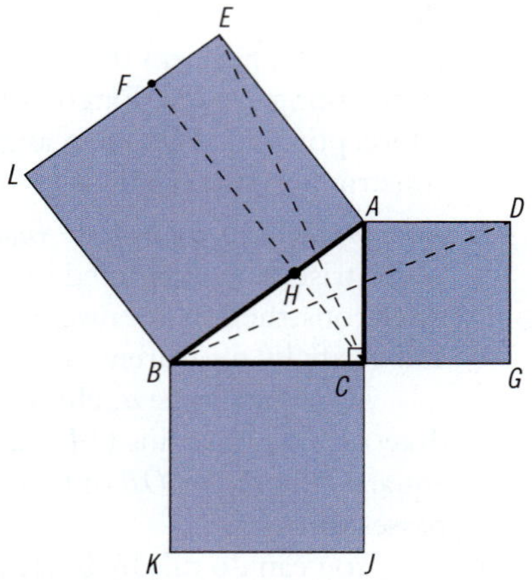

FIGURE 3.91 ■ Euclid's Proof of the Pythagorean Theorem

copies of the given right triangle. How can you show that the square on the hypotenuse is equal to the sum of the squares on the legs?

47. *Euclid's Proof.* Euclid proves the Pythagorean theorem in Book 1, Proposition 47. See if you can reproduce Figure 3.91 and explain the steps.

   Construct a square on each side of the given right triangle. Draw altitude *CH* and extend to *F*. Draw segments *BD* and *CE*. The triangle *ABD* is identical to triangle *ACE*, except for a rotation. If you turn one of these triangles by 90°, it will cover the other. Square *ACGD* has twice the area of triangle *ABD*, because they have the same base (*AD*) and the same height (*AC*). Similarly, rectangle *AEFH* has twice the area of triangle *ACE*. Because triangles *ABD* and *ACE* are equal, square *ACGD* equals rectangle *AEFH*.

   In the same way, you can show that square *BCJK* equals rectangle *BLFH*. Therefore, Square *ACGD* + Square *BCJK* = Rectangle *AEFH* + Rectangle *BLFH* = Square *ABLE*, which proves the theorem.

48. *Pythagorean Triples.* A set of three integers that satisfy the Pythagorean theorem are called *Pythagorean triples*. One such set is for the familiar 3-4-5 triangle. Here are several formulas for Pythagorean triples. Try them with various values and check your results. Do this by calculator or use a spreadsheet.

42 PYTHAGOREAN TRIPLES

a. Attributed to Plato: For any *m*,

$$(2m)^2 + (m^2 - 1)^2 = (m^2 + 1)^2$$

b. Rule of Pythagoras: *m* is an odd integer.

$$m^2 + \left(\frac{m^2 - 1}{2}\right)^2 = \left(\frac{m^2 + 1}{2}\right)^2$$

c. Rule of Plato: *m* is an even integer.

$$m^2 + \left(\frac{m^2}{4} - 1\right)^2 = \left(\frac{m^2}{4} + 1\right)^2$$

d. Rule of Masères: *m* and *n* are any integers, and *m* > *n*.

$$(2mn)^2 + (m^2 - n^2)^2 = (m^2 + n^2)^2$$

49. *Pythagorean Triples Related to Fibonacci Numbers.* Livio points out an interesting connection between Pythagorean triples and Fibonacci numbers. Take any four consecutive Fibonacci numbers, for example,

$$1, 2, 3, 5.$$

Then compute as follows:

a. The product of the outer terms 1 and 5:

$$1(5) = \mathbf{5}$$

b. Twice the product of the inner terms 2 and 3:

$$2(2)(3) = \mathbf{12}$$

c. The sum of the squares of the inner terms 2 and 3:

$$2^2 + 3^2 = \mathbf{13}$$

   This gives the Pythagorean triple **5, 12, 13**. Also note that the last number (13) is also a Fibonacci number. Try this with other consecutive Fibonacci numbers, and see if it works there.

50. *Fermat Point.* Devise a way to find the Fermat point, as shown in Figure 3.92. *Hint:* Start by drawing an equilateral triangle on each side of the original triangle and connecting the vertices of each to the vertices of the original triangle. Then drag a corner of the triangle to change its shape. By doing this, can you determine what kinds of triangles do not have a Fermat point?

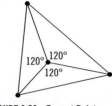

FIGURE 3.92 ■ Fermat Point

38   FERMAT POINT

Some triangles have a point such that lines drawn from it to each vertex form three 120° angles.

SOURCES

Aaboe, Asger. *Episodes from the Early History of Mathematics.* New York: Random House, 1964.

Calter, Paul. *Technical Mathematics.* New York: Wiley, 2007.

Coxeter, H. S. M. *Introduction to Geometry,* 2nd ed. New York: Wiley, 1989.

Coxeter, H. S. M., et al. *Geometry Revisited.* Washington, D.C.: Mathematical Association of America, 1967.

Euclid. *The Thirteen Books of the Elements.* New York: Dover, 1956.

Eves, Howard. *An Introduction to the History of Mathematics.* New York: Holt, 1953.

Heath, Sir Thomas. *A History of Greek Mathematics.* New York: Dover, 1981. First published in 1921.

Loomis, E. S. *The Pythagorean Proposition,* 2nd ed. Ann Arbor, MI: Edwards, 1940.

March, Lionel. *Architectonics of Humanism.* Chichester, UK: Academy Editions, 1998.

Olson, Alton T. *Mathematics Through Paper Folding.* Reston, VA: National Council of Teachers of Mathematics, 1975.

NOTES

1. Aaboe, p. 87.
2. Euclid V2, p. 96.
3. This exercise is from Olson, p. 23.
4. Olson, p. 13.

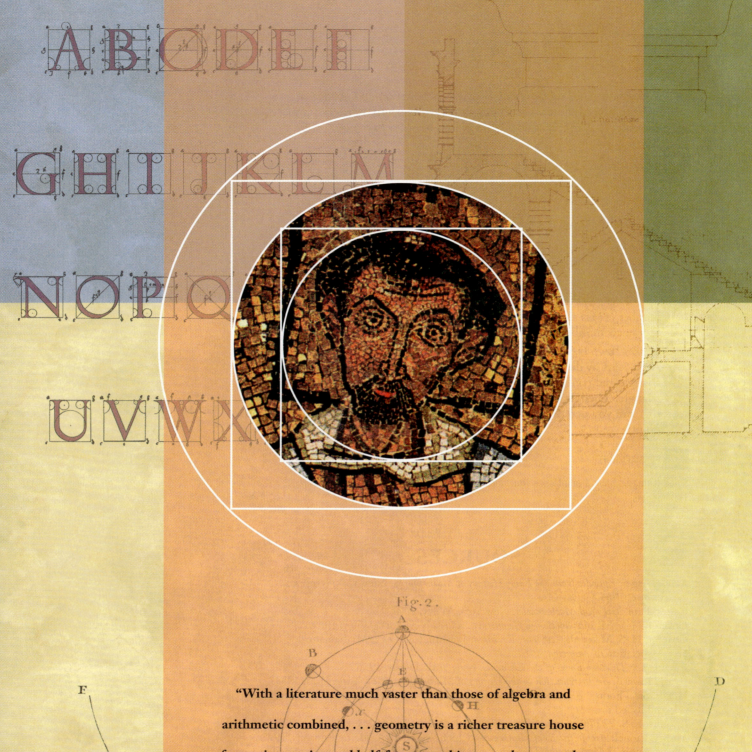

"With a literature much vaster than those of algebra and

arithmetic combined, . . . geometry is a richer treasure house

of more interesting and half-forgotten things . . . than any other

division of mathematics."

E.T. BELL

FIGURE 4.1 ▪ Pope John VII, c. 705, Mosaic Fragment (detail)
©Vatican Museum

4

Ad Quadratum and the Sacred Cut

I n this chapter, we'll journey from the three-sided triangle to the four-sided square and rectangle, still confining ourselves to the two-dimensional plane. We'll examine the square and its diagonal, the so-called *ad quadratum* figure, and the root-2 rectangle. We'll describe the Pythagorean method of alternation for finding the incommensurate diagonal of a square, and we'll examine the principle of *gnomonic growth*. In addition, we'll look at *the sacred cut*, a geometric construction involving the square, and the so-called *Roman rectangle*, which is thought to be central to the ancient Romans' system of proportions. We'll also give the geometry of other four-sided figures: parallelograms, rhombuses, and trapezoids.

Just as we saw previously with the triangle, both the square and the rectangle had significance to early artists, architects, and builders. Note the square halo behind the head of Pope John VII in Figure 4.1. Although rare, the square halo is sometimes used in Christian symbolism for someone who is still alive and

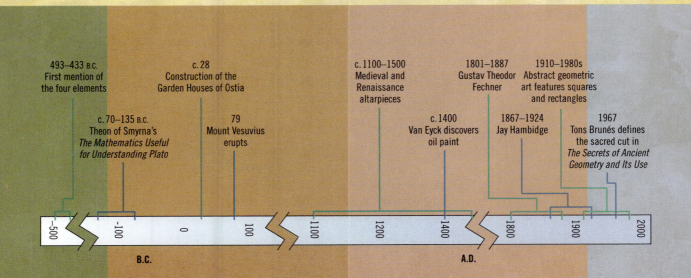

493–433 B.C.
First mention of
the four elements

c. 70–135 B.C.
Theon of Smyrna's
*The Mathematics Useful
for Understanding Plato*

c. 28
Construction of the
Garden Houses of Ostia

79
Mount Vesuvius
erupts

c. 1100–1500
Medieval and
Renaissance
altarpieces

c. 1400
Van Eyck discovers
oil paint

1801–1887
Gustav Theodor
Fechner

1867–1924
Jay Hambidge

1910–1980s
Abstract geometric
art features squares
and rectangles

1967
Tons Brunés defines
the sacred cut in
*The Secrets of Ancient
Geometry and Its Use*

-500   -100   0   100   1100   1200   1400   1800   1900   2000

B.C.   A.D.

who is probably headed for sainthood. In this chapter, we'll explore square- and rectangular-shaped frames, how and why those particular shapes were used to make frames, and how the chosen shape affected the enclosed picture or relief sculpture. We'll briefly track the development of paintings having no frames, open frames, complex frames, or multiple frames, and we'll take a look at altarpieces. Further, we'll explore several rectangular formats: the square, the double square, overlapping squares, and the golden rectangle. In addition to the outside shape of a picture, we'll examine some compositional devices, such as axes of symmetry, zones, diagonals, and microthemes.

Finally, we'll discuss various researchers' advanced theories about how systems of proportions based on the square were used in the domestic Roman architecture of Ostia, Pompeii, and Herculaneum; in Renaissance architecture; and in the ground plans of churches. We'll view these claims with the same critical eye we used earlier when we examined systems of proportions said to be based on the musical ratios or the golden ratio.

THE SQUARE

The *tetragon* and *quadrangle* are also names for the quadrilateral.

Recall that a polygon is a plane figure bound by straight line segments. A *quadrilateral* is a polygon with four sides, such as the square. Other quadrilaterals include the rectangle, rhombus, parallelogram, and trapezoid. We'll study the square in this section and the other quadrilaterals in later sections.

Let's begin by revisiting the simple formulas for the perimeter and area of the square, which you learned in earlier mathematics courses.

46 PERIMETER AND AREA OF A SQUARE

a. The perimeter of a square is the sum of the four sides.

b. The area of a square is the square of one side.   ●

■ **EXAMPLE:** What is the perimeter and area of a square of side 5?

● **SOLUTION:**

$$\text{Perimeter} = 5 + 5 + 5 + 5 = 20$$
$$\text{Area} = 5^2 = 25$$ ■

Using what we learned from our discussion of symmetry in Chapter 3, "The Triangle," we know that the square has four axes of symmetry, as shown in Figure 4.2. They are the vertical and horizontal lines and the two diagonal lines through the center. The square also has fourfold rotational symmetry about its center.

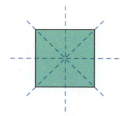

FIGURE 4.2 ■ Symmetry of the Square

Diagonal of a Square

In Chapter 1, "Music of the Spheres," we noted that the diagonal of a square is *incommensurable* with the side of that square, meaning that there is no smaller length that will divide into both side and diagonal an exact number of times.

The diagonal d of a square of side s can be found using the Pythagorean theorem:

$$d^2 = s^2 + s^2$$
$$= 2s^2$$
$$d = \sqrt{2}s.$$

Therefore, the diagonal is $\sqrt{2}$, or about 1.414, times the length of a side.

Another way to calculate the diagonal of a square is to use a method called *alternation*, which is demonstrated in Figure 4.3. This method is presented in

Theon of Smyrna's book *The Mathematics Useful for Understanding Plato*, the same book in which we found the *Ten Sets of Four Things*. We show the method here because it gives the diagonal of a square of side 1, and it provides an interesting series of numbers that we'll use later.

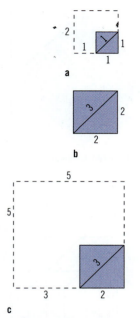

To use the method of alternation, start with a square of side 1 (Figure 4.3a) and assume that its diagonal is 1. Because the diagonal of a square cannot equal its side, we know that side 1 is too small to be the diagonal; we are just using it as a starting value. Make a second square (Figure 4.3b) with a side that equals the old side plus the old diagonal:

$$\text{new side} = 1 + 1 = 2.$$

Assume that the new diagonal equals the preceding diagonal plus the two preceding sides:

$$\text{new diagonal} = 1 + 1 + 1 = 3.$$

Now we have a diagonal that is too large because the square of the diagonal of a square cannot exceed the sum of the squares of two of its sides. Make a third square (Figure 4.3c) with a side that equals the preceding side plus the preceding diagonal, or $(2 + 3 = 5)$. Let the new diagonal equal the preceding diagonal plus the two preceding sides, or $(3 + 2 + 2 = 7)$.

Keep repeating the procedure, where each new square has a side equal to the preceding square plus the preceding diagonal, and where you assume that its diagonal equals the preceding diagonal plus two preceding sides. You'll find that the diagonal alternates between being too small and being too large but keeps getting closer to the true value. In modern terminology, this is called a method of *successive approximations* or *iteration*. The computation is said to be *converging* on a value.

FIGURE 4.3 ■ Method of Alternation

The following table gives the values obtained for the first nine squares. Notice that the ratio of diagonal to side, which is approaching the value of the square root of 2, is also shown. The square root of 2 is 1.414211, by calculator, rounded to six decimal places.

Alternation Square	Side	Diagonal	Diagonal/Side
1	1	1	1.000000
2	2	3	1.500000
3	5	7	1.400000
4	12	17	1.416667
5	29	41	1.413793
6	70	99	1.414286
7	169	239	1.414201
8	408	577	1.414216
9	985	1393	1.414213

Of interest for our study of Roman architecture is the sequence of numbers giving the sides and diagonals of successive squares:

$$1, 1, 2, 3, 5, 7, 12, 17, 29, 41, 70, 99, \ldots$$

We'll return to this sequence of numbers, which we'll refer to as the *alternation sequence*.

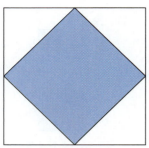

FIGURE 4.4 ■ Ad Quadratum

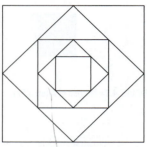

FIGURE 4.5 ■ Nested Squares

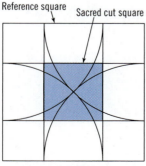

FIGURE 4.6 ■ The Sacred Cut

Root-2 or Ad Quadratum Sequence

When we connect the midpoints of the sides of a square to form an inner square so that one square is set diagonally inside another, we get a design known as the *ad quadratum* figure. Figure 4.4 illustrates an ad quadratum figure in which the side of the larger square equals the diagonal of the smaller square.

As we just demonstrated, the diagonal of any square is $\sqrt{2}$ times its side. Therefore, starting with a square of side 1, we know that the diagonal of that square is $\sqrt{2}$. If $\sqrt{2}$ is taken as the side of the next larger square, its diagonal is $\sqrt{2}$ times $\sqrt{2}$, or just 2. With *successive iterations*, we obtain the root-2 or ad quadratum sequence:

$$1, \sqrt{2}, 2, 2\sqrt{2}, 4, \ldots$$

This is a geometric progression with a common ratio of $\sqrt{2}$. Therefore, we find each successive number in the sequence by multiplying the preceding number by $\sqrt{2}$, and we find each preceding number by dividing by $\sqrt{2}$. As such, if you were to start with a value of, say, 41, and work in both directions, you would get the following sequence of numbers, each rounded to the nearest whole number:

$$\ldots, 29, 41, 58, 82, \ldots$$

The ad quadratum sequence will appear again later in this chapter when we look for ratios in Roman architecture. You'll see why we chose 41 as our starting value.

We can also find the ad quadratum sequence geometrically, using the construction shown in Figure 4.5. Citing it as one "among the many very useful theorems of Plato," Vitruvius mentions a method for finding a square of *twice the area* of a given square. "Suppose there is a place or a field in the form of a square and we are required to double it. . . . Nobody can find this by means of arithmetic . . . [so] let a diagonal be drawn from angle to angle of that square . . . [and then] taking this diagonal as the length, describe another square."[1]

The areas of the nested squares in Figure 4.5 are found by squaring the sides:

$$1, 2, 4, 8, 16, \ldots$$

Notice that the area of each square is double the preceding square. This is another geometric progression, this time with a common ratio of 2. Therefore, Vitruvius' two fields are incommensurable as to their sides, but they *are* commensurable as to their areas. They are sometimes said to be *commensurable in power*.

The Sacred Cut

In addition to the ad quadratum figure, the root-2 rectangle and the method of alternation, there is yet another geometric construction involving the square. It is given by Italian Renaissance architect Sebastian Serlio (Book 1, Chapter 1, Folio 11) as a method for constructing an octagon. It is sometimes called the *sacred cut*, a name coined by the Danish engineer Tons Brunés in his book *The Secrets of Ancient Geometry and Its Use*. In the book, Brunés claims that the sacred cut is found in the layout of many ancient buildings, including the Parthenon.

Figure 4.6 illustrates how the sacred cut is made.[2] Starting with a square that we call the *reference square*, swing an arc from each corner with radius equal to half the square's diagonal. From the point where each arc cuts the square, draw a line perpendicular to that side of the square. Those four lines will define a center square called the *sacred cut square*. This construction can be extended inward by repeating the construction on the sacred cut square. It can also be extended outward. Try it for yourself: Extend each of the four arcs in Figure 4.6 into a complete circle. You'll see the result in Figure 4.7. The next larger sacred cut square is circumscribed about the four circles.

Figure 4.7 shows a sacred cut square of side *s*. The distance *ab* is equal to the radius of one of the circles, and so is half the diagonal of the sacred cut square, or $s\sqrt{2}/2$. The side *S* of the (larger) reference square is then

$$S = 2ab + s = 2s\left(\frac{\sqrt{2}}{2}\right) + s$$
$$= s(\sqrt{2} + 1).$$

Conversely, if the side of the reference square is *S*, then the side *s* of the (smaller) sacred cut square is

$$s = S(\sqrt{2} - 1).$$

If you start with a square of side 41, as you did for the ad quadratum sequence, calculate in both directions, and include the sides $(S - s)/2$ of the corner squares, you get the *sacred cut sequence*:

$$\ldots, 5, 7, 12, 17, 29, 41, 70, \ldots$$

Do some of these numbers look familiar? You saw them earlier in the ad quadratum sequence and the alternation sequence, and you'll see them again in some Roman houses.

Some Designs Based on the Square

The ad quadratum figure is often used for architectural decoration, as you can see in Figures 4.8 and 4.9.

As we saw earlier with the triangle, several four-branched designs related to the square can also be found in architectural history. Figure 4.10, the Solomon's Knot, is another ancient four-armed figure often seen in pavements. The ovals have neither beginning nor end, symbolizing immortality and eternity. This knot is also called the *endless knot* or *lover's knot*. Figure 4.11 illustrates the four-lobed *quatrefoil*, similar to the three-lobed trefoil we encountered in Chapter 3. The quatrefoil is a common shape for windows in Gothic architecture, and we'll see it again, along with other foliations, in Chapter 7, "Circular Designs in Architecture."

■ EXERCISES ● THE SQUARE

1. Define or describe the following:

polygon	quadrilateral	square
rectangle	perimeter	alternation
ad quadratum	sacred cut	

2. Find the perimeter and area of a square with a side equal to
   a. 3.5 in.   b. 77.3 cm

3. Find the side and perimeter of a square with an area equal to
   a. 55.6 cm²   b. 186 ft²

4. Verify that the areas of successive squares in the root-2 sequence form a geometric progression:

   $$1, 2, 4, 8, \ldots$$

5. Construct the ad quadratum figure by hand or by a computer graphics program.

6. Find the diagonal of a square graphically by the method of alternation.

Why did Brunés call the cut *sacred*? You'll find out in Chapter 8, "Squaring the Circle."

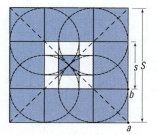
FIGURE 4.7 ■ Extending the Sacred Cut

FIGURE 4.8 ■ Pavement from Herculaneum

FIGURE 4.9 ■ Design on Pisa Duomo

FIGURE 4.10 ■ Solomon's Knot in Pompeii

FIGURE 4.11 ■ A Quatrefoil

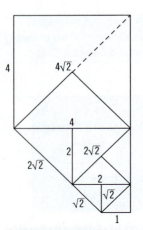

FIGURE 4.12 ■ Constructing the Root-2 Sequence

7. Use a spreadsheet to calculate the diagonal of a square by alternation. Do 20 iterations.

8. Make a geometric construction of the root-2 or ad quadratum sequence by connecting midpoints of adjacent sides of a square.

9. Make a geometric construction of the root-2 or ad quadratum sequence by drawing lines through the corners of a square parallel to the diagonals.

10. Make a geometric construction of the root-2 or ad quadratum sequence by repeatedly constructing a square on the diagonal of a square, as in Figure 4.12.

11. Make a sacred cut construction. Then extend the construction to larger and smaller squares.

THE RECTANGLE

Let's begin this section by reviewing the formulas for finding the perimeter and area of the rectangle and examining some of its unique features.

47 PERIMETER AND AREA OF RECTANGLE

a. The *perimeter* is the sum of the four sides, or twice the sum of width and length.

b. The *area* is the product of the width and the length. ●

■ **EXAMPLE:** Find the perimeter and area for a rectangle of width 5 units and length 8 units.

● **SOLUTION:**

$$\text{Perimeter} = 2(5 + 8) = 26 \text{ units}$$
$$\text{Area} = 5(8) = 40 \text{ square units}$$ ■

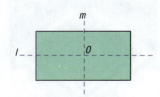

FIGURE 4.13 ■ Symmetry of the Rectangle

In Figure 4.13, we see a rectangle that has two axes of mirror symmetry, l and m. It also has twofold rotational symmetry about center O.

A *root rectangle* is a rectangle with an irrational number as the ratio of length to width:

$$\sqrt{2}, \sqrt{3}, \sqrt{5}, \ldots$$

Root rectangles are easy to construct, starting with a square and swinging arcs equal in length to the diagonals, as in Figure 4.14. (Another construction is explained in the exercises for this section.)

FIGURE 4.14 ■ Construction of the Root Rectangles

Similar Rectangles

Just as we can have similar triangles, we can have similar rectangles. Similar rectangles are two rectangles that have the same ratio of length to width. Figure 4.15 shows a simple procedure for constructing a similar rectangle within another rectangle. First, draw diagonal *BD* in rectangle *ABCD*. Construct a perpendicular to *BD* through *C*, and extend it to *AB* at *E*. Finally, construct a perpendicular to *AB* at *E*.

You can have similar figures of any shape and similar solids of any shape. You'll learn about their properties in later chapters.

We can demonstrate that the two rectangles *ABCD* and *BCFE* are similar to each other by using Equation 41, which states, "In a right triangle, the altitude drawn to the hypotenuse forms two right triangles that are similar to each other and to the original triangle." When we apply Equation 41 to Figure 4.15, we can see that altitude *GC* to right triangle *BCD* creates another right triangle *BCG* that is similar to triangle *BCD*. Therefore, angle *BCG* is equal to angle *BDC*, so triangle *BCE* is also similar to triangle *BCD*. Because corresponding sides of similar triangles are in proportion, you can say

$$EB : BC = BC : CD.$$

This means that rectangle *BCFE* is similar to the original rectangle, and it is called the *reciprocal* of the original rectangle. The remainder *AEFD* of the original rectangle is called a *gnomon*.

Repeating the construction in Figure 4.15 produces a figure called the *whirling rectangles* or *whirling gnomons* as shown in Figure 4.16. Note that all the gnomons in the figure have the same shape. No matter how many gnomons you add (at either the larger end or the smaller end), the appearance of the figure remains unchanged. You can't tell whether you're looking at an enlarged picture of gnomons the size of an atom or a reduced picture of gnomons the size of a football field. They are *self-similar under magnification*, which is also a property of fractals as you'll see in Chapter 13, "Fractals." Connecting the vertices of the gnomons with a smooth curve gives a *logarithmic spiral*. There's more about this in Chapter 9, "The Ellipse and the Spiral."

As you learned in Chapter 3, the sides of similar triangles are in proportion to each other, and the areas of similar triangles are proportional to the squares of corresponding sides. Because any rectangle can be subdivided into two triangles, these properties can be applied to rectangles as well. For instance, if the lengths of two similar rectangles are in the ratio of 4 : 1 (a scale factor of 4), then the widths are also in the ratio of 4 : 1. The areas, however, are proportional to the squares of corresponding sides, so the area of the larger rectangle is 4^2 or 16 times the area of the smaller rectangle.

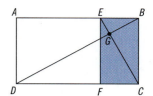

FIGURE 4.15 ■ Constructing the Reciprocal and Gnomon of a Rectangle

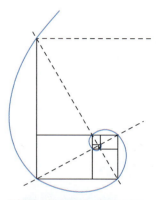

FIGURE 4.16 ■ Whirling Rectangles

The Golden Rectangle

Recall from Chapter 2, "The Golden Ratio," that dividing a line segment in the golden ratio enabled us to create a golden rectangle with length equal to Φ times its width. An interesting property of the golden rectangle is that when you construct a reciprocal, the resulting gnomon is a *square*, as shown in Figure 4.17. The reciprocal is, of course, another golden rectangle. If you begin the whirling rectangle construction with a golden rectangle, which has square gnomons, you will produce *whirling squares*.

Summary of Special Rectangles

For reference, let's summarize the various rectangles in a table and include a few rectangles with proportions that are the musical ratios and the Roman rectangle, which you'll learn about later in this chapter.

The table on the following page relates the shapes of various important rectangles and relates some to the musical ratios we found in Chapter 1, along with their Greek and Latin names. It provides a valuable catalog that we can refer back to when we later encounter such shapes in art and architecture.

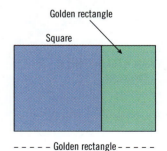

FIGURE 4.17 ■ A Golden Rectangle

Rectangle Name(s)	Ratio	Approximate Decimal Value
Square	1 : 1	1.000
Tone, Tonus, Sesquioctavus	8 : 9	1.125
Fourth, Diatesseron, Sesquitertia	3 : 4	1.333
Root-2	$1 : \sqrt{2}$	1.414
Fifth, Diapente, Sesqualtera	2 : 3	1.500
Golden	$1 : \Phi$	1.618
Root-3	$1 : \sqrt{3}$	1.732
Double-Square, Duplus, Octave	1 : 2	2.000
Root-5	$1 : \sqrt{5}$	2.236
Roman	$1 : (1 + \sqrt{2})$	2.414

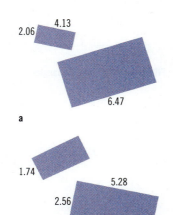

a

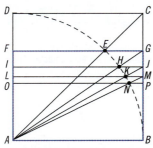

b

FIGURE 4.18 ▪ Two Pairs of Rectangles

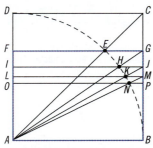

FIGURE 4.19 ▪ Constructing the Root Rectangles

▪ EXERCISES ● THE RECTANGLE

1. Define or describe the following:

   similar rectangles   reciprocal of a rectangle   gnomon of a rectangle
   golden rectangle   whirling rectangles

2. Find the perimeter and area of each rectangle:
   a. 5 × 9   b. 3.84 × 7.73   c. 293 × 837

3. How many 9-inch square tiles are needed to cover a pavement measuring 48 ft by 12 ft?

4. What is the cost of paving a rectangular pathway 312 ft long and 6.15 ft wide, at $55.75 per sq. yd?

5. Find the cost of tiling a rectangular garden pool that is 15 ft long, 12 ft wide, and 1.5 ft deep, if the tiles for the sides cost $12.50 per sq. ft and the tiles for the bottom cost $8.85 per sq. ft.

6. A solar collector array consists of six rectangular panels, each 45.3 in. by 92.5 in. Each has a blocked rectangular area (needed for connections) measuring 4.70 in. by 8.80 in. Find the total collecting area.

7. Draw several rectangles of different shapes. Construct the reciprocal for each.

8. Prove that the reciprocal to a rectangle is similar to the original rectangle.

9. Prove that the gnomon in a golden rectangle is a square.

10. Find the missing parts for the pairs of similar rectangles in Figure 4.18.

11. *Root Rectangles.* Repeat the construction for the root rectangles shown in the text.

12. *Root Rectangles in a Square.* Figure 4.19 shows another construction for the root rectangles, contained within a square. Repeat this construction and explain why it works.

    Draw square *ABCD*. With center at *A*, swing arc *BD*. Draw diagonal *AC*, cutting the arc at *E*. Draw *FEG*, parallel to *AB*. *ABGF* is now a root-2 rectangle. Draw *AG*, cutting the arc at *H*. Draw *IHJ*, parallel to *AB*. *ABJI* is a root-3 rectangle (1: $\sqrt{3}$). Continue for as many root rectangles as needed.

OTHER QUADRILATERALS

A *quadrilateral*, as mentioned earlier, is a polygon that has four sides. We've already considered the square and the rectangle; now let's examine the others. The familiar shapes shown in Figure 4.20 are all quadrilaterals. With the exception of the general quadrilateral, the equation for determining the area of each shape is also given.

A *parallelogram* is a quadrilateral in which opposite sides are parallel and equal and the diagonals bisect each other. To obtain a formula for the area of a parallelogram, imagine the shaded triangle in Figure 4.21 being cut off and attached at the opposite end of the parallelogram. This will give you a rectangle of height h and length b, with an area that is bh. These properties are summarized in Statement 48.

48 PARALLELOGRAM

a. Opposite angles of a parallelogram are equal.

b. The diagonals of a parallelogram bisect each other.

c. The area equals the product of the perpendicular distance between two opposite sides and the length of one of those sides.   ●

■ **EXAMPLE:** Find the area of a parallelogram, two sides of which are 8.44 cm apart and separated by a distance of 7.41 cm.

● **SOLUTION:** By Equation 48,

$$\text{Area} = 8.44 \times 7.41 = 62.5 \text{ cm}^2.$$ ■

A *rhombus* is a parallelogram with equal sides; therefore, the properties of the parallelogram given in Statement 48 apply to the rhombus as well. The rhombus has additional properties, which are given in Statement 49.

49 RHOMBUS

a. The diagonals of a rhombus bisect each other at *right angles*.

b. The diagonals of a rhombus bisect the interior angles of the rhombus.   ●

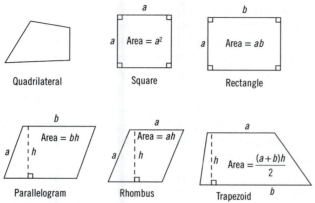

Quadrilateral

Square $\text{Area} = a^2$

Rectangle $\text{Area} = ab$

Parallelogram $\text{Area} = bh$

Rhombus $\text{Area} = ah$

Trapezoid $\text{Area} = \dfrac{(a+b)h}{2}$

FIGURE 4.20 ■ Quadrilaterals

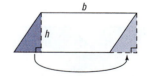

FIGURE 4.21 ■ Area of a Parallelogram

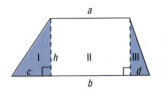

FIGURE 4.22 ■ A Trapezoid

A *trapezoid* has two parallel sides, called *bases*, and its *altitude* is the distance between the bases. To determine the area of a trapezoid, add the areas of the regions labeled I, II, and III in Figure 4.22:

$$\text{Area} = \frac{ch}{2} + ah + \frac{dh}{2}$$

$$= h\left(\frac{c + 2a + d}{2}\right)$$

$$= h\left[\frac{a + (a + c + d)}{2}\right]$$

$$= h\left(\frac{a + b}{2}\right)$$

because $a + c + d = b$.

50 **TRAPEZOID**

The area of a trapezoid equals the altitude times the average of the bases.   ●

■ **EXAMPLE:** What is the area of a trapezoid with bases of length 3.48 cm and 5.53 cm and an altitude of 2.89 cm?

● **SOLUTION:**

$$A = 2.89\frac{(3.48 + 5.53)}{2} = 13.0 \text{ cm}^2$$

■ **EXERCISES ● OTHER QUADRILATERALS**

1. Define or describe the following:

   quadrilateral   parallelogram
   rhombus   trapezoid

2. Find the perimeter and area of each quadrilateral in Figure 4.23.

3. Make a paper cut-out of a parallelogram. Then fold along the diagonals. How are the diagonals related? Can you prove this?

4. Make a paper cut-out of a parallelogram. Then, by cutting off a piece of the parallelogram and pasting that piece in a new location, arrive at a formula for its area.

5. *Euclid Book I, Proposition 85.* Euclid wrote that "two parallelograms that have the same base and that lie between the same two parallel lines have the same area." Demonstrate this theorem by making a CAD drawing, such as the one in Figure 4.24, in which you drag the parallelograms between parallels to show that the areas remain the same.

6. Make a paper cut-out of a rhombus. Then fold along the diagonals. How are the diagonals related? Can you prove this?

7. Make a paper cut-out of a trapezoid. Then fold it to make a crease that is parallel to and midway between the two parallel sides. With further cutting and pasting, can you arrive at a formula for the area of a trapezoid? Can you prove your formula?

8. *Golden Rhombus.* A *golden rhombus* is a rhombus in which the diagonals are in the golden ratio. Construct such a rhombus and add it to your growing collection of "golden" figures.

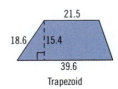

1.85
1.44
1.85

Rhombus

a

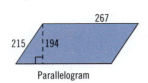

21.5
18.6 15.4
39.6

Trapezoid

b

267
215 194

Parallelogram

c

FIGURE 4.23 ■ Some Quadrilaterals

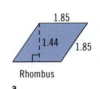

FIGURE 4.24 ■ Areas of Parallelograms

THE RECTANGULAR FORMAT IN PAINTING

*"Composition reveals itself when, as we inevitably do, we see a painting
or sculpture or building as an arrangement of definable shapes organized
in a comprehensive structure."*[3]

RUDOLPH ARNHEIM

In this section, we'll explore square- and rectangular-shaped paintings. When we use the word *frame* in this section, we are not referring to the wooden or metal construction around a painting; we are talking about the outer shape of the painting: square, round, rectangular, and so forth. The shape of a frame is often called the *format*. To gain a better understanding of the use of frames, let's begin by investigating three early examples.

Types of Frames

Horizontal bands of decoration, such as the *frieze* shown in Figure 4.25, are examples of *partial frames*, which are closed at the top and bottom and open at the ends. In the figure, the architrave is the horizontal beam that spans from column to column, and the frieze is the horizontal member directly above it. The frieze is often found on temples in classical architecture and is the ideal format for depicting marches and processions. Other partial frames—again, closed at the top and bottom and open at the ends—include scrolls and tapestries such as the famous *Bayeux Tapestry*. Depicting the Battle of Hastings and William the Conqueror's invasion of England, this work is open at the ends to suggest a progression of events.

A *mural* is a painting on a wall or ceiling, and a *fresco* is a mural painted on fresh, moist plaster. Murals and frescoes might be open, or they might have clearly defined outlines, such as those shown in the image of the Scrovegni Chapel in Figure 4.26. Sometimes they include a time progression as well as a spatial one

FIGURE 4.25 ■ Doric Entablature

©Scala/Art Resource, NY

FIGURE 4.26 ■ Scrovegni Chapel Frescoes, Giotto, c. 1305

FIGURE 4.27 ■ Baptistery North Doors, Ghiberti, 1403–1424

FIGURE 4.28 ■ *Deposition,* Fra Angelico, c. 1440

in which the individual pictures are similar to the frames in a movie film or the pages in a biography. The frescoes in the Scrovegni Chapel depict a series of religious events in the life of the Virgin Mary. Similarly, the Baptistery North Doors, shown in Figure 4.27, depict scenes from the New Testament.

Frames have taken all sorts of elaborate shapes, including lancets, pointed arches, and quatrefoils. Perhaps the most elaborate frames of all were the medieval and Renaissance altarpieces, with their ranges of pointed peaks. Most of these works were painted on wood panels, the standard for altarpieces until at least the 1520s.

Altarpieces, such as the one shown in Figure 4.28, may look old-fashioned to us now, but according to Jacob Burckhardt, a Swiss historian of art and culture writing in the 1800s, the great tradition of modern European easel painting started with the Italian Renaissance altarpiece.[4] Altarpieces were on the cutting edge of Italian painting, and the most advanced methods and materials were used to produce them. Burckhardt writes, "There were no rules for the design of an altarpiece. They were probably not even specified by clergy or given a clear function by the Church, so it was a place where artists could experiment with new methods and materials. Altarpieces show continual change, both in shape and subject matter, and were regularly replaced by the latest model, the old altarpiece often cut up and pieces scattered in various galleries."

Why Frame a Picture?

Of course, not all pictures are framed or partially framed. Cave paintings are examples of unframed pictures. Why frame a picture? Consider these reasons:

- *To separate inside from outside.* A frame asserts that the painting is a world of its own and not part of the surrounding world.

- *For visual control.* Separating inside from outside lets the painter control the composition. The meanings of the objects in a painting depend partly on their surroundings. Therefore, if the surroundings are controlled, the meanings can be controlled too.

- *To provide a window.* When you study *perspective,* you'll see that a frame can function *as a window* through which you can view the world.

■ *For portability.* This is arguably the strongest reason. As we've just seen, many early pictures were fixed in place—painted on a wall or an altarpiece, or carved on a building or temple. During the Renaissance, however, people wanted to buy and own paintings and other art objects. Paintings and small sculptures could be made in a studio with no particular patron or client in mind. Portable art could be sold to whomever desired it, and it could eventually be resold and become part of the *art market.*

Format

You've seen what frames can do, but how can the shape of a frame, or the format, enhance a work of art? Let's begin by examining the rectangular format, or *quadro.* When the need for portability led artists to prefer using canvas over wood panels, those artists also needed easily constructed frames. As a result, the use of the rectangular format increased.

When an artist decides to use a rectangular frame, he must decide on the *orientation* and the *proportions* of that rectangle. The terms *portrait* and *landscape* are familiar to us today because computers and printers offer those orientations to distinguish between vertical and horizontal formats. Deciding which format to use for a given picture is usually easy. A landscape or seascape, for example, normally calls for a horizontal format; a standing figure or waterfall calls for a vertical format. Perceptual psychologist Rudolf Arnheim points out that orientation can affect how the contents of a painting are viewed. A standing figure painted using a portrait format appears taller and thinner. We've intentionally cropped the *Birth of Venus* in Figure 4.29 to demonstrate this principle. However, in the original landscape *Birth of Venus,* which is shown in Figure 4.30, the height of a standing figure is accentuated. The contrast with the horizontal frame makes it stand out even more.

MATERIALS

Canvas was widely used in Venice for large-scale painting during the late 1400s. It was cheaper than wood and more portable, especially for large works. This made canvas more suitable for the growing art market of the 1500s.

At about the same time, there was a shift from *egg tempera* paint to oil paint. To make paint, an artist had to grind pigment from minerals or plants to a powder and add a liquid to make a paste. Raw egg, which had been used in Egypt since at least 2000 B.C., was the liquid used to make egg tempera. Artists continued to use it throughout the Renaissance.

Egg tempera worked well enough, but it dried very quickly. The Flemish painter Jan Van Eyck wanted to produce smoother color transitions in his work, so he substituted oil for egg and created a slower drying mixture. As such, Van Eyck is sometimes called *the discoverer of oil painting.* The switch to oils and to large-scale works widened the use of canvas, which came into common use during the 1500s.

FIGURE 4.29 ■ *Birth of Venus,* Botticelli, c. 1482–1485 (cropped into a portrait format)

FIGURE 4.30 ■ *Birth of Venus,* Botticelli, c. 1482–1485

Once an artist determines whether to use a portrait or a landscape orientation, she needs to decide what proportions to use. Rectangular formats include the ones used for movie films, wide-screen movies, and video. Photographic film, writing paper, and computer monitors also use rectangular formats.

Let's consider the proportions of some of our special rectangles—including the golden rectangle and the root-2 rectangle—as formats. We'll also look at the double-square rectangle and rectangles formed from overlapping squares.

Some Rectangular Formats

The golden rectangle is supposed to be the most aesthetically pleasing rectangle, and its proportions have occasionally been used for frames. Remember that a golden rectangle is one in which the ratio of the long dimension to the short dimension is Φ, or the golden ratio. The *Birth of Venus* approximately uses the golden ratio in its frame and even in the proportions of Venus herself. Salvador Dali's *Last Supper* also uses a ratio of about 1.6 to 1. Georges Seurat's *A Sunday on La Grande Jatte* is said to have the shape of a golden rectangle; actually, its ratio is closer to 1.5 to 1 based on dimensions from the Art Institute of Chicago, which owns the painting. Claims about the beauty of the golden rectangle are based mostly on the work of Gustav Fechner. When Fechner presented people with rectangles of white cardboard on a dark table, they tended to prefer proportions approaching the golden ratio (1.62 to 1).[5] About 76% of the respondents picked rectangles with ratios from 0.57 to 0.67 (where 0.62 is the golden ratio).[6] However, when he measured frames in a museum, Fechner found that the favorite ratios were approximately 5 : 4 (or 1.25 to 1) for vertical pictures and 4 : 3 (or 1.33 to 1) for horizontal. In other words, artists used frames that were more compact than the golden rectangle, with the vertical more so than the horizontal.

Why did artists prefer a shape that was squarer than the golden rectangle? Arnheim says that these preferences may reflect the requirements of the composition. Another possibility is reflected in the title of Arnheim's book *The Power of the Center*. As the eye moves away from the center of a picture, the objects that are encountered are perceived as being less important. Also, a squarer format is consistent with rest, repose, dignity, and timelessness—concepts that artists often want their paintings to convey. A square format conveys stillness and serenity, a calmness and dignity that you'll see again in the round format. This property makes it ideal for subjects such as the Madonna, as in Donatello's *Madonna of the Clouds* shown in Figure 4.31.

It appears then that most rectangular pictures were made with no particular proportions in mind. If you look long enough, pictures of any desired shape can be found among the millions of pictures in existence. However, a few shapes appear often enough to deserve mention.

We occasionally see a painting with a *double-square* format, that is, a rectangle with a width that is twice the height. An example of the double-square is *The Family of Darius before Alexander*, by Veronese, which was painted in 1565–1570. A more common rectangle is formed by *overlapping squares*, which we see in Giotto's *Saint Francis Undergoing the Test by Fire Before the Sultan*, shown in Figure 4.32. In this picture, the sides (of the square) are equal to the short side of the outer rectangle; also, the crossings of the diagonals fix the base of the throne and the height of the partition in the background.[7] The outline is nearly a golden rectangle, so removing squares would create an approximate golden rectangle on each side of the throne.

Finally, in Figure 4.33 we have a representation of the *root-2 rectangle*. You were introduced to root rectangles earlier and were shown how to construct

FIGURE 4.31 ■ *Madonna of the Clouds,* Donatello, c. 1427

GUSTAV THEODOR FECHNER
(1801–1887)

Gustav Fechner was a German physicist, philosopher, and experimental psychologist who did important work studying sensations and the stimuli that produce them. He devised experiments for measuring sensations in relation to the magnitude of stimuli. He deduced what is now known as *Fechner's Law*, which says that the intensity of a sensation increases as the *logarithm* of the stimulus. He also sought to determine, by actual measurement, which shapes and dimensions are most pleasing.

FIGURE 4.32 ■ *Saint Francis Undergoing the Test by Fire Before the Sultan*, Giotto, c. 1320s

FIGURE 4.33 ■ *Flagellation of Christ*, Piero della Francesca, 1455

them. We'll discuss them again when we discuss architectural proportions later in this chapter. For now, consider Piero's *Flagellation of Christ*, which is a famous example of a root-2 rectangle in painting. The painting's width is approximately $\sqrt{2}$ times its height.[8]

Compositional Devices

Sometimes a painting is subdivided into distinct regions, or *zones*, each with its own characteristics. Identifying these zones can enhance your appreciation of a painting. In Fra Angelico's *The Annunciation* (Figure 4.34), the vertical columns of the loggia divide the painting into vertical panels. The loggia itself is subdivided vertically into two zones: the celestial, represented by the archangel Gabriel, and the earthly, represented by Mary. The work also contains an interesting *time* division. Gabriel and Mary represent the present, and the portrait of Christ to come, in the round carving above the middle column, represents the future. A very similar painting by Fra Angelico shows Adam and Eve in a left-most panel, representing the past.

FIGURE 4.34 ■ *The Annunciation*, Fra Angelico, 1435–1445

FIGURE 4.35 ■ Line Drawing after *Apollo Pursuing Daphne*, Tiepolo, 1755–1760

The *main diagonal* is another compositional device used to make a picture more dynamic. Sometimes such a diagonal is used to split the picture into zones, as in Figure 4.35. Here the diagonal shows the god Apollo on one side and the mortals on the other. As in *The Annunciation*, the scene reflects the celestial versus the earthly.

Unlike the rectangle, the diagonal of a square is also an *axis of symmetry*. In *Stigmatization of St. Francis*, as shown in Figure 4.36, Giotto used the diagonal of the cliffs to separate the terrestrial from the celestial. The earthly region is below the diagonal and the heavenly region is above it.

In Chapter 1, you were introduced to Raphael's *School of Athens*. Here, we'll consider one of his earlier works. *The Knight's Dream*, shown in Figure 4.37, is notable for how it depicts zones and what Arnheim calls a *microtheme*. A microtheme is a miniature version of the main theme, acted out near the center of the picture, often using the hands. In *The Knight's Dream*, the tree acts as a vertical axis,

FIGURE 4.36 ■ *Stigmatization of St. Francis*, Giotto, c. 1320

FIGURE 4.37 ■ *The Knight's Dream*, Raphael, 1504–1505

separating two women, who represent virtue on the left and beauty on the right. The microtheme is the choice between the book (presumably a Bible) held by the modest maiden on the left and the flowers held by the other, more glamorous maiden on the right. This choice echoes the larger choice between virtue and beauty, a popular theme in art.

Edvard Munch's *The Sick Girl* (Figure 4.38 is a representational sketch) gives us a powerful depiction of two zones. A strong diagonal separates the realm of the sick daughter from that of the grieving mother. The mother's head lies near the dividing line between these zones, in what Arnheim calls a *compositional latch*. "She is trying to breach the barrier. But closeness does not foster communication; her bent head hides her face."[9] The microtheme is at the very center, where the mother reaches for the daughter's hand but can no longer touch it.

FIGURE 4.38 ■ Line Drawing after *The Sick Girl* by Edvard Munch, Original, 1896

Jay Hambidge and the Root Rectangles

The paintings we've been discussing come from the early history of art. However, the geometrical principles used in them have carried forward to today, as we'll see in the coming sections. The start of the twentieth century saw great changes in art with Cubism, Futurism, Constructivism, and other movements. To many people art, like science, suddenly became incomprehensible, even alien. A reaction to this new, "modern" art wasn't surprising, and an artistic movement surrounding the concept of *dynamic symmetry* emerged as part of that reaction. The most important figure in this movement was Jay Hambidge.

According to Hambidge, *dynamic symmetry* means "commensurable in power." He used the word *symmetry* differently than we have used it elsewhere in this book; he used it to mean *proportion*. In Hambidge's view, dynamic symmetry is a system of proportions based on the root rectangles, with sides that are not commensurable but with areas that are. The construction of the root rectangles (Figure 4.14) is the very first figure in Hambidge's book *The Elements of Dynamic Symmetry*.

Hambidge's writings do not state very clearly what he meant by *static symmetry*, which he also called *passive symmetry*, although presumably it is the opposite of dynamic symmetry. "Static symmetry, as the name implies, is a symmetry which has a sort of fixed entity or state."[10] It is based on the regular polygons: the equilateral triangle, square, pentagon, hexagon, and octagon. However, he wrote, "At any rate there is no question of the superiority of the dynamic over the static."[11] He claimed this superiority based on his findings of the dynamic in plants and the human figure, and in Egyptian and Greek art. "The basic principles underlying the greatest art so far produced in the world may be found in the proportions of the human figure and in the growing plant."[12] The idea was that when an artist or architect designed something—a painting, a sculpture, a building, and so on—that design would be more pleasing if the artist used ratios that involved irrational numbers rather than rational numbers.

Dynamic symmetry caught the attention of many artists, and it was seen as a rational way to return to sanity after what many thought were the excesses of the new art. Those attracted to dynamic symmetry were of a classical turn of mind, leaning toward classical concepts of order, harmony, beauty, and truth. It was popular in art design and education from 1920 to 1940 and was used for industrial design in the 1920s and 1930s. Some of the twentieth-century artists who were influenced by dynamic symmetry were Maxwell Armfield, Robert Henri, and George Bellows. Others who claimed to have used dynamic symmetry include Barnett Newman, Mark Rothko, Max Weber, Dorothea Rockburne, Kenneth Noland, Maxfield Parish, and Rutherford Boyd.

JAY HAMBIDGE (1867–1924)
Jay Hambidge was born in Ontario, Canada, and was, at various times, a reporter, an illustrator, a lecturer, and a teacher. He studied at the Art Student's League in New York. A turning point in his life came when he was sent to Agrigento in Sicily to draw the temples. There he felt he had rediscovered a system of proportions based on the golden ratio. Convinced that he had found the basic principles underlying both art and nature, Hambidge invented a theory of art and design that he called *dynamic symmetry*.

Recall from earlier in this chapter Vitruvius' layout of fields, which were not commensurable as to their sides but were commensurable as to their areas—that is, they were *commensurable in power*.

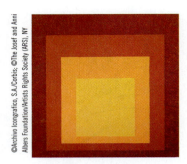

FIGURE 4.39 ■ *Homage to the Square, Attuned*, Josef Albers, 1958

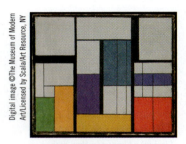

FIGURE 4.40 ■ *Composition II, Indigo Violet Derived from Equilateral Triangle*, Georges Vantongerloo, 1921

FIGURE 4.41 ■ *Tableau 3 with Orange, Red, Yellow, Black, Blue and Grey*, Piet Mondrian, 1921

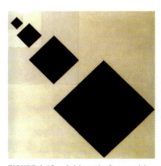

FIGURE 4.42 ■ *Arithmetic Composition 1*, van Doesburg[15]

However, some artists and critics saw dynamic symmetry as a hindrance to emotion and imagination. Applied zealously, they said, it could lead to a loss of spontaneity and could cause rigidity and brittleness. For the mathematically minded, however, the early twentieth century was an interesting chapter in American art.

The Rectangle in Mid-Twentieth-Century Abstract Geometric Art

Not surprisingly, geometric art frequently uses the rectangle, not just as a frame, but as a compositional element. Starting at about 1910 and continuing into the 1980s, such painting involved many artists, including van Doesburg, Mondrian, Vantongerloo, Klee, Diller, MacLaughlin, Nicholson, von Weigand, Diebenkorn, Novros, Sander, Schoonhaven, Aneuskiewics, Andre, Jenson, Kelly, and Malevich. Josef Albers did an entire series called *Homage to the Square*, one of which is shown in Figure 4.39. Also shown are paintings by Vantongerloo (Figure 4.40), Mondrian (Figure 4.41), and van Doesburg (Figure 4.42).

Perhaps the best-known compositions based on squares and rectangles are by Piet Mondrian. According to Arnheim, Mondrian wanted to eliminate any reference to the pull of gravity, at least in the painting shown in Figure 4.41.[13] His painting shows "a perfect balance between horizontal and vertical, a weightless universe, homogeneous and endless, beholden to no one and to nothing."[14] Also, there is no explicit reference to the center. Each shape has its own center, but no shape is strong enough to act as a center for other shapes to organize around. Mondrian's preference for an even distribution of weights probably made him prefer to paint on the horizontal surface of a table rather than on an easel.

ARCHITECTURAL PROPORTIONS BASED ON THE SQUARE AND THE RECTANGLE

We've already discussed three of the main systems of proportion used in architecture. Chapter 1 presented a system based on the Pythagorean musical ratios. Chapter 2 discussed a system based on the golden ratio. Earlier in this chapter, we introduced a system of proportions based on the square, which apparently was used extensively by the Romans.

Now we'll examine several ways to derive a system of proportions from the square, including

- The *root-2* rectangle
- The *ad quadratum* construction
- The Pythagorean *method of alternation* for finding the diagonal of a square
- The principle of *gnomonic growth*
- The *sacred cut*

How the Sacred Cut and Root-2 Rectangle Are Related

Mathematics professor Jay Kappraff has pointed out that four sacred cuts subdivide a square into nine units, as shown in Figure 4.43. There are five squares, one in the center and the four corners, and four root-2 rectangles. The rectangle that is formed when we combine a corner square with the adjacent root-2 rectangle is what Kappraff calls a *Roman rectangle*.[16]

Kappraff has also pointed out some interesting additive properties for the square, the root-2 rectangle, and the Roman rectangle:

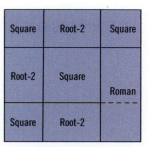

FIGURE 4.43 ■ Kappraff's Subdivided Square

- By definition, adding a root-2 rectangle to a square gives a Roman rectangle. Conversely, subtracting a square from a root-2 rectangle gives a Roman rectangle.

- Adding two squares to a Roman rectangle gives a larger Roman rectangle.

- Adding two root-2 rectangles gives a larger root-2 rectangle. Conversely, halving a root-2 rectangle gives two smaller root-2 rectangles.

These properties enable the figures to be combined in a variety of ways, giving the flexibility needed for an architectural layout. The additive properties shown in Figure 4.44 are the very additive properties we were seeking in a system of architectural proportions (recall the section titled "Systems of Architectural Proportions" in Chapter 1).

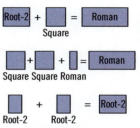

FIGURE 4.44 ■ Additive Properties of the Squares and Rectangles

The Garden Houses of Ostia[17]

We've created a lot of interesting constructions so far, but what do they have to do with actual buildings? Let's examine two ways in which these constructions were used: first, to supply the actual layout for the ground plan, and second, as a source for a sequence of numbers that were used in building design. The first example will show how the sacred cut was used primarily as a ground plan but also was used as a source for a number sequence. We'll limit our examples to *domestic* Roman architecture, or the design of dwellings, although it is claimed the use of geometry can be found in Roman city planning and monumental architecture as well.

The remains of the Garden Houses are located west of Rome, at the mouth of the Tiber River. This is the location of the ancient Roman port of Ostia. The Garden Houses were a planned apartment complex, complete with shops and gardens. Donald and Carol Watts wrote, "In the second-century Garden Houses of Ostia a rigorous geometry prevailed. From the overall plan to individual mosaics, a particular pattern, rich in philosophical suggestiveness, underlay the design. . . . " See Figure 4.45. The Watts analyzed the Garden Houses and found a single controlling idea: the sacred cut.

The geometry of the Garden Houses is based on three successive sacred cuts, as can be seen in Figure 4.46. Donald and Carol Watts describe the geometry as follows: "(a) A square (the outermost square) roughly congruent with the perimeter of the complex encloses a circle that touches the corners of the courtyard. (b) Sacred cuts of the east and west sides of this reference square determine the position of the outer walls of the courtyard buildings. (c) The second reference square, concentric with the first, is defined by the width of the courtyard and the positions of the fountains; the sacred cuts of its east and west sides guide the placement of the shared walls along the spines of the courtyard buildings. (d) The third reference square is the sacred cut square of the second, and its cuts define the innermost walls of the courtyard buildings. (e) The buildings are precisely five times as long as the final sacred cut square, and their width is equal to its diagonal. (f) A superposition of all the sacred cuts shows how they unfold from a common center, thereby emphasizing the major east-west axis of the complex. When the three reference squares are superimposed, they nest

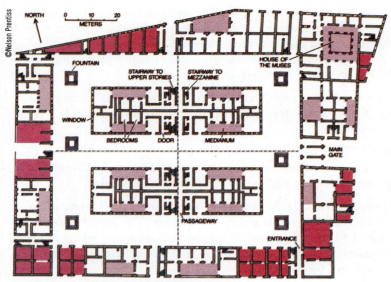

FIGURE 4.45 ■ Ground-Floor Plan, Garden Houses of Ostia

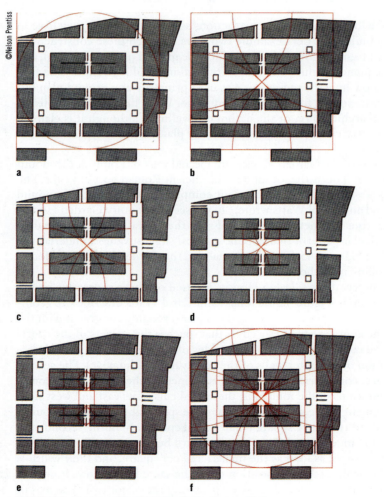

FIGURE 4.46 ■ Geometry of the Floor Plan, Garden Houses of Ostia

concentrically inside one another, and their sacred cuts appear to unfold like the layers of an onion."

As shown in Figure 4.47, "Apartment plans in the courtyard buildings are regulated by the sacred cuts of a square whose sides are 41 Roman feet long (a Roman foot is 0.297 meters, or roughly $11\frac{1}{2}$ inches) and whose diagonal is equal to the interior width of the buildings (58 feet). Two back-to-back apartments, separated by a 2-foot shared wall, are shown. The dimensions established by successive sacred cuts of the square appear throughout the apartments; the number 7 and its multiples are particularly frequent."

The sacred cut was apparently used to supply the ground plan. However, it appears that $\sqrt{2}$ was also used as the source for a sequence of proportional, whole-number dimensions that regulate the plan of the apartments. Recall the sacred cut sequence from earlier in this chapter, where each larger square was $(\sqrt{2} + 1)$ times the side of the preceding square, and each smaller square was the side of the preceding square divided by $(\sqrt{2} + 1)$. If you generate a sacred cut sequence starting with a square that is 41 Roman feet on a side and you round to the nearest whole number, you'll get (working both forward and backward and including the corner squares, as we did on p. 97) the following sequence:

$$\ldots, 5, 7, 12, 17, 29, 41, 70, \ldots$$

Rather than being generated arithmetically, this sequence could be obtained geometrically by a sacred cut construction using stretched cords on the ground.

Now recall the method of *alternation* given earlier in this chapter. We used it to approximate the diagonal of a square. The sides and diagonals gave us the following sequence of numbers:

$$\ldots, 5, 7, 12, 17, 29, 41, 70, \ldots$$

These are the same numbers as in the preceding sacred cut sequence. These are also the numbers used in the Garden Houses of Ostia. The Watts say, "By basing the plan on a sequence of numbers that approximate the irrational square root of 2, the architect of the Garden Houses was making a philosophical statement akin to that of squaring the circle (which is equivalent to approximating the irrational number π). In both instances the sacred cut is the means of expressing the irrational and undefinable by the rational and definable."

Pompeii and Herculaneum[*]

For our second example, let's go to southern Italy in the first century A.D., to the coast south of Naples. We'll examine the claim that the sacred cut sequence and the ad quadratum sequence provided dimensions for domestic structures.

Here Mount Vesuvius (Italian *Vesuvio*, from the Oscan word *fesf*, "smoke") rises to a height of 4190 ft; it is the only active volcano on the European mainland. See Figure 4.48. Vesuvius erupted on August 24 in A.D. 79. The top of the mountain blew off, and the cities of Pompeii, Herculaneum, and Stabiae were buried—not by lava, but by ashes and mud. Most of the inhabitants survived, but about 2000 people were killed. The mud went westward toward Herculaneum, while the ash rained southeast toward Pompeii.

Pompeii was founded about 600 B.C. by the Oscans, and it became a Roman colony in 80 B.C. Pompeii was a favorite resort for wealthy Romans and a center

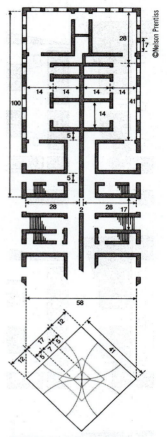

FIGURE 4.47 ■ **Analysis of the Floor Plan, Garden Houses of Ostia**

[*]All the information, quotations, and figures in this section are from the paper by Carol Martin Watts in *Nexus: Architecture and Mathematics*.

FIGURE 4.48 ■ Pompeii, with Vesuvius in the Background

for trade. Its population was about 20,000 at the time of the eruption. Pompeii lay undisturbed beneath the ashes until 1748, when excavations began. The wet ashes had formed a seal around Pompeii, which preserved many buildings. The victims were encased in shells formed from the wet ashes, which remained after the bodies themselves had decomposed.

Herculaneum is named for the mythical Greek hero Hercules. According to mythology, Hercules founded the city. Like Pompeii, Herculaneum was a popular resort area for wealthy Romans. When Mount Vesuvius erupted, Herculaneum was buried in mud about 65 ft deep. Figure 4.49 gives a sense of this. Systematic excavations of the ruins were begun at about the same time as those in Pompeii, and they have continued ever since. The excavators found elaborate villas, a theater, marble and bronze sculptures, paintings, and an extensive library of papyrus rolls in the Villa of the Papyri. The design of the old Getty Museum in Malibu, California, is based on the Villa of the Papyri.

FIGURE 4.49 ■ Herculaneum, Showing the Depth of the Mud

Let's look at a single house in Herculaneum, the House of the Tuscan Colonnade, which was a typical house or *domus* for the average patrician. According to Carol Martin Watts, "The domus was one of [the Romans'] earliest building types and remained the archetype for other buildings. As such, it embodies the basic characteristics of Roman design. . . ." In her study of the *domus*, she discovered an underlying geometry. "Analysis of houses at Pompeii and Herculaneum suggests that two simple geometric systems, both based on the square, underlie the design of the Roman house at all scales. [See Figure 4.50.] The placement of major elements of the house, such as the atrium, peristyle, tablinum, and other major spaces appear to be related to a regulating square and its sacred cut and/or ad quadratum."

According to Watts, "At Pompeii and Herculaneum the houses use a limited number of proportions and of regulating square sizes. The most common

House	Diagram	Number of Squares	Size of Regulating Square
Labyrinth, Pompeii	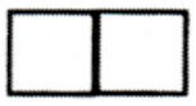	2	116 Oscan ft.
L. Ceius Secundus, Pompeii	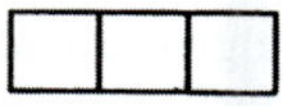	3	34 Oscan ft.
Tragic Poet, Pompeii	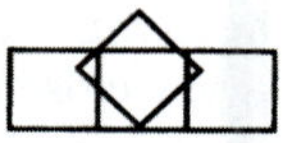	3 width=diag.	34 Roman ft.
M. Lucretius Fronto, Pompeii	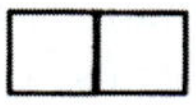 or 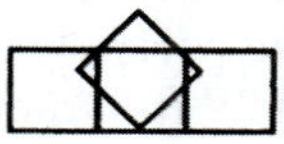	2 or 3 with width=diag.	58 Oscan ft. or 41 Oscan ft.
Vettii, Pompeii		1	116 Oscan ft.
Faun, Pompeii	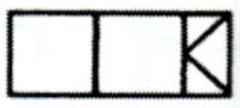	2 1/2	116 Oscan ft.
Sallustius, Pompeii		1	116 Oscan ft.
Carbonized Furniture, Herculaneum	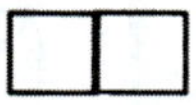 or 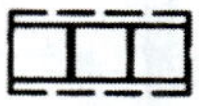	2 or 3	41 or 24 Roman ft.
Samnite, Herculaneum	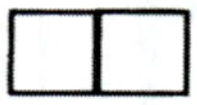	2 (to original lot line)	41 Oscan ft.
Tuscan Colonnade, Herculaneum	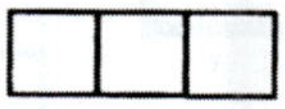	3	41 Oscan ft.
Wooden Partition, Herculaneum	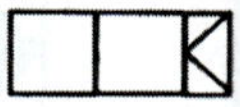	2 1/2	58 Oscan ft.
Bicentenary, Herculaneum	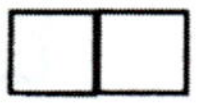	2	70 Oscan ft.

FIGURE 4.50 ■ Table of Proportions and Regulating Square Size for Case Study of Houses at Pompeii and Herculaneum

proportion of width to depth was 1 : 3, followed by 1 : 2. There were a few occurrences of 1 : $2\frac{1}{2}$ and 1 : 1. These can be expressed as squares as shown here. In some cases the entire depth of the house could be divided into an exact number of squares, and the diagonal of these squares was the width of the house."

Watts speculates that, after a simple preliminary drawing was made, the regulating square would be laid out on the site using ropes and stakes. "Such a method would allow the plan to be laid out without the use of dimensions, although dimensions in whole numbers of feet frequently occur in house plans . . . An important result of such a design method is to insure commensurable proportions throughout the house."

In the House of the Tuscan Collonade shown in Figure 4.51, we find the same sequence of numbers that we found when we wrote the sacred cut sequence for the Garden Houses:

$$\ldots, 5, 7, 12, 17, 29, 41, 70, \ldots$$

We also find numbers from the ad quadratum sequence. If we use the number 41 as the side of our initial square, and we find each successive larger square by multiplying by $\sqrt{2}$ and each successive smaller square by dividing by $\sqrt{2}$, we get the following numbers, rounded to the nearest whole number:

$$\ldots, 29, 41, 58, 82, \ldots$$

Watts writes, "The site is three squares deep, with each square 41 Oscan ft. The sacred cut of this 41 ft square gives the dimensions 17 and 12. Seventeen plus 12 equals 29, part of the ad quadratum sequence from 41. The diagonal of 41 is 58. The plan indicates where these dimensions are found within the house. The width of the atrium is very close to 29 Oscan feet. The width of the tablinum is 17 Oscan ft. The peristyle with its porticoes is 41 ft. square, and the diagonal of the peristyle colonnade is also 41. The peristyle width within the colonnades is 29 ft. Other spaces use dimensions of 17, 12, 7

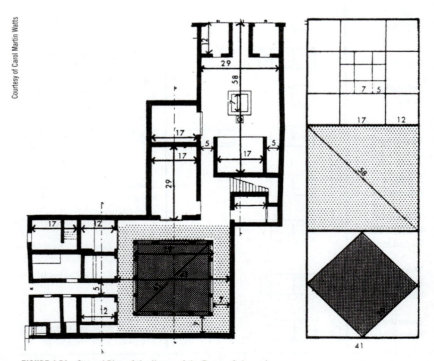

Courtesy of Carol Martin Watts

FIGURE 4.51 ■ Ground Plan of the House of the Tuscan Colonnade

and 5 Oscan ft." In addition to the dimensions of the ground plan, Watts found similar geometrical relationships in the heights of walls and in the smaller-scale details of the rooms.

The Root-2 Rectangle in Architecture

Vitruvius mentions using the diagonal of a square as one way to find the dimensions of the atrium of a Roman house, "by using the width [of the atrium] to describe a square . . . [and] drawing a diagonal in this square, and giving the atrium the length of this diagonal line."[18] We used this method earlier when we constructed the root rectangles. We got a rectangle with its long side equal to the diagonal of the square and with its width equal to a side of the original square. Here we'll review a few examples of the root-2 rectangle in architecture.

   According to architect Kim Williams, the ground plan of the Pantheon in Rome is circumscribed by a root-2 rectangle, as we can see in Figure 4.52. If a circle is drawn extending from face to face of the niches on the north-south axis of the Pantheon, and a square is circumscribed about that circle, the diagonal of that square gives the length of the enclosing rectangle.[19] Rudolph Wittkower mentions the use of the root-2 rectangle by Renaissance architects Francesco di Giorgio Martini and Sebastian Serlio, and he says, "As far as we can see this is the only irrational number widely propagated in Renaissance theory of architectural proportion."[20] Recall from Chapter 1 that Renaissance architects used *integer* ratios, especially the musical ratios.

   In his *Four Books of Architecture*, Palladio suggested the shapes that should be used for rooms. Most of the proportions were musical ratios, as we saw in Chapter 1, but he included one in which the "length will be the diagonal line of the square."[21] This, of course, is the root-2 rectangle. Later, he described his *Corinthian Atrio*, "whose length is the diagonal line of the square of its breadth."[22]

   Finally, Kim Williams extensively measured Michelangelo's *Medici Chapel* (the New Sacristy) in San Lorenzo, Florence. She concluded that Michelangelo used the root-2 rectangle at least twice in the design of the floor plan. Note the diagonals in Figure 4.53.[23]

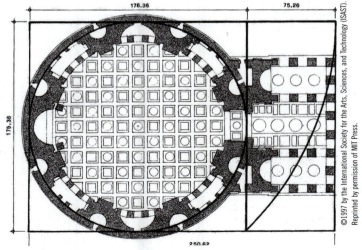

FIGURE 4.52 ■ The Pantheon Ground Plan in a Root-2 Rectangle

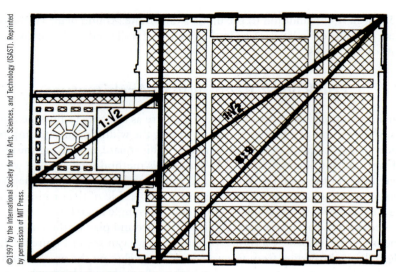

FIGURE 4.53 ■ Floor Plan of the Medici Chapel

The Cruciform Church

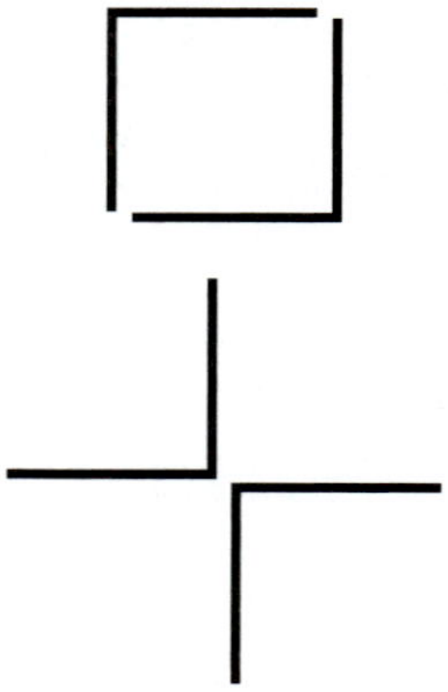

The square and the rectangle are, of course, obvious choices for the ground plan of a building, and their use may also have provided systems of proportions. However, the *cross*, a geometric figure related to the square, has also been used to design ground plans. These fundamental shapes are presented in Figure 4.54. Of all the various kinds of crosses, we'll look at just two: the Latin cross, with its stake and crossbeam, and the equilateral Greek cross. Both were used as ground plans for churches.

The variety of crosses is bewildering, with numerous decorative variations. Figure 4.55 depicts just a few. Here are some details:

a. *Latin or Roman Cross.* These are the most familiar crosses, signifying crucifixion.

b. *Tau Cross.* This cross is named for its shape, the Greek letter *tau*. St. Francis used this as his signature. It's also called St. Anthony's cross, the Egyptian cross, *Crux commissa*, and the Old Testament cross.

c. *Greek Cross.* For the ancients, this equal-armed cross symbolized the four indestructible elements and, therefore, permanence.

FIGURE 4.54 ■ Square and Cross
A square can be transformed into a cross by a simple translation of two adjacent sides.

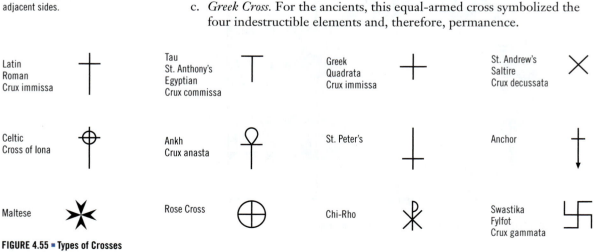

FIGURE 4.55 ■ Types of Crosses

d. *Saint Andrew's Cross.* This is like the Greek cross, turned a quarter-revolution. St. Andrew is said to have been martyred on one of these.

e. *Celtic Cross.* The Celtic cross includes a circular element.

f. *Ankh or Key of the Nile.* This was an Egyptian symbol of life and also a symbol of the Coptic Christians in Egypt.

g. *Saint Peter's Cross.* St. Peter is said to have died on an inverted cross, hence the name.

h. *Anchor Cross.* A cross disguised as an anchor was an early Christian symbol of hope.

i. *Maltese Cross.* Like four spearheads point-to-point, this cross was the symbol of the Knights of Malta.

j. *Rose Cross.* The cross from which the Rosicrucians, or Order of the Rose cross, got their name.

k. *Chi-Rho Monogram.* This is an abbreviation of the Greek letters χ (*chi*) and ρ (*rho*), the first letters of χρΙΣΤΟΣ, which is Greek for *Christos*. When a horizontal arm is added, Burckhardt calls it an eight-spoked universal wheel.[24] The vertical and horizontal lines refer to the four cardinal directions, the axial cross of the heavens. The slanted lines depict the intercardinal directions. Together they make up what Burckhardt calls the *cosmic cross*, indicating the eight major compass directions.

l. *Swastika.* This symbol was derived from the sun wheel and was used by early Christians as a disguised cross. It is also called the *Crux gammata* because it can be considered to be made up of four of the Greek letter Γ (*gamma*).

Carl Jung points out that up to about A.D. 1000 the equilateral cross or Greek cross was the usual form, but that over time the center bar moved upward until the cross took the Latin form that it has today. To Jung, this symbolized a movement away from the earthly toward the more spiritual sphere.[25] This upward movement reached its climax in the Middle Ages.

During this Gothic period, the Latin cross was generally used for the ground plan of buildings, like the representation of Chartres Cathedral in Figure 4.56, although baptisteries were centrally symmetrical. This form was an outgrowth of the rectangular Roman basilica, with the addition of transepts. Its shape was also compared to that of the human figure, and in Figure 4.57 we see how Francesco di Giorgio Martini illustrated a cruciform ground plan with an inscribed human figure. Again, to quote Jung, "With the Renaissance the upward movement went into reverse . . . man turned back to earth and nature . . . art broke away from the religious subjects of the middle ages and embraced the whole visible world [and] . . . in contrast to the soaring gothic cathedrals, there were more circular ground plans. The circle replaced the Latin cross."

Wittkower further writes, "For the Renaissance architect the circular or polygonal church was the man-made echo or image of God's universe." Wittkower mentions Leon Battista Alberti, who in his *Ten Books of Architecture*[26] wrote, "It is manifest that nature delights principally in round figures." Alberti also extolled the hexagon and other polygons. Wittkower says that Alberti was attracted to the "central" church (one of circular or polygonal shape) by classical structures: Sto. Stefano Rotondo, Sta. Constanza, the octagonal baptistery near the Lateran, and the twelfth-century S. Giovanni in Florence, thought to be a Roman temple turned into a church and now a baptistery. Alberti maintained that the staggering beauty in a church has a purifying effect and produces a state of innocence pleasing to God. No geometrical form, he felt, is

> The cross may symbolize a *crossroad*, that place where things meet and from which a change in direction is possible.

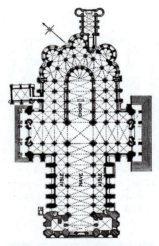

FIGURE 4.56 ■ Ground Plan of Chartres

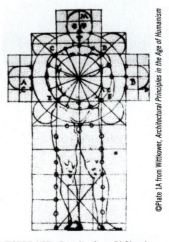

FIGURE 4.57 ■ Drawing from Di Giorgio

©Plate 1A from Wittkower, *Architectural Principles in the Age of Humanism*

better suited for this plan than the circle or forms derived from it—that is, shapes having rotational symmetry.

Of the round temple, Palladio writes, "it is the only one that is simple, uniform, equal, strong, and capacious. . . . [It has] neither beginning nor end, nor are they to be distinguished one from the other, . . . having parts similar to one another, and all participating of the figure of the whole; . . . the extreme being found in all its parts, equally distant from the middle, it is exceeding proper to demonstrate the unity, the infinite essence, the uniformity, and the justice of God."[27]

Again quoting Wittkower, "With the Renaissance revival of the Greek mathematical interpretation of God and the world, and invigorated by the Christian belief that Man as the image of God embodied the harmonies of the universe, the Vitruvian figure inscribed in a square and a circle became a symbol of the mathematical sympathy between microcosm and macrocosm. How could the relation of man to God be better expressed than by building the house of God in accordance with the fundamental geometry of square and circle?"[28]

The round church may have been the ideal, but polygonal and Greek cross churches are much more prevalent than round ones. Why is this so? Perhaps because

- All polygons are closely related to the circle and share many of its properties.
- The circular dome above a polygonal or Greek cross church carried the symbolism of the circle, regardless of the shape below.
- The Greek cross shape was more appealing than round or polygonal churches because it is, in fact, a cross.

One appeal of the Greek cross is that, unlike the Latin cross, it is *centrally symmetrical*. This property of central symmetry closely relates it to circular and regular polygons, which were extolled during this period.

Wittkower notes that from 1485 to 1527 there was a profusion of centralized churches, the first of which was Giuliano da Sangallo's S. Maria delle Carceri in Prato, in the Tuscan region of Italy. The basic layout is shown in Figure 4.58. Of this church, Wittkower writes, "Four short arms are joined to the crossing which . . . is based on the two elementary figures of square and circle. The ratios are simple and therefore as evident as possible. The depth of the arms, for instance, is half their length and the four end walls of the cross are as long as they are high, i.e., they form a perfect square."[29]

Wittkower thinks that Donato Bramante planned St. Peter's as a Greek cross shape, but that "although Bramante's plan underwent many and decisive changes, it remained a tremendous stimulus to architects all over Italy, and churches with a high dome over a Greek cross arose everywhere."[30] Wittkower lists these churches, including the church of S. Maria della Steccata in Parma, shown in Figure 4.59.

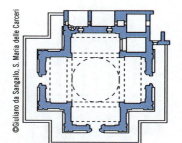

FIGURE 4.58 ■ S. Maria delle Carceri, Prato, Giuliano da Sangallo, begun 1485

FIGURE 4.59 ■ S. Maria della Steccata, Parma

THE EARTHLY AND THE MUNDANE

The square, with its parallel and perpendicular sides, is often associated with earthly existence and the works of humankind. Plato equated the cube, with its square faces and square corners, to the element *earth*. In this section, we'll explore how the square is associated not only with the earth as a whole, but with its features—its winds, rivers, and so on.

Earthly Groups of Four

The movements of celestial bodies may have been part of the reason why the number *four* became identified with the *four directions:* toward the sunrise, toward the sunset, and the two directions perpendicular to these. Four is the number of the *cardinal points* (north, south, east, and west) and the directions in which a person can move (right, left, forward, and back).

The Old Testament contains many references to the four directions: "And he shall . . . gather together the dispersed of Judah from the four corners of the earth"[31] and "An end, the end is come upon the four corners of the land."[32]

The first mention of the *four elements* probably comes from Empedocles (c. 493–433 B.C.), Greek philosopher, statesman, poet, and disciple of Pythagoras. Empedocles asserted that all things are composed of four primal elements: earth, air, fire, and water. He believed that the creation of new matter is not possible and only changes in the combinations of the four existing elements can occur. In art, the elements are sometimes portrayed as gods. *Air* may be represented by one of the four wind gods (Boreas, Zephyr, Notus, or Eurus). *Earth is* frequently represented by Ceres (Greek Demeter), the goddess of agriculture. *Fire* is represented by Vulcan (Greek Hephaestus) at his forge. *Water* is often represented by Neptune (Greek Poseidon). Other representations can be found as well.

Rudolph Koch gives the very basic symbols for the elements shown in Figure 4.60. (He also developed a set based on the circle.)

The *four continents* or *four parts of the world* may be represented by a woman, a river, or certain animals. For example, *Africa* is often portrayed as a woman with black skin, perhaps wearing a coral necklace. Other symbols include the crocodile, lion, snake, elephant, scorpion, the sphinx, and the Nile. *The Americas* wears a feathered headdress and holds a bow and arrows, with the alligator and the Plate River in South America as other symbols. *Asia* is adorned with flowers and jewels, perhaps with a camel, rhino, elephant, palms, perfumes, or the river Ganges. *Europe*, queen of the world, wears a crown and holds a scepter. A bull or horse may be depicted, and the river is the Danube.

The Old Testament is a rich source of numerical symbolism. The *four rivers* are mentioned in Genesis 2:10. "And a river went out of Eden . . . and parted . . . into four heads. The . . . first [is] Pison . . . which compasses the whole land of Havilah . . . the second [is] Gihon . . . that compasses the whole land of Ethiopia . . . the third [is] Hiddekel . . . that goes toward the east of Assyria . . . and the fourth [is] Euphrates that goes eastward to Assyria." The four rivers are usually represented by a fountain or by river gods. The fountain in Figure 4.61

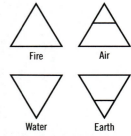

Fire   Air

Water   Earth

FIGURE 4.60 ■ **Symbols for the Four Elements**

The *Rape of Europa* is a very popular art motif. Zeus fell in love with the maiden Europa, disguised himself as a bull, and abducted her to Crete. She eventually gave birth to the continent of Europe and, some say, Minos, who became king of Crete. For more illustrations, see *Rape of Europa* (Giordano, 1686), Titian's *Rape of Europa*, Tiepolo's *Apollo and the Four Continents*, and Rubens's *Four Parts of the World.*

©Mary Ann Sullivan

FIGURE 4.61 ■ **Fountain of the Four Rivers, Bernini, Piazza Navona, Rome**

FIGURE 4.62 ■ *Aeneas and the Sibyl Enter Charon's Boat*

depicts the Nile, the Rio de la Plata, the Ganges, and Danube rivers. (It does not depict any Old Testament rivers.) From left to right, we see the river gods of the Danube, the Ganges, and the Nile.

The *four rivers of Hades*—Acheron (woe), Styx (sorrow), Phlegethon (fire), and Cocytus (wailing)—are important elements from mythology. They play an important role in Plato's *Phaedo*, Dante's *Inferno*, and Virgil's *Aeneid*. The Aeneid is represented in Figure 4.62. When Aeneas and the Sibyl entered Charon's boat to cross at the junction of the Acheron and the Cocytus, they were initially refused passage because they had not been duly buried; Charon yielded at the sight of the golden bough that Aeneas held.

To Plato and Dante, these four rivers represented the four stages of punishment awaiting a person after death. The newly dead who drank from another river, the *Lethe*, lost all memory of their past existence.

In the *Aeneid*, Virgil wrote about Aeolus, the King of the Winds, who lived on an island just north of Sicily. Aeolus ruled the *four winds*, which he kept locked in a cave. The four winds were considered to be gods who controlled the fates of sailors. They were Boreas, the north wind (Latin, Aquilo); Zephyr, the west wind (Latin Favonius); Notus, the south wind (Latin Auster); and Eurus, the east wind (also Eurus in Latin). The winds are usually portrayed in art as heads blowing with puffed cheeks, as in Botticelli's *Birth of Venus* (Figure 4.30). They are sometimes bodiless or blowing a horn.

The winds are also mentioned in the Old Testament: "I bring the four winds from the four quarters of heaven, and will scatter them toward all those winds."[33] They also appear in the New Testament: "And after this I saw four angels standing on the four corners of the earth, holding the four winds of the earth, that the wind should not blow on the earth. . . ."[34]

The number four is associated not only with the winds and waters and other features of Earth, but also with humankind. In the Middle Ages, much was made of the *four humours*, fluids contained in the body: *phlegm*, *blood*, *bile* or *choler*, and *black bile*. The four humours were associated with types of personalities or temperaments. *Phlegmatic* was associated with a calm, stolid temperament and was considered unemotional. A *sanguine* temperament was benign, gentle, cheerful, and optimistic. A *choleric* personality was easily angered, bad-tempered, and irritable. A *melancholic* individual was moody.

Other groups of four associated with the Earth are the four times of day and the four seasons. We'll discuss them in Chapter 11, "The Spheres and Celestial Themes in Art and Architecture."

Correspondences

According to the scholar Vincent Hopper, the idea that the members of any group defined by the same number were related, such as the seven planets and the seven days of the week, came from astrology.[35] Italo Calvino wrote, "The ancients saw the microcosm and the macrocosm in [terms of] correspondences, between psychology and astrology, between humours, temperaments, solids, planets, and constellations. . . . " The following table lists the traditional correspondences between our groups of four: elements, humours, solids, properties, zodiac signs, cardinal points, river gods, seasons, and times of day. Items in this table may vary in different places and times, but this listing gives the general idea.

Element	Humour	Solid	Property	Zodiac Signs	Cardinal Point	River God	Season	Time of Day
Water	Phlegmatic	Icosahedron	Cold	Cancer, Scorpio, Pisces	North	Cocytus	Winter	Night
Air	Sanguine	Octahedron	Hot	Gemini, Aquarius, Libra	South	Acheron	Spring	Morning
Fire	Choleric	Tetrahedron	Moist	Aries, Leo, Sagittarius	East	Phlegethon	Summer	Midday
Earth	Melancholic	Cube	Dry	Taurus, Virgo, Capricorn	West	Styx	Autumn	Dusk

Similar information is provided in the famous illuminated manuscript called *Les Très Riches Heures du Duc de Berry*, which was painted by the Limbourg brothers in the early 1400s. Figure 4.63 is a page from this medieval Book of Hours. (A Book of Hours was an almanac that related seasonal activities and celestial motion to Biblical passages and specific prayers.) This picture associates the

©Réunion des Musées Nationaux/Art Resource, NY

FIGURE 4.63 ■ **Zodiacal Man**

humours with the qualities hot, cold, moist, and dry. It shows three zodiac signs in each of the four corners. Each corner is also a cardinal point of the compass.

lower right	north	septentrionalia	Phlegmatic	sluggish, dull
lower left	south	meridionalia	Sanguine	hopeful, confident
upper left	east	orientalia	Choleric	hot-tempered, passionate
upper right	west	occidantalia	Melancholic	sad, depressed, melancholy

SUMMARY

After presenting the geometry of the square, the rectangle, and the other quadri-laterals, we briefly traced the development of shapes of paintings. We discussed open frames, as in the frieze; multiple frames, often showing a time sequence; and the altarpiece. We learned that an emerging art market required portability, which encouraged the use of canvas, which in turn may have encouraged the use of the rectangular format. We explored several rectangular formats, the square, and overlapping squares, and came to see that the golden rectangle was not a very important one. We learned that, indeed, if we looked hard enough, we could find many pictures having almost any arbitrarily chosen shape. We saw how some artists used axes of symmetry, zones, diagonals, and microthemes to strengthen the compositions of their paintings.

We discussed four reasons to use the closed frame: to separate the world of the picture from the rest of the world, to control the composition of the picture, to aid portability, and to use as a *window frame* through which to view the world. The notion that a picture is a window through which we look will be expanded when we discuss the development of perspective in Chapter 12, "Brunelleschi's Peepshow and the Origins of Perspective."

We learned that the desirability of using a system of proportions, rather than picking each dimension with no regard to the others in a structure, was clearly stated by Vitruvius. To our earlier systems of architectural proportions based on the Pythagorean musical ratios and the golden ratio, we added another one, based on the square.

It's hard to say whether the use of these particular proportions was truly widespread, because we have no written record verifying their use. Quoting Hersey, "So here's an important point: in the absence of documentation, there is absolutely no way of proving that [an architect] worked from the one shape and not from another. To make geometric analyses by trying to get inside the head of a long-departed architect is folly . . . the shapes an architect chooses are a matter of fact; how he arrived at them is a matter of opinion."[36]

Vitruvius is vague about the Roman's use of certain proportions. Aside from the Romans' urge to impose order on the world, there are more pragmatic reasons for them to use an established system of proportions. Such a system provided a way to ensure proportional relations without having to do calculations. All constructions could have been done with a straightedge and a compass (both done with stretched cord) and were very easy to use for on-site layout.

In the next chapter, we'll explore the role of other polygons. As with the triangle and the square, we'll explain some of the symbolism associated with each, the so-called *sacred geometry*, and some art motifs connected with the number of sides of each polygon. We'll use the sacred cut to construct an octagon. We aren't finished with frames, either. We'll discuss round and elliptical frames in later chapters. Finally, the square will reappear in Chapter 8, "Squaring the Circle."

EXERCISES AND PROJECTS

1. Define or describe the following:

   frieze   format   orientation
   microtheme   zones   cruciform
   correspondences   compositional latch Solomon's knot

2. Give four reasons for framing a picture.

3. What characteristic does the square format possess that the rectangular format does not?

4. What sort of subject matter lends itself to the square format?

5. Give examples of how the square format symbolizes the earthly and mundane.

6. Do the construction, drawing a rectangle similar to a given rectangle, as in Figure 4.15. Verify by measurement that the new rectangle has the same proportions as the original.

7. Find and photograph any ad quadratum figures in your school or neighborhood, in windows, pavements, and so forth.

8. Make the sacred cut construction and photocopy it onto card stock. Cut the copies apart, creating an assortment of squares, root-2 rectangles, and Roman rectangles. Try to reassemble some of the pieces into an architectural layout, either plan or elevation. Do these modules exhibit the additive properties claimed?

9. Repeat Fechner's experiments and decide for yourself if the golden rectangle is the most pleasing shape. Draw or cut out a number of rectangles, including the golden rectangle and the others mentioned in this chapter. Poll at least 35 students as to which one they think is the most pleasing. Do a simple statistical analysis of your findings. Do you find a statistically significant preference for any particular shape?

10. Make a graphic arts design based on the ad quadratum figure.

11. Make a graphic design or a work of art based on the sacred cut and its extensions, such as the tee shirt shown in Figure 4.64.

12. Arrange the squares and rectangles from the sacred cut construction into a graphic design or work of art. Use color.

13. Create a graphic design based on whirling rectangles.

14. Make a painting or a relief sculpture based on the square or rectangle.

15. Make a painting or a relief sculpture that uses axes of symmetry as a main compositional device.

16. Make a painting or a relief sculpture that contains a microtheme.

17. Make a painting or a relief sculpture that uses a strong diagonal as a main compositional device.

18. Make a wall sculpture based on the so-called *God's Eye*, a design that incorporates both the cross and the square (shown in Figure 4.65). Starting at the center, wind a piece of string or yarn around two crossed sticks, varying their color to obtain a series of concentric squares or rhombi.

19. Create a graphic design using the principles of dynamic symmetry. See Edwards, *Pattern and Design with Dynamic Symmetry*, for ideas.

20. *Team Project.* Assemble a team. Using ropes and pegs, lay out a large root-2 rectangle on your school lawn or ball field.

21. *Team Project.* Using ropes and pegs, make a large sacred cut construction on your school grounds.

FIGURE 4.64 ■ A Student's Tee Shirt Design

FIGURE 4.65 ■ God's Eye

22. *Team Project.* Using ropes and pegs, lay whirling rectangles on your school grounds. Then use a heavy rope or garden hose to connect vertices of the rectangles to produce a logarithmic spiral.

23. Design a building floor plan based on ad quadratum or the sacred cut. Lay it out on a flat field using stretched cords.

24. Make a model of a small building with proportions based on the ad quadratum or sacred cut sequences.

25. *The Rectangle Paradox.* Cut the pieces for the rectangle paradox shown in Figure 4.68 from thin wood or cardboard to make a puzzle. Use it to amaze friends and relatives.

26. *Jigsaw Puzzle.* Paste a picture on one side of a piece of thin wood or cardboard. Then cut the sheet apart into shapes representing as many kinds of quadrilaterals that you can, labeling each on the back, to make a jigsaw puzzle for children.

27. From an art book or at a museum, take a sampling of rectangular frames, noting whether they are in landscape format or portrait format. For each frame, show how the format correlates to the subject matter.

28. Obtain a photocopy of Botticelli's *Birth of Venus* and crop it from a landscape format into a portrait format. (Use a scanned image on the computer, or just use scissors and a photocopied image.) Discuss how the change from landscape to portrait orientation changes the appearance of the figure of Venus. Repeat the experiment with other images.

29. Find as many standard rectangular picture formats as you can (for example, movies, camera film, TV, computer monitor). For each format, measure width and height and compute ratio. Write a paper on your findings.

30. Make a golden rectangle viewing template. Use it at a museum to see how many pictures are nearly in the golden ratio, how many deviate, and to what degree.

31. In a large art book or museum, look for pictures that contain examples of axes, zones, and microthemes. Summarize your findings in a short paper.

32. Search for paintings in which the artist has used a strong diagonal. Discuss how it affects the composition and whether the picture is divided into zones.

33. Visit a museum and take a random sampling of frame shapes. Record the shape, orientation, dimensions, date, and country of origin. Then do a simple statistical analysis. Find the frequency distribution by period or country and find the mean. For rectangular frames, compute the ratio of length to width, do a frequency distribution, and find the mean and standard deviation. Contact a mathematics instructor for references regarding the statistics needed.

34. Do the project in Exercise 33, but search for paintings in a large art book instead of going to a museum.

35. Repeat the project in Exercise 33, but instead of paintings choose common objects around the home—tables, windows, books, magazines, and appliances. Do you find any preferences for the proportions used?

36. Write a book report on any book listed in the sources at the end of this chapter.

37. *Paper Suggestions.* As usual, use these suggestions as a starting point for developing your own topic for a short paper or a term paper.

    ■ Systems of proportions such as those given in this chapter, discussing whether they were actually used in the construction of Roman houses and backing up your opinion with facts

- The role of the frame in art, including modern works in which the frame is violated, and how the artist has used this violation to make a statement
- The use of axes of symmetry and the diagonal in pictorial composition
- The notion of the *Tree of Life* and its relationship to the cross
- Jay Hambidge and the dynamic symmetry movement
- The history and development of painting materials
- The altarpiece and its importance in art history
- The work of Gustav Fechner
- The eruption of Mount Vesuvius and the subsequent work to unearth Pompei and Herculaneum
- The change in shape of church ground plans over the centuries and the reasons for those changes
- The use and significance of the *square halo* in Christian art, and how it relates to other symbolism of the square mentioned here
- Traditional quilt patterns using squares, such as the ad quadratum layout in *Indiana Puzzle* (pattern # 1367 in Rhemel's *The Quilt I.D. Book*)
- Various groups of four and their representation in art. See Appendix E for ideas.
- The Quadrivium of knowledge: the four subjects needed for a bachelor's degree in the Middle Ages—Arithmetic, Music, Geometry, and Astronomy
- The Four Gospels and Four Evangelists
- The Four Horsemen of the Apocalypse

38. Make a quilt based on squares and rectangles, using a pattern from a quilt book, or, better, a design of your own, as in the student's quilt shown in Figure 4.66.

39. The composer Carl Nielson wrote a symphony called *Four Temperaments*. Can you find other works of music based on quantities of four?

40. Make an oral presentation on any of the previous projects.

41. Bake a square cake. In class, perform the sacred cut on the cake and distribute the pieces.

42. *A Puzzle.* A mother bakes a rectangular cake for her twins' birthday, with half to go to each twin. Before she can serve the cake, the father steals a rectangular piece from somewhere in the cake, as shown in Figure 4.67. How can the mother cut the remaining cake into two exactly equal pieces with only one straight cut?

Mathematical Challenges

43. Demonstrate that the ratio of the side of the sacred cut square to the side of the reference square is $\sqrt{2} - 1$.

44. Demonstrate that the ratio of the side of the reference square to the side of the sacred cut square is $\sqrt{2} + 1$.

45. *The Rectangle Paradox.* A square is cut apart as shown in Figure 4.68a, and its pieces are rearranged as the rectangle shown in Figure 4.68b. Note that the area of the rectangle is *one square unit greater* than that of the square! Where did the extra area come from?

46. *Golden Rectangle in a Square.* Construct a square and divide each side by the golden ratio, as shown in Figure 4.69. The figure created by connecting the division points of adjacent sides appears to be a golden rectangle. Can you prove that it is?

FIGURE 4.66 ■ A Student's Quilt

FIGURE 4.67 ■ Cake with a Piece Missing

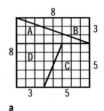

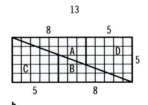

FIGURE 4.68 ■ The Rectangle Paradox

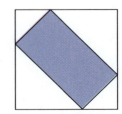

FIGURE 4.69 ■ Golden Rectangle in a Square

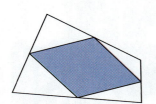

FIGURE 4.70 ■ A Midpoint Quadrilateral

1	12	23	9	20
8	19	5	11	22
15	21	7	18	4
17	3	14	25	6
24	10	16	2	13

FIGURE 4.71 ■ A 5 × 5 Magic Square

47. *Midpoint Quadrilateral.* Construct any quadrilateral. Then connect the midpoints of the sides of the quadrilateral to form another quadrilateral (Figure 4.70), the *midpoint quadrilateral.* What can you say about the midpoint quadrilateral? Can you prove your assertion?[37]

48. In Figure 4.70, connect two opposite vertices of the original quadrilateral. How is the diagonal related to the sides of the midpoint quadrilateral? Can you prove this?

49. In Figure 4.70, compute the area of the midpoint quadrilateral and that of the original quadrilateral. If you are using The Geometer's Sketchpad or a similar program, the computer can find the area for you. What is the ratio of the two areas? Will they always have that ratio, regardless of how the original quadrilateral is dragged? Can you prove this?

50. *Magic Squares.* A magic square is a square array in which the sum of the numbers is the same along any row, column, or diagonal, and in which no number is repeated.
    a. Design your own 3 × 3 magic square.
    b. Design a magic square 4 × 4 or larger.
    c. Research and write a paper on magic squares, such as the one in Figure 4.71.

51. *Geometric Algebra.* The product of two quantities (for example, x and y) can be represented by the area of a rectangle having sides x and y. Similarly, the expression $2x^2 + 5x + 2$ can be thought of as the area of a rectangle having sides $(x + 2)$ and $(2x + 1)$. From this simple idea, a system of *geometric algebra* can be developed. Write a paper on geometric algebra, using as your starting point Heath's summary in his translation of Euclid's *Elements.*[38]

SOURCES

Arnheim, Rudolph. *The Power of the Center.* Berkeley: University of California Press, 1988.

Bouleau, Charles. *The Painter's Secret Geometry.* New York: Harcourt, 1963.

Burckhardt, Jacob. *The Altarpiece in Renaissance Italy* (1898), ed. and trans. P. Humfrey. Cambridge, UK: Cambridge University Press, 1988.

Burckhardt, Titus, *Chartres and the Birth of the Cathedral.* Ipswitch, UK: Golgonooza, 1995.

Coxeter, H. S. M. *Introduction to Geometry,* 2nd ed. New York: Wiley, 1989.

Hambidge, Jay. *The Elements of Dynamic Symmetry.* New York: Dover, 1967.

Kappraff, Jay. "Musical Proportions at the Basis of Systems of Architectural Proportion Both Ancient and Modern." In *Nexus: Architecture and Mathematics,* Kim Williams, ed. (Fucecchio, Florence: Edizioni dell'Erba, 1996), pp. 115–133.

Lawlor, Robert. *Sacred Geometry.* New York: Thames & Hudson, 1982.

Levey, Michael. *Early Renaissance.* New York: Penguin, 1967.

Pope-Hennessy, John. *The Study and Criticism of Italian Sculpture.* Princeton, NJ: Princeton University Press, 1980.

Vitruvius. *The Ten Books on Architecture.* New York: Dover, 1960.

Watts, Donald J., and Carol Martin Watts. "A Roman Apartment Complex." In *Scientific American,* Vol. 255, No. 6 (December 1986), pp. 132–139.

Watts, Carol Martin. "The Square and the Roman House: Architecture and Decoration at Pompeii and Herculaneum." In *Nexus: Architecture and Mathematics,* Kim Williams, ed. (Fucecchio, Florence: Edizioni dell'Erba, 1996), pp. 167–181.

Watts, Donald J., and Carol Martin Watts. "The Role of Monuments in the Geometrical Ordering of the Roman Master Plan of Gerasa." *Journal of the Society of Architectural Historians,* LI, No. 3, September 1992, pp. 306–314.

Williams, Kim. "Michelangelo's Medici Chapel: The Cube, the Square, and the Root-2 Rectangle." *Leonardo,* Vol. 30, No. 2, 1997, pp. 105–112.

Wittkower, Rudolf. *Architectural Principles in the Age of Humanism.* New York: Random House, 1965.

NOTES

1. Vitruvius, IX, Intro, pp. 4–5.
2. This construction is described in Serlio, Chapter 1, Folio 11.
3. Arnheim, p. 1.
4. J. Burckhardt, *The Altarpiece in Renaissance Italy*, p. 81.
5. Arnheim, p. 63.
6. Markowsky says, "[T]his can hardly be viewed as overwhelming evidence for the importance of the golden ratio in esthetics," and he holds that Fechner's testing was limited in that he offered only ten choices. Markowsky, p. 13.
7. Bouleau, p. 45.
8. For a detailed geometric analysis of this painting, see Kemp's *The Science of Art*, p. 30.
9. Arnheim, p. 107.
10. Hambidge, *Elements*, p. xiii.
11. Hambidge, *Elements*, p. xiii.
12. Hambidge, *Elements*, p. xi.
13. Arnheim, p. 104.
14. Arnheim, p. 105.
15. For an analysis of this painting, see Marion Walter, "Looking at a Painting with a Mathematical Eye," *For the Learning of Mathematics*, Vol. 21, No. 2 (2001).
16. Kappraff, p. 121.
17. The material and the quotations in this section are from Donald and Carol Watts's article in *Scientific American*.
18. Vitruvius, VI, III, p. 3.
19. *Leonardo*, Vol. 30, p. 108. In this article, Williams also shows how the root-2 rectangle is related to the musical ratios.
20. Wittkower, p. 108.
21. Palladio, Book I, Chapter XXI.
22. Palladio, Book II, Chapter VI.
23. *Leonardo*, Vol. 30, Figure 5. Calter and Williams have measured the elevations in the Medici Chapel using a theodolite and a trigonometric method. We hope to determine whether Michelangelo used the root-2 rectangle in elevation as well as in plan. See Calter and Williams, "Measuring Up to Michelangelo: A Methodology," *Nexus III, Architecture and Mathematics*, p. 23.
24. Titus Burckhardt, p. 15.
25. *Man and His Symbols*, p. 273.
26. Alberti, Book VII, Chapter IV.
27. Palladio, Fourth Book, Chapter II.
28. Wittkower, p. 16.
29. Wittkower, p. 20.
30. Wittkower, p. 26.
31. Isaiah 11:12.
32. Ezekiel 7:2.
33. Jeremiah 49:36.
34. Revelation 7:1.
35. Hopper, p. 90.
36. George Hersey, p. 14.
37. This exercise and the next two are from *Teaching Mathematics with the Geometer's Sketchpad*.
38. Euclid, Vol. 1, p. 372.

"A mathematician, like a painter or a poet, is a maker of patterns."[1]

G. H. HARDY

FIGURE 5.1 ■ Self-Portrait (detail), Dürer, c. 1498
©Archivo Iconografico, S.A./CORBIS

5

Polygons, Tilings, and Sacred Geometry

Polygons are not new to us. We explored triangles in Chapter 3, "The Triangle," and squares, rectangles, and other quadrilaterals in Chapter 4, "Ad Quadratum and the Sacred Cut." In this chapter, we'll explore polygons in general, and we'll calculate their interior angles, exterior angles, and area. We'll also extend our discussions of similar figures and symmetry, ideas that were introduced in Chapters 3 and 4.

Next, we'll discuss polygons with sides numbering from five (the pentagon) up to eight (the octagon). With each figure, we'll introduce some geometric figures that are *not* polygons, such as *polygrams* (including the five-pointed star). Some of these figures will transport us to the realm of *sacred geometry*.

Finally, we'll have a brief introduction to *tilings*, the art of covering a plane surface with polygons. Most of our tilings will use regular polygons, but we'll also discuss *Penrose tilings*, which cover the plane with irregular polygons.

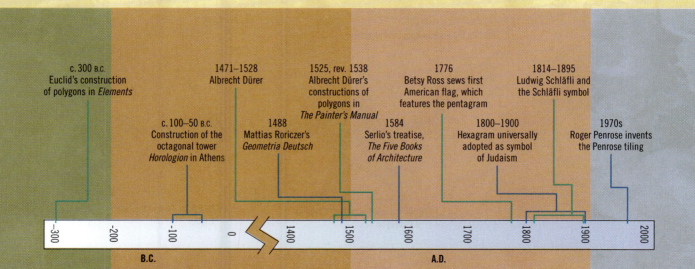

c. 300 B.C.
Euclid's construction
of polygons in *Elements*

c. 100–50 B.C.
Construction of the
octagonal tower
Horologion in Athens

1471–1528
Albrecht Dürer

1488
Mattias Roriczer's
Geometria Deutsch

1525, rev. 1538
Albrecht Dürer's
constructions of
polygons in
The Painter's Manual

1584
Serlio's treatise,
*The Five Books
of Architecture*

1776
Betsy Ross sews first
American flag, which
features the pentagram

1800–1900
Hexagram universally
adopted as symbol
of Judaism

1814–1895
Ludwig Schläfli and
the Schläfli symbol

1970s
Roger Penrose invents
the Penrose tiling

-300 -200 -100 0 1400 1500 1600 1700 1800 1900 2000

B.C.　　　　　　　　　A.D.

POLYGONS

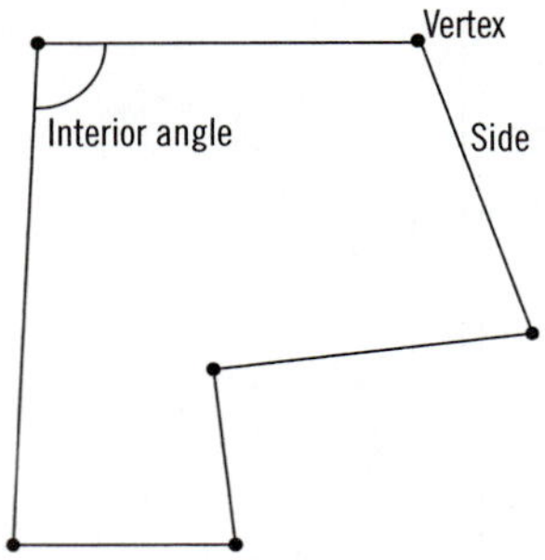

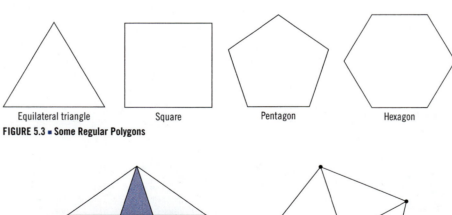

FIGURE 5.2 ■ A Polygon

A *polygon*, as seen in Figure 5.2, is a plane figure bounded by straight line segments, which are called the *sides* of the polygon. The word comes from the Greek words *poly*, which means "many," and *gon*, which means "angle." Some polygons, such as the pentagon, have unique names. Others, like the 16-gon, are just indicated by their number of sides. A general polygon is also called an *n-gon*. The sides of a polygon intersect at points called the *vertices*. The angle between two sides is called an *interior angle* or a *vertex angle*.

A *regular polygon* is one in which all the sides and interior angles are equal. Figure 5.3 shows us four regular polygons. A *regular* polygon of *n* sides and *n* vertices has *n-fold rotational symmetry* about its center. A regular polygon also has *n* axes of mirror symmetry.

The center, or *centroid*, of a polygon (or of any plane area) is the point where all the area can be considered to be concentrated. It is analogous to the *center of gravity* of a solid; however, because a plane area has no weight, the word *centroid* is used instead. The centroid of a regular polygon or a circle is at its center; the centroid of a triangle is at the intersection of its medians. As we saw in Chapter 3, if a polygon is cut from some uniform material and hung from a point, it will swing to where its centroid is directly below the point of suspension. (See Figure 3.26.)

You can draw a poly*gram* by connecting the vertices of a polygon. Therefore, you can draw a pentagram in a pentagon, as shown in Figure 5.4, a hexagram in a hexagon, an octagram in an octagon, and so forth. A polygram is also called a *star-polygon*. By extension, a *pentagram* and a *star-pentagon* are the same thing; a hexagram and a star-hexagram are the same thing, and so on.

Interior Angles of a Polygon

Let's derive a formula for the sum of the interior angles of any polygon, regular or irregular. First, place a point *P* anywhere inside the polygon, as shown in Figure 5.5,

| Equilateral triangle | Square | Pentagon | Hexagon |

FIGURE 5.3 ■ Some Regular Polygons

FIGURE 5.4 ■ A Pentagram Drawn in a Pentagon

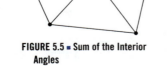

FIGURE 5.5 ■ Sum of the Interior Angles

and then connect it to each vertex. If the polygon has *n* sides, doing this will form *n* triangles.

Because the sum of the interior angles of each triangle is 180°, the sum of all angles in all these triangles is then 180*n*. From this, you must subtract the angles around point *P* because they are not interior angles of the polygon. Their sum is, of course, 360°. This gives $180n - 360 = 180(n - 2)$.

25 SUM OF THE INTERIOR ANGLES OF A POLYGON

$$\text{Sum} = 180°(n - 2),$$

where *n* is the number of sides in the polygon.   ●

■ **EXAMPLE:** Find the sum of the interior angles in a polygon having nine sides.

● **SOLUTION:** From Equation 25, with *n* = 9,

$$\text{sum} = 180(9 - 2) = 180(7) = 1260°.$$ ■

■ **EXAMPLE:** Find angle θ in Figure 5.6.

● **SOLUTION:** The polygon shown has seven sides, so *n* = 7. Therefore,

$$\text{the sum of the angles} = (7 - 2)(180°) = 900°.$$

Adding the six given interior angles gives us

$$278° + 62° + 123° + 99° + 226° + 43° = 831°.$$

Therefore,

$$\theta = 900° - 831° = 69°.$$ ■

FIGURE 5.6 ■ **An Irregular Polygon of Seven Sides**

You can compute the vertex angles for the regular polygons simply by finding the sum of the angles and then dividing that sum by the number of sides. The following chart gives the sum of the angles, for *any* polygon of *n* sides, and the vertex angle (rounded to two decimal places) for regular polygons only. (A regular polygon can't have angles less than 60°, because a polygon can't have fewer than three sides.)

Interior Angles of a Regular Polygon		
Sides (*n*)	Angle Sum 180 (*n* − 2)	Vertex Angle Sum/*n*
3	180	60.00
4	360	90.00
5	540	108.00
6	720	120.00
7	900	128.57
8	1080	135.00
9	1260	140.00
10	1440	144.00
11	1620	147.27

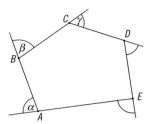

FIGURE 5.7 ■ Exterior Angles of a Polygon

Exterior Angles of a Polygon

An *exterior angle* of a polygon is the supplement of the interior angle. To find the sum of the exterior angles, imagine the polygon in Figure 5.7 as if it were laid out on a field. If you were walking in the direction from *A* to *B*, you would have to turn through an angle β to walk from *B* to *C*. At *C*, you would have to turn through angle γ to proceed to *D*, and so forth. Therefore, a total of *n* turnings would return you to your original direction (*A* to *B*), and you would have turned a total of one revolution, which gives us Statement 26.

26 SUM OF THE EXTERIOR ANGLES OF A POLYGON

Sum of exterior angles = 360° ●

■ **EXAMPLE:** Find angle θ in Figure 5.8.

● **SOLUTION:**

$$\theta = 360° - 88.4° - 51.2° - 136.1° + 41.2° - 96.6° = 28.9°$$

Note that the 41.2° angle is taken as positive because it is turning in a direction opposite from the others. ■

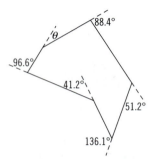

FIGURE 5.8 ■ An Irregular Polygon of Six Sides

Similar Polygons

Like similar triangles and similar rectangles, other similar polygons are figures that have the same shape, but are scaled up or down by some scale factor, and perhaps are rotated or flipped. Because any polygon can be subdivided into triangles, the characteristics of similar triangles also apply to other similar polygons.

94 CORRESPONDING DIMENSIONS OF SIMILAR FIGURES

Corresponding dimensions of plane or solid similar figures are in proportion. The ratio of their lengths is the scale factor. ●

95 AREAS OF SIMILAR FIGURES

Areas of similar figures are proportional to the squares of corresponding dimensions, the square of the scale factor. ●

Therefore, if one pentagon has a side that is 2.5 times larger than a corresponding side of a similar pentagon, as in Figure 5.9, then every side is 2.5 times larger than the corresponding side of the smaller pentagon. If the smaller pentagon has an area of 100 cm², then the larger pentagon has an area $(2.5)^2$—or 6.25 times greater, or 625 cm².

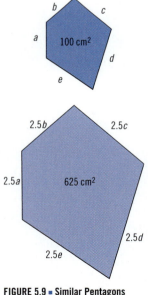

FIGURE 5.9 ■ Similar Pentagons

■ **EXERCISES • POLYGONS**

1. Define or describe the following terms:

   polygon   regular polygon   star-polygon   polygram
   similar polygons   interior angle   exterior angle

2. Figure 5.10 shows two polygons.
   a. Find angle θ in Figure 5.10a.
   b. Find angle θ in Figure 5.10b.

3. Figure 5.11 shows two polygons.
   a. Find angle θ in Figure 5.11a.
   b. Find angle θ in Figure 5.11b.

4. Figure 5.12 shows two similar polygons.
   a. Find the dimensions x and y.
   b. Find the area of the larger polygon.

5. Figure 5.13 shows two similar polygons.
   a. Find the dimensions x and y.
   b. Find the area of the smaller polygon.

6. Find angle θ in the irregular building lot shown in Figure 5.14.

7. In Figure 5.15, find the angle θ between side AB of a building lot and the boundary line L.

8. A scale model of a building has a scale factor of 1 : 20 (1 ft on the model equals 20 ft on the building). What dimension on the building would correspond to 0.55 ft on the model?

9. The façade of the model in Exercise 8 has an area of 2.75 sq. ft. What is the area of the façade on the building?

10. A certain floor plan is drawn at one-tenth scale. The area of a room measures 95 sq. in. on the floor plan. Find the area of the room on the actual building.

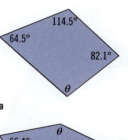

a

b

FIGURE 5.10 ■ Figures for Exercise 2

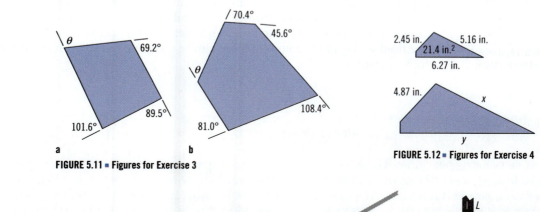

a

FIGURE 5.11 ■ Figures for Exercise 3

b

FIGURE 5.12 ■ Figures for Exercise 4

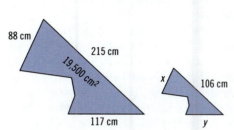

FIGURE 5.13 ■ Figures for Exercise 5

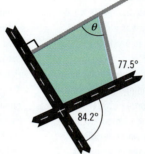

FIGURE 5.14 ■ A Building Lot

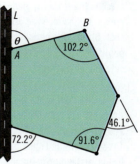

FIGURE 5.15 ■ Another Building Lot

FIGURE 5.16 ■ A Nonconvex Polygon

11. Show that the formula for the sum of the interior angles of a polygon (Equation 25) holds for a nonconvex polygon, such as the one shown in Figure 5.16.

12. Making only straight cuts, how would you divide a square cake into nine slices so that each person gets the same amount of cake and the same amount of icing?[2] Explain why the solution works.

PENTAGONS AND PENTAGRAMS

"Geometry is the right foundation of all painting."

ALBRECHT DÜRER

A *pentagon* is a polygon that has five sides. A *regular* pentagon has five equal sides and five equal interior angles. Several methods can be used to construct a pentagon; in this section, however, we will discuss only one method—the most fundamental one.

The approximate construction of a pentagon given in Figure 5.17 is the second of two pentagon constructions given in Dürer's *The Painter's Manual*.[3] To begin, lay out line segment *AB* to form one side of the pentagon. Draw circles of radius *AB* about *A* and about *B*, locating *C* and *D*, and draw line *CD*. Next, draw a circle of radius *AB* about *C*, locating *E*, *G*, and *F*, and draw *EFH* and *GFI*. Finally, from *I* and *H*, draw arcs of radius *AB* to obtain point *J*. Check that *C*, *F*, *D*, and *J* lie on a straight line.

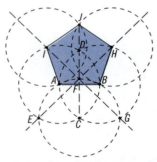

FIGURE 5.17 ■ Dürer's Approximate Construction of the Pentagon

This construction is also given in *Geometria Deutsch*, a German book of applied geometry for stonemasons and craftsmen. It uses a figure called the *vesica*, which we'll cover in Chapter 7, "Circular Designs in Architecture." It is called a *rusty compass* construction, which means that it can be completed with one fixed setting of the compass. It can also be completed using a circular object, such as a jar lid.

ALBRECHT DÜRER (1471–1528)

Albrecht Dürer, the third of eighteen children, was born in Nuremberg. His father was a goldsmith and his son's first art teacher. Dürer was apprenticed in 1486 to a painter and print-maker, and he received extensive instruction in making woodcuts. During this period, southern Germany was a center for publishing, so it was common for painters to be skilled at woodcuts and engravings.

In Nuremberg from 1495 to 1505, Dürer produced many of the works that established his reputation, including his series the *Apocalypse* (1498), the engravings *Large Fortune* (1501–1502), and *Adam and Eve* (1504). A self-portrait he painted around 1498 is shown in Figure 5.1. Many of these works demonstrate his understanding of human proportions based on passages from the works of Vitruvius.

Like Piero and Leonardo, Dürer was an artist who wrote about mathematics. A natural geometrician, Dürer knew about Euclid; he had bought a copy of Euclid's *Elements* in 1507.[4] He was also familiar with Archimedes, Hero, Sporus, Ptolemy, and Apollonius.

He wrote a treatise on geometry, *Unterweysung der Messung mit dem Zirckel un Richscheyt* (1525, rev. 1538), translated as *Instruction in Measurement with Compass and Ruler, in Lines, Planes, and Solid Bodies*. The English translation is called *The Painter's Manual*, and we will refer to it as such. Dürer was an admirer of Vitruvius and praises him in his book.

The Painter's Manual is a book for practical use and not a treatise on pure mathematics. Dürer wanted it to be understood by artists and artisans, and he used their language. The work also familiarized mathematicians with "workshop geometry."

You can easily construct a pentagram from a regular pentagon, either by connecting its vertices, as in Figure 5.4, or by extending its sides. You can also draw a pentagram without lifting your pen, as illustrated in Figure 5.18. Finally, the pentagram can also be made up of five *A*s, as in Figure 5.19; hence its other name—*pentalpha*.

The regular pentagon and the regular pentagram have five axes of mirror symmetry, which are shown in Figure 5.20. Each figure also has *fivefold* rotational symmetry about the center.

Symbolism of the Pentagram

Because of its intrinsic association with the number five, the pentagram has been used to represent many things. Five is the number of fingers on a human hand or the number of toes on a human foot, and the four limbs plus the head add up to five. Therefore, five is the symbol for the flesh. The pentagram became a symbol for man as the microcosmos, as shown in Figure 5.21; the five points matched the two arms, two legs, and head. Man has five senses, and four of them are regularly symbolized in art: as a musical instrument for hearing, a mirror for sight, fruit for taste, and flowers or perfume for smell. The sense of touch has no consistent representation in art.

The Pythagoreans called five the *pentad*. It was the masculine marriage number, uniting the first female number (two) and the first male number (three) by addition. It was called *incorruptible* because multiples of five end in five. Five is also the number of the Platonic solids, and we'll explore the Platonic solids in Chapter 10, "The Solids."

FIGURE 5.18 ■ **The Pentagram Drawn as Endless Loop**

FIGURE 5.19 ■ **The Five *A*s (the Pentalpha)**

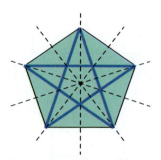

FIGURE 5.20 ■ **Symmetries of the Pentagon and Pentagram**

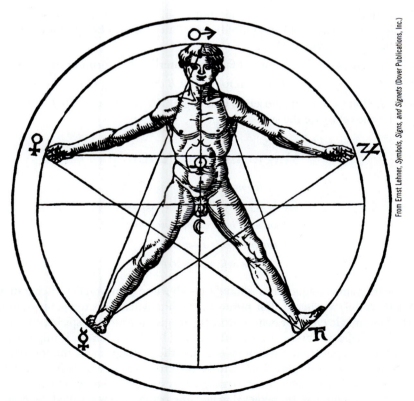

From Ernst Lehner, *Symbols, Signs, and Signets* (Dover Publications, Inc.)

FIGURE 5.21 ■ **Symbolic Representation of *Man as Microcosmos*, Agrippa (1486?–1535)**

FIGURE 5.22 ■ Tombstone Showing Symbol of the Order of the Eastern Star (the Women's Auxiliary of the Freemasons)

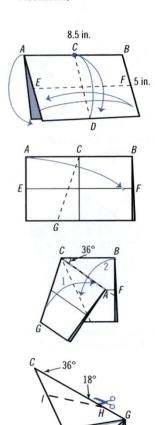

FIGURE 5.23 ■ Cutting an Approximate Pentagram

The star pentagram was a badge of recognition among the Pythagoreans, as was the *Tetraktys* mentioned in Chapter 1, "Music of the Spheres." It may have been especially important to the Pythagoreans because, as you will see, it contains the golden ratio. The construction of the star pentagram was supposed to have been a jealously guarded secret. According to legend, Hippocrates of Chios was rejected by the Pythagoreans for divulging the construction of the pentagram to someone outside the group. Like the Pythagoreans, other select groups used the pentagram as a symbol by which to recognize their members. Note, for example, the Freemasons' Flaming Star and the symbol of the Order of the Eastern Star (a representation of the latter can be seen in Figure 5.22). The U.S. Department of Veteran Affairs has recently added the Wicca symbol, a pentagram inscribed in a circle, to the list of approved religious symbols that may be engraved on veterans' headstones. The pentagram was also said to appear on the shield of Sir Gawain, one of the Arthurian Grail Knights.

The pentagram appears on many flags, including the American flag. According to Betsy Ross's daughter, George Washington and some of his contemporaries came to her mother's upholstery shop in Philadelphia in June of 1776 with a rough draft of a flag. Their design contained stars with six points. Ross showed them how to make a five-pointed star by folding a piece of paper and making a single cut with the shears, as in Figure 5.23. Upon seeing her example, Washington changed the design to include the pentagram. This story was later discredited, but it is interesting to note that George Washington studied engineering, geometry, trigonometry, and surveying. His family crest contained stars and stripes, and some say that the crest may have influenced the flag design. Also, Washington was a Freemason.

You can try folding a pentagram for yourself.

1. Begin by folding an 8.5 × 10 in. (not 11 in.) sheet of paper in half, along line *AB*. Crease (fold and unfold) centerlines *CD* and *EF*. (Point *C* will be the center of the star.)

2. Fold the paper at *C* so that *A* falls on centerline *EF*, forming edge *CG*.

3. Fold *CG* up to edge *CA*; then fold point *B* down along *CG*.

4. Cut along a line *HI*, chosen so that angle *CHI* is half angle *HCI*. Unfold the cut-off to see the star.

Magic and the Pentacle

"*. . . A pointed cap, such as of yore*
Clerks say that Pharaoh's Magi wore.
His shoon were marked with cross and spell,
Upon his breast a pentacle . . . "[5]

SIR WALTER SCOTT

The pentagram is frequently a symbol in both white (benevolent) witchcraft and black (harmful) witchcraft. For purposes of white witchcraft, the pentagram was part of the magic circle drawn to protect a magician from evil spirits, and it was carved on cribs or lintels to ward off evil. For black witchcraft, it might have a more sinister appearance and purpose; note the pentagram on the forehead of the figure shown in Figure 5.24. Unlike the cross, the pentagram was used almost exclusively for magic, but that magic was strengthened by reference to the cross and the stigmata. In medieval magic, the pentagram was called the *pentacle*. The pentacle appears in the *Tarot*, a deck of cards mainly used for fortune-telling

FIGURE 5.24 ■ *Baphomet,* Eliphas Levi, 1896

FIGURE 5.25 ■ A Tarot Ace of Pentacles

and introduced into Europe between 1095 and 1270 by Crusaders or by Gypsies. Figure 5.25 shows us the Ace of Pentacles from a Tarot deck.

In *Faust* by Goethe (1749–1832), Mephistopheles is prevented from leaving Faust's study because a pentagram or *wizard's foot* (*drudenfuss* in German) was engraved on the lintel.[6]

> "I must admit I cannot go!
> A trifling hindrance makes me stay . . .
> That Wizard's Foot upon your sill below . . . "

Faust then asks how he entered in the first place.

> "What, the pentagram is in your way?
> Come, confess, you son of Hell,
> How did you enter if you cannot leave?
> A spirit like you, how could one deceive?"

Mephistopheles explains,

> "Look carefully—it's not drawn well,
> The outer angle, that one nearest me,
> Is slightly open, as you see."[7]

Pentagons and Pentagrams in Architecture

The pentagram and pentagon appear occasionally in architecture (for example, the Pentagon in Washington, D.C., the Palazzo Farnese, and the papal palace at Caprarola, Italy). Serlio, discussing diverse forms of temples, shows the pentagonal ground plan of Figure 5.26 in Folio 5 of Book 5, Chapter 14. (Serlio's book also shows plans for hexagonal and octagonal temples.) The pentagram is sometimes a feature in rose windows, as shown in Figure 5.27.

The quincunx is another design based on the number five. It is not a polygon, but it is an arrangement of five items within a square, with one of the things placed at each corner and one in the center. The five dots on a die provide a familiar example. A quincunx church is one with five domes—one at each corner and one, usually larger, at the center, as shown in Figure 5.28. You can find more examples of the quincunx in Chapter 6, "The Circle," and Chapter 7, "Circular Designs in Architecture."

> The five-lobed *cinquefoil*, along with other foliations, will be illustrated in Chapter 7.

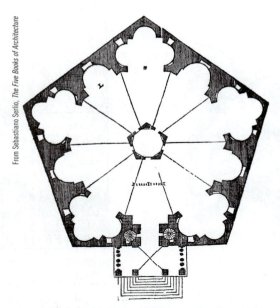

From Sebastiano Serlio, *The Five Books of Architecture*

FIGURE 5.26 ■ The Floor Plan for a Pentagonal Temple from Serlio

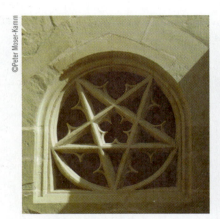

©Peter Moser-Kamm

FIGURE 5.27 ■ (a) Window in the Cloisters of Hauterive; (b) Pentagram in a Rose Window

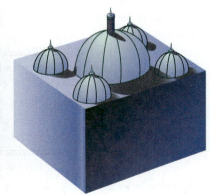

FIGURE 5.28 ■ A Quincunx Church

Golden Triangles in the Pentagon and Pentagram

Recall the golden triangle from Chapter 3. A golden triangle is an isosceles triangle in which the ratio between leg and base is the golden ratio (Φ). There are two types of golden triangles, acute and obtuse, depending on whether you take the ratio of leg to base or the ratio of base to leg, as shown in Figure 3.31. In the *acute* golden triangle, the base angles are 72° and the remaining angle is 36°. In the *obtuse* golden triangle, the base angles are 36° and the remaining angle is 72°. Both types of angles are found in the pentagon and the pentagram.

In Figure 5.29, each interior angle is

$$\alpha = 180(5 - 2)/5 = 108°.$$

Then the base angle of triangle *ABC* is

$$\beta = (180 - 108) \div 2 = 36°.$$

Therefore, triangle *ABC* is an obtuse golden triangle. Now,

$$\Phi = 108 - 36 = 72°.$$

Therefore,

$$\gamma = 180 - 2(72) = 36°.$$

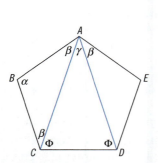

FIGURE 5.29 ■ Angles in the Pentagon

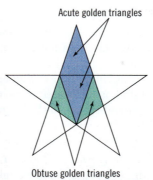

FIGURE 5.30 ■ Golden Triangles in a Pentagon and Pentagram

As such, triangle *ACD* is an acute golden triangle. Using the angles in the pentagon, you can find the angles in the pentagram. Figure 5.30 shows that each "arm" of a pentagram is an acute golden triangle. Further, a pentagon can be subdivided into two obtuse and one acute golden triangle.

Figure 5.31 shows another way to dissect a pentagon into golden triangles and smaller pentagons, and two more ways to dissect golden triangles into pentagons and smaller golden triangles. We'll use golden triangles for a special kind of tiling later in this chapter.

■ EXERCISES ● PENTAGONS AND PENTAGRAMS

1. Define or describe the following terms:

pentagon	pentagram	rusty compass construction
pentacle	quincunx	pentalpha

2. Check for the golden ratio in the pentagon by using the golden ratio dividers made in Chapter 2, "The Golden Ratio."

3. Make a pentagon by tying an overhand knot in a strip of paper and flattening the knot.

4. Fold a pentagram using Betsy Ross's method.

5. Make a pentagram using either of these methods:
   a. By extending the sides of a pentagon
   b. By connecting the vertices of a pentagon

6. Draw a pentagon using Dürer's approximate method, which is shown in the text. Devise a way to check the accuracy of the method.

7. Construct a pentagon using the method in Figure 5.32. Do it by hand or with The Geometer's Sketchpad. Check it for accuracy.
   a. Draw line segment *AB* to form one side of the pentagon, and extend.
   b. Erect a perpendicular to *AB* at *A*.
   c. From *A*, draw an arc of radius *AB*, locating point *C*.
   d. Bisect *AB* with point *D*.
   e. From *D*, draw an arc with radius *DC*, locating points *E* and *F*.
   f. From *A*, draw an arc of radius *AF*, and from *B*, draw an arc with radius *AF*. Mark point *G* at the intersection of these two arcs. Mark point *H* at the intersection of arc *EG* and arc *BC*.
   g. From *B*, draw an arc with radius *AB*. Mark point *I* where this arc intersects arc *GF*.
   h. Connect *A*, *H*, *G*, *I*, and *B* to form a pentagon.

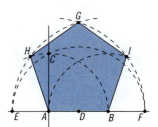

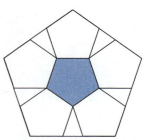

FIGURE 5.31 ■ Golden Triangles in a Pentagon and Pentagons in Golden Triangles

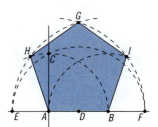

FIGURE 5.32 ■ Constructing the Pentagon, Given One Side

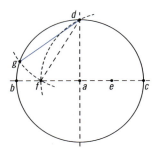

FIGURE 5.33 ■ Dürer's First Construction of the Pentagon

8. *Dürer's First Method for Constructing the Pentagon.*[8] Repeat the construction in Figure 5.33, and devise some way to evaluate its accuracy.
   a. Draw a circle with center *a*, diameter *bac*, and a radius *ad* perpendicular to *bac*.
   b. Locate midpoint *e* of segment *ac*.
   c. With *e* as the center, draw arc *df*.
   d. Segment *df* is then the length of one side of the inscribed pentagon. Draw it as chord *dg*.
   e. Step chord *dg* around the circle to complete the construction.

9. *Euclid's Pentagon and Decagon Construction.* Euclid gives a construction for inscribing a pentagon in a circle in Book IV, Proposition 11.[9] This is shown in Figure 5.34. According to the translator T. L. Heath, these methods were probably developed by the Pythagoreans.[10] In Chapter 3, you learned how to construct acute and obtuse golden triangles, following Euclid IV, 10. You saw that the angle opposite the base of the acute golden triangle was 36°, or one-tenth of a revolution. To inscribe a regular decagon in a circle, you simply construct an acute golden triangle *AOB* with its equal angles *A* and *B* on the circle and its 36° angle at the center of the circle. The base of the triangle is then one of the ten sides of the decagon. For a regular pentagon, connect every other vertex of the decagon. Find this construction in Euclid, and reproduce it yourself by hand or with a computer drafting program. Then evaluate its accuracy.

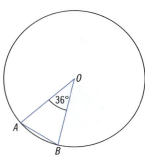

FIGURE 5.34 ■ Inscribing a Regular Decagon in a Circle

10. Put one acute triangle and two obtuse golden triangles together to form a pentagon.

11. Only two dimensions are needed to construct a pentagon: 1 unit and Φ units. Given these two dimensions, use a compass and a straightedge to construct a pentagon.

12. Use your regular pentagon to construct a regular decagon.

13. Construct a set of nested pentagons and pentagrams, as in Figure 5.35. What sequence is formed by the dimensions of corresponding parts?

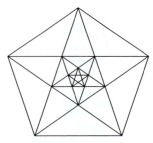

FIGURE 5.35 ■ Nested Pentagons and Pentagrams

OTHER POLYGONS

Geometry, nature, art, and architecture offer many polygonal shapes. In the following sections, we'll briefly explore hexagons, heptagons, and octagons. Of course, we'll also discuss their corresponding hexagrams, heptagrams, and octagrams.

Hexagons and Hexagrams

". . . nature is sometimes delighted with Figures of six sides; for Bees, Hornets, and all other Kinds of Wasps have learnt no other Figure for building their Cells in their Hives, but the Hexagon."

LEON BATTISTA ALBERTI

FIGURE 5.36 ■ Hexagon Made from Six Equilateral Triangles

A *hexagon* is a polygon that has six sides. A *regular* hexagon has six equal sides and is closely related to the equilateral triangle, as you can see in Figure 5.36.

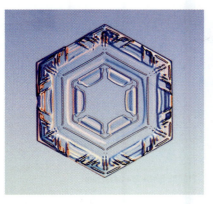
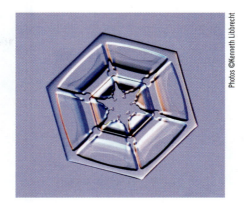

FIGURE 5.37 ■ Examples of Snow Crystals

The hexagon is found in nature in the honeycomb, in some crystals such as basalt, and of course, as we see in Figure 5.37, in snowflakes.

The hexagon is popularly used in architectural decoration partly because it is so easy to draw. In fact, it can be drawn using *rusty compass constructions*, which can be made with a forked stick. The ease of construction is due to the fact that *the side of the hexagon is equal to the radius of the circumscribed circle*. To inscribe a regular hexagon in a circle, simply step the radius around the perimeter six times, as shown in Figure 5.38. This is the same construction given in Dürer's *The Painter's Manual*, page 143, and it follows Euclid Book IV, Proposition 15.

In Figure 5.38, angle γ is one-sixth of a revolution, or 60°. Because the sum of the angles of any triangle must be 180°, you have

$$\alpha + \beta = 180° - \lambda = 120°.$$

Because triangle *AOB* has two sides equal to r, it is isosceles, and hence $\alpha = \beta$. Therefore,

$$2\alpha = 120°$$

$$\alpha = \beta = 60°.$$

Hence, triangle *AOB* is equilateral, and side *AB*, as well as the other sides of the hexagon, equals the radius r of the circumscribed circle.

Six circles will fit around a seventh circle of the same diameter, as in Figure 5.39. As noted, the radius of a circle exactly divides the circumference

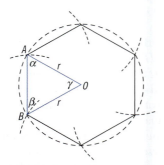

FIGURE 5.38 ■ Inscribing a Regular Hexagon in a Circle

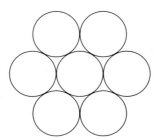

FIGURE 5.39 ■ Six Circles around a Seventh

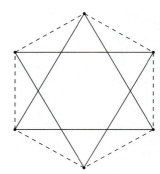

FIGURE 5.40 ■ Six-Petalled Rose　　　**FIGURE 5.41 ■ The Hexagram**

into six parts. Drawing circles centered on these six points gives the *six-petalled rose* illustrated in Figure 5.40, a figure that Keith Critchlow says is culturally universal.[11]

Connecting alternate points of a hexagon gives a *hexagram*, a six-pointed star as illustrated in Figure 5.41. The regular hexagon and hexagram both have six axes of mirror symmetry. Each also has sixfold rotational symmetry about its center.

Hexagons and Hexagrams in Art and Architecture

Figure 5.26 showed a pentagonal ground plan from Serlio; he also gave a hexagonal ground plan in Book 5, Chapter 14, Folio 6. In addition to this, the hexagon and hexagram occur in many other places. The hexagram is often called a *Solomon's Seal*. (Solomon's name is sometimes given to the pentagram and octagram as well.) Jewish legends tell that God gave King Solomon a magic ring that enabled him to control demons.[12] Joseph Campbell says that King Solomon used a seal to imprison monsters and giants in jars.[13] In the *Arabian Nights*, or *The Book of a Thousand Nights and a Night*, Scheherazade tells the story of the fisherman who found a brass bottle in his net, "whose mouth is made fast with a leaden cap, stamped with the seal-ring of our Lord Sulayman son of David."[14] The fisherman opened the bottle and out popped a genie.

Muslims gave the hexagram mystical associations, especially with the cosmos. The seven regions of the hexagram became identified with the seven celestial bodies: Mercury, Venus, Earth, Mars, Jupiter, Saturn, and the Moon. Other cultures, both earlier and later, took the hexagram a few steps further: When a hexagram is placed within a circle, the twelve outer regions are associated with the signs of the zodiac. See Figure 5.42.

The hexagram has symbolic meaning as the Hebrew *Magen David* (Shield of David), or *Mogen David*. The name *Magen David* refers to God as the protector of David. In a religious or cultural context, the hexagram is often referred to today as the Star of David. The Star of David didn't have any religious significance for Judaism in the Middle Ages, and it was found on some medieval cathedrals as well as synagogues. Kabbalists popularized the use of the symbol as a protection against evil spirits. It became a general sign of Judaism from the seventeenth century on, and it was almost universally adopted by Jews in the nineteenth century, even though there is no mention of this symbol in the Bible or in the Talmud. The Star of David is seen on the Israeli flag, synagogues, Jewish tombstones, and so forth. Under Nazism, Jews were forced to wear a yellow Star of David.

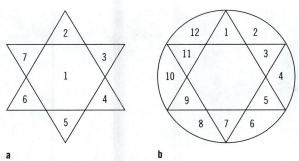

FIGURE 5.42 ■ (a) The Hexagram and (b) the Zodiac

If you mark the six vertices of a hexagon and the six vertices of the surrounding hexagram, and then mark the center, you create the figure known as *The Twelve Surrounding One*. This figure has been interpreted as the twelve tribes of Israel surrounding Moses, the twelve disciples surrounding Jesus, and the twelve prophets surrounding Mohammed.

Joseph Campbell notes that the hexagram can also be viewed as two overlapping Pythagorean *Tetraktys*. Campbell writes, "In the Great Seal of the U.S. there are two of these interlocking triangles. We have thirteen points, for our original thirteen states, and six apexes: one above, one below, and four to the four quarters. The sense of this might be that from above or below, or from any point of the compass, the creative word may be heard, which is the great thesis of democracy."[15]

Recall, from Chapter 4, the symbols for the four elements in Figure 4.60. Figure 5.43 shows a way to relate them to the hexagram. The hexagram can also be drawn as two separate triangles. Sometimes, one is shown dark and the

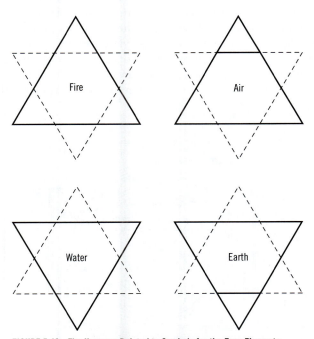

FIGURE 5.43 ■ The Hexagon Related to Symbols for the Four Elements

FIGURE 5.45 ■ Symbol on a Church Façade in Venice

FIGURE 5.44 ■ Hexagram as Two Separate Equilateral Triangles

other is shown light, as in Figure 5.44. In this case, the hexagram is meant to be a symbol for the union of dark and light and the *reconciliation or union of opposites*, a theme you will encounter often. Carrying this theme a step further, in Figure 5.44 the triangles are shown *interwoven*. The theme of union or reconciliation of opposites may also be symbolized graphically by interweaving a triangle with a trefoil or with a circle, as in Figure 5.45.

Many eastern meditation figures, called *yantras*, are geometric in design (for example, the *mandala* that you'll learn about when we get to the circle). Figure 5.46 shows the *Sri Yantra*, which is based on equilateral triangles and the hexagon. This design is composed of nine linked triangles. According to Carl Jung, the yantra is another symbol of the union of opposites. In *Man and His Symbols*, Jung says, ". . . a very common yantra motif is the two interpenetrating triangles . . . one point-upward, the other point-downward. . . . Traditionally, this shape symbolizes the union of Shiva and Shakti, the male and female divinities . . . In terms of psychological symbolism, it expresses the union of

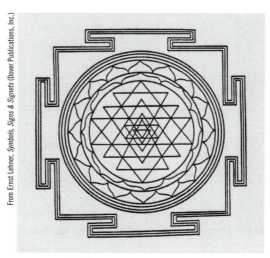

From Ernst Lehner, *Symbols, Signs & Signets* (Dover Publications, Inc.)

FIGURE 5.46 ■ The *Sri Yantra*, Nepal, c. 1700

FIGURE 5.47 ■ **Design on S. Croce, Florence**

FIGURE 5.48 ■ **Design on Pisa Duomo**

FIGURE 5.49 ■ ***Moses Cupola,* S. Marco, Venice**
From F. M. Hessemer, *Historic Designs and Patterns in Color*
(Dover Publications, Inc.)

FIGURE 5.50 ■ **Inlaid Design at Pompeii**

opposites—the union of the personal, temporal world of the ego with the non-personal, timeless world of the non-ego. Ultimately, this union is the fulfillment and goal of all religions: It is the union of the soul with God."[16]

Designs using the hexagram are common in architecture. Consider the hexagrams on the façade of S. Croce and on the Duomo in Pisa—shown in Figures 5.47 and 5.48, respectively. Figure 5.49 shows an exquisite example of a six-petalled rose, in mosaic. The marvelous design from Pompeii shown in Figure 5.50 is made up of a central hexagon surrounded by squares, equilateral triangles, and rhombi.

Heptagons and Heptagrams

A *heptagon* is a seven-sided polygon. Heptagons and heptagrams are rare in art, architecture, and nature. We include them for completeness, for the significance of the number seven and for their interesting constructions.

The construction of a regular heptagon is different from the construction of other geometric shapes. A heptagon cannot be constructed in the usual manner, with a compass and a straightedge. To construct it, you must mark your straightedge. Such methods were known to the ancient Greeks, who called them *neusis constructions*. Archimedes is credited with a trisection of an angle that you may recall from the exercises in Chapter 3. He also developed an exact construction for the heptagon,[17] but it is long and complex. The construction in Figure 5.51 details a somewhat simpler approximate construction from Geometria Deutsch. Begin by drawing radius *OA* in a circle. Draw chord *BC* perpendicular and equal in length to *OA*. (How would you do this if you didn't know in advance the location of point *D*?) Mark the intersection point at *D*. *OD* is then the approximate side of the inscribed heptagon, which is then stepped off around the circle.

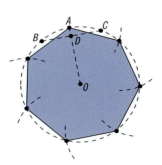

FIGURE 5.51 ■ **Approximate Construction of the Heptagon**

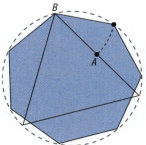

FIGURE 5.52 ■ Dürer's Approximate Construction of the Heptagon

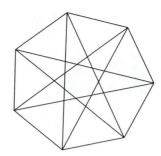

FIGURE 5.53 ■ Heptagrams

FIGURE 5.54 ■ Australian Flag

Figure 5.52 illustrates another heptagon construction from Dürer's *The Painter's Manual*. In this instance, you construct an equilateral triangle in a circle and bisect one side at *A*. *AB* is then the length of a side of the heptagon, and that length can be stepped around the circle seven times.

To form a heptagram, connect the vertices of the heptagon. As shown in Figure 5.53, you can draw two kinds of heptagrams, the first by connecting every other vertex, and the second by connecting every third vertex. You can draw either of these without lifting your pen.

Seven-pointed stars are rare, although heptagonal stars appear on the flags of Australia, Jordan, and the Georgian Republic. One example, the Australian flag, is shown in Figure 5.54. Figure 5.55 shows a rare seven-pointed design. In Figure 5.56, you can see that the heptagon and heptagram have seven axes of mirror symmetry. The heptagon also has sevenfold rotational symmetry about its center.

The Significance of Seven

Seven is one of the main astrological numbers. Seven is also identified with the seven planets known in early times. In contrast to today's usage, the sun and moon were grouped with the five known planets at that time (Mars, Jupiter, Saturn, Mercury, and Venus). The days of the week are named for the sun, the moon, the planet Saturn, and the gods of Teutonic mythology (*Wotan*, the supreme deity; his wife, *Frigg*; *Thor*, god of thunder; and *Tiw*, the war god). Further, medieval alchemy identified seven metals.

As you saw with the number four, the number seven has its *correspondences*. Each planet is identified with a specific day of the week and a specific metal.

FIGURE 5.55 ■ Seven-Pointed Design from the Pisa Baptistery

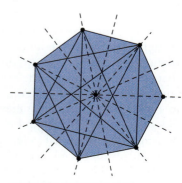

FIGURE 5.56 ■ Symmetries in the Heptagon and Heptagram

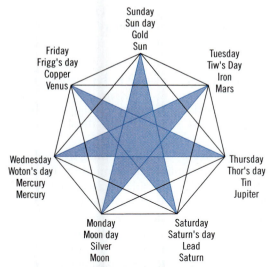

FIGURE 5.57 ■ Days, Planets, Metals, and the Heptagon

Figure 5.57 illustrates the associations. Therefore, the planets, the days of the week, and various metals are associated with the heptagon and the two kinds of heptagrams.

Octagons and Octagrams

You can construct a regular octagon with a compass or simply by folding paper. To construct an octagon within a circle, first inscribe a square to get four vertices, and then extend the bisectors of its sides to obtain four more vertices. You can also construct an octagon using the sacred cut, which was introduced in Chapter 4 and is shown in Figure 5.58. First, make the sacred cut by placing a compass point at each corner of a square and striking an arc that passes through the center of the square and intersects the two adjacent sides. Connect the points where the arcs cut the square, and you have an octagon.

There are three ways to draw an eight-pointed star, or *octagram*, from an octagon. One way is to connect every other vertex, as shown in Figure 5.59, to get two separate squares, or simply extend the sides of the octagon. If the points of one square indicate the cardinal directions, then the other square defines the so-called *intercardinal points*. The squares are often shown *interlaced*, as in Figure 5.60 (recall Figure 5.45, where two triangles were interwoven to make a hexagram). A second way to make an eight-pointed star is to connect every *third* vertex with a continuous line, as in Figure 5.61. A third way is to connect

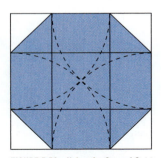

FIGURE 5.58 ■ Using the Sacred Cut to Construct an Octagon

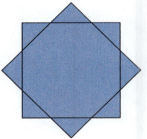

FIGURE 5.59 ■ Octagram Formed from Two Squares

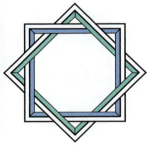

FIGURE 5.60 ■ Octagram as Two Interlaced Squares

FIGURE 5.61 ■ Octagram Drawn Without Lifting the Pen

FIGURE 5.62 ■ Octagrams in an Italian Pavement

From F. M. Hessemer, *Historic Designs and Patterns in Color* (Dover Publications, Inc.)

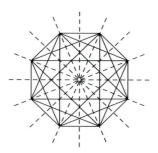

FIGURE 5.63 ■ Symmetries in the Octagon and Octagram

every *fourth* vertex, as shown in Figure 5.62. The octagon and the various octagrams each have eight axes of mirror symmetry; see Figure 5.63. They also have eightfold symmetry about their centers.

Octagons in Architecture

In addition to his pentagonal and hexagonal designs, Serlio gives an octagonal ground plan.[18] Figures 5.64 through 5.68 display some other octagonal structures.

Because Jesus is said to have risen from the grave eight days after his entry into Jerusalem, in Christian symbolism the number eight represents resurrection and rebirth. Similarly, the eight-sided octagon is a symbol of baptism, the spiritual rebirth of a person. Many baptismal fonts are octagonal in shape. We'll encounter the baptismal font in Figure 5.64 again, because it contains the first existing example of linear perspective, in a bronze relief by Donatello.

The high balcony in the Pisa Baptistery shown in Figure 5.65 affords a spectacular view of the octagonal font and pavement below. The baptistery in Figure 5.66 is located in the center of Florence, the birthplace of the Italian Renaissance. Created by Andrea Pisano in 1336, it is named *San Giovanni*, for John the Baptist, whose life story is carved on the south doors. The Baptistery of San Giovanni is the oldest church in Florence. The walls of the octagonal building date from about the seventh century A.D., and its foundations are even older. You'll see this baptistery again in Chapter 12, "Brunelleschi's Peepshow and the Origins of Perspective," when we examine it as the site of Brunelleschi's *peepshow*, the first known demonstration of a perspective painting.

Octagonal buildings other than baptisteries are not common. One curious example is the Castel del Monte in Apulia, shown in Figure 5.67, which not only is octagonal in shape but also has an octagonal tower at each vertex.[19] Octagonal barns are also rare examples of octagons in architecture. Another especially noteworthy octagonal structure is the *Dome of the Rock*, shown in Figure 5.68. Also called the *Mosque of Omar*, this holy site is located on the Temple Mount in Jerusalem, the site of Solomon's Temple and its successors. Muslims believe that the Prophet Muhammad ascended through the heavens to Allah from this site. For Jews, it is here that Abraham prepared to sacrifice his son Isaac. The Dome of the Rock shrine was built between A.D. 685 and 691, and it is the oldest Islamic monument still in existence. It has a wooden dome surrounded by an octagonal arcade of 24 piers and columns; these are surrounded by octagonal outer walls.

FIGURE 5.64 ■ Baptismal Font in Siena

FIGURE 5.65 ■ Baptismal Font in Pisa

FIGURE 5.66 ■ The Florence Baptistery

FIGURE 5.67 ■ Castel del Monte

FIGURE 5.68 ■ Dome of the Rock

FIGURE 5.69 ■ Tower of the Winds

Earlier we studied the *four* winds, but in his *Ten Books on Architecture* Vitruvius says there are eight. "Some have held that there are only four winds: Solanus from due east, Auster from the south; Favonius from due west; Septentrio from the north. But more careful investigators tell us that there are eight."[20] Vitruvius goes on to show how to lay out streets to avoid the disagreeable effects of some of the eight winds. Vitruvius also mentions the *Horologium* (Timepiece) or *Horologion*, an octagonal marble tower (see Figure 5.69) in Athens built about 100−50 B.C. by Andronicus of Cyrrhus. It is still standing at 42 ft high and 26 ft in diameter. Each of its eight sides faces a compass point and bears a relief carving representing one of Vitruvius' eight winds. At one time, the tower is known to have also contained eight or nine sundials, as well as a weathervane atop the tower that contained a *clepsydra* (water clock) to record the time.

■ EXERCISES ● OTHER POLYGONS

1. Define or describe the following terms:

hexagon	hexagram	heptagon
heptagram	octagon	octagram
six-petalled rose	Solomon's Seal	Magen David
yantra	correspondences	intercardinal points

2. In *The Painter's Manual*, Dürer gives constructions for the triangle, square, pentagon, hexagon, heptagon, and octagon. Study and reproduce his constructions for the hexagon, heptagon, and octagon.[21]

3. Create a hexagram from a hexagon by connecting the hexagon's alternate points.

4. Create a heptagram from a heptagon in each of the following ways:
   a. Connect every other vertex.
   b. Connect every third vertex.

5. Create an octagram from an octagon in each of the following ways:
   a. Connect every other vertex.
   b. Connect every third vertex.
   c. Connect every fourth vertex.

TILINGS WITH POLYGONS

"The regular division of the plane . . . is the richest source of inspiration
I have ever struck, nor has it dried up yet."[22]

M. C. ESCHER

Tiling refers to the complete covering, without gaps, of a plane surface by plane figures. The words *tile* and *tiling* can be used both as nouns and as verbs. For example, a *tile* is one of the pieces used to *tile* a plane. The act of *tiling* a plane results in a *tiling* of the plane. Tilings are also called *tessellations*, from *tesserae*, which is Latin for "tiles."

Tilings can range from the very simple to the incredibly complex. For example, draw lines on a sheet of paper, straight or curved, from any edge to any other edge, and subdivide the sheet into distinct regions, as shown in Figure 5.70. You have just produced a tiling, although it may not be attractive or easy to make with its odd-shaped tiles. Jigsaw puzzles, the pattern of cracks in a crazed pot glazing, and the pattern of dried mud are all examples of tilings. Even simple squares can produce an infinite number of tile patterns. Figure 5.71 shows how you can create such tilings simply by shifting one row of squares relative to another.

To keep our exploration of tilings manageable in this chapter, we need to impose some restrictions on the kinds of tilings we'll consider. Even within a narrower range of possibilities, an amazing variety of exquisite tilings are possible. This section will be just a brief introduction. First, we'll limit ourselves to tiling with *regular* polygons: the equilateral triangle, square, regular pentagon, and so forth. Second, we'll require that the polygons be *congruent*, meaning that they are all the same size. Finally, we'll look only at *edge-to-edge* tilings, where adjoining tiles share an edge and two vertices. Figure 5.72a shows what we mean by an edge-to-edge tiling, while Figure 5.72b does not.

Regular Tilings

Let's first consider tilings made from *regular* polygons. To simplify things even further, we'll impose a fourth restriction: Our regular tilings will have just one kind of regular polygon. When that restriction is imposed, only three tilings are possible. Here's why: In a regular tiling, several congruent regular polygons have to meet at a point and fill the plane around that point with no gaps or overlaps. To prove that only three are possible given this fourth restriction, let's try different numbers of polygons around the point.

FIGURE 5.70 ■ A Freehand Tiling

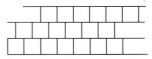

FIGURE 5.71 ■ A Square Tiling

a

b

FIGURE 5.72 ■ Edge-to-Edge Tiling

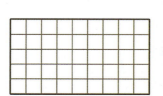
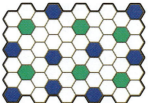
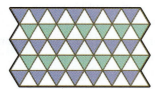

FIGURE 5.73 ■ The Three Regular Tilings

■ For *three* polygons, each would have interior angles of 360 ÷ 3, or 120°. These are *hexagons*.

■ For *four* polygons, the interior angles would be 360 ÷ 4, or 90°. These are, of course, *squares*.

■ For *five* polygons, the interior angles would be 360 ÷ 5, or 72°. *There is no such regular polygon.*

■ For *six* polygons, the interior angles would be 360 ÷ 6, or 60°. This gives *equilateral triangles*.

■ For *seven* polygons, the interior angles would be 360 ÷ 7, or 51.43°.

■ Earlier you saw that a regular polygon cannot have a vertex angle less than 60°, so tilings with seven or more interior angles are not possible.

Therefore, we can conclude that only equilateral triangles, squares, and regular hexagons will produce a regular tiling. Examples of the three are shown in Figure 5.73. Regular tilings, as in Figures 5.74 and 5.75, are commonly used in pavements. The hexagonal tiling is sometimes used to create the illusion of a cube, as shown in Figure 5.75. Just connect every other vertex to the center, forming three diamonds, and shade each diamond differently.

You can denote a regular polygon by its number of sides—three for a triangle, four for a square, and so forth. One shorthand way to describe the polygons surrounding each vertex in a tiling is to give the number of each polygon in sequence, either clockwise or counterclockwise. Using this method, the regular tilings would be 333333, 4444, and 666. Each set of numbers is referred to as a *vertex net*. Another way to designate a regular polygon is with the *Schläfi symbol* {p, q}, named after the nineteenth-century German mathematician Ludwig

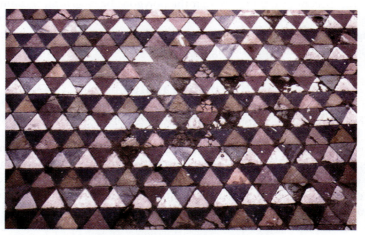

FIGURE 5.74 ■ A Regular Tiling with Equilateral Triangles from Pompeii

FIGURE 5.75 ■ A Hexagonal Tiling from Pompeii

Schläfi.[23] Here p represents the number of sides possessed by each polygon, and q is the number of such polygons about the vertex. Using this notation, 333333 is {3, 6}, 4444 is {4, 4}, and 666 is {6, 3}.

Semiregular Uniform Tilings

Let's relax that last restriction and allow *more than one kind* of regular polygon around each vertex. In this new set of possibilities, the angles around each vertex must still add up to 360°, as with the regular tilings. By trial and error, you can determine that there are 21 ways to arrange regular polygons around a vertex, including the three regular polygons we already mentioned. Of the eighteen remaining ways, only eight can be extended to tile the entire plane. These eight, shown in Figure 5.76, are called the *semiregular* or *Archimedean* tilings. The Islamic tiling shown in Figure 5.77 is an example of a 6363 tiling.

The semiregular tilings, together with the three regular tilings, make up what are called the *uniform* tilings, tilings of the entire plane that contain multiples of *just one* kind of vertex. Each vertex must have the same tiles in the same order. In other words, the entire tiling must have only one kind of vertex net. A tiling is said to be *k-uniform* if it contains *k* different kinds of vertices. Therefore, the uniform tilings in Figures 5.73 through 5.76 that contain just one kind of vertex are called *one-uniform tilings*. Tilings that use two kinds of vertex nets are called *two-uniform tilings*, and so on.

Symmetry of a Pattern

We've discussed symmetry in regard to single plane figures. The same ideas can be extended to a tiling or a pattern. As with a plane figure, a tiling can have axes of mirror symmetry. Placing a mirror along any such axes won't change the appearance of the tiling. A tiling with no mirror axis is called *enantiomorphic*. It has a left-hand and right-hand version. Of the uniform tilings we've discussed, all of them have mirror symmetry except 63333.

A pattern can also have *rotational symmetry* about one or more centers. Turning the pattern about such a center will leave the pattern unchanged, except perhaps for a mirror reflection.

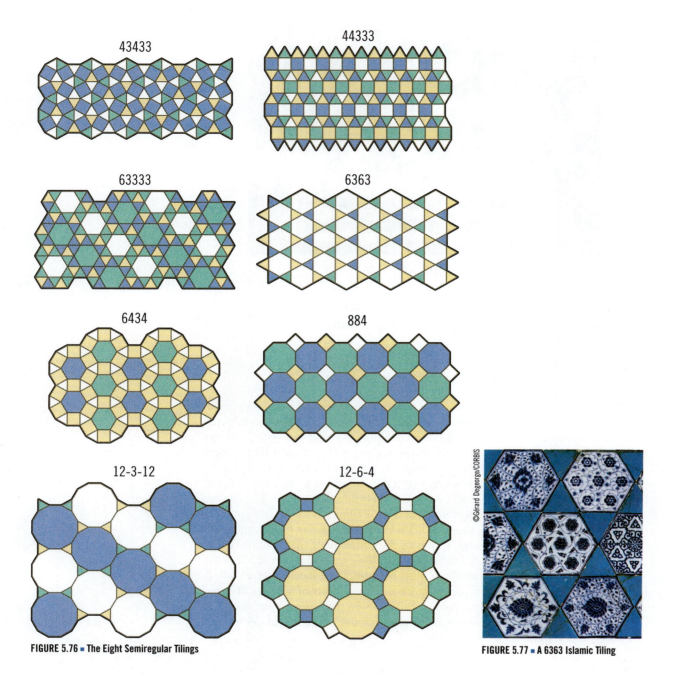

43433

44333

63333

6363

6434

884

12-3-12

12-6-4

FIGURE 5.76 ■ The Eight Semiregular Tilings

FIGURE 5.77 ■ A 6363 Islamic Tiling

©Gérard Degeorge/CORBIS

Penrose Tilings

A tiling is called *periodic* if a translation of a portion of the tiling will cause it to match another portion. Otherwise, the tiling or pattern is called *nonperiodic* or *aperiodic*. The first tiling we made with randomly drawn lines—see Figure 5.73— is nonperiodic. However, the challenge is to construct a nonperiodic tiling with as few tile shapes as possible. The *Penrose Tiling*, invented by Roger Penrose in the late 1970s, is the most well-known such tiling. It uses only two kinds of tiles, and it will tile a plane without repeating the pattern the way ordinary tilings do.

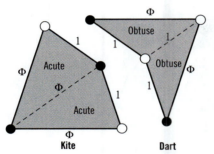
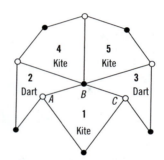

FIGURE 5.78 ■ Kites and Darts   **FIGURE 5.79 ■ Making a Penrose Tiling**

There are two sets of Penrose tiles: One uses *kites and darts* and the other uses *diamonds* or *rhombi*. The first Penrose tiling we'll create will use kites and darts. A kite is made from two acute golden triangles, and a dart is made from two obtuse golden triangles, as shown in Figure 5.78.

As with any tiling, we need some rules to construct a Penrose tiling. Let's keep our earlier restriction that these must be edge-to-edge tilings in which adjoining tilings share an edge and two vertices. Also, to ensure a nonrepeating pattern, mark the vertices of the kites and darts with black dots and white dots, as shown in the figure. When the tiles are placed together, the black dots can touch only black dots, and the white dots can touch only white dots. Lastly, the tiled surface must have no spaces or overlapping tiles.

As with regular and semiregular tilings, let's explore the ways in which tiles can be arranged about a single vertex *B*, as in Figure 5.79. Start with a single tile, either kite or dart. Let's choose a kite, tile 1. Choose a second tile—say, a dart. Following the rules for construction, join tile 2 to tile 1 along side *AB*. Because *A* is a white dot and *B* is a black dot, the second tile (2) can only fit as shown. Choose another dart, tile 3, to fit along side *BC*. The two kites (tiles 4 and 5) can fit between tiles 4 and 5. Therefore, vertex *B* is surrounded with no gaps or overlaps, and none of the rules for construction has been violated.

The pattern we obtained in the preceding example is just one of seven possible arrangements of tiles about a vertex. Figure 5.80 illustrates all seven. Continuing on from here, the combinations and possibilities are almost unlimited. Beginning with the tiling in Figure 5.79, we can move outward to another vertex and surrounding it with our choice of tiles, as long as we follow the rules for construction, to get Figure 5.81. We can even use color to enhance the tilings, as in Figure 5.82.

> You can also mark the tiles by notching their edges or drawing arcs from the vertices.

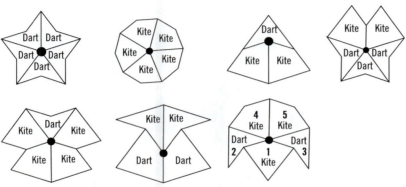

FIGURE 5.80 ■ Seven Ways to Arrange Kites and Darts about a Vertex

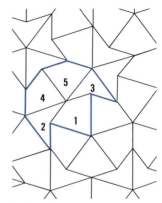

FIGURE 5.81 ■ Continuation of the Tiling

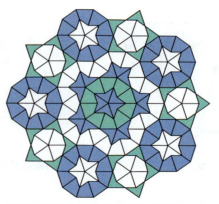

FIGURE 5.82 ■ A Penrose Tiling Made with Kites and Darts

Another pair of tiles that will produce a nonperiodic tiling are *diamonds* or *rhombi*, as shown in Figure 5.83. Like the kites and darts, the diamonds are made up of acute and obtuse golden triangles. The matching rules here are that a dark blue area can abut only a blue area on the adjoining tile and that green must abut green. Figure 5.84 shows one of the ways three fat and two thin diamonds may be fitted together in accordance with the matching rule, and Figure 5.85 shows a complete tiling.

Not surprisingly, the four Penrose tiles are closely related to the pentagon and pentagram. In Figure 5.86, a kite is shown lightly shaded, and a dart is shown with heavier shading. Can you also locate a fat and a thin diamond within the figure?

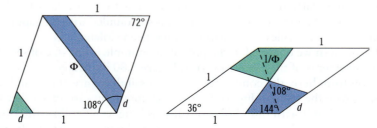

FIGURE 5.83 ■ Fat and Thin Diamond Tiles

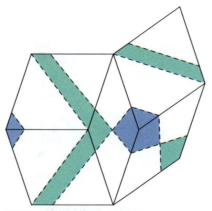

FIGURE 5.84 ■ Fitting Fat and Thin Diamonds

FIGURE 5.85 ■ A Penrose Tiling Made with Diamonds

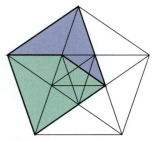

FIGURE 5.86 ■ Kites and Darts in the Pentagon

■ EXERCISES ● TILINGS WITH POLYGONS

1. Define or describe the following terms:

tiling	tessellation	edge-to-edge tiling
regular tilings	vertex net	semiregular tilings
Archimedean tilings	uniform tilings	two-uniform tilings
periodic tilings	nonperiodic tilings	aperiodic tilings
Penrose tilings	kites	darts

2. What restrictions apply when you are making a regular tiling?

3. What restrictions apply when you are making a semiregular tiling?

4. What restrictions apply when you are making a Penrose tiling with kites and darts?

5. What regular polygons will produce a regular tiling of the plane?

6. Make cardboard cut-outs, photocopies, or block prints of the regular polygons, from triangles to octagons. Make duplicates of each. Try to tile the plane using only one kind of regular polygon. In your own words, explain why some will work and some will not.

7. Using a computer drafting program, create a design within an equilateral triangle, a square, or a hexagon. Then copy the design repeatedly to produce a regular tiling.

8. Start with triangles at a vertex, and then remove and replace them to produce semiregular tilings.

9. Sketch the tiles around one vertex if the vertex net is the following:
   a. 6363
   b. 43433
   c. 884
   d. 6434

10. Show that all triangles and all quadrilaterals will tile the plane.

11. Make cardboard cut-outs, photocopies, or block prints of regular polygons, from triangles to octagons. Make duplicates of each. Use these in various combinations to tile the plane.

12. Using your cut-outs, try to reproduce the eight semiregular tilings.

13. Repeat the preceding two exercises using a computer drafting program.

14. Construct a kite and a dart. Make photocopies, block prints, or cardboard cut-outs. Use them to make a Penrose tiling.

15. Using your cut-outs, see if you can verify the seven possible ways to surround a vertex with kites and darts.

16. Repeat the preceding two exercises using a computer drafting program.

17. The ratio of kites to darts in a Penrose tiling is approximately Φ. Verify this amazing fact by counting the number of kites and of darts in your tiling or any other Penrose tiling.

SUMMARY

In this chapter, we finished the study of polygons that we began with triangles and quadrilaterals. We covered the pentagon, hexagon, heptagon, and octagon, and, by extension, the decagon, dodecagon or 12-gon, and so forth. We studied some of the numerical symbolism associated with some of the figures. Because geometric figures are visual, they have more impact than their associated numbers. The Cross and the Star of David are two examples. We also introduced regular, nonregular, and nonperiodic tilings. Although we've covered polygons, we haven't quite finished with them. You'll meet them again as the faces of the Platonic solids, in Chapter 10.

If you take a square and cut off its corners, you'll get an octagon. Cutting off the eight corners of an octagon creates a hexadecagon, or 16-gon. Cutting again will create a 32-gon. If you continue to add sides to the polygon, you'll eventually get the ultimate geometric figure: the most sacred of all in geometry (and the subject of our next chapter), "The Circle."

EXERCISES AND PROJECTS

1. In Book VII, Chapter IV, of *The Ten Books of Architecture*, Alberti shows how to construct regular polygons of six, eight, ten, and twelve sides. Read his instructions, try them, and report your findings.

2. You can double the sides of a regular polygon by bisecting its sides and cutting off its corners. Use this method to construct a decagon, a 12-gon, a 14-gon, and a 16-gon.

3. Construct a set of nested octagons and octagrams, as in Figure 5.87. What sequence is formed by the dimensions of corresponding parts?

4. Dürer gives constructions for the triangle, square, pentagon, hexagon, heptagon, and octagon.[24] Study his constructions, reproduce some, and write a short paper on your findings.

5. Repeat the preceding project with Dürer's constructions for polygons with nine, eleven, and thirteen sides.[25]

6. Repeat the preceding project with Dürer's constructions for five-, six-, seven-, and eight-pointed stars.[26]

7. Find examples of tilings on your campus or in your neighborhood. Take photos or rubbings. Classify them and give their vertex nets. Describe your findings in writing.

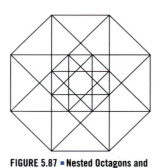

FIGURE 5.87 ◾ Nested Octagons and Octagrams

8. Fold a hexagon as shown in Figure 5.88: Starting with equilateral triangle *ABC*, fold vertices *A*, *B*, and *C* to center *O*.

9. Fold the hexagon from the preceding exercise along its axes of symmetry. With it still folded, cut notches and holes of various shapes. Unfold it to display a paper snowflake.

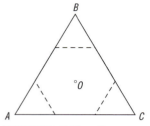

FIGURE 5.88 ■ Folding a Hexagon

10. Cut a circle from paper, fold in quarters vertically, then again horizontally, making a 4 × 4 grid. Mark the circumference where it crosses the grid. Connect these points in various ways to make the familiar regular polygons.

11. Devise a construction for the Islamic Pentagonal Seal, as shown in Figure 5.89.

12. A 4-5-4 triangle can be used to approximately construct a heptagon. Draw such a triangle and figure out how to use it to make the heptagon.

13. *The Enneagon.* Dürer shows how to construct a nine-sided regular polygon, the *enneagon* or *nonagon*.[27] Do this construction, as illustrated in Figure 5.90, and evaluate its accuracy. Locate points *a*, *b*, and *c* equally spaced on the circumference of a circle. From each, draw an arc passing through the center *o* of the circle, getting what Dürer called *fish bladders*. Extend *oa* to *d* and trisect segment *od*. (Chapter 1 explained how to trisect a line segment.) Through the lower trisection point *g*, draw *egf* perpendicular to *od*. Draw a circle with center *o* and radius *oe*.

   The segment *ef* is now one side of the enneagon inscribed in the smaller circle.

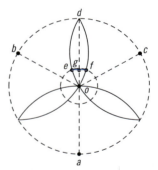

FIGURE 5.89 ■ Islamic Pentagonal Seal

14. Make a work of art or a screensaver based on the material in this chapter, perhaps a design using polygons or a tiling.

15. Make a "stained glass window" based on the material in this chapter. Use actual glass or rigid plastic cut-outs, or simply paint a sheet of plastic. Temporarily install your "stained glass" over a window in your classroom.

16. *Team Project.* On a level field on campus, drive pegs to locate the corners of an irregular polygon with four or more sides. Connect the corners with string. Then find the area of this polygon as follows: Subdivide the polygon into triangles using string, measure the sides and compute the area of each triangle, and add to obtain the total area of the original polygon.

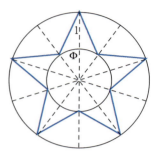

FIGURE 5.90 ■ Dürer's Construction of the Enneagon

17. Make a Penrose tiling, as in the student project shown in Figure 5.91. Use either kites and darts, or diamonds.

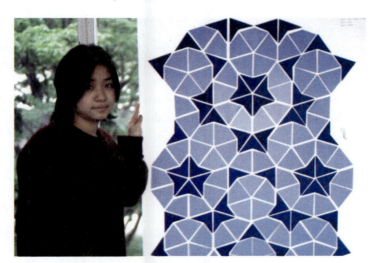

FIGURE 5.91 ■ A Student's Penrose Tiling

FIGURE 5.92 ■ A Student's Inlaid Wooden Plaque

18. Make an inlaid design, as in the student's wooden plaque shown in Figure 5.92.

19. Research Islamic tilings, and make one of your own in a similar style. Write a short paper describing your work.

20. In *The Painter's Manual*, Dürer includes instructions for tiling the plane with various polygons.[28] Read his material and report on your findings.

21. Analyze one of Escher's tilings. Find the basic unit, trace it, and see if you can use it to reproduce the entire design. Design a similar tiling of your own.

22. Use a computer drawing program to design wrapping paper for small gifts. Use a color printer to print some sheets.

23. *Jigsaw Puzzle*. Design a regular, semiregular, or Penrose tiling. Paste a picture on one side of the tiling, and cut the tiles apart to make a jigsaw puzzle.

24. Make a ceramic pot or jar and decorate it with tile patterns.

25. *Kaleidoscope*. Research the kaleidoscope. Write a short paper describing its construction and its principle of operation. Make a working model.

26. Take a frontal digital photograph of your face or a classmate's face. Then, using graphics software, erase the left half of the face and replace it with a mirror image of the right half. What conclusions can you draw about the symmetry of the face?

27. Research *flexagons*, polygons that alter their shape when flexed. Make some flexagons and demonstrate them to your class. Describe your findings in writing.

28. Write a report on any book listed in the sources at the end of this chapter.

29. The following are a few suggestions for short papers or term papers:
    - The representation of halos in art, including polygonal halos
    - Symmetry in Islamic ornament
    - The Castel del Monte
    - The tilings in the Alhambra
    - The Dome of the Rock
    - The shapes in E. A. Abbot's *Flatland*, both for inhabitants and for their buildings
    - Traditional quilt patterns using the polygons, such as the octagonal *Lone Star* (pattern #3781 in Rhemel's *The Quilt I.D. Book*)

- Kepler's work on tilings
- Italian pavements (See Williams's *Italian Pavements in Space* to get started.)
- The tilings of M. C. Escher
- The *passion-flower* (passiflora), which symbolizes events in the Passion of Jesus, once used by missionaries to teach about the crucifixion
- The plant called *Solomon's Seal* (genus *Polygonatum*). When the stem is detached from the root, a scar in the shape of a seal (the *Solomon's Seal*) is left.
- Groups of seven in art: the seven metals, the seven days of the week, the seven planets, the seven ages of man, the seven days of creation, the seven liberal arts, the seven virtues, and the seven vices (See Appendix E for ideas.)

30. Make an oral presentation to your class on any of the previous projects.

31. Use sheet metal to make cookie cutters in the shape of the regular polygons. Use the cookie cutters to make polygonal cookies. Decorate the cookies with different colored icings and sprinkles, and arrange them into tiling patterns. Share with your classmates, but be sure to photograph your design first.

32. Repeat the preceding project with Penrose tilings.

Mathematical Challenges

33. *Transformation of Areas.* A problem of interest to the Pythagoreans was how to transform, with compass and straightedge, a polygon into a triangle or square of equal area. This shape is represented in Figure 5.93. Draw an irregular polygon and use the following method to construct, with compass and straightedge, a triangle of equal area:
    - Given polygon *ABCDE*, draw line *AC*.
    - Draw a parallel to *AC* from *B*, cutting *DC*, extended, at *F*.

    Because triangles *ABC* and *AFC* have the same base *AC* and the same altitude drawn to this base, they are equal in area. That means that quadrilateral *AFDE* has the same area as the original pentagon, but it has one less side. Repeat the process until you find a triangle that has the same area as the original polygon.

34. On page 171 of *The Painter's Manual*, Dürer explains transformation of areas. Compare his method to the one listed in Exercise 33, and write a few paragraphs of explanation.

35. *The Five-Disk Problem.* Five disks, each with radius one unit, are placed as in Figure 5.94. Their points of intersection form the vertices of a regular pentagon, and the circles all pass through *O*, the center of the pentagon. The five disks together cover a circular area of radius *OA*. Find *OA*.

36. *The Pythagorean Theorem Generalized.* The Pythagorean theorem states that the area of a *square* constructed on the hypotenuse of a right triangle equals the sum of the areas of the *squares* on the other two legs. What would happen if we replaced the *squares* in that statement with *regular polygons*?
    a. Use a computer drafting program to construct regular pentagons on the three sides of a right triangle (Figure 5.95), measure their areas, and see if the Pythagorean theorem, generalized, holds.

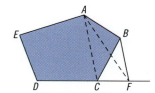

FIGURE 5.93 ■ **Transformation of Areas**

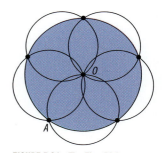

FIGURE 5.94 ■ **The Five-Disk Problem**[29]

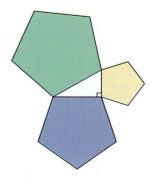

FIGURE 5.95 ■ **The Pythagorean Theorem Generalized**

    b. Try it again with regular hexagons.

    c. Write a short paper about your findings. See Putz and Sipka, "On Generalizing the Pythagorean Theorem," *The College Mathematics Journal*, Vol. 34, No. 4, Sept. 2003.

SOURCES

Aaboe, Asger. *Episodes from the Early History of Mathematics*. New York: Random House, 1964.

Boles, Martha, and Newman, Rochelle. *The Golden Relationship: Art, Math & Nature*. 4 Vols. Bradford, MA: Pythagorean Press.

Campbell, Joseph, with Bill Moyers. *The Power of Myth*. New York: Doubleday, 1988.

Coxeter, H. S. M. *Introduction to Geometry*, 2nd ed. Wiley, 1989.

Dürer, Albrecht. *The Painter's Manual*. Walter Strauss, trans. New York: Abaris, 1977. Original 1525.

Escher, M. C. *The Graphic Works of M. C. Escher*. New York: Ballantine, 1960.

Euclid. *The Thirteen Books of the Elements*. New York: Dover, 1956.

Eves, Howard. *An Introduction to the History of Mathematics*. New York: Holt, 1953.

Fisher, Sally. *The Square Halo*. New York: Abrams, 1995.

Grünbaum, Branko, et al. *Tilings and Patterns: An Introduction*. New York: Freeman, 1989.

Hargittai, István, ed. *Fivefold Symmetry*. New York: World Scientific, 1991.

Jones, Lesley, ed., *Teaching Mathematics and Art*. Cheltenham, UK: Stanley Thornes (Publishers), 1991.

Jung, Carl G., et al. *Man and His Symbols*. New York: Dell, 1964.

Kappraff, Jay. *Connections: The Geometric Bridge between Art and Science*. New York: McGraw-Hill, 1990.

Livio, Mario. *The Golden Ratio*. New York: Broadway Books, 2002.

March, Lionel. *Architectonics of Humanism*. Chichester, UK: Academy, 1998.

Olson, Alton T. *Mathematics Through Paper Folding*. Reston, VA: National Council of Teachers of Mathematics, 1975.

Rehmel, Judy. *The Quilt I.D. Book: 4,000 Illustrated and Indexed Patterns*. New York: Prentice-Hall, 1986.

Runion, Garth E. *The Golden Section*. Palo Alto, CA: Seymour, 1990.

Venters, Diana, et al. *Mathematical Quilts*. Emeryville, CA: Key Press, 1999.

NOTES

1. Hardy, G. H. *A Mathematician's Apology*. London: Cambridge, 1940.
2. From Coxeter's *Introduction to Geometry*.
3. Dürer, p. 147.
4. Kemp, p. 55.
5. Sir Walter Scott describing a wizard in his romantic narrative poem *Marmion* (1808).
6. Faust, Part I, line 1394.
7. Trans. by Alice Raphael, 1930.
8. Dürer, *The Painter's Manual*, p. 145. It is also found in Serlio, Book 1, Chapter 1, Folio 11. This method is originally from "After Ptolemy," *Almagest* Book 1, Chap. 9.
9. Euclid, Vol. 2, pp. 100–104.
10. Euclid, Vol. 2, p. 97.
11. Critchlow, p. 85.
12. Ben-David, Calev, "Ring of the King," *The Jerusalem Report*, Oct. 1995, p. 56.

13. Campbell, p. 27.
14. *Tales from the Arabian Nights*. Translated and annotated by Richard F. Burton. Avenel, 1978.
15. Campbell, p. 27.
16. Jung, p. 267.
17. Aaboe, p. 88.
18. Serlio, Book 5, Chap. 14, Folio 7.
19. See Heinz Götze, "Friedrich II and the Love of Geometry," *The Mathematical Intelligencer*, Vol. 17, No. 4, 1995. Also, Heinz Götze, Friedrich II and the Love of Geometry, *Nexus: Architecture and Mathematics*, 1996.
20. Vitruvius, Book 1, Chap. VI, para. 4.
21. Dürer, pp. 143–146.
22. *The Graphic Works of M. C. Escher*, p. 9.
23. Coxeter, *Introduction*, p. 61.
24. Dürer, pp. 143–146.
25. Dürer, p. 149.
26. Dürer, pp. 153–155.
27. Dürer, p. 149.
28. Dürer, pp. 157–169.
29. Huntley, p. 45.

"The only perfect form is the circle."

BARUCH SPINOZA

FIGURE 6.1 ▪ *Slavia* (detail), Alfons Mucha, 1908
©2006 Mucha Trust. Used with permission.

6

The Circle

When we studied polygons, we started with those having three sides and then added one side at a time until we got to the octagon. Now, let's increase the number of sides to 10, 100, 1000, 1,000,000, and more. As we add sides, the polygons will look more and more like circles. Because all regular polygons are embraced by the circle, the circle is considered a *symbol of unity*. It is also a *symbol of infinity*, without beginning or end and with an infinite number of sides. The circle, with all points equidistant from the center, is a *symbol of democracy*, and it is the preferred formation for an assembly of equals—for example, the council circle, the campfire circle, and King Arthur's round table. It is the easiest geometric figure to draw accurately, even with a string and a peg or with a forked stick. The circle is usually considered the ultimate geometric symbol.

To see as many properties of the circle as we can, we'll divide our exploration across three chapters. In this chapter, we'll address the basics by defining the circle and its parts and reviewing how its circumference and area are computed. We'll also define *radian measure* and determine arc lengths, central angles,

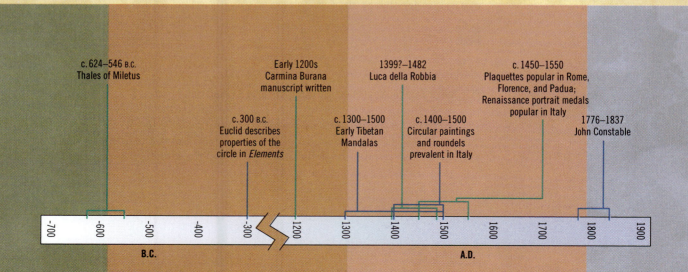

c. 624–546 B.C.
Thales of Miletus

Early 1200s
Carmina Burana
manuscript written

1399?–1482
Luca della Robbia

c. 1450–1550
Plaquettes popular in Rome,
Florence, and Padua;
Renaissance portrait medals
popular in Italy

c. 300 B.C.
Euclid describes
properties of the
circle in *Elements*

c. 1300–1500
Early Tibetan
Mandalas

c. 1400–1500
Circular paintings
and roundels
prevalent in Italy

1776–1837
John Constable

-700   -600   -500   -400   -300   1200   1300   1400   1500   1600   1700   1800   1900

B.C.   A.D.

and sector areas. Next, we'll combine the circle with the straight line, in the form of tangents, secants, and chords. Then we'll move from mathematics to art and architecture. We'll explore the round format, both in painting and relief sculpture, and the symbolism of the circle as it applies to common art motifs.

In Chapter 7, "Circular Designs in Architecture," we'll combine the circle with triangles and other polygons and show examples of its use in architecture, from the rose window to the flying buttress. In Chapter 8, "Squaring the Circle," we'll combine circle and square to arrive at the theme of this book: *squaring the circle.*

GEOMETRY OF THE CIRCLE

Let's start with a review of some geometry of the circle that you probably remember from earlier courses. Statement 51 defines the circle and its parts, and Figure 6.2 illustrates them.

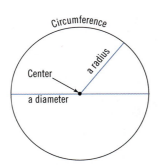

FIGURE 6.2 ■ Diagram of a Circle

51 CIRCLE DEFINITIONS

A *circle* is a plane curve consisting of all points at a given distance (called the *radius*) from a fixed point (called the *center*). The *diameter* is twice the radius. The diameter cuts the circle into two *semicircles.* ●

Circumference and Pi (π)

Wrap a strip of paper around a circular object, such as a jar lid, and mark the point where it starts to overlap. This is illustrated in Figure 6.3. The length from the mark to the end of the strip is the *circumference* of the circular object.

FIGURE 6.3 ■ Measuring the Circumference of a Circle

53 CIRCUMFERENCE

a. The *circumference* of a circle is its total length. This term is also used to mean the circle itself, rather than its *interior.*

b. $C = \pi d = 2\pi r$

The circumference of a circle equals π times its diameter. ●

Now measure the diameter of the lid, and divide that number into the circumference. The quotient is a bit larger than 3, regardless of the size of the lid. The ratio of the circumference C of a circle to its diameter d is the same *for all circles.* It is denoted by the Greek letter π (pi).

52 Pi (π)

$$\pi = \frac{Circumference}{Diameter} = \frac{C}{d} \cong 3.1416$$

Pi is the ratio of the circumference of any circle to its diameter. ●

Pi (π) is an *irrational number* with the approximate value 3.1416. Your calculator stores pi to more decimal places than you will ever need. Look for a key marked π. You can use the definition of π to find the circumference of a circle.

> Refer back to Chapter 1, "Music of the Spheres," if you have forgotten about irrational numbers.

■ **EXAMPLE:** What is the circumference C of a circle having a diameter of 5.54 in.?

● **SOLUTION:** By Equation 53b,

$$C = 5.54\pi \cong 17.4 \text{ in.}$$ ■

■ **EXAMPLE:** Find the radius r of a circle having a circumference of 854 cm.

● **SOLUTION:** By Equation 53b,

$$r = \frac{854}{2\pi} \cong 136 \text{ cm.}$$ ■

> Standard practice is to round your answers to as many significant digits as used in the original data.

Arcs and Sectors

At this point, you need to learn a few more definitions so that you can perform the constructions in this chapter and the next. Figures 6.4 and 6.5 illustrate the definitions that are defined in Statements 55, 57, and 58.

55 **ANGLES IN A CIRCLE**

a. A *central angle* is one whose vertex is at the center of the circle.

b. An *inscribed angle* is one whose vertex is on the circle.   ●

57 **ARC**

An *arc* is a portion of the circle between two points on the circle. The smaller arc is called the *minor arc*, and the longer arc is called the *major arc*.   ●

58 **SECTOR**

A *sector* is a plane region bounded by two radii and one of the arcs intercepted by those radii.   ●

In Book III, Proposition 20, Euclid describes the relationship between a central angle and an inscribed angle, as shown in Figure 6.6. This is Statement 60.

60 **AN INSCRIBED AND A CENTRAL ANGLE SUBTENDING THE SAME ARC**

If an inscribed angle ϕ and a central angle θ subtend the same arc, the central angle is twice the inscribed angle.

$$\theta = 2\phi$$ ●

A semicircle has the property listed in Statement 56. Dürer used this property to draw a right angle.[1] Euclid, in Book VI, Proposition 13, uses this property to find *a mean proportional* or *geometric mean*, a construction given in Chapter 1. This property is sometimes called *Thales' theorem* after the ancient Greek philosopher/mathematician Thales (pronounced Thālēs) of Miletus, who is said to have invented deductive mathematics.

56 **ANGLE INSCRIBED IN A SEMICIRCLE**

Any angle inscribed in a semicircle is a right angle.   ●

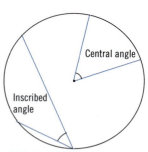

FIGURE 6.4 ■ **A Central Angle and an Inscribed Angle**

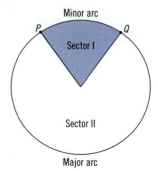

FIGURE 6.5 ■ **Arcs and Sectors**

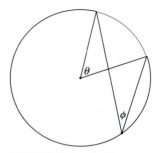

FIGURE 6.6 ■ **An Inscribed Angle and a Central Angle**

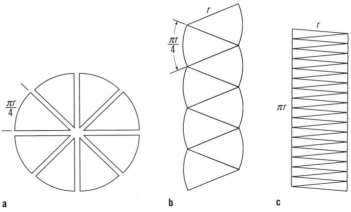

FIGURE 6.7 ■Finding the Area of a Circle

Area of a Circle

Draw a circle on a sheet of paper, and cut it along radial lines into eight equal sectors, as shown in Figure 6.7a. The length of each arc will be one-eighth the circumference of the circle, or $\pi r/4$. Arrange the sectors as shown in Figure 6.7b. The area of this figure is equal to the area of the original circle, because nothing has been either added or removed.

Next, mentally divide each sector along imaginary radial lines into smaller sectors—say, 100 of them—and fit them together as before. This is shown in Figure 6.7c. Your figure should be starting to look like a *rectangle* having an approximate length of πr and an approximate width of r. Again, mentally subdivide each sector into 1000 sectors, a million sectors, and even more. While it's physically impossible, you can do it *mathematically*! We say that as the number of subdivisions approaches infinity, our figure *approaches* a rectangle with width r, height πr, and an area A equal to πr^2.

> The concept of a limiting value is central to calculus. You would say, "The limit of A, as the number of subdivisions approaches infinity, is equal to πr^2."

54 AREA OF A CIRCLE

$$A = \pi r^2$$

The area of a circle (that is, its interior) equals π times the square of the radius.   ●

■ **EXAMPLE:** What is the area of a circle having a radius of 3.75 in.?

● **SOLUTION:**

$$A = \pi(3.75)^2 = 44.2 \text{ sq. in.}$$ ■

■ **EXAMPLE:** If the area of a circle is 583 cm^2, what is its diameter?

● **SOLUTION:** By Equation 54,

> As usual, we carry *one more* significant digit in the *intermediate* steps than is used for the final answer.

$$r^2 = \frac{583}{\pi} = 185.6.$$

Take the square root:

$$r = \sqrt{185.6} = 13.62.$$

Then double the radius to get the diameter:

$$d = 2(13.62) = 27.2 \text{ cm.}$$ ■

■ EXERCISES ● GEOMETRY OF THE CIRCLE

1. Draw a circle. From memory, draw and label the following:
   a. A diameter
   b. A radius
   c. The circumference
   d. An arc
   e. A sector
   f. A central angle
   g. An inscribed angle

2. In your own words, explain the *meaning* of π, not just its value.

3. Find the circumference of a circle having a radius of 5.83 in.

4. Find the circumference of a circle having a diameter of 66.4 cm.

5. Find the diameter of a circle having a circumference of 69.3 m.

6. Find the radius of a circle having a circumference of 384 in.

7. Find the area of a circle having a radius of 44.3 in.

8. Find the area of a circle having a diameter of 55.8 mm.

9. Find the area of a circle having a circumference of 8.48 ft.

10. Find the radius of a circle having an area of 63.8 sq. cm.

11. Find the circumference of a circle having an area of 743 sq. in.

12. The circumference of a certain Doric column having a circular cross section is measured at its base using a tape measure and found to be 3.55 m. Find the diameter of the column at that point.

13. Find the perimeter of the window represented in Figure 6.8, which has a shape that is a rectangle surmounted by a semicircular arch.

14. Find the open area of the window in Exercise 13.

15. A circular frame with an inside diameter of 28.5 in. and a width of 3.15 in. is to be covered with gold leaf. Calculate the area of the front surface of the frame (ignoring the edge).

16. A round window is to have an open area of 250 sq. ft. What diameter should it have?

17. During the Renaissance and even later, the rectangle surmounted by a semicircle (for example, Piero della Francesca's *Baptism of Christ* in Figure 8.26) was a popular frame shape. The reported dimensions vary a bit. For this exercise, let's say the rectangular portion is 116 cm wide and 109 cm high.
    a. What is the total perimeter of the picture?
    b. What is the area of the picture?

18. Figure 6.9, an example of a circular rosette, has an inner ring diameter of 1.05 m and an outer diameter of 3.75 m. Find the area of the rosette between the two rings.

19. Find the circumference and area of a circular fountain that is 4.85 m in diameter.

20. Use the property of a semicircle (as was shown in Figure 1.17) to find, by construction, the mean proportional between 4 and 7. Check your answer by computation.

21. The hemispherical dome represented in Figure 6.10 is supported by concrete members, one of which is shown. Use Statement 56 to find dimension *AB*.

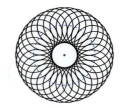

5.72 ft

3.65 ft

FIGURE 6.8 ■ A Window

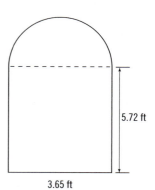

FIGURE 6.9 ■ A Rosette

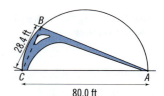

28.4 ft

B

C

A

80.0 ft

FIGURE 6.10 ■ A Hemispherical Dome

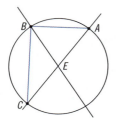

FIGURE 6.11 ■ Constructing a Right Angle

22. *Constructing a Right Angle.*[2] Try the method illustrated in Figure 6.11. Draw any two lines that intersect at a point *E*. From *E*, swing a circle using any compass opening, intersecting the original lines at *A*, *B*, and *C*. *AB* will be perpendicular to *BC*. Explain why this works.

RADIANS

Chapter 3, "The Triangle," introduced angular measure, and we have been using *degrees* as the unit of measure for angles since its introduction. However, for some purposes, such as computing the lengths of arcs and areas of sectors, the *radian*, another angular unit, is a more convenient unit for measuring circles. Here's an easy way to draw an angle of one radian. In any circle, lay off an arc with length equal to the radius of the circle. (Imagine a jar lid of radius *r* and a piece of string of length *r*; lay the string along the perimeter of the lid.) Connect the ends of the arc to the center of the circle. The central angle formed is defined as *one radian*. Figure 6.12 illustrates a radian, which is defined in Statement 61.

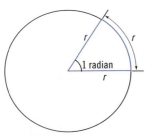

FIGURE 6.12 ■ The Radian

61 THE RADIAN

a. One *radian* (rad) is the central angle subtended by an arc equal in length to the radius of the circle.

b. Radian Measure: The *radian measure* of a central angle *θ* is the ratio of the arc *s* it subtends to the radius *r* of the circle.

$$\theta = \frac{s}{r}$$

   ●

An arc equal to two radii will subtend a central angle of two radians, an arc of three radii will subtend an angle of three radians, and so forth.

■ **EXAMPLE:** Find the central angle *θ* subtended by an arc of length 44.5 cm in a circle of radius 37.3 cm.

● **SOLUTION:**

$$\theta = \frac{s}{r} = \frac{44.5 \text{ cm}}{37.3 \text{ cm}} = 1.19$$

In the previous equation, the units (cm) in the numerator *cancel* the same units in the denominator, leaving radians as a *dimensionless ratio*. However, mathematicians usually write *rad* after an angle in radians just as a label, even though it is not a true dimension like degrees. Also, they say "*θ* equals 1.19," rather than the more correct "the measure of *θ* is 1.19." Because working with radians doesn't require us to deal with dimensions, radians can be useful for making many calculations.   ■

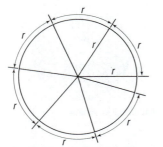

FIGURE 6.13 ■ Laying Off Radii Around the Circumference

Many calculators have built-in conversions between degrees and radians. Check yours.

Angle Conversions

Now let's lay off arcs of length *r* completely around the circle, as shown in Figure 6.13. Six radii are nearly enough to fit around the circumference. In fact, because the circumference is equal to 2π radii (from Statement 53b), the radius should fit around the circle 2π (about 6.28) times, giving one revolution or 360 degrees. This gives us a way to convert from radians to degrees to revolutions and so on.

21 UNITS OF ANGULAR MEASURE

$$1 \text{ revolution} = 360° = 2\pi \text{ rad}$$
$$1° = 60 \text{ minutes}$$
$$1 \text{ minute} = 60 \text{ seconds}$$

●

■ **EXAMPLE:** Convert 1 radian to degrees, to one decimal place.

● **SOLUTION:** Given that 2π rad $= 360°$, we can calculate the following:

$$1 \text{ rad} = (1 \text{ rad}) \frac{360°}{2\pi \text{ rad}} = 57.3°.$$

Therefore, 1 radian is nearly equal to 60°.   ■

■ **EXAMPLE:** Convert 45.8° to radians.

● **SOLUTION:**

$$45.8° = (45.8°) \frac{2\pi \text{ rad}}{360°} = 0.799 \text{ rad}$$

■

Arc Length

Equation 61b ($\theta = s/r$) relates the central angle, the arc length, and the radius. Knowing any two of these three quantities enables you to find the third. For example, multiplying both sides by r gives the arc length defined in Statement 62. Just remember that *the angle must be in radians* for this formula to work. If the given angle is in degrees, convert it to radians.

62 ARC LENGTH

$$s = r\theta$$

The length of an arc of a circle equals the radius times the central angle (in radians) that it subtends.   ●

■ **EXAMPLE:** Find the arc length intercepted by a central angle of 48.4° in a circle of radius 5.34 in.

● **SOLUTION:** Converting the angle to radians gives the following:

$$\theta = 45.8° = (45.8°) \frac{2\pi \text{ rad}}{360°} = 0.7994 \text{ rad}.$$

Therefore,

$$s = r\theta = 5.34 \text{ in. } (0.7994 \text{ rad}) = 4.27 \text{ in.}$$ ■

■ **EXAMPLE:** Find the radius of a circle in which an arc of length 66.8 cm subtends a central angle of 1.33 rad.

● **SOLUTION:** Solving Equation 62 for r and substituting values gives the following:

$$r = \frac{s}{\theta} = \frac{66.8 \text{ cm}}{1.33 \text{ rad}} = 50.2 \text{ cm.}$$

■

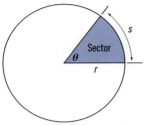

FIGURE 6.14 ■A Circle with Radius and Sector

Area of a Sector

Radian measure allows us to derive a neat formula for the area of a sector, as shown in Figure 6.14. Start with this proportion: The ratio of the area A_s of a sector to the area πr^2 of the full circle equals the ratio of the central angle θ of the sector to the central angle 2π of the full circle.

$$\frac{A_s}{\pi r^2} = \frac{\theta}{2\pi}$$

Now multiply both sides by πr^2:

$$A_s = \frac{\theta}{2\pi}(\pi r^2) = \frac{\pi r^2 \theta}{2\pi}.$$

When you cancel π, you get the formula in Statement 59.

59 AREA OF A SECTOR

$$A_s = \frac{r^2 \theta}{2}$$

■ **EXAMPLE:** What is the area of a sector that has a central angle of 0.885 rad and a radius of 8.35 cm?

● **SOLUTION:**

$$A_s = \frac{(8.35 \text{ cm})^2 (0.885 \text{ rad})}{2} = 30.9 \text{ cm}^2$$

■ EXERCISES ● RADIANS

1. In your own words, state what a *radian* is.
2. Find the central angle, in radians, subtended by an arc of length 4.83 in a circle of radius 5.86.
3. Find the central angle, in degrees, subtended by an arc of length 55.8 in a circle of radius 49.7.
4. Find the arc length subtended by a central angle of 0.885 rad in a circle of radius 184.
5. Find the arc length subtended by a central angle of 33.8° in a circle of radius 69.3.
6. Find the radius of a circle in which a central angle of 1.66 rad subtends an arc of length 55.9.
7. Find the radius of a circle in which a central angle of 44.9° subtends an arc of length 883.
8. A radian is called a *dimensionless ratio*. What does that mean?
9. Besides a radian, name as many other dimensionless ratios as you can.
10. Convert the following angles to radians:
    a. 60.5°   b. 165°   c. 5.58°
11. Convert the following angles to degrees:
    a. 1.95 rad   b. 5.83 rad   c. 0.884 rad

12. Find the area of a sector having a central angle of 0.772 rad in a circle of radius 6.93.

13. Find the area of a sector having a central angle of 55.6° in a circle of radius 77.6.

14. Find the length *AB* of the flying buttress illustrated in Figure 6.15.

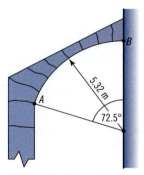

FIGURE 6.15 ■ A Flying Buttress

LINES INTERSECTING A CIRCLE

So far we've discovered some interesting facts about the circle itself. However, in many designs, the circle touches lines, other circles, or polygons—and knowing how the circle relates to these other figures can be useful. In this section, we'll cover circles intersected by lines, and in the next chapter, we'll study how circles interact with polygons and other circles.

Figure 6.16 shows some straight lines touching a circle. In the figure, the tangent *T* touches the circle at *P*, and the secant line *S* cuts the circle at two points *A* and *B*. Line segment *AB* is a chord. Statement 63 defines some of the elements displayed in the figure.

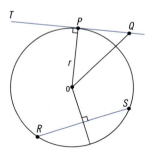

FIGURE 6.16 ■ Tangent, Secant, and Chord

63 TANGENT, SECANT, AND CHORD

a. A line that touches a circle at just one point is called a *tangent* to the circle.

b. A line that cuts across a circle at two points is called a *secant* to the circle.

c. A *chord* is the portion of a secant that joins two points on the circle. ●

These terms are the same when applied to curves other than the circle. Statements 64 and 65 are two theorems about chords and tangents.

64 PERPENDICULAR BISECTOR OF A CHORD

The perpendicular bisector of a chord passes through the center of the circle. ●

65 TANGENT AND RADIUS

A tangent to a circle is perpendicular to the radius drawn through the point of contact. ●

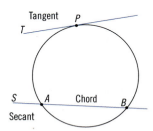

FIGURE 6.17 ■ A Chord and a Tangent

Why must Statement 65 be true? In Figure 6.17, line *T* is tangent to the circle at *P*. If on line *T* you select any point *Q*, *OQ* must be longer than *OP* because *Q* lies outside the circle. Therefore, *OP* is the shortest line that can be drawn from *O* to *T*. Because the shortest distance between a point and a line is perpendicular to the line, then the radius *OP* must be perpendicular to *T*.

■ **EXAMPLE:** Find the distance *x* in Figure 6.18.

● **SOLUTION:** By Statement 65, you know that the angle between the plane and the radius drawn to the point of contact is a right angle. This enables you to use the Pythagorean theorem.

$$x^2 = (24.3)^2 + (58.3)^2 = 3989$$

$$x = 63.2 \text{ mm}$$ ■

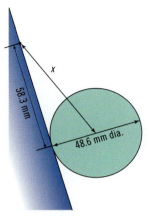

FIGURE 6.18 ■ A Disk Suspended by a String

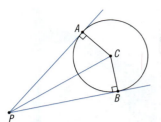

FIGURE 6.19 ■ Two Tangents Drawn to a Circle

Our second theorem—Statement 66—concerns two tangents to a circle drawn from an external point, as illustrated in Figure 6.19. Can you prove the theorem? See the exercises at the end of this section.

66 TWO TANGENTS DRAWN TO A CIRCLE

Two tangents drawn to a circle from a point outside the circle make equal angles with a line drawn from the circle's center to the external point. The distances from the external point to each point of tangency are equal.   ●

The next relationship between tangents and secants comes from Euclid, Book III, Proposition 36. Statement 67 is illustrated in Figure 6.20.

67 A TANGENT AND A SECANT DRAWN TO A CIRCLE

For a tangent and a secant drawn to a circle from a point outside the circle, the product of the portions of the secant equals the square of the tangent. Therefore,

$$(OP)(OQ) = (OT)^2.$$ ●

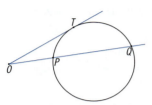

FIGURE 6.20 ■ A Circle with a Tangent and Secant Drawn from a Point Outside

Turning now from tangents to chords, we have the theorem in Statement 68, which is illustrated in Figure 6.21.

68 INTERSECTING CHORDS

If two chords in a circle intersect, the product of the parts of one chord equals the product of the parts of the other chord. Here,

$$ab = cd.$$ ●

■ **EXAMPLE:** Find x in Figure 6.22.

● **SOLUTION:** Applying Equation 68, you can write

$$16.3x = 10.1(12.5)$$

$$x = \frac{10.1(12.5)}{16.3} = 7.75.$$ ■

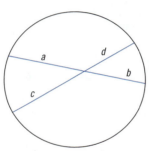

FIGURE 6.21 ■ Intersecting Chords

In Book III, Proposition 3, Euclid gives the relationship between a chord and a diameter, as discussed in Statement 69 and illustrated in Figure 6.23.

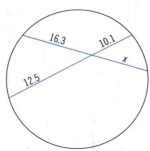

FIGURE 6.22 ■ Circle with Intersecting Chords

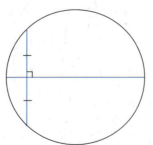

FIGURE 6.23 ■ Intersecting Chord and Diameter

69 INTERSECTING CHORD AND DIAMETER

If the diameter of a circle bisects a chord, it is perpendicular to that chord. Conversely, if the diameter of a circle is perpendicular to a chord, it bisects the chord. ●

Further, from Euclid, Book III, Proposition 21, you have Statement 70, which is illustrated in Figure 6.24.

70 ANGLES SUBTENDED BY A CHORD

A chord subtends equal angles from any point on the circle that are on the same side of the chord. Here,

$$\theta = \phi.$$ ●

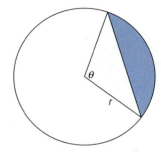

FIGURE 6.24 ■ Angles Subtended by a Chord

Finally, a chord subdivides a circle into two regions called *segments*, as shown in Figure 6.25 and defined in Statement 71.

71 SEGMENT OF A CIRCLE

A *segment of a circle* is a plane region bounded by a chord of a circle and the arc cut off by that chord. As with arcs and sectors, there is a *major* (larger) segment and a *minor* (smaller) segment. (We say "segment *of a circle*" to avoid confusion with *line* segment.) ●

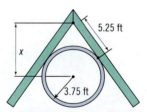

FIGURE 6.25 ■ Segment of a Circle

■ EXERCISES ● LINES INTERSECTING A CIRCLE

1. Draw a circle. From memory, draw and label the following:
   a. A chord
   b. A tangent
   c. A secant
   d. A segment

2. If $b = 3.84$, $c = 6.83$, and $d = 5.28$, find the length of segment a in Figure 6.21.

3. If $b = 186$, $c = 163$, and $d = 158$, find the length of segment a in Figure 6.21.

4. Figure 6.26 shows a circular window under the eaves of a roof. Find the distance x.

5. Figure 6.27 shows a round window in a dormer the shape of an equilateral triangle. Find the radius r of the window.

6. Figure 6.28 shows a circular design, 2.04 m in diameter, located in the angle of a roof. Find distances AB, AD, and AC.

7. Figure 6.29 shows some of the internal framing in a rose window. Find AB, given that $BC = 7.38$ ft, $DB = 4.77$ ft, and $BE = 5.45$ ft.

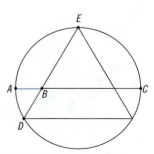

FIGURE 6.26 ■ Design for a Circular Window

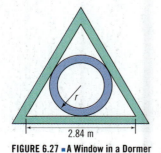

FIGURE 6.27 ■ A Window in a Dormer

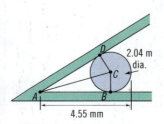

FIGURE 6.28 ■ A Circular Design

FIGURE 6.29 ■ A Rose Window

THE ROUND FORMAT IN ART

We've already discussed the square and rectangular formats in art. Now we'll discuss the round format. A *tondo* is simply a round painting. It is also called a *desco da parto*, or painting on a platter. In Chapter 5, "Polygons, Tilings, and Sacred Geometry," you saw the importance of the center of a square painting, where the four axes of symmetry intersect. The center is even more important in a circular painting that has an infinite number of axes of symmetry, both for mirror symmetry and for rotation, and where all points on the circumference are equidistant from that center.

Another characteristic of the circular form is *detachment*. A frame is used to help define a picture as a closed entity, separate from its surroundings, and make it a world of its own. According to Rudolph Arnheim, a round frame does this best because it strengthens the center, making the picture more self-referential, while a rectangular frame does it less well because its sides strengthen the reference to the vertical and horizontal of the outside world.[3] This detachment makes the tondo a natural choice for religious pictures. Its separation from the secular and vulgar—the worldly and common—makes it the all-time favorite format for the Madonna. When describing the picture of the Madonna presented in Figure 6.30, Jacob Burckhardt wrote, "it contains . . . the entire philosophy of the round picture so clearly that any unhampered look can realize what this most difficult and most beautiful of formats . . . means for representation."

We saw that the axes of a square are often used to divide a picture into zones. Although a tondo has no axes, it too can be divided into zones. In Figure 6.31, the vertical tree divides the round picture into two zones of *good* and *evil*. On the left, the tree of life bears a crucifix and Mary feeds the righteous; an angel stands to her left. On the right, the tree bears a skull. The serpent picks fruit from the

FIGURE 6.30 ■ *Madonna della Sedia* (Seated Madonna), Raphael, c. 1516

FIGURE 6.31 ■ *Adam and Eve,* Museo dell'Opera del Duomo, Florence, Italy, Fifteenth Century

tree, which is the Tree of the Knowledge of Good and Evil, and gives it to Eve to feed to the wicked. A skeleton stands on the right, opposite the angel. Adam has fallen to the ground behind the tree. Four small circles are interwoven around this large circle to form a design called a *quincunx*. (We'll see more examples of this design in Chapter 7.)

A *roundel* is a circular relief sculpture. Like circular paintings, the roundel became popular in the fifteenth century. Like the tondo, it was a favorite shape for religious figures, especially the Madonna, and for celestial and classical themes.

THE DELLA ROBBIA FAMILY

Della Robbia was an important name in Renaissance terra cotta sculpture. Luca Della Robbia (1399?–1482) founded a family studio in Florence, which mainly produced works in glazed terra cotta. Terra cotta (literally "baked earth" in Italian) is a kiln-fired reddish clay. In its unglazed form, it is used to make bricks. A finer grade is often glazed and used for sculpture and pottery.

Three of the family's best-known works are shown in Figures 6.32 through 6.34. Luca Della Robbia's most important sculptures in terra cotta are the roundels of Apostles in the Pazzi Chapel, created around 1442. One of these is shown in Figure 6.32. Luca's nephew Andrea (1435–1525) took over the workshop after his uncle's death. Andrea's best-known works are the charming roundels of foundlings in swaddling clothes on the façade of the *Ospedale degli Innocenti* in Florence. See Figure 6.33. Copies of these roundels may be seen on the Children's Hospital, Boston.

Andrea's son Giovanni (1469–1529), in turn, took charge of the studio upon the death of his father. Giovanni's younger brother Girolamo (1488–1566) was trained in Andrea's studio and eventually moved to France.

Hexagonal frames, such as those used for the scene by Luca Della Robbia shown in Figure 6.34, were also popular in the Middle Ages and the Renaissance. This is an *allegory to music*, one of a series depicting the seven liberal arts that we'll discuss in Appendix E. Much of our discussion about the round frame applies to the hexagonal shape as well.

FIGURE 6.33 ■ *The Foundling,* Andrea Della Robbia, c. 1487

FIGURE 6.32 ■ *St. James the Evangelist,* Luca Della Robbia

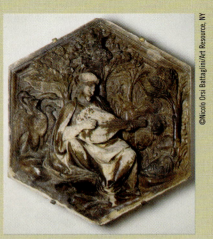

FIGURE 6.34 ■ *Orpheus Playing to the Animals,* Luca Della Robbia, c. 1438

The Italian Plaquette[4]

Tondi and roundels are examples of the loosening connection between artworks and their locations. More works of art became portable objects made for the art market, and the ultimate example of that is the *plaquette*. Plaquettes were small, single-sided bronze reliefs. Most were made from around 1450 to 1550, often in Rome, Florence, and Padua. Many were rectangular or oblong, but most were round, usually less than 6 inches in diameter. Some themes put forth on plaquettes were religious, but many were drawn from mythology, as is the figure of Mars by Moderno shown in Figure 6.35. With many plaquettes, the aim was to portray principles of good conduct: courage, rectitude, constancy, decency, restraint, and virtue. Moderno was perhaps the most prominent and prolific sculptor of plaquettes. Riccio (1470?–1532) was also active in making plaquettes.

One notable type of plaquette was the Renaissance Portrait medal inspired by antique Roman coins and purely commemorative in nature.[5] Its virtual inventor and greatest practitioner was Pisanello (Antonio Pisano, active c. 1415–1455). The typical design of these medals included a profile portrait on one side and an allegorical allusion or an emblem intended to portray some characteristics of the person portrayed on the reverse side. An inscription usually appeared around the rim.

These medals also often bore the family *impressa*, a sort of trademark, logo, or insignia. For example, the reverse side of the medal of Marsilio Ficino, the great Florentine Neoplatonic philosopher, bore only the single word *Plato*. Rulers and princes and other wealthy citizens were not the only personalities portrayed in portrait medals. Scholars, humanists, and even teachers were commemorated. Pisanello made the medal shown in Figure 6.36, which portrays Vittorina da Feltre, who had a school in Mantua. Its reverse shows a pelican feeding its young by pecking at its own breast. The encircling inscription reads *Pater Mathematicus et omnis Humanitatis*, which, loosely translated, means "Father of Mathematics and all Humanities."

Michael Levey writes of the Renaissance portrait medal, "At once a new artifact but with antique echoes, personal, naturalistic and yet allusive, easily portable yet particularly durable, the portrait medal is a perfect symbol of Renaissance endeavor and achievement."[6] Here, a few inches of metal held the promise of eternity. It is a perfect example of the Egyptian definition of the word *sculptor* as "one who keeps alive."

©2006 Board of Trustees, National Gallery of Art, Washington, D.C.

FIGURE 6.35 ■ *Mars Surrounded by Trophies,* Moderno

FIGURE 6.36 ■ Medal of Vittorino da Feltre, Pisanello

THE CIRCLE AS A SYMBOL IN ART

"My heart leaps up when I behold a rainbow in the sky"...[7]

WILLIAM WORDSWORTH

As we did with other geometric figures, let's journey from the mathematical to the symbolic. The circle is often taken as a *symbol of unity*, because all of the regular polygons are embraced by the circle. For instance, all the inhabitants of Edwin A. Abbot's *Flatland* were two dimensional.[8] Soldiers and lower classes of workmen were isosceles triangles; the middle class were equilateral triangles; professionals were squares; gentlemen were pentagons; nobility were hexagons, rising to the honorable title of *polygonal*. Priests were *circles*. Even in this fantasy, the circles were ranked higher than the polygons. The circle was considered to be the supreme shape.

The circle can also be a *symbol of democracy*. The circle dance portrayed in Figure 6.37, like Arthur's Round Table or a council circle, is also democratic, where no one has a more prominent position than anyone else. Another well-known

FIGURE 6.37 ■ *Circle Dance*, Goya

depiction of a circle dance is Poussin's *Ballo della Vita Humana*. Erwin Panofsky describes that painting as "a kind of humanized Wheel of Fortune . . . where Poverty joins hands with Labour . . . Wealth . . . Luxury, [and] dance to the lyre of Time."[9]

The Mandala[10, 11]

"The contemplation of a mandala is meant to bring an inner peace,
a feeling that life has again found its meaning and order."[12]

M. L. VON FRANZ

FIGURE 6.38 ■ Wheel of Dharma

The circle symbolizes *infinity*. That property of *endlessness* has prompted its extensive use as a symbol in religion. For example, the *Wheel of Dharma* symbolizes Buddha's teaching as it continues to spread endlessly. The eight spokes you see in Figure 6.38 represent the *Eightfold Path of Buddha*.

Perhaps the most beautiful examples of the circle in art can be seen in the Indian or Tibetan mandala. See Figure 6.39. In Sanskrit, *mandala* literally means "circle and center"[13] or Holy Circle, and it is essentially a vehicle for concentrating the mind. In their book *The Secret of the Golden Flower*,[14] Carl Jung and Richard C. Wilhelm explore the therapeutic value of this function of the mandala.[15] Mandalas were mentioned briefly in Chapter 5, and we will see the mandala again in Chapter 8.

FIGURE 6.39 ■ Mandala of Thanka, Tibetan

The Sun and the Rainbow

Throughout the ages, the circle has been used to symbolize the sun, sometimes radiating rays of fire. Figure 6.40 is an ancient Aztec depiction of the sun and its rays, among other symbols. The circle has further heavenly associations in the rainbow, which appears to touch both heaven and earth at the same time. Just as with other halos around the sun and the moon, rainbows are full circles. From our earthly vantage point, we see only an arc of that circle, and the earth blocks the remainder. The earliest reference to the rainbow in the Old Testament comes after the great flood:

> "I set my bow in the cloud, and it shall be a sign
> of the covenant between me and the earth."[16]

FIGURE 6.40 ■ Aztec Sunstone

Iris was the Greek goddess of the rainbow, a messenger of the gods, like Hermes.[17] She descended to earth on a rainbow, which touched both realms, representing a communication between the heavenly and the earthly. She appears often in Virgil's *Aeneid*. The following is one of many examples: "The maiden Iris hurried on her way, along her rainbow with a thousand colors. . . . "[18] In Biblical references, the rainbow was often used as the Lord's throne and in scenes of the Last Judgment. In Figure 6.41, the rainbow is shown as an oval surrounding the Lord on His throne. The connection to the rainbow probably came from the Old Testament, in Ezekiel 1:26:

> "Like the appearance of the bow that is in the cloud on the day of rain,
> so was the appearance of the brightness round him."

In the New Testament we read,

> ". . . and round the throne was a rainbow. . . ."[19]

When rainbows appear in landscape paintings, they tend to give a spiritual feeling to the scene by connecting the earth with the heavens. The landscapes of George Inness, for example, often have a sense of divine presence. Rainbows had a spiritual significance for Inness, and he sometimes included them in his paintings, as in Figure 6.42.[20] A double rainbow appears in John Constable's painting *Hampstead Heath with Rainbow*. Constable was interested in rainbows

FIGURE 6.41■ *Last Judgment* (Scrovegni Chapel), Giotto, 1305

FIGURE 6.42 ■ *The Rainbow*, George Inness, c. 1878
Photography © Robert Wallace. Reproduced by permission of the Indianapolis Museum of Art. Gift of George E. Hume.

his entire life, and he tried to capture their transient effects. They later became, for him, a metaphysical and emblematic symbol. Around 1830, he became interested in the science of rainbows and made geometric diagrams describing them. Constable's *Salisbury Cathedral*, J. M. W. Turner's *Buttermere Lake*, and Jean-François Millet's *Spring* are also notable paintings that include rainbows.

The Ring

The ring has long been a symbol for eternity, union, and delegated authority, and it has been deemed to have magical powers. As a symbol for eternity, the ring is, of course, used for betrothal and marriage. Betrothal rings were an old Roman custom; wedding rings came into use later. A bishop's ring signifies his union with the Church,[21] and a nun's ring signifies her marriage to Christ. Figure 6.43 shows St. Catherine receiving a ring.

Related to the ring by shape and meaning, the serpent devouring its own tail, as in Figure 6.44, also represents eternity, infinity, and the cyclic all-devouring nature of the universe. It is also the alchemical symbol for chemical change. It is called the *ouroboros* (or *uroboros*), Greek for "tail-devourer."

The ring as a symbol of authority or status comes from the use of the *signet ring*, once used to make an impression or sign in sealing wax, indicating authenticity. Moreover, it is a symbol of *designated* or *delegated* authority, because a ring is easily passed to someone else.[22]

In Rome, wearing rings of various metals was strictly regulated. Citizens wore rings made of iron, and slaves were forbidden to wear rings at all. Church rings showed the ecclesiastical office of the wearer. The Papal ring worn by a particular pope—or *Fisherman's Ring*, bearing the image of St. Peter fishing—is broken after a pope's death. Romans wore rings dedicated to the goddess Salus (Hygeia) and engraved with a pentagram and a coiled snake to ensure good fortune. In the Middle Ages, rings made of nails from coffins or church doors were

FIGURE 6.43 ■ *The Mystic Marriage of St. Catherine of Alexandria*, Giovanni Francesco Guercino

FIGURE 6.44 ■ Serpent Feeding on Its Own Tail

popular talismans for curing cramps and other disorders. Rings have often been thought to be endowed with magical powers. Egyptians wore rings of jasper or bloodstone to ensure success in battle or other struggles. The Koran says Solomon's magic ring gave him power over his enemies and transported him to a celestial sphere where he could rest from the cares of state. The ring of power plays a central part in J. R. R. Tolkien's *Lord of the Rings:*

> "Three rings for the Elven-kings under the sky,
> Seven for the Dwarf-lords in their halls of stone,
> Nine for Mortal Men doomed to die,
> One for the Dark Lord on his dark throne
> In the Land of Mordor where the Shadows lie.
>
> One Ring to rule them all, One Ring to find them,
> One Ring to bring them all, and in the darkness bind them
> In the Land of Mordor where the Shadows lie."

The Wheel

In the form of a wheel, the circle becomes the symbol of *mobility*. Consider Figure 6.45—a page from the illuminated manuscript of the Winchester Bible—which illustrates a strange passage from the Book of Ezekiel in the Old Testament. In our day, the passage is popular with flying saucer enthusiasts:

"Now as I looked . . .

I saw a wheel upon the earth beside the living creatures . . .
their construction being as it were a wheel within a wheel . . .
The four wheels had rims and they had spokes,
and their rims were full of eyes round about.
And when the living creatures went, the wheels went beside them,
and when the living creatures rose from the earth, the wheels rose.
Wherever the spirit would go, they went, and the wheels rose along with them;
for the spirit of the living creatures was in the wheels."[23]

This passage was probably the basis for a well-known gospel hymn:

"Ezekiel saw the wheel, way up in the middle of the air,
Ezekiel saw the wheel, way in the middle of the air,
And the little wheel runs by faith,
and the big wheel runs by the grace of God,
Tis a wheel in a wheel, way in the middle of the air."

From "Ezekiel Saw the Wheel"

Because it can *turn*, the wheel has often been associated with chance and fortune.[24] The roulette wheel is a modern example. The wheel of fortune in

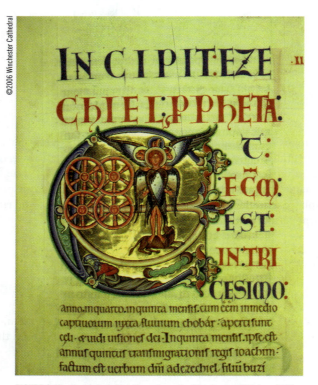

©2006 Winchester Cathedral

FIGURE 6.45 ■ Ezekiel's Initial (Page from the Winchester Bible), c. 1165

FIGURE 6.46 ■ Wheel of Fortune from the *Carmina Burana* Manuscript

Figure 6.46 (from the *Carmina Burana*, a collection of poems and songs, most written in the early thirteenth century) contains these words:

Regnabo,   Regno,   Regnavi,   Sum sine regno.
I shall reign,   *I reign,*   *I reigned,*   *I don't reign.*

Fortuna, or *Lady Luck*, is a common theme in art. You've seen her portrayed with a wheel, but she is frequently shown standing on a sphere or a soap bubble, indicating chance or uncertainty. She may be shown blindfolded and holding a cornucopia, symbolizing the favors she may capriciously hand out. Dice are sometimes present, and billowing drapery is often used to indicate the variable winds of chance.

> The symbolism of the sphere will be discussed in more depth in Chapter 11, "The Sphere and Celestial Themes in Art and Architecture."

SUMMARY

In our quest to explore the intersection of the fields of geometry, art, and architecture, we examined the geometric figure that is one of the easiest to construct and perhaps the most significant. Building on our study of the regular polygons in an earlier chapter, we kept increasing the number of sides, examining ever more complex shapes until, with an infinite number of sides, we arrived at the simplest figure. After reviewing some circle geometry and building on our earlier work with the rectangular and square formats, we discussed the round format and how it is used for painting (the tondo) and in sculpture (the roundel and the portrait medal).

We then explored the circle and circular arcs in art. We saw that the circle has a diverse set of associations. In different places and at different times, the circle has represented eternity, unity, democracy, infinity, perfection, mobility, chance, punishment, union, and authority. It was a sun symbol and appeared in art as the rainbow, the halo, the ring, the wheel, and the mandala (a vehicle for meditation and prayer).

Our exploration of the circle has not ended. In the next chapter, we'll see the role of the circle and circular arcs in Gothic tracery (arches, windows, rosettes, and so forth) and become familiar with the *vesica* or *mandorla*, made from circular arcs. After that, we'll combine the circle with the square for the highly symbolic concept of *squaring the circle*.

EXERCISES AND PROJECTS

1. Define or describe the following terms:

   mandala   tondo
   roundel   plaquette

2. Measure the diameters of an assortment of circular items (soup cans, jar lids, etc.). Then, use a strip of paper to measure the circumference of each item. Divide each circumference by the corresponding diameter. Write a paragraph or two about your findings.

3. Use a computer drafting program to draw a circle. On-screen, display the diameter of the circle, the circumference of the circle, and the ratio of circumference C to diameter d. Then *drag* the circle to different sizes, noting the ratio of C/d. What do you observe?

4. Use a computer drafting program to draw a circle one unit in radius (called a *unit circle*). Draw two radii. Display on-screen (a) the length of the arc that is intercepted by the two radii and (b) the angle between the radii; show both in radians. Then drag one radius so that the central angle and the arc change in size. What do you observe?

The following paper-folding projects are from Olson's *Mathematics Through Paper Folding*.

5. Fold the diameter of a circle by folding a circle onto itself.

6. Find the center of a circle by folding two intersecting diameters.

7. Fold a chord, and then fold a diameter perpendicular to that chord.

8. By folding two chords and the perpendicular bisector of each, find the center of an incomplete circle that contains the center of the complete circle.

9. Fold the bisector of a central angle.

10. Fold an angle inscribed in a semicircle. Show that it is a right angle.

Complete the constructions in Exercises 11 through 14 using The Geometer's Sketchpad or another CAD program, or, preferably, complete the constructions by hand using straightedge and compass. Try to figure out how to produce them before reading the instructions given in the program.

11. Construct the center of a circle, given an arc of the circle. (First, draw two chords in the arc. Then intersect the perpendicular bisectors of these chords at the center of the arc and hence of the circle.)

12. Construct a circle through three given points. (First, connect the points with straight lines, which are then chords of the circle. Draw perpendicular bisectors of these chords to find the center of the circle. From this center, draw a circle through the three points, using the distance from the center to any point as the radius.)

13. Construct a circle circumscribed about a given triangle. (Consider the vertices of the triangle as three given points on the circle, and proceed as in Exercise 12.)

14. Construct a circle inscribed inside a given triangle. (The bisectors of the angles of the triangle will meet at the center of the inscribed circle. From this center, draw a perpendicular to any side of the triangle. The length of this perpendicular will be the radius of the inscribed circle).

15. *The Butterfly.*[25] Use CAD software to perform the following steps: Draw a chord *PQ* in a circle, and find its midpoint *M*. (See Figure 6.47.) Draw any other two chords *AB* and *CD* through *M*. Draw chords *AD* and *BC*, cutting the original chord *PQ* at *x* and *y*. Demonstrate that *M* bisects *xy*. Drag *P*, *Q*, *A*, or *C* to see if *M* continues to bisect *xy*.

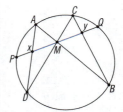

FIGURE 6.47 ■ The Butterfly

16. Find some round paintings in an art history book. After examining the paintings, answer the following questions:
    a. Does it seem true that the round format was used mostly for sacred themes?
    b. Can you identify a microtheme (this concept was introduced in Chapter 5) near the center of a round painting or sculpture? If so, what does it signify?
    c. Can you find examples of instances where an artist included vertical and/or horizontal elements to orient the work to the outside world?
    d. Can you identify any *zones* (again, recall our discussion in Chapter 5) in a round work of art? If so, what is the significance of each zone?

17. Make a design based on the circle. Use it to make a graphic design, a pillow quilt, a screensaver, an inlaid design using cutout pieces, a tee shirt, or for some use of your own.

18. One piece of software for exploring the circle is "Geometry of the Circle with The Geometer's Sketchpad," by James King. If the software is available to you, try it, and write a short paper on your findings.

19. Report on any book listed in the sources at the end of this chapter.

20. Write a paper or make a class presentation on some topic relating to this chapter. Come up with your own topic, using these suggestions to jog your imagination:

    ■ The history of pi
    ■ The history of methods that have been used to compute pi
    ■ The importance of the circle to the Lakota Sioux tribe of North America
    ■ The pervasive use of the *wheel of fortune* theme in medieval art
    ■ Isidore of Seville's method of visualizing various concepts as wheels or *rotae*
    ■ The *vanitas* art motif
    ■ The rainbow in art, literature, or the Old Testament
    ■ The Italian plaquette

- The Renaissance portrait medal
- Traditional quilt patterns using circles or parts of circles, such as *True Lover's Buggy Wheel* or *Wheel of Chance* (This is pattern #1662 in Rhemel's *The Quilt I.D. Book.* Rhemel devotes an entire chapter to circular designs.)
- The *coin* as an example of the round format
- How the circle sometimes symbolized *punishment*, as illustrated by the *Catherine Wheel* and the *millstone* and by the Greek myth of *Ixion*

21. Make a pizza incorporating as many circular geometric features as you can. Serve it to your class.

Mathematical Challenges

22. Prove that any angle inscribed in a semicircle is a right angle. *Hint:* In Figure 6.48, prove that $a^2 + b^2 = (2r)^2$.

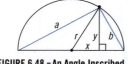

FIGURE 6.48 ■ **An Angle Inscribed in a Semicircle**

23. *Dürer's Approximate Method for Finding the Circumference of a Circle.*[26] Complete the construction in Figure 6.49. Evaluate its accuracy, and explain why it works. Draw three circles, centered on the same line and tangent to each other. Subdivide an end circle into seven equal parts. (Chapter 5 explained how to do this.) Add one of these parts to the end of the line to get point *B*. The length of *AB* is the approximate circumference of one of the circles.

24. *Finding a Square Root, Geometrically.* To geometrically find the square root of some number *n*, follow the steps shown in Figure 6.50. Complete the construction, and explain why it works. Lay out *n* and 1, end-to-end, to form line *AB*. Draw a semicircle on *AB*. Erect a perpendicular to *AB* at *C*, and extend to the semicircle at *D*. *CD* equals the square root of *n*.

25. Figure 6.51 shows what is called an *equilateral pointed arch*, where the radius of each curved portion equals the width of the span. Find the length of each curved portion.

26. *The area under the arch is equal to twice the area of sector ABDC minus the area of triangle ABC* (see Figure 6.51). Make paper cut-outs and use them to verify this statement.

27. Using the statement in Exercise 26, find the area under the given arch (see Figure 6.51).

Area of a Segment of a Circle. The formula for the area of a segment is found in Statement 72. Use Statement 72 to solve the following four exercises. Put your calculator in RADIAN mode, and use the SIN key.

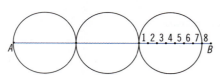

FIGURE 6.49 ■ **Dürer's Method for Finding the Circumference of a Circle**

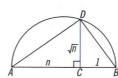

FIGURE 6.50 ■ **Constructing a Square Root**

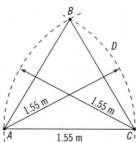

FIGURE 6.51 ■ **An Equilateral Pointed Arch**

72 AREA OF A SEGMENT OF A CIRCLE

$$\text{Area} = \frac{1}{2}r^2(\theta - \sin\theta),$$

where θ is in radians.

28. A chord in a 24.8-cm-diameter circle subtends a central angle of 1.22 radians. Find the area of the segment cut off by the chord.

29. A chord in a 128-cm-diameter circle subtends a central angle of 1.55 radians. Find the area of the segment cut off by the chord.

30. A chord in a 36.9-in.-diameter circle subtends a central angle of 2.23 radians. Find the area of the segment cut off by the chord.

31. Find the area of the church window shown in Figure 6.52. Take the diameter of the window as 9.6 ft and the length of the flat section as 7.4 ft. *Hint:* You will need to use trigonometry to find the central angle from the given information.

32. Prove that two tangents drawn to a circle from a point outside the circle are equal in length, and that they make equal angles with a line drawn from the circle's center to the external point. *Hint:* Revisit Statement 66 in the earlier section of this chapter titled "Lines Intersecting a Circle."

33. Figure 6.53 shows a square window surmounted by a circular arch. Find the following:
    a. The radius r of the arch
    b. The length s of the curved underside of the arch
    c. The total open area of the window, including the area of the square and the area under the arch

34. Use the following method to find the area under the pointed arch in Figure 6.51: (1) Find the area of the equilateral triangle. (2) Find the area of the segment bounded by one branch of the arch and one side of the triangle. (3) Add the triangle area to the area of the two segments.

FIGURE 6.52 ■ A Church Window in Bethel, Vermont

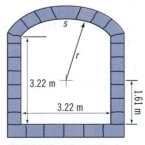

3.22 m

3.22 m

1.61 m

FIGURE 6.53 ■ A Square Window Topped by a Circular Arch

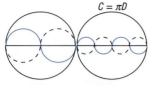

$C = \pi D$

FIGURE 6.54 ■ Yin-Yang Symbol

35. Figure 6.54 shows some elements of the Yin-Yang symbol (depicted completely in Appendix E, Figure E.4). Note that the sum of circumferences of two inner circles equals that of the outer circle. If D is the diameter of the outer circle, then $\pi D/2 + \pi D/2 = \pi D$. If you continue subdividing the diameter into smaller and smaller circles, the sum of their circumferences will still equal that of the outer circle. Eventually, the small circles will be indistinguishable from the diameter, giving the paradox that the diameter of a circle appears to become equal to its circumference. Repeat this construction and write a short paper discussing this paradox.

SOURCES

Aaboe, Asger. *Episodes from the Early History of Mathematics*. New York: Random House, 1964.

Arnheim, Rudolph. *The Power of the Center*. Berkeley: University of California Press, 1988.

Calter, Paul. *Technical Mathematics*. New York: Wiley, 2007.

Coxeter, H. S. M. *Introduction to Geometry*, 2nd ed. New York: Wiley, 1989.

Coxeter, H. S. M., et al. *Geometry Revisited*. Washington, D.C.: Mathematical Association of America, 1967.

Jung, Carl G., et al. *Man and His Symbols*. New York: Dell, 1964.

Levey, Michael. *Early Renaissance*. New York: Penguin, 1967.

NOTES

1. Dürer, *The Painter's Manual*, p. 136.
2. From *Geometria Deutsch*, in Roriczer, p. 114.
3. Arnheim, p. 56.
4. Pope-Hennessy, p. 192.
5. Levey, pp. 73–77.
6. Levey, p. 74.
7. *The World in Literature*, Vol. 2, p. 392.
8. Abbot, p. 7.
9. Panofsky, p. 93.
10. Argüelles, José and Miriam, *Mandala*.
11. Campbell, Joseph, *The Power of Myth*.
12. *Man and His Symbols*, p. 230.
13. Arguelles, p. 13.
14. New York: Harcourt Brace, 1931.
15. Arguelles, p. 13.
16. Genesis 9:13.
17. Gomm, p. 120.
18. *Aeneid*, V, 609.

19. Revelation 4:3.
20. Janson, p. 715.
21. Ferguson, p. 153, p. 178.
22. Hall, p. 264.
23. Ezekiel 1:15–22.
24. Campbell, p. 118.
25. From Coxeter, *Geometry Revisited*, p. 45.
26. From *Geometria Deutsch*, in Dürer's *The Painter's Manual*, p. 419.

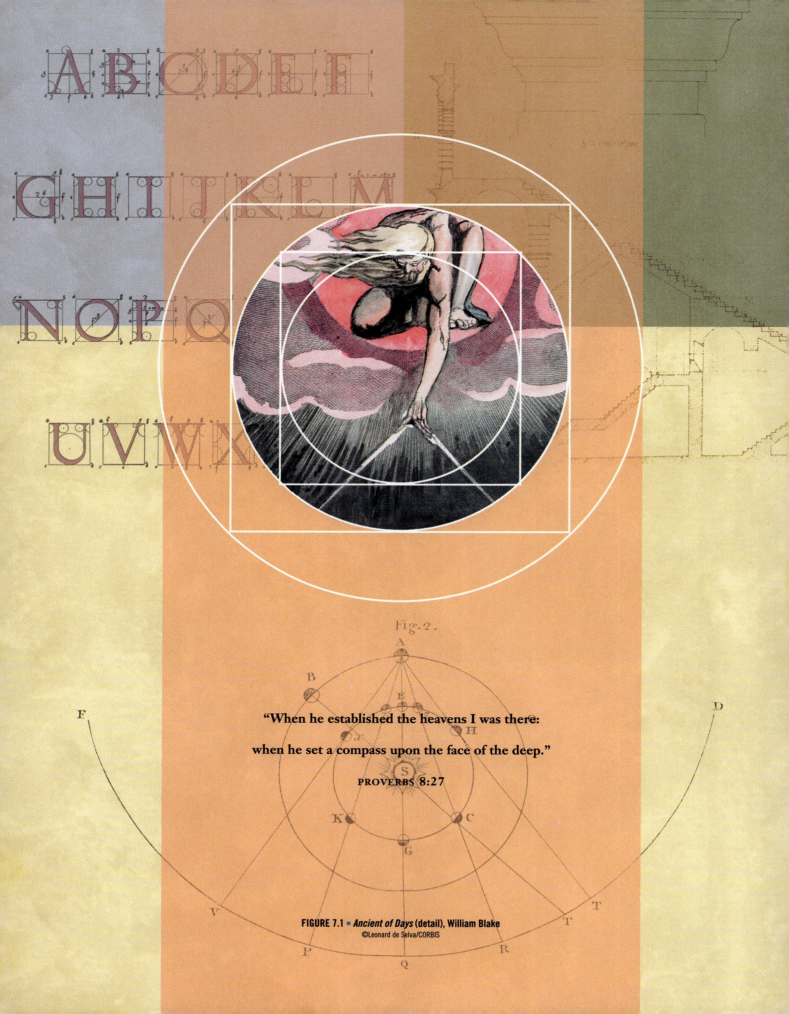

"When he established the heavens I was there:

when he set a compass upon the face of the deep."

PROVERBS 8:27

FIGURE 7.1 ■ *Ancient of Days* (detail), William Blake
©Leonard de Selva/CORBIS

7

Circular Designs in Architecture

uilding on our work with the circle in the preceding chapter, let's explore the geometry of intersecting circles and circles inscribed in and circumscribed about polygons. In this chapter, we'll become involved in the fascinating world of *Gothic tracery*, the ornamental openwork in Gothic windows. We'll see how the circle was prominent in such tracery, especially in the rose window, and also in circular designs such as inlaid designs and rosettes, the pointed arch, and the vault. We'll construct what will appear to be incredibly complex Gothic designs. (Given the great variety of these designs, we'll limit ourselves to some of the simplest.) Finally, we'll examine figures with circular boundaries, such as the *vesica*, and we'll study their use as art motifs.

Speaking of construction, circles are often symbolized by the instruments used to draw them, and compasses or dividers appear frequently in art. They are an attribute of astronomy, one of the seven liberal arts; of Euclid, the personification of astronomy; and of Urania, the muse of astronomy. Architects such as

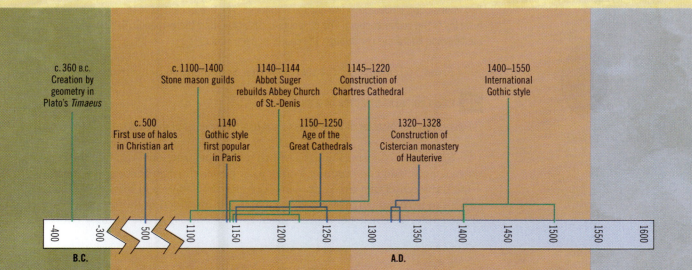

c. 360 B.C. Creation by geometry in Plato's *Timaeus*	c. 1100–1400 Stone mason guilds	1140–1144 Abbot Suger rebuilds Abbey Church of St.-Denis	1145–1220 Construction of Chartres Cathedral		1400–1550 International Gothic style	
	c. 500 First use of halos in Christian art	1140 Gothic style first popular in Paris	1150–1250 Age of the Great Cathedrals	1320–1328 Construction of Cistercian monastery of Hauterive		

-400 -300 500 1100 1150 1200 1250 1300 1350 1400 1450 1500 1550 1600

B.C.   A.D.

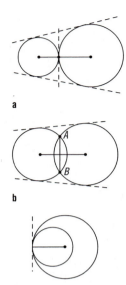

a

b

c

FIGURE 7.2 ■ **Intersecting Circles**

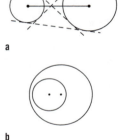

a

b

c

FIGURE 7.3 ■ **Nonintersecting Circles**

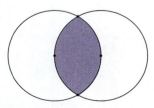

FIGURE 7.4 ■ **The Vesica**

Brunelleschi and Palladio and navigators such as Columbus are often portrayed holding compasses. In Figure 7.1, William Blake shows the Creator reaching down from the heavens with a compass in the act of creation. (For another representation of such construction, see *God the Geometer*, Figure 13.47.)

MORE CIRCLE GEOMETRY

In the last chapter, we discussed tangents, secants, and chords for a single circle. In this chapter, we'll extend those ideas to pairs of circles.

Pairs of Circles

Two circles in a plane can intersect at one or two points, as in Figure 7.2, or not intersect at all, as in Figure 7.3. The line connecting the centers of two circles is called the *line of centers*, shown as a solid line in both Figures 7.2 and 7.3. When one circle is inside another and both have the same center, as in Figure 7.3c, they are said to be *concentric* circles; otherwise, they are called *eccentric* circles, as in Figures 7.2c and 7.3b. Lines tangent to both circles are called *common tangents*; these are shown as broken lines in the figures. Pairs of circles can have zero, one, two, three, or four common tangents.

Circles that touch at just *one* point (Figures 7.2a and 7.2c) are said to be *tangent* to one another.

74 TANGENT CIRCLES

Two tangent circles have three common tangents. The line of centers passes through the point of contact and is perpendicular to the common tangent at that point.　●

If one circle is inside the other, the three common tangents coincide.

For circles that touch at *two* points (Figure 7.2b), the line joining the points of contact is called the *common chord*. Figure 7.4 illustrates a special set of intersecting circles, one in which the circles are equal in diameter. The portion common to both circles is called a *vesica pisces* or simply a *vesica*, or a *mandorla*. We will have much more to say about the vesica later in this chapter. The following statements apply to circles with common chords.

75 CIRCLES INTERSECTING AT TWO POINTS

Two intersecting circles have two common tangents. The line of centers is the perpendicular bisector of the common chord. The converse is also true, that the perpendicular bisector of the common chord passes through the center of each circle.　●

From Statement 75, it is also clear that Statement 76 holds true.

76 PERPENDICULAR BISECTOR OF A CHORD

The perpendicular bisector of a chord passes through the center of the circle.　●

Statement 76 forms the basis of two useful constructions—finding the center of a circle and constructing a circle that passes through three given points. To find the center of a circle, draw any two chords, as shown in Figure 7.5. Construct the perpendicular bisector of each. Their intersection locates the center of the circle. To construct a circle given three points *A*, *B*, and *C*, draw chords

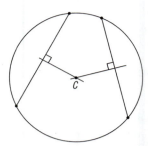

FIGURE 7.5 ■ Finding the Center of a Circle

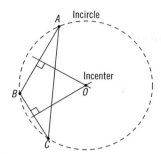

FIGURE 7.6 ■ Constructing a Circle Given Three Points

AB and BC, as shown in Figure 7.6. Construct the perpendicular bisector of each. Their intersection locates the center of the circle. To confirm this, draw chord AC. Its perpendicular bisector should intersect at the same point O, the center of the circle.

Inscribed Circles

In this section, we'll further explore the relationship between inscribed circles and polygons. A circle is said to be *inscribed in* a polygon if that circle is tangent to every side of the polygon. The polygon, on the other hand, is said to be *circumscribed about* the circle. The inscribed circle is called the *incircle*, and its center is called the *incenter*.

> Triangles and circles inscribed in and circumscribed about each other were discussed in Chapter 3, "The Triangle."

The incenter of a circle inscribed in a regular polygon is simply the center of the regular polygon. To find the incenter of a circle that is inscribed in a triangle, use Statement 66, which states, "Two tangents drawn to a circle from a point outside the circle make equal angles with a line drawn from the circle's center to the external point." From this, we have Statement 77.

77 INCENTER OF A TRIANGLE

The incenter of a circle inscribed in a triangle is at the intersection of the bisectors of the angles of the triangle. ●

A circle can be inscribed in any regular polygon, but not in every irregular polygon. A circle can, however, be inscribed in any triangle. Because a side of the polygon is a tangent to the circle, a radius of the incircle drawn to its point of contact is perpendicular to that side, as shown in Figure 7.7.

Let's find the radius r of a circle inscribed in a triangle having sides a, b, and c, as in Figure 7.8. Draw a radius from O to each point of contact between the circle and the triangle, noting that each radius is perpendicular to that side. Triangle AOB has an altitude r and a base c, so its area is $\frac{1}{2}rc$. Similarly, the areas of triangles BOC and AOC are $\frac{1}{2}ra$ and $\frac{1}{2}rb$. The area of triangle ABC is

$$\text{Area} = \frac{1}{2}ra + \frac{1}{2}rb + \frac{1}{2}rc$$

$$= \frac{1}{2}r\,(a + b + c)$$

$$= \frac{1}{2}rP,$$

where P is the perimeter of the triangle $(a + b + c)$. Solving for r gives Equation 78.

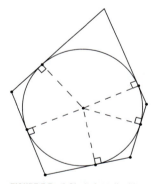

FIGURE 7.7 ■ A Circle Inscribed in a Polygon

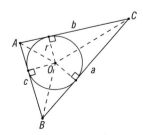

FIGURE 7.8 ■ Incircle of a Triangle

78 RADIUS OF THE INCIRCLE OF A TRIANGLE

$$r = \frac{2(\text{Area})}{P}$$

The radius of a circle inscribed in a triangle equals twice the area of the triangle divided by the perimeter of the triangle.

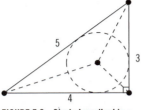

FIGURE 7.9 ■ Circle Inscribed in a 3-4-5 Right Triangle

■ **EXAMPLE:** Find the radius of the circle inscribed in the 3-4-5 right triangle shown in Figure 7.9. This incircle is represented by a dashed line in the figure.

● **SOLUTION:** The area of the triangle is

$$\text{Area} = \frac{1}{2}(4)(3) = 6 \text{ square units}$$

and

$$P = 3 + 4 + 5 = 12 \text{ units.}$$

Therefore, the radius of the incircle is

$$r = \frac{2(6)}{12} = 1 \text{ unit.}$$

■

Circumscribed Circles

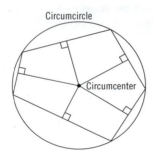

FIGURE 7.10 ■ Circle Circumscribed About a Polygon

A circle is said to be *circumscribed about* a polygon if it passes through every vertex of the polygon. This is apparent in Figure 7.10. Such a circle is called a *circumcircle*; the polygon is then said to be *inscribed in* the circle. A circle cannot be circumscribed about every polygon, but it can be circumscribed about every *regular* polygon. A circle can also be circumscribed about any triangle.

The center of a circumscribed circle is the same as the center of the regular polygon, and it is called the *circumcenter*. You can locate the circumcenter by recalling that the perpendicular bisector of a chord of a circle passes through the center of that circle. Because each side of the inscribed polygon is a chord of the circle, Statement 79 holds true.

79 CIRCUMCENTER OF A POLYGON

If a polygon has a circumcircle, its center (the circumcenter) is at the intersection of the perpendicular bisectors of the sides of the polygon.

■ **EXAMPLE:** Graphically circumscribe a circle about a triangle.

● **SOLUTION:** Construct the perpendicular bisector to each side of the triangle. Their point of intersection, the circumcenter of the triangle, is the center of the circumscribed circle. The result should look like Figure 7.11.

■

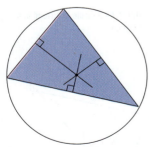

FIGURE 7.11 ■ A Circle Circumscribed About a Triangle

The problem of inscribing a regular polygon of *n* sides in a circle is no different than one of dividing the circle into *n* equal parts; you can do it easily with a protractor by dividing 360° by *n* and repeatedly measuring off this angle. However, it is more interesting and more instructive to inscribe polygons using only compass and straightedge.

You've already seen how to inscribe a triangle in a circle with a compass and straightedge. To inscribe a *square* in a circle, simply draw a diameter and

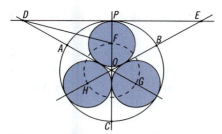

FIGURE 7.12 ■ Inscribing *n* Tangent Circles Within a Circle

construct another diameter perpendicular to the first. Connect the points of intersection of the diameters and the circle. Chapter 5, "Polygons, Tilings, and Sacred Geometry," explained how to inscribe a pentagon, hexagon, heptagon, octagon, and enneagon. To obtain a decagon, simply bisect the angles of the pentagon. For a 12-gon, bisect the angles of a hexagon, and so forth. In general, bisecting the angles of any regular polygon will give another polygon having twice the number of sides.

Another useful geometric technique for drawing Gothic traceries is to inscribe a given number of circles within a circle so that each inscribed circle is tangent to the outer circle and to its neighbors. You can use the following procedure to inscribe any number *n* of circles. Figure 7.12 illustrates the procedure for *n* = 3 circles.

> Rose windows are one prominent architectural shape that is frequently subdivided into 12 parts. (12 is one of the most important astrological numbers.)

1. Subdivide the circle into *n* equal parts using the methods of the preceding section. Draw *n* radii, *OA*, *OB*, and *OC* in your illustration.

2. Bisect the central angle between two adjacent radii *OA* and *OB*, cutting the circle at *P*.

3. Draw a tangent at *P* that intersects the extended radii at *D* and *E*, forming isosceles triangle *ODE*.

4. Inscribe a circle in triangle *ODE* using Statement 77, which says that the center of a circle inscribed in a triangle is at the intersection of the angle bisectors of the triangle. You already have the angle bisector *OP*, so bisect either angle *D* or *E* to locate *F*.

5. Locate the centers *G* and *H* of the other inscribed circles on the lines *DAO* and *EBO*, extended, at a distance *OF* from *O*.

6. Draw the inscribed circles with radius *FP*.

■ EXERCISES ● MORE CIRCLE GEOMETRY

1. Define or describe the following terms:

tangent circles	intersecting circles	concentric circles
incenter	eccentric circles	common tangent
common chord	circumcenter	inscribed circle
circumscribed circle	inscribed polygon	incircle
vesica	circumscribed polygon	line of centers

2. Explain why is it possible to circumscribe a circle about any triangle.

3. Find the radius of a circle inscribed in a right triangle with legs that are 385 cm and 422 cm.

4. Find the radius of a circle inscribed in a triangle with an area of 5.84 sq. ft and a perimeter of 13.8 ft.

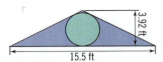

FIGURE 7.13 ■ A Round Window

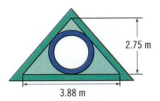

FIGURE 7.14 ■ A Round Glass Window

5. Find the radius of the round window in Figure 7.13. Assume that the triangle is isosceles.

6. Find the area of the round glass window in Figure 7.14. Assume that the frame about the round window is 9 cm wide and that the triangle is isosceles.

7. Figure 7.15 is a design on a famous church by Alberti. From the photograph, estimate the proportions of the triangle. Then do a construction to insert the circle into the triangle.

Complete the constructions in Exercises 8 through 30 using a compass and straightedge or using The Geometer's Sketchpad or other CAD program. Do not use a protractor.

8. In a given circle, inscribe a triangle that is similar to a given triangle (Euclid IV, 2).

9. About a given circle, circumscribe a triangle that is similar to a given triangle (Euclid IV, 3).

10. Inscribe a circle in a given triangle (Euclid IV, 4).

11. Circumscribe a circle about a given triangle (Euclid IV, 5).

12. Inscribe a square in a given circle (Euclid IV, 6).

13. Circumscribe a square about a given circle (Euclid IV, 7).

14. Inscribe a circle in a given square (Euclid IV, 8).

15. Circumscribe a circle about a given square (Euclid IV, 9).

16. Inscribe a regular pentagon in a given circle (Euclid IV, 11).

17. Circumscribe a regular pentagon about a given circle (Euclid IV, 12).

18. Inscribe a circle in a given regular pentagon (Euclid IV, 13).

19. Circumscribe a circle about a regular pentagon (Euclid IV, 14).

20. Inscribe a regular hexagon in a given circle (Euclid IV, 15).

21. Inscribe a circle in a regular hexagon.

22. Circumscribe a circle about a regular octagon.

23. Inscribe a regular octagon in a circle.

24. Inscribe three tangent circles in a circle.

25. Inscribe four tangent circles in a circle.

FIGURE 7.15 ■ A Design on Santa Maria Novella, Florence

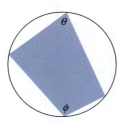

FIGURE 7.16 ■ A Quadrilateral Inscribed in a Circle

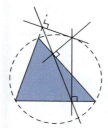

FIGURE 7.17 ■ Simpson Line

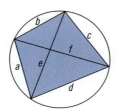

FIGURE 7.18 ■ Ptolemy's Theorem

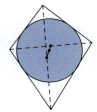

FIGURE 7.19 ■ Newton's Theorem

26. Inscribe five tangent circles in a circle.

27. Inscribe six tangent circles in a circle.

28. Is it possible to inscribe a circle in any convex polygon? Try to draw a convex polygon for which it is impossible to inscribe a circle.

29. Is it possible to circumscribe a circle about any convex polygon? Try to draw a convex polygon for which it is impossible to circumscribe a circle.

30. Inscribe 12 tangent circles in a circle.

31. Demonstrate that the sum of opposite angles of any quadrilateral inscribed in a circle, as in Figure 7.16, is equal to two right angles (Euclid III, 22).

32. *Simpson Line.* By construction—such as you see in Figure 7.17—demonstrate that the feet of the perpendiculars from the sides of a triangle to a point on the circumcircle lie on a straight line, which is called the *Simpson line.*

33. *Ptolemy's Theorem.* By construction, demonstrate that if a quadrilateral is inscribed in a circle, the sum of the products of two opposite sides is equal to the product of the diagonals. Therefore, in Figure 7.18, $ac + bd = ef$.

34. *Newton's Theorem.* By construction, such as you see in Figure 7.19, demonstrate that if a circle is inscribed in a quadrilateral, the midpoints of the diagonals and the center of the circle lie on a straight line.[1]

35. Two intersecting circles have a common chord PQ, which intersects the common tangent XY at Z. You can see a representation of this in Figure 7.20. Prove that Z is the midpoint of XY.

36. Choose points A', B', and C' somewhere on the sides of given triangle ABC. Draw a circumcircle about each small triangle, $AB'C'$, $BA'C'$, and $CA'B'$. Demonstrate by construction, such as you see in Figure 7.21, that the three circumcircles meet at a common point.

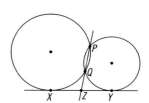

FIGURE 7.20 ■ Intersecting Circles

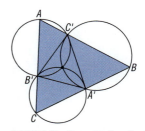

FIGURE 7.21 ■ Figure for Exercise 36

THE CIRCLE IN GOTHIC ARCHITECTURE

"But thou hast arranged all things by measure and number and weight."

THE WISDOM OF SOLOMON, BOOK XI, VERSE **20**

The importance of the circle in the Middle Ages was often symbolized by pictures of God holding a pair of compasses, as depicted in Figure 7.1. The art historian Ernst Gombrich credits the opening quote, which is a passage from the

Old Testament (Proverbs 8:27) that we'll repeat here, as the inspiration for these portrayals:[2]

> "When he established the heavens I was there:
> when he set a compass upon the face of the deep."

Creation by geometry is not a new idea. In *Timaeus*, Plato described the creation of the world out of triangles and the five Platonic solids. We'll show numerous examples of the circle in Gothic architecture, but first let's set the stage by pointing out the role of numbers and geometry during that period.

<aside>Chapter 11, "The Sphere and Celestial Themes in Art and Architecture," describes the Ptolemaic arrangement of the world as a set of concentric circles.</aside>

Numbers and Geometry in the Medieval Cathedral

According to Vincent Hopper, medieval thinkers believed numbers to be of divine origin.[3] One reason for this, according to Umberto Eco, is the triad of terms *numerus, pondus, mensura* (number, weight, and measure) in the *Book of Wisdom of Solomon*, from the Old Testament Apocrypha.[4] Because numbers were so important, incorporating them into the monuments of the church on earth was considered a necessity. Vincent Hopper further writes: "The frozen eloquence of the medieval cathedral is to a great extent the eloquence of number by which the very sum of pillars, gates, or windows was meaningful . . . the altar steps were always three or a multiple of three . . . At the consecration, the central door is sprinkled thrice with holy water . . . 12 candles are lighted, triple thanks offered to heaven, 3 solemn knockings on the door, each of the 4 arms of the cruciform plan are sprinkled thrice, as are the 7 altars . . . 12 priests carry crosses through the 4 parts of the church. The mass consists of 7 parts, or offices . . . the episcopal procession is led by 7 acolytes, indicating the 7 gifts of the holy spirit, followed by the pontiff and 7 subdeacons and 7 deacons, (7 pillars of wisdom), 12 priors (apostles) . . . The sign of the cross is made 3 times . . . ."[5] All of this shows a determination to find numerical significance in all things.

Apparently, the importance of number symbolism was matched by a dedication to geometry. Kenneth Clark points out that "to medieval man geometry was a divine activity."[6] According to Painton Cowen, churches have been built on geometric principles since early Christian times. "Geometry was the basis of all Gothic cathedrals, everything being created from basic relationships. Geometry and arithmetic were traditional studies, but the rediscovery of Euclid's *Elements* by Adelard of Bath [who was connected to the Chartres school] gave the subject new life."[7]

The ground plan of a church was usually cruciform, the baptismal font was almost always octagonal, as was the baptistery itself, and the circle was used everywhere. One obvious appearance of the circle was in the halo, as seen in Figure 7.22. Halos were used for the sun gods Mithras, Apollo, and Helios long before the Middle Ages, and they began appearing in Christian art in the fifth century or so. The halos of saints and other holy persons were usually circular; triangular halos were reserved for God the Father. Square halos were used for living persons, and hexagonal halos occasionally were used for allegorical figures.

The Greek cross within a circle (*cruciform nimbus*) was often used in Christian symbolism. In the San Giovanni ceiling shown in Figure 7.22, the seated figure is in the shape of a Greek cross within a circle. This, as well as the halo, can be taken as an example of *squaring the circle*, an idea we'll cover in detail in the next chapter. Note the stylized rainbow in this picture, and recall our discussion of the rainbow as religious symbol in Chapter 6, "The Circle." Body halos, or *aureoles*, are described later in this chapter when we discuss the vesica. After the Renaissance, the use of halos lost popularity.

FIGURE 7.22 ■ Florence Baptistery Ceiling (detail)

The Masons

The secrets of geometry and building were transmitted by the medieval guilds of builders and masons. A mason who reached the second degree of initiation received a personal seal, which he kept for life. When traveling outside his home guild, he had to be able to "prove" his seal—that is, to construct it within a circle and explain his construction when challenged.

Building techniques were carefully guarded secrets. The following resolution, for example, was adopted in 1459 at a meeting of architects and builders in Regensburg, in Bavaria in the southeast of Germany: "Item: no worker, master, polisher, craftsman; no one, no matter what he is called, unless he belongs to our trade organization, shall be taught how to build or erect structures from a ground plan."[8]

The stone mason guilds (their members were called *operative* masons) eventually died out, but many of their ideas and symbols were adopted by the Freemasons, who are active to this day. We've already discussed the Freemasons in connection with the pentagram. The square and compass of the Freemason's symbol, as shown in Figure 7.23, offer yet another suggestion of the squaring of the circle.

FIGURE 7.23 ■ Freemason's Symbols from Tombstone

The Gothic Style

The word *Gothic* pertains to the Goths, a Germanic tribe that had nothing to do with the kind of architecture we now refer to as *Gothic*. Originally, *Gothic* was a term of derision for any "barbarian" style not Greek or Roman. About 1140, the style developed and began to be popular around Paris. It was followed by the Age of the Great Cathedrals, 1150–1250. By 1250, the Gothic style had spread over most of Europe, and by 1400 an International Gothic style prevailed. By 1450 the Gothic style was shrinking, and by 1550 it was gone.[9]

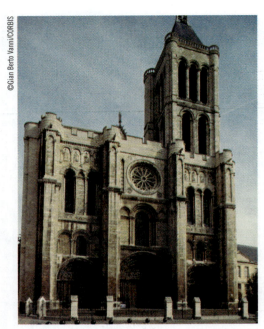

©Gian Berto Vanni/CORBIS

FIGURE 7.24 ■ Abbey Church of St.-Denis (1140–1144)

The rebuilding of the royal Abbey Church of St.-Denis, which was located outside Paris and named for the patron saint of France, pinpoints the origin of the Gothic style in architecture. A photograph of the church is shown in Figure 7.24. Anthony Janson writes of this church, "The new spirit that strikes us at St.-Denis is the emphasis on strict geometric planning and the quest for luminosity." Abbot Suger (1081–1151), who wanted to make St.-Denis the spiritual center of France, rebuilt and enlarged it. Janson paraphrases the abbot and writes, "Harmony (that is the perfect relationship among parts in terms of mathematical proportions or ratios) is the source of all beauty. . . ."[10]

A year after the construction of St.-Denis was completed, rebuilding began on Chartres Cathedral [more formally, the Cathedral of Our Lady of Chartres; in French, *Cathédrale Notre-Dame de Chartres*].[11] With Chartres, Figure 7.25, the medieval passion for numbers and geometry seemed to reach its peak. According to Cowen, "The scholars at Chartres were clearly fascinated by numbers and . . . geometry . . . as a key to understanding nature . . . their preoccupation with numbers led to a trend of almost reducing theology to geometry." Look closely at Figure 7.26 to see if you can discern any of the following details. The right-hand door of the west façade displays the seven liberal arts, personified by Pythagoras, Euclid, Aristotle, Cicero, Ptolemy, Boetius, and Priscian. The left-hand door of the west façade also has a familiar subject: the signs of the zodiac. In addition to numbers and geometry, the architectural design of Chartres was strongly influenced by astrology. That influence can be seen in the towers of the sun and moon, a sundial, an astronomical clock, a zodiac window, and the doorway shown in Figure 7.26. Three large rose windows are subdivided by the astrological number 12.

Umberto Eco adds, "The School of Chartres remained faithful to the Platonic heritage of the *Timaeus*, and developed a kind of 'Timaeic' cosmology. . . . For the School of Chartres, the work of God was order, opposite of the primeval

Chapter 11 will discuss the influence of astrology in greater detail.

©Adam Woolfitt/CORBIS

FIGURE 7.25 ■ Chartres Cathedral (1145–1220)

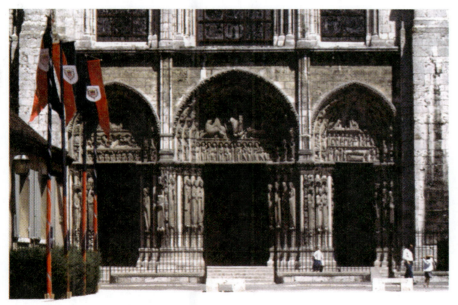

FIGURE 7.26 ■ Chartres Cathedral, West Façade

chaos."[12] Clark writes, "Chartres was the centre of a school of philosophy devoted to Plato, and in particular to his mysterious book called the *Timaeus*, from which it was thought that the whole universe could be interpreted as a form of measurable harmony. So perhaps the proportions of Chartres reflect a more complex mathematics than one is inclined to believe."[13]

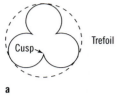

Cusp — Trefoil

a

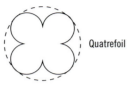

Quatrefoil

b

Cinquefoil

c

FIGURE 7.27 ■ Foliations

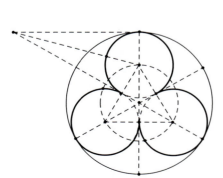

FIGURE 7.28 ■ Trefoils and Quatrefoils, Episcopal Church in Bethel, Vermont

ROUND WINDOWS

"The ultimate expression of the medieval love of geometry and of the circle is the rose window."

PAINTON COWEN

In tracery, a *foil* is a circular lobe tangent to the inner side of a circle. Figures 7.27 and 7.28 present examples of foils. The points where the lobes meet are called *cusps*. A design containing foils is called a *foliation*. The term *foliation* is also used for a leaf-like decoration. A design having more than five foils is called a *multifoil*.

To construct foliations, you subdivide a circle into a number of equal parts, as we did earlier in this chapter. To construct a trefoil window based on tangent circles and a variation of that trefoil window, first inscribe three tangent circles within the outer circle, as explained previously. Next, locate the point of tangency between each pair of circles by drawing the line of centers. Remove the portions of the circles inside the points of tangency, forming three cusps. Finally, remove the portion of each of the three inner circles between the point of tangency with the outer circle and the point of tangency with the neighboring circle, as shown in Figure 7.29.

To add more detail, you can insert three more small circles into each of the three inner circles. Each of these small circles should pass through the center of an inner circle, be tangent to that circle, and be tangent to either the outer circle or to a neighboring inner circle. From the center *A* of an inner circle, draw a circle with half the radius of the inner circle, as shown in Figure 7.30. The centers of the small circles will be on this circle. Next, draw the line of centers *AB* to locate point *C*. Similarly, the line of centers *OA* will locate *D*. Locate *E* at a distance *OA* from *O*. Finally, mark the points of intersection of the smallest circles and complete the arcs as shown in the figure.

Figure 7.31 illustrates the same window with some framing added. This produces a paisley-like design.

Windows in the Cloisters of Hauterive[14]

The Cistercian monastery of Hauterive was built between 1320 and 1328. It is located in Switzerland, about 10 km from Fribourg. According to Benno Artmann, the architect at Hauterive was interested in theoretical geometry, especially in the subdivisions of a circle given in Book IV of Euclid's *Elements*. Artmann calls the cloisters of Hauterive "a commentary to Euclid carved out of stone." He makes the following observations on these windows, which are shown

FIGURE 7.29 ■ Trefoil Window

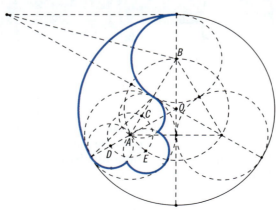

FIGURE 7.30 ■ Variation on the Trefoil Window

FIGURE 7.31 ■ The Finished Window

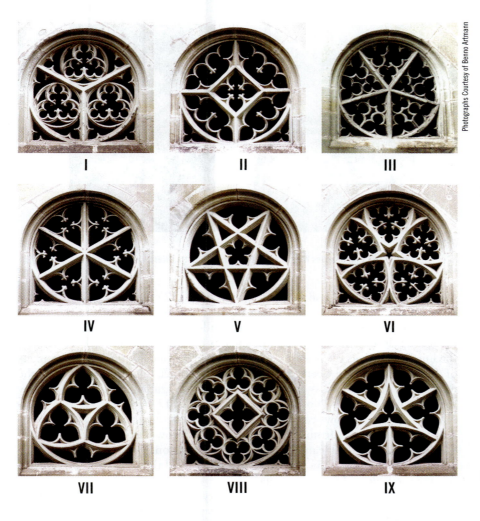

I II III

IV V VI

VII VIII IX

FIGURE 7.32 ■ Rose Windows at Hauterive

in Figure 7.32. "First observe that the architect stresses interest in the circle by selecting round arches. He approaches this topic systematically by subdividing the circle in three (window I), four (window II), five (window III), [and] six (window IV) equal parts. . . . Five-fold symmetry is repeated by the pentagram in window V and again in the beautiful and ingeniously constructed rose in window VI. The subject of triangles is taken up and iterated in window VII. Observe that window I has the sequence 3-9-27 built into it. . . . Four and eight are combined in window VIII."

Figure 7.33 illustrates how the main features of Hauterive window IX were constructed. Artmann says of this window, "Three- and six-fold symmetries are built into this window, which has a very simple construction once you view the circle as the incircle of an equilateral triangle." You can follow Artmann's lead and duplicate his construction. To begin, circumscribe an equilateral triangle about a circular window. Draw an arc from each vertex of the triangle with radius equal to half the side of the triangle. Last, draw three radial lines from the circle's center toward each vertex of the triangle.

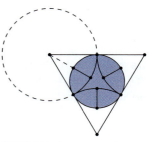

FIGURE 7.33 ■ Construction of Hauterive Window XVI

Rose Windows at Chartres

The rose window is also called the *Catherine window* and the *wheel window*. According to Painton Cowen, the astrological number 12 is one of the most common numbers in rose windows.[15] At Chartres, there are three large rose

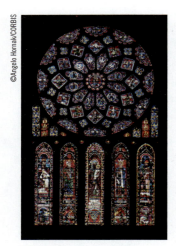

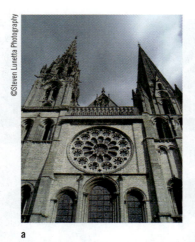

FIGURE 7.34 ■ Chartres North Window

a

b

FIGURE 7.35 ■ (a) Chartres West Window, Exterior; (b) Chartres West Window, Interior

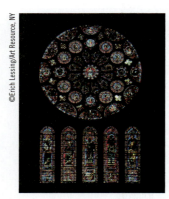

FIGURE 7.36 ■ Chartres South Window

The notion of complexity down to the smallest detail is a property of a *fractal*, which we'll discuss in Chapter 13, "Fractals."

windows, north, south, and west, each divided into 12 segments. The north window, the *Rose de France*, is displayed in Figure 7.34. It is dedicated entirely to the Madonna, who is at the center and surrounded by doves, angels, and thrones from the celestial hierarchy. The 12 squares contain the kings of her ancestry, the line of David, as recorded by St. Matthew.[16] Beyond are the last 12 prophets of the Old Testament. Seen from different directions, the four doves imply the four points of the compass, or the whole world.[17] Figures 7.35 and 7.36 display the other two windows. Note again the multiples of 12.

Some rose windows contain stained glass, while others do not. A rose window may look grand from outside a cathedral. However, when viewed from inside with sunlight streaming through them, the ones with stained glass are magnificent. Figure 7.35a and 7.35b are pictures of the same window, viewed from the outside and from the inside.

According to Painton Cowen, geometry is used in rose windows in three different ways.[18] First, it is "*manifest geometry* that makes the most immediate impact on the eye; the web of complexity and precision within which every space is defined by a yet smaller geometric figure—a trefoil, quatrefoil, rosette, or spherical triangle. And within these . . . can be seen an even finer pattern woven into the glasswork . . . right down into every fiber and corner of the cosmic rose." Then there is *hidden geometry*, the secret geometry of the relationships and proportions of the parts, and finally, *symbolic geometry*, a kind of shorthand, where geometric figures represent different things.

Given such layers of meaning, you might, while sitting in a cathedral and gazing at a rose window, think of some of the following: the theme shown in the window; the suffering of St. Catherine on the torture wheel; the unpredictable turns of the *rota fortuna*; the turning of the world; the circles of heaven; the beauty of the rose and of nature; the sun; the infinite complexity of the universe; the shapes chosen by the designers of the window and what they mean; and the particular relationships and proportions between parts and the numbers of elements.

CIRCULAR ORNAMENTATION IN WALLS AND PAVEMENTS

The rose window wasn't the only circular feature used in medieval and Renaissance art and architecture. Circular designs were used extensively in façades, and especially in the interior pavements, as can be seen in Figure 7.37.

a S. Martino, Lucca

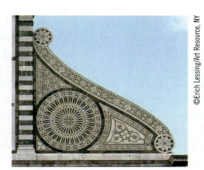

b S. Maria Novella, Florence

c Pompeii

d S. Maria Novella, Florence

e Pisa Duomo

FIGURE 7.37 ■ Some Circular Designs

a

b

FIGURE 7.38 ■ Rosettes in Architecture

Rosettes are a common pavement design featured in façades. In common usage, the word *rosette* can mean an ornament made of fabric, gathered at the center to resemble a rose, or any circular design.[19] The term *rosette* is also used in architecture and in mathematics. In architecture or art, a rosette can be a round design, carved or painted, and it usually has a floral motif.[20] Note Figure 7.38. In mathematics, a rosette or rose is the graph of certain curves in polar coordinates. See Figure 7.39. We'll explore the mathematical version of the rosette in Chapter 9, "The Ellipse and the Spiral."

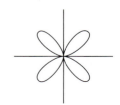

FIGURE 7.39 ■ Rosettes in Mathematics

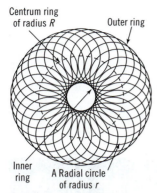

FIGURE 7.40 ■ A Circular Rosette
with Radial Circle Smaller than
Centrum Ring (*r* < *R*)

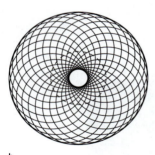

a

b

FIGURE 7.41 ■ (a) A Circular Rosette,
Radial Circle Equal to Centrum
Ring (*r* = *R*); (b) A Circular
Rosette, Radial Circle Larger than
Centrum Ring (*r* > *R*)

In this discussion, we'll use the word *rosette* to mean a geometric pattern obtained when copies of a mathematical curve are positioned at equal intervals about a central point, forming a circular design. When the curve is a circle, the design is the *circular rosette*; when the curve is a logarithmic spiral, the design is a *logarithmic rosette*. In the section on the hexagon in Chapter 5, we first discussed the *six-petalled rose*. Look again at Figure 5.40. In general, if you draw circles of the same size, equally spaced around another circle, you'll get a circular rosette. Now let's call each of those small circles a *radial circle* and denote its radius by *r*. We'll call the circle joining the centers of the radial circle the *centrum ring* of radius *R*. The center of the centrum ring is the *center of the rosette*. The empty *inner ring* and the *outer ring* of the rosette are concentric with the centrum ring. Figure 7.40 shows a circular rosette in which the radius *r* of each radial circle is smaller than the radius *R* of the centrum ring.

Figure 7.41 shows two more common rosettes. In Figure 7.41a, the radius of each radial circle is equal to the radius of the centrum ring. In Figure 7.41b, the radius of each radial circle is greater than the radius of the centrum ring. In each case, the radius of the inner ring is equal to the difference between the radius *R* of the centrum ring and the radius *r* of each radial circle. Also, the radius of the outer ring is the sum of the radius of the centrum ring and that of a radial circle.

The rosette's appearance is also affected by the number of radial circles used; more circles make the rosette *denser*. This rosette in Figure 7.42 is identical to the one in Figure 7.40, except that it has 90 radial circles instead of 30.

To construct a rosette, you must choose the radii of the various rings as well as the number of radial circles.

■ **EXAMPLE:** Construct a rosette with a centrum ring of radius 5, containing 18 radial circles of radius 4.

● **SOLUTION:** Draw the centrum ring and a single radial circle with the center on the centrum ring, as shown in Figure 7.43. The outer ring will have a radius of 9, and the inner ring will have a radius of 1. Draw more radial circles spaced by an angle of 360° ÷ 18 or 20°.　　　　　　　　■

The outer ring of the full circular rosette is the circle tangent to the radial circles at their outer limits. If you crop the rosette at some smaller radius, in particular at some radius where the radial circles intersect, you'll get what is sometimes called a *starburst* pattern, as in Figures 7.44 and 7.45. The starburst

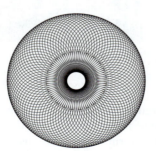

FIGURE 7.42 ■ Rosette with
90 Radial Circles

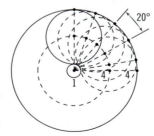

FIGURE 7.43 ■ Constructing a
Rosette

FIGURE 7.44 ■ Starburst Pattern

FIGURE 7.45 ■ Rosette in Marble
Pavement, San Marco, Venice

©From F. M. Hessemer, *Historic Designs and Patterns in Color*, Dover Publications, Inc.

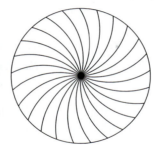

FIGURE 7.46 ■ Fan Pattern

FIGURE 7.47 ■ Guilloche in San
Marco, Venice

©Steven Lunetta Photography

pattern in Figure 7.45 comes from a pavement in Venice. One final design, called the *fan* pattern, is obtained when just one arc of each radial circle is used, as shown in Figure 7.46.

In addition to the rosette, the serpentine design, or *guilloche*, shown in Figure 7.47, is a popular pavement design. It is often shown in a four-lobed version, a kind of *Solomon's Knot*, as in Figure 7.48, or displayed in a five-lobed variation, the *quincunx*, as shown in Figure 7.49. A third option is to give the guilloche a circular arrangement, as in Figure 7.50.

To construct a Solomon's Knot, just follow these steps (the guilloche and quincunx are constructed using similar methods):

1. Inscribe four tangent circles in a circle, as shown earlier in this chapter.

2. Draw an inner circle and an outer circle concentric with each of these circles. Choose radii for them that will make the intertwining strands as fat or as thin as you like.

3. Draw the lines of centers to locate the points of tangency between inner and outer circles—points such as *A* and *B* in Figure 7.51.

4. Where one strand crosses another, decide whether it goes *over* that strand or whether it goes *under* that strand. Darken the appropriate arcs to complete the design.

The Solomon's Knot was introduced in Chapter 4, and the Quincunx was introduced in Chapter 5.

FIGURE 7.48 ■ A Rosette and a
Solomon's Knot in Pavement

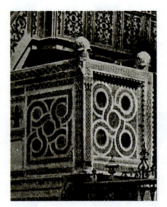

FIGURE 7.49 ■ Quincunx in the Choir
of Santa Maria in Cosmedin, Rome

FIGURE 7.50 ■ A Circular Guilloche in
the Vatican Museum

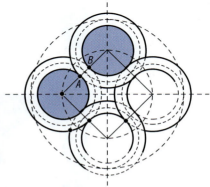

FIGURE 7.51 ■ Construction of the Solomon's Knot

ARCHES AND VAULTS

"An arch consists of two weaknesses which, leaning one against the other, make a strength."

LEONARDO DA VINCI

Two problems that concerned builders over the ages were how to span an opening in a wall and how to span an overhead space between four columns, especially with brittle masonry. A solution to the first problem was the arch and to the second was the vault. We will explore both the arch and the vault in this section.

The Arch

One of the most common architectural features based on the circle is the *arch*. An arch is a curved structural member that is used to span an opening and to support the load above. The arch is used over doorways and windows, for ceiling vaulting, and for buttresses. Arches can take a great variety of forms, as shown in Figure 7.52.

When we're talking about arches, some of the terminology refers to arches in general, and some terminology refers specifically to the masonry arch. Figure 7.53 illustrates the general terms. A point from which an arch rises from its vertical support is known as a *spring point*. The line connecting the spring points is called the *spring line*. The top of the arch is called the *crown*, which is flanked by the *haunches*. The *rise* is the distance from spring line to crown, and the *span* is the width of the opening. The inner curve of an arch is called the *intrados*, and the outer curve is called the *extrados*.

As you can see in Figure 7.54, in *masonry* construction a series of wedge-shaped blocks are set side-by-side to form the arch. Each block is called a *voussoir*. The central voussoir is called the *keystone*. The roughly triangular space above each haunch is called a *spandrel*. Each side of the arch rests on a *pier* or *abutment*, from which it may be separated by a masonry member called an *impost*.

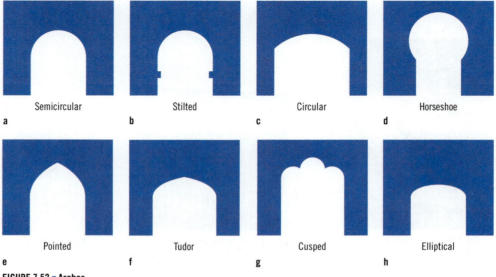

Semicircular a Stilted b Circular c Horseshoe d

Pointed e Tudor f Cusped g Elliptical h

FIGURE 7.52 ■ Arches

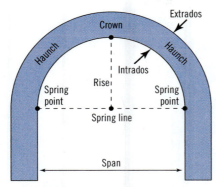

FIGURE 7.53 ■ Arch Terminology

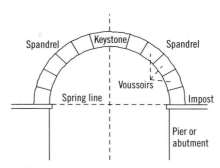

FIGURE 7.54 ■ Masonry Arch Terminology

During construction, an arch requires support from below until the last stone has been set and the mortar has hardened. Such temporary wooden support is called *centering*. Arches can be made of any structural material—wood or steel as well as masonry. For masonry construction, the arch has advantages over the horizontal beam, or *lintel*. It can span wider openings with the use of lighter blocks, rather than with a heavy stone lintel. An arch can also carry a much greater load than a lintel can. A downward force on an arch tends to press the voussoirs together instead of apart, and these in turn create an outward force on the piers, tending to push them apart. This thrust must be counteracted by heavy masonry, by buttressing, or by placing arches in a row, called an *arcade*, so that each arch counteracts the thrust of its neighbors. The arches at the end of the arcade must, of course, be buttressed.

The rise of a round arch is automatically half its span. The pointed arch, also called the *Gothic* arch, is more versatile than the round arch because its rise is independent of its span. It also has less outward thrust than the round arch. Symbolically, it appears to be pointing upward to heaven, as do the spires of a church.

First, let's draw the simplest pointed arch: the *equilateral* pointed arch. From each spring point, swing an arc with a radius equal to the span, as illustrated in Figure 7.55. The two arcs will intersect at the crown *B*, on the centerline of the span.

Next, let's construct a general pointed arch, one whose radius is not equal to the span. To construct a pointed arch of given span and rise, begin by connecting the spring point *A* with the crown *B*, as shown in Figure 7.56. Next, draw the perpendicular bisector *L* to line *AB*. Locate point *C* where *L* intersects the spring line. Point *C* may fall within the span of the arch, on a spring point, as with the equilateral arch, or outside. Finally, draw arc *AB* with center at *C* and radius *AC*. This gives half the arch. The other half is simply the mirror image of arc *AB* about the centerline. Can you explain why this construction works?

Our next constructions involve inserting a rose into an arch, as shown in Figure 7.57. This window was built by the architect Jean d'Orbais in the cathedral of Reims during the years 1211–1221; its construction is based on the equilateral triangle. When the rose has the same diameter as the width of the window, as within the Reims window, and passes through the spring points, the method used to insert the rose is simple. When the rose is inscribed in the arch and is tangent to the spring line, as in Figure 7.58, the method used to insert it is much more complex. In both the simple case and the complex case, the diameter of the rose is fixed by the size of the arch. A more general case occurs when the rose touches the arch but not the spring line. When this happens, as seen in

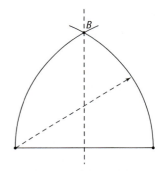

FIGURE 7.55 ■ Equilateral Pointed Arch

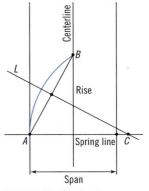

FIGURE 7.56 ■ Drawing the Pointed Arch

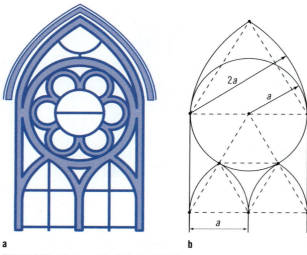

a   b

FIGURE 7.57 ■ Window at the Cathedral of Reims

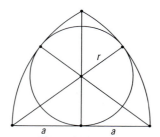

FIGURE 7.58 ■ A Circle Inscribed in an Equilateral Pointed Arch

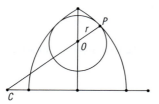

FIGURE 7.59 ■ Inserting a Rose into a Pointed Arch

Figure 7.59, the designer can choose either the radius of the rose or the location of its center.

To insert a rose into a pointed arch, first let *C* be the center of one arc of the arch, as in Figure 7.59. It is, as we have seen, on the spring line. If the location *O* of the rose's center is given, draw a line from *C* through *O*, cutting the arch at *P*. This locates the point of contact *P* between rose and arch, and gives the radius *r* of the rose. If, instead, the radius *r* of the rose is given, swing an arc from *C* with radius *R*−*r*, where *R* is the radius of the arch. This arc will intersect the window centerline at *O*, giving the center of the rose. Extending that line will give the point of contact *P* between rose and arch.

Figure 7.60 shows an example of a complete window from a church built in Strasbourg around 1185. Let's draw a complete window by following these steps, based on Figure 7.61. Begin by drawing an equilateral main arch of radius *2a*, the width of the window. Next, draw the two equilateral midsized arches, each of radius *a*. Then draw the four smallest equilateral arches, each of radius *a*/2. Because the radius *AP* of the arch is *2a*, and the radius *AQ* of each midsized arch is *a*, the diameter of the large rose must be *AP* minus *AQ* or *a*. Locate its center *O* at a distance *OA* from spring point *A*, where *OA* = *a* + *a*/2, or *3a*/2. Lines from the center of the main rose to the spring points locate the centers of the two midsized roses. Find their diameters in the same way you found the diameter of the main rose.

FIGURE 7.60 ■ Strasbourg North Window

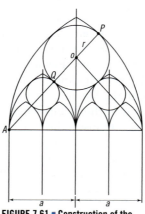

FIGURE 7.61 ■ Construction of the Strasbourg North Window

FIGURE 7.62 ■ Flying Buttresses at Notre Dame

A *basilica* is a large building with a rectangular floor plan and an open hall, called the *nave*, which extends from end to end. It is usually flanked by side aisles. In Gothic cathedrals, these side aisles were sometimes spanned with half-arches that took the lateral thrust of the main vault. If you can imagine an aisle with the roof removed, the series of half-arches spanning the aisle would be what are called *flying buttresses*, as shown in Figure 7.62. These exterior half-arches supported the walls of the nave at points of greatest stress, allowing builders to reduce the thickness of those walls. These buttresses, in turn, conducted stresses to heavier buttresses or piers below.

The Vault

One of the main structural challenges in building, especially in masonry, is how to span the overhead space between four columns. In medieval Western Europe, this feat was accomplished almost exclusively with the *vault*. Here we'll give a highly simplified account of the development of the vault.

Early basilica churches were usually roofed over with horizontal wooden beams. Because the wooden beams were prone to catching fire, masonry roofing was desired. However, horizontal masonry would collapse of its own weight. Solving this problem led to the development of the *vault*, which is simply a masonry ceiling or roof that uses an arrangement of arches to support the weight. According to Burckhardt, fire resistance was not the only virtue of the vault. "In medieval art there is no form that has only a practical purpose and does not . . . serve as a spiritual expression . . . the covering vault always possessed the meaning of the vault of Heaven. This is proved by the pictures with which it was decorated."[21] (*The Dome of Heaven* is discussed in Chapter 11.)

The simplest and easiest way to cover a long building is with a semicircular construction known as a *barrel vault*. A barrel vault is basically a very deep arch, as you can see in Figure 7.63. The earliest barrel vaults used the round arch, which provided a semicylindrical covering over a space bounded by parallel walls. While later barrel vaults used the pointed arch rather than the round, we'll assume that the barrel arch is round, unless otherwise noted.

The barrel vault, like the arch, exerts a downward force from the weight of the masonry in and above the arch, as well as an outward thrust from the wedge action of the stones in the arch. This outward thrust was resisted in several ways. One way was to support the vault along its entire length with heavy walls that had just a few small openings. This, of course, blocked light and inhibited circulation.

The addition of ribs spanning the nave increased the strength of the walls enough to allow more openings, giving rise to the *banded* barrel vault. This subdivided the vault into *bays*. This also allowed the vault to be constructed one bay at a time, rather than all at once from end to end. This method was more efficient and economical, as formwork from a finished bay could be moved to the next bay, and finishing work could start on a completed bay before the entire vault was completed.

When two barrel vaults of equal span intersect at right angles, they form a *groin vault*, as shown in Figure 7.64. (In this case, it is also called a *cross vault*.) The lines of intersection are called *groins*. The groin vault is square in design. A number of groin vaults can be constructed in a row to produce a long aisle, and they can be arranged to cover areas of any size, in what is called a *grid system*. Most of the weight of the groin vault lies on pillars at the four corners, allowing the walls to be more open than is possible with a barrel vault.

The groin vault, which used round arches, had two problems. First, if the arches on each of the four sides were round, then the diagonal arch had to be *flattened*. Otherwise, it would rise higher than the four side arches. This flattening weakened the arches, which had to be made heavier and stiffer to carry the load. Conversely, if the diagonal arches were full semicircles, the four side arches had to be *stilted*, that is, raised on straight sides in order to rise to the same height as the diagonal arches. Geometrically, that was the only way they could span the greater diagonal distance between pillars while rising to the same height, the full semicircular arch always having a rise equal to half its span. This problem was overcome by the use of the pointed arch, whose rise, as we have seen, is independent of its span.

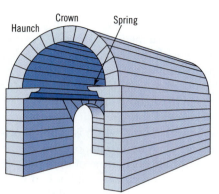

FIGURE 7.63 ■ A Barrel Vault with Round Arches

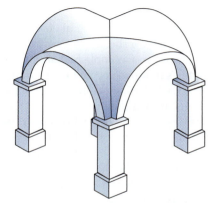

FIGURE 7.64 ■ Groin Vault with Round Arches

The second problem was that to build a groin vault, a form had to be made on which to pour or lay the entire vault. This required complex scaffolding from the ground up. The problem was overcome by the invention of the *rib vault*. Builders first made *ribs*, structural members spanning the four sides of the vault diagonally with each rib requiring relatively simple centering. See Figure 7.65. The masonry of the vault was then laid on this skeleton of ribs.

The rib vault with pointed arches became an important element in the development of Gothic architecture. But, as John Fichten points out, writers have seen the adoption of the pointed arch as "the basis of distinction between Romanesque and Gothic architecture . . . the pervading structural characteristic of the latter era . . . [but] there were many occasions on which the Gothic builders retained the semicircular arch."[22] He also points out that the great majority of diagonal ribs in Gothic rib vaults are, in fact, easy-to-make round arches.

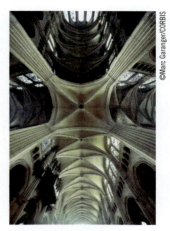

FIGURE 7.65 ■ Rib Vaulting at Chartres

THE VESICA

The *vesica* is another geometric figure that is made from circular arcs, as in Figure 7.66. As we discussed earlier in this chapter, the vesica is created when two equal circles intersect. The vesica is the portion that is common to both circles. The vesica is a very common symbol in art history, and you'll see that it has had several meanings at various times and places. Notice in Figure 7.67 that the construction of the vesica also gives two *crescents* and that half the vesica is a *pointed arch*.

The vesica is properly called the *vesica pisces* or the *mandorla*. *Mandorla*, which means "almond" in Italian, refers to the vesica's almond shape, which is a powerful symbol because a nut is the seed from which a tree grows. The almond is often mentioned in the Old Testament. Exodus 25 says that the bowls of the menorah were to be made in the shape of almonds. Further, from Proverbs 9, the rod of Aaron (Moses' older brother) was made of almond wood. It bloomed, signifying the Lord's choice of Aaron as leader of the tribe of Levi.

> ". . . behold, the rod of Aaron for the house of Levi brought forth buds, and bloomed blossoms, and yielded almonds."[23]

Vesica pisces literally means "fish bladder." If the ends on one side are extended, the result is a symbol for the fish. The vesica is also a variation of the zodiac sign for Pisces, as shown in Figure 7.68. The fish symbol is closely associated with Christianity, and fish are often mentioned in the New Testament. In the story of "The Loaves and Fishes," Jesus took seven loaves and a few fish, broke them into pieces, and fed 4,000 hungry people.[24] Simon Peter and Andrew were fishermen. In "The Drought of Fishes," Simon Peter, having caught nothing, was told by Jesus to lower his net, and it miraculously came up full; this story is portrayed in Duccio's *The Calling of the Apostles Peter and Andrew*. A fish sometimes

The word *vesica* is sometimes used only when each circle passes through the *center* of the other. However, that condition is not part of the definition we'll use here.

FIGURE 7.66 ■ Vesica on the Cararra Duomo

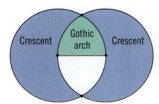

FIGURE 7.67 ■ Vesica, Arch, and Crescents

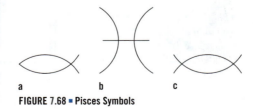

FIGURE 7.68 ■ Pisces Symbols

FIGURE 7.69 ■ Car Medallion

appears on the table in paintings of the Last Supper. Fish also figure in the stories of "Jonah and the Whale," Antony of Padua who preached to the fishes, and Tobias who carried a fish while accompanied by the angel Raphael. The story of Peter finding money in a fish's mouth (Matthew 17:24–27) is portrayed in Masaccio's *The Tribute Money*, which is shown in Chapter 12. Another reason for Jesus' association with the fish is that *fish* in Greek is written ΙΧΘΥΣ or Ichthus. These are also the initial letters in Greek of the words **J**esus **Ch**ristos **Th**eou **U**ios **S**oter, or *Jesus Christ God's Son Savior*. The fish symbol shown in Figure 7.69 eventually came to stand for Christ himself.

During the Middle Ages, the vesica was frequently used as an *aureole*, a field of radiance surrounding the entire body and a sort of *body halo* surrounding holy figures. Figure 7.70 shows the vesica or mandorla over a doorway in the Duomo in Florence. The central figure is surrounded by an aureole. This practice continued into the Renaissance. It is the most common use of the vesica in art.

As an aureole for a standing person, the vertical vesica seems a logical choice. However, sometimes the vesica is *horizontal*. Such uses give the impression of an *eye*, usually representing the omnipresent, all-seeing Divinity. On the first day of Rosh Hashana, Jews are encouraged to visit a body of water containing live fish and recite prayers to cast away sins. Because a fish's eyes never close, they symbolize God's unceasing watchfulness. As such, an eye framed by a triangle can be a symbol of God. Just such an image appears on the United States dollar bill. A detail of this image is shown in Figure 7.71. About the eye on the pyramid on the dollar bill, Joseph Campbell writes, "When you're down on the lower levels of the pyramid you will be on either one side or the other. But when you get to the top, the points all come together, and there the eye of God opens."[25]

The *crescent* is a geometric figure related to the vesica. The remaining portions from the construction of a vesica (look again at Figure 7.67) were crescents, or *lunes*. The crescent moon has long been a religious symbol and eventually became the symbol of the Byzantine Empire, supposedly because the sudden appearance of the moon saved the city of Byzantium (Constantinople) from a surprise attack. The Ottoman Turks had been using the crescent symbol since

The *lune* or crescent will be discussed in greater detail when we cover the moon in Chapter 11.

FIGURE 7.70 ■ Porta della Mandorla

FIGURE 7.71 ■ Eye on Pyramid on Dollar Bill

the mid-1300s, but rather than representing the moon it may have been two horns or claws, joined base-to-base. The crescent has come to be a symbol of Islam in general, and it is seen today on the flags of many predominantly Islamic countries. It is the symbol of the Red Crescent, the Muslim equivalent of the Red Cross.

SUMMARY

In this second chapter on the circle, we expanded our exploration of geometric theorems by showing intersecting circles and circles inscribed in and circumscribed about polygons. We then used these new theorems to construct a variety of Gothic designs and Gothic traceries. After a brief overview of Gothic architecture, we constructed foliations, rose windows, rosettes, Solomon's Knots, guilloches, arches, vaults, vesicas, and crescents. The aim was not, of course, to train you to be a medieval builder, but to increase your understanding of geometry by showing how the circle might have been used in a very interesting period in architecture. We have one more chapter on the circle: the symbolic *squaring of the circle*, which is next.

EXERCISES AND PROJECTS

1. Define the following terms:

Gothic style	foliations	trefoil
quatrefoil	cinquefoil	rosette
Solomon's Knot	guilloche	quincunx
arch	*intrados*	*extrados*
spring line	spring points	*voussoir*
keystone	spandrel	centering
mandorla	crescent	*lune*
aureole	vault	barrel vault
cross vault	groin vault	rib vault
flying buttress		

2. Demonstrate by construction that for two intersecting circles, the line of centers is the perpendicular bisector of the common chord.

3. Demonstrate by construction that for two tangent circles, the line of centers is perpendicular to the common tangent.

4. Demonstrate by construction that a tangent is perpendicular to the radius drawn to the point of contact.

5. Demonstrate by construction that the perpendicular bisector of a chord passes through the center of the circle.

6. Using compass and straightedge only, divide a circle into 3, 4, 5, 6, 8, and 12 parts.

7. Using The Geometer's Sketchpad, or a similar program, draw two circles and their line of centers. Construct a tangent where the line of centers crosses each circle. Then drag one circle into different positions relative to the first—not touching, tangent, intersecting, and so on. Write a few paragraphs on your findings.

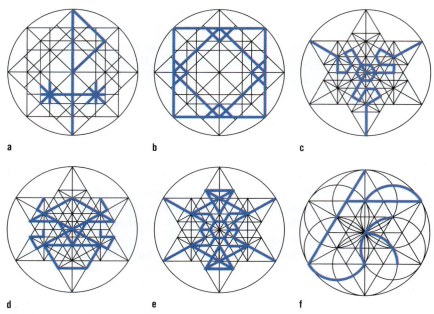

FIGURE 7.72 ■ Some Mason's Marks[26]

8. Design your own "mason's mark." Like a medieval mason, be prepared to construct it if requested. Figure 7.72 should give you some ideas.

9. Select a window from the Cloisters of Hauterive shown in Figure 7.32, and perform a geometrical reconstruction.

10. Using the Hauterive windows as a source of ideas, design your own "Hauterive window."

11. To construct the bifoil window in Figure 7.73, follow these steps. Inscribe two tangent circles inside the outer circle. Remove half of each inner circle to form the yin-yang design. Subdivide the radius *AC* into four equal parts. From each of the three points *B*, *D*, and *E*, draw a circle with radius one-quarter that of the original circle. Draw a perpendicular to *AC* at *D*, locating point *F*. With center at *F*, draw a circle with radius *DF*. Remove portions of the circles, as shown, to complete the design. (Note the similarity to the Chinese yin-yang symbol. See Appendix E, Figure 4.)

For many of the following constructions derived from undimensioned photographs, you may need to make assumptions about the missing dimensions. Most of the photographs appear distorted from being necessarily photographed from below. You must compensate for that distortion. For all these constructions, draw only the centerlines of the members, ignoring their widths.

12. Figure 7.74 shows a quatrefoil, or *quadrillobate*, medallion. It is also called a *Florentine*, a stylized flower from which the city of Florence gets its name. Construct this figure.

13. Reconstruct the window in Figure 7.75, which shows three curvilinear triangles in a circle.

14. Reconstruct both the upper and lower windows of Figure 7.76.

15. Reproduce the quatrefoil window in Figure 7.77. Then use your drawing as a starting point for a design of your own.

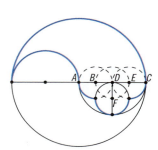

FIGURE 7.73 ■ Constructing a Bifoil Window

©Alinari/Art Resource, NY

FIGURE 7.74 ■ Baptistery South Doors (detail), Andrea Pisano

FIGURE 7.75 ■ Church Window in Montpelier, Vermont

FIGURE 7.76 ■ Church Façade in York, Maine

FIGURE 7.77 ■ Quatrefoil Window in Barre, Vermont

FIGURE 7.78 ■ Window at Wesleyan
University in Middletown, Connecticut

FIGURE 7.79 ■ Chartres South Window
(Exterior)

©Steven Lunetta Photography

16. The window in Figure 7.78 combines both three-fold and four-fold symmetry. Construct this window.

17. Construct the Chartres South rose window shown in Figure 7.79. All of the proportions are fixed by the geometry of the window except for the diameter of the central circle, which you should estimate.

18. Design a roof that is a circular arc, like the one in Figure 7.80. Design a façade that harmonizes with the roof line. Make a model.

19. Construct a window that uses all circular arcs, such as the one in Figure 7.81.

20. Construct either of the semicircular windows in Figure 7.82. Assume that the outer frames are semicircular, even though they appear foreshortened in the photograph. Then design a semicircular window of your own.

21. Construct an equilateral, a lancet, and an obtuse pointed arch.

FIGURE 7.80 ■ A Roof Line That Is a Circular Arc

FIGURE 7.81 ■ Window in Montpelier, Vermont

22. Using The Geometer's Sketchpad or a similar program, draw a rectangular window and half a pointed arch. Mirror that half across the centerline of the window to complete the arch. Then drag the center on the circle, forming the arch along the spring line and causing the arch to change shape. Write a few paragraphs on your findings.

23. *Team Project.* Find a large, flat floor or pavement. On it, lay out a full-sized pointed arch. Use a stretched cord for a compass, as if this were to be used as a template for cutting the stones of an actual arch.

24. The window in Figure 7.83 combines a number of geometrical elements: quatrefoil, vesica, ad quatratum, and so forth. Design your own window that also combines elements we have covered in a pleasing and interesting design. Then make a model from translucent material.

25. Figure 7.58 shows a circle inscribed in an equilateral pointed arch. Find r, and do the construction.[27]

26. Inscribe a rose in a pointed arch of radius R, given:
    a. The radius r of the rose
    b. The distance h from the spring line to the rose center

a

b

FIGURE 7.82 ■ Semicircular Windows

FIGURE 7.83 ■ A Window in Barre, Vermont

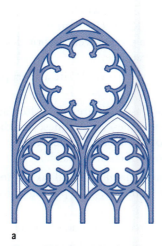

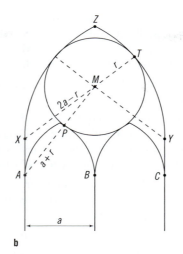

a

b

FIGURE 7.84 ■ Haina North Window

27. Most Gothic windows have the main arch, one or more roses, and from two to eight vertical panels. Draw the Reims window shown in Figure 7.57, which has two vertical panels.

28. In Figure 7.84, we see a variation of the construction of the Reims window—this is from the north window of the church of Haina. The two smaller parts of this window are repetitions of the first Reims tracery. The upper part of the window, however, is a more subtle design based on circles in contact. Draw this window, given the width $2a$ of the window, the radius r of the large rose, and the spring line ABC of the two small arches, as follows. Draw the small arches of radius a, from centers A, B, and C. Locate the center M of the rose on the centerline of the window at a distance $a + r$ from A. Line AM locates the point of tangency P on a small arch. Draw the large rose with center at M with radius $r = MP$. Locate the spring points X and Y for the outer arch. Because the radius XT of that arch equals the width $2a$ of the window, the distance MX must be $2a - r$. Draw the outer arch, with centers at X and Y and radius $2a - r$.

29. Reconstruct the window shown in Figure 7.85.

30. Construct the mandorla window from Cararra, shown in Figure 7.66. Assume that each of the three small circles has half the diameter of the large circle.

31. Construct a vesica by drawing two circles of the same radius, each passing through the center of the other. Complete the construction shown in Figure 7.86, and demonstrate that triangles ABC, BDE, ABD, ADF, and CEF are all equilateral.

32. Construct a circular rosette, first choosing the radii of the centrum ring and the number and radius of the radial circles. If you use a CAD program, use the ROTATE command to draw all the circles after the first.

33. Construct a guilloche, and embellish it with your own designs.

34. Construct a quincunx, and embellish it with your own designs.

35. Make a round arch by stacking dominoes. Use a can or an oatmeal box for centering.

36. Make a groin vault by intersecting two half-cylinders. Use a cardboard oatmeal box or other cylinders.

FIGURE 7.85 ■ Window in York, Maine

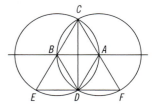

FIGURE 7.86 ■ Construction from Lawlor[28]

37. Make a graphic design based on the rosette, rose window, a foliation, the vesica, guilloche, quincunx, or Solomon's Knot. Use your design as decoration for wrapping paper, a greeting card, a screensaver, a fabric design, a pillow, a tee shirt, a quilt, or a dress, for example.

38. Research the hex sign, and design one of your own. Write a short paper on your findings.

FIGURE 7.87 ■ A Student's Rose Window

39. Design your own complete window, such as in Figure 7.87. After laying out the geometrical construction, give thickness to the lines, as if they were cut stones. Cut the "stones" out of styrofoam, foamcore, or cardboard. Insert stained "glass" windows of colored plastic.

40. Visit a cathedral, or search your campus or neighborhood taking note of any architectural features covered in this chapter. Take photographs and write a report on your findings.

41. Do a report on any book listed in the sources at the end of this chapter.

42. Ken Follett's *The Pillars of the Earth* contains many references to medieval building features. Read the book and extract these features in a book report.

43. Write a short paper or a term paper on some topic from this chapter. Come up with your own topic, using these suggestions to jog your imagination:
    - Number symbolism in the medieval cathedral
    - Medieval building techniques
    - The Gothic style in architecture
    - The development of the rose window
    - The development of the arch and/or the vault
    - *Circle-packing*, circles in a circle, circles in a square, and circles in an equilateral triangle. See the entry on "Circle Packing" in the *CRC Concise Encyclopedia of Mathematics*, 2nd ed., p. 429.
    - On the close-packing of circles and their appearance in Buddhist art. See Tarnai, et al. "Circle Packings and the Sacred Lotus," *Leonardo*, Vol. 36, No. 2 (2003), pp. 145–150.
    - The halo in art
    - The history of the Masons
    - Knot patterns, such as the Solomon's Knot, including Leonardo's interest in knots
    - The rosette. See Paul L. Rosin, "Rosettes and Other Arrangements of Circles," *Nexus Network Journal*, Autumn 2001.
    - The extensive appearance of the vesica in medieval art

44. Make an oral presentation on any of the projects or papers in this section.

45. Make a rose-window pizza, and share it with your class.

Mathematical Challenges

46. Some technical students may be familiar with the methods of descriptive geometry. If you have these skills, use them to draw the intersection of two cylinders. Do this by hand or with a CAD program. Relate your drawing to the cross vault.

47. The radius of a circle inscribed in a triangle equals twice the area of the triangle divided by the perimeter of the triangle. Starting with this,
   a. Replace *area* with the expression given by Hero's formula (Equation 30), replace *perimeter* with the sum of the sides, and derive the following formula for the radius in terms of the sides of a triangle.

81 RADIUS OF A CIRCLE INSCRIBED IN A TRIANGLE

$$r = \frac{\sqrt{s(s-a)(s-b)(s-c)}}{s}$$

where s is half the perimeter of the triangle;

$$s = \tfrac{1}{2}(a + b + c)$$ ●

   b. Use Equation 81 to find the radius of a circle inscribed in a triangle with sides that are 194 cm, 285 cm, and 314 cm.

48. The radius r of a circle circumscribed about a triangle with sides a, b, and c is given by Equation 82. Use this formula to find the radius of a circle circumscribed about a triangle with sides that are 683 in., 736 in., and 884 in.

82 RADIUS OF A CIRCLE CIRCUMSCRIBED ABOUT A TRIANGLE

$$r = \frac{abc}{4\sqrt{s(s-a)(s-b)(s-c)}},$$

where s is half the perimeter of the triangle.   ●

49. Find the radii of each of the three circular designs in the pediment in Figure 7.88.

50. Figure 7.89 illustrates the *nine-point circle*.
   a. Demonstrate by construction that there is a circle that will pass through the three midpoints M of the sides of a triangle, the three points F where each altitude touches its base, and the three midpoints J of the line segments joining each vertex to the orthocenter.
   b. How is the center of this "nine-point-circle" related to the *Euler line* from Chapter 3?
   c. How is the nine-point circle related to the *orthic triangle* of Chapter 3?

51. Write an expression for the radius of the incircle of the right triangle ABC shown in Figure 7.90.

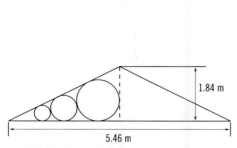

FIGURE 7.88 ▪ **Three Circular Designs**

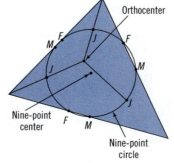

FIGURE 7.89 ▪ **Nine-Point Circle**

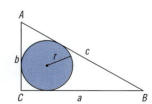

FIGURE 7.90 ▪ **Circle Inscribed in a Right Triangle**

52. Find the radii of the circles in Figure 7.91. Take the side of each triangle as 1 unit.

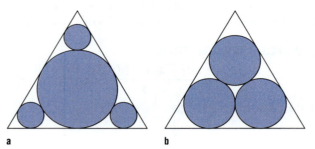

a   b

FIGURE 7.91 ■ Circles Inscribed in an Equilateral Triangle

SOURCES

Artmann, Benno. "The Cloisters of Hauterive." *Nexus: Architecture and Mathematics*. Kim Williams, ed. Fucecchio (Firenze). Edizione dell'Erba, 1996.

Calter, Paul. *Technical Mathematics*, 5th ed. New York: Wiley, 2007.

Cowen, Painton. *Rose Windows*. London: Thames and Hudson, 1979.

Coxeter, H. S. M., et al. *Geometry Revisited*. Washington, D.C.: Mathematical Association of America, 1967.

Coxeter, H. S. M. *Introduction to Geometry*, 2nd ed. New York: Wiley, 1989.

Eco, Umberto. *Art and Beauty in the Middle Ages*. New Haven, CT: Yale University Press, 1986.

Fichten, John. *The Construction of Gothic Cathedrals*. Oxford, UK: Oxford University Press, 1961.

Ghyka, Matila. *The Geometry of Art and Life*. New York: Dover, 1977.

Hopper, Vincent. *Medieval Number Symbolism*. New York: Columbia University Press, 1938.

Janson, H. W. *History of Art*, 5th ed. Revised by Anthony F. Janson. New York: Harry N. Abrams, Inc., 1995.

Williams, Kim. *Italian Pavements, Patterns in Space*. Houston: Anchorage, 1997.

Williams, Kim. "Spirals and the Rosette in Architectural Ornament." *Nexus Network Journal*, Vol. 1 (1999), pp. 129–138.

NOTES

1. This construction and the two that follow are from Sergey Markelov, "Circles and Parabolas," *The Mathematical Intelligencer*, Vol. 21, No. 1, Winter, 1999.
2. *The Story of Art*, p. 366.
3. Hopper, p. 98.
4. Eco, p. 19.
5. Hopper, pp. 114–115.
6. Clark, *Civilization*, p. 52.
7. Cowen, p. 91.
8. From the Introduction to Dürer's *The Painter's Manual*. With the publication of his book, Dürer intended to help pierce this veil of secrecy.
9. Janson, 5th ed., p. 330.
10. Janson, 5th ed., p. 332.
11. Richter, p. 91.

12. Eco, p. 33.
13. *Civilization*, p. 52.
14. *Nexus*, p. 22. The window numbers here are different from those in Artmann's paper.
15. Cowen, p. 92.
16. Matthew 1:1–17.
17. For a detailed geometrical analysis of this window, see Cowen, p. 125.
18. Cowen, p. 92.
19. Much of the material in this section is from Williams's book and MAA article.
20. Harris, p. 470.
21. Titus Burckhardt, p. 29.
22. Fichten, p. 78.
23. Numbers 17:8.
24. Matthew 15:36.
25. Joseph Campbell, *The Power of Myth*, p. 25.
26. Ghyka, p. 121.
27. *Nexus 96*, p. 17.
28. Lawlor, p. 33.

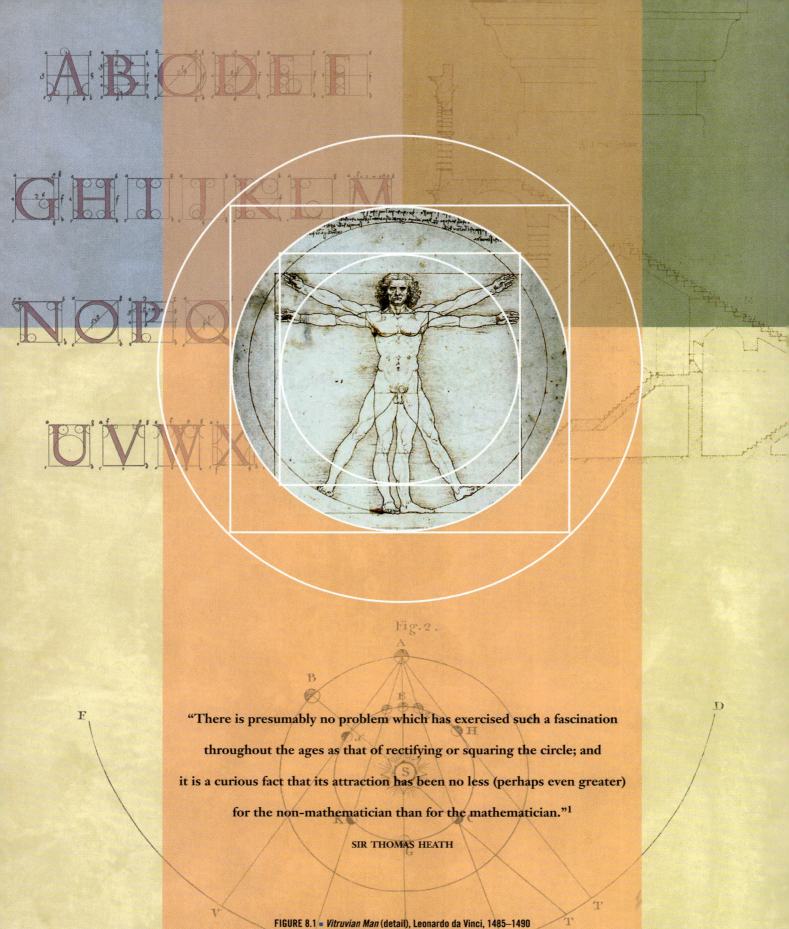

"There is presumably no problem which has exercised such a fascination

throughout the ages as that of rectifying or squaring the circle; and

it is a curious fact that its attraction has been no less (perhaps even greater)

for the non-mathematician than for the mathematician."[1]

SIR THOMAS HEATH

FIGURE 8.1 ■ *Vitruvian Man* (detail), Leonardo da Vinci, 1485–1490

8

■

Squaring the Circle

●

In Chapter 6, "The Circle," you learned that the circle was regarded as the ultimate geometric figure—perfect, infinite, and representing the divine. In earlier chapters, you saw that the square often represented the earthly and mundane. In this short chapter, you'll see that combining the two figures can have special significance: the reconciliation of the heavenly and infinite with the earthly and finite.

First, we'll explore the geometry connected with squaring the circle. Then, building on our earlier exploration of the Great Pyramid (recall Chapter 2, "The Golden Ratio"), we'll examine the claim that the pyramid contains the squaring of the circle. After that, we'll explore some other examples from art and architecture that appear to exhibit the idea of squaring the circle, in spirit if not geometrically correct.

Here, we're again venturing into the misty world of sacred geometry. Much of the material in this chapter is highly speculative and without documentation. As such, it calls for a generous amount of skepticism. In spite of that, the material should be motivating for mathematicians, artists, and architects alike.

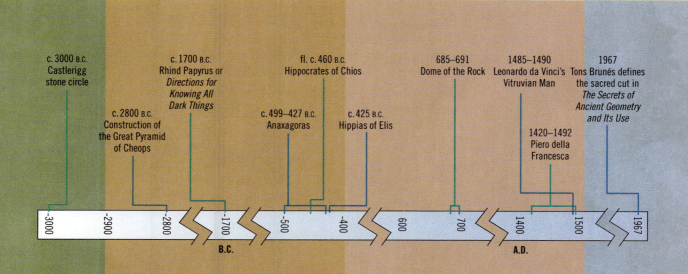

c. 3000 B.C.
Castlerigg
stone circle

c. 2800 B.C.
Construction of
the Great Pyramid
of Cheops

c. 1700 B.C.
Rhind Papyrus or
*Directions for
Knowing All
Dark Things*

c. 499–427 B.C.
Anaxagoras

fl. c. 460 B.C.
Hippocrates of Chios

c. 425 B.C.
Hippias of Elis

685–691
Dome of the Rock

1420–1492
Piero della
Francesca

1485–1490
Leonardo da Vinci's
Vitruvian Man

1967
Tons Brunés defines
the sacred cut in
*The Secrets of
Ancient Geometry
and Its Use*

-3000 -2900 -2800 -1700 -500 -400 600 700 1400 1500 1967

B.C.   A.D.

THE GEOMETRY OF SQUARING THE CIRCLE

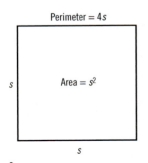

Perimeter = 4s

s

Area = s^2

s

a

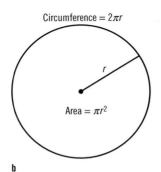

Circumference = $2\pi r$

r

Area = πr^2

b

FIGURE 8.2 ■ **Perimeter and Area of a Square and a Circle**

The problem of squaring the circle is to construct, using only a compass and a straightedge, a square with a perimeter that is exactly equal to the circumference of a given circle, or to construct a square with an area that is exactly equal to the area of a given circle. Squaring of the circle is also called *quadrature* of the circle. It is one of three famous problems of early mathematics. The other two are trisecting an angle and duplicating a cube, each with the limitation that only a compass and straightedge can be used. Over the centuries, there have been many attempts to solve these problems, and there have been many approximate solutions. In the nineteenth century, however, the solution was proved to be impossible. Fortunately, according to Howard Eves, the time spent in the search for a solution was not wasted. "The energetic search for the solution to these three problems profoundly influenced Greek geometry and led to many fruitful discoveries. . . ."[2]

Squaring the Circle by Calculation

In this chapter, you'll see that there are many approximate ways to square the circle. Although these methods are not exact, they do convey the same symbolic meaning of squaring the circle. Before we begin, let's review how to find the exact perimeter and area of a square and the circumference and area of a circle, so that we can later compare the results of geometric constructions with these exact values. These formulas were introduced in Chapter 4, "Ad Quadratum and the Sacred Cut," and Chapter 6, "The Circle," and are listed below for your reference. Take a look at Figure 8.2, which illustrates these concepts.

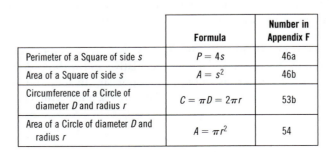

	Formula	Number in Appendix F
Perimeter of a Square of side s	$P = 4s$	46a
Area of a Square of side s	$A = s^2$	46b
Circumference of a Circle of diameter D and radius r	$C = \pi D = 2\pi r$	53b
Area of a Circle of diameter D and radius r	$A = \pi r^2$	54

Now, let's "square" a circle, not with a compass and straightedge but by calculation. For simplicity, let's use a circle of radius 1 unit, called a *unit circle*. The circumference C of a unit circle can be stated as follows:

$$C = 2\pi r = 2\pi(1) = 2\pi.$$

Let s be the side of a square having a perimeter equal to the circumference of the circle. Set the perimeter of the square, $4s$, equal to the circumference of the circle

$$4s = 2\pi$$
$$s = \pi/2 \approx 1.5708.$$

Now, let's turn to areas:

$$\text{Area of a unit circle} = \pi r^2 = \pi(1)^2 = \pi.$$

For a square of side *s* of equal area

$$s^2 = \pi$$
$$s = \sqrt{\pi} \approx 1.772.$$

Figure 8.3 shows a unit circle, a square of equal perimeter having a side of $\pi/2$ units, and a square of equal area having a side of $\sqrt{\pi}$ units. Of course, we're not limited to the unit circle. For *any* circle, the square with a perimeter that equals the circle's circumference has a side $\pi/2$ times the circle's radius. Note that each square has a side *less* than the diameter of the circle and a diagonal *greater* than the diameter of the circle.

Squaring the circle by calculation, as we just did, gives us the exact side of the square having an area (or perimeter) equal to that of a given circle. The method we used was not a geometric method using only compass and straightedge. Therefore, it does not constitute a legitimate solution to the quadrature problem. It does, however, give us exact values with which to evaluate the accuracy of the approximate methods to follow.

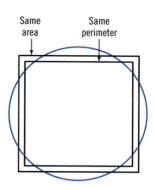

FIGURE 8.3 ■ Squaring the Circle

The Rhind Mathematical Papyrus

Let's consider a very early Egyptian method for squaring the circle. We find it in the Rhind Papyrus, a page of which is shown in Figure 8.4. The formal title of the Rhind Papyrus is *Directions for Knowing All Dark Things*. It is a collection of 85 problems in geometry and arithmetic, and it has been called *the first math book*.[3] It was written about 1700 B.C. by a priest named Ahmes, who copied and edited it from another papyrus dating from about 1800 B.C. In 1858, a Scotsman named A. Henry Rhind bought a large section of the papyrus in Luxor, Egypt. (This section eventually wound up in the British Museum. Some missing pieces turned up in the New York Historical Society 50 years later.)

The papyrus was originally 18 ft long and 13 in. high, and it contains a lot of material. Here, however, we'll look only at the instructions given for finding the area of a circle. The Papyrus states, "Take the diameter of the circle, subtract

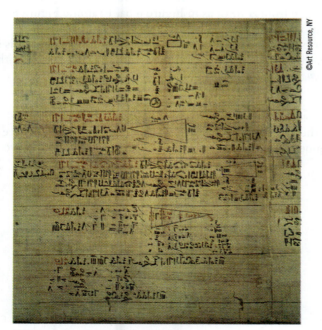

©Art Resource, NY

FIGURE 8.4 ■ Page from the Rhind Papyrus

its ninth part, and square the result to get the area." For a circle of diameter d, subtracting a ninth part gives

$$d - \frac{d}{9} = \frac{9d}{9} - \frac{d}{9} = \frac{8d}{9}.$$

When you square the result, you get

$$\text{Area of a circle} \approx \left(\frac{8d}{9}\right)^2.$$

For a circle of diameter 1, you get

$$\text{Area} \approx \left(\frac{8}{9}\right)^2 \approx 0.779012 \quad \text{using the Papyrus formula}$$

and

$$\text{Area} = \frac{\pi d^2}{4} \approx 0.78540 \quad \text{using Equation 54.}$$

This is a difference of about 0.8%.

Gay Robins and Charles Shute speculate that before the Egyptians devised the previous mathematical method, they may initially have squared the circle as illustrated in Figure 8.5. Draw a square and mark the one-quarter points along its sides. Draw a circle through the points that are one-quarter unit from each corner, as shown in the figure.[4] That circle then has approximately the same area as the square.

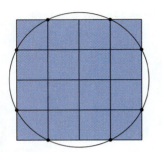

FIGURE 8.5 ■ A Geometric Method for Squaring the Circle

Hippocrates' Lunes

Recall that when we constructed the vesica in Figure 7.67 of Chapter 7, "Circular Designs in Architecture," the remaining portions were crescents, or lunes. Hippocrates of Chios devised a successful way to square the lune. Asger Aaboe has said that this problem "doubtless grew out of attempts at squaring the circle." To square the lune according to Hippocrates, draw a semicircle on the diagonal AC of a square $ABCD$. Then, with D as center and AD as radius, draw a 90° circular arc from A to C, as in Figure 8.6a. Examine Figure 8.6 to see how the steps unfold.

Either of the segments I is a 90° segment of a circle, and so is segment II. Therefore, they are similar. Similar figures, however, have areas with ratios that equal the *square* of their linear ratio. Therefore,

$$\frac{\text{segment I}}{\text{segment II}} = \left(\frac{AB}{AC}\right)^2 = \frac{(AB)^2}{(AC)^2}.$$

However, the last ratio is $\frac{1}{2}$, because AC is the diagonal of a square of side AB. Hence, segment II is twice segment I or equal to the sum of the two segments I.

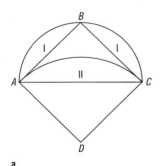

a

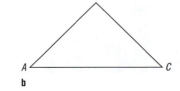

b

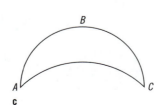

c

FIGURE 8.6 ■ Hippocrates' Lunes

If you remove two segments I or one segment II from the semicircle, you must end up with the same area—because in either case, you have removed the same amount. However, in the first instance, you obtain the triangle *ABC*, as in Figure 8.6b, and in the second, you obtain the crescent or lune *ABC*, as in Figure 8.6c. The triangle and the lune must, therefore, have the same area. As such, you are able to *square* the lune. The area of the lune in Figure 8.7 is half the area of the square *ABCD*.

Hippocrates' successful squaring of the lune was just one of several early Greek attempts to square the circle with compass and straightedge. Anaxagoras (c. 499–c. 427 B.C.), a contemporary of Hippocrates, was the first Greek known to work on the quadrature problem. He is said to have occupied himself with the problem while in prison, but what he contributed is not known. Hippias of Elis (c. 425 B.C.) discovered the curve now known as the *quadratix*. It is used to solve both the famous quadrature and angle-trisection problems we referred to earlier (also see Exercises and Projects at the end of this chapter). In Chapter 9, "The Ellipse and the Spiral," you'll see how Archimedes squared the circle by means of a spiral.

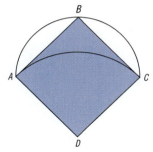

FIGURE 8.7 ■ Squaring the Lune

■ EXERCISES ● THE GEOMETRY OF SQUARING THE CIRCLE

1. Define or describe the following terms:

   quadrature   lune   crescent

2. What does it mean, geometrically, to square the circle?

3. Using The Geometer's Sketchpad, construct a circle concentric with a square. Display the perimeter and area of each on the screen. Then drag a corner of the square while watching the displayed numbers. Note where the perimeters of square and circle are the same and where the areas are the same.

4. a. Show by calculation that the circumference of a circle is approximately equal to three times its diameter plus $\frac{1}{5}$ the side of the inscribed square.
    b. To what approximate value of π does this lead?

5. Refer to Figure 8.8 as you do the following construction for finding the approximate circumference of a circle: Given a circle of diameter *AB*, draw a tangent at *B*. Locate *C* so that angle *BOC* is 30°. On the tangent, lay off *CD* equal to three times the radius *r*. Draw *AD*. When you've completed the construction, show by measurement that 2*AD* is approximately equal to the circumference of the circle. Find the value of π that this implies.

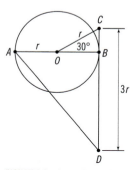

FIGURE 8.8 ■ Approximate Circumference of a Circle

6. Refer to Figure 8.9 on the next page as you complete the following construction, developed by Robert Lawlor,[5] which connects the vesica, golden ratio, Yin-Yang symbol, pentagon, and squaring of the circle.
   a. Draw *x* and *y* axes.
   b. Draw a circle from the origin of radius 1, cutting the *y*-axis at *C* and *F*.
   c. Inscribe two tangent circles of radius $\frac{1}{2}$. This forms the Yin-Yang symbol.
   d. Draw a line from *C* through the small circle center *D* and extend to *E*.
   e. Draw an arc from *C* through *E*, and a similar arc from *F*, tangent to the small circles. This forms a *vesica*.
   f. Connect *F* to the points *A* and *B* where the upper arc of the vesica cuts the circle. This gives two sides of a *pentagon*.

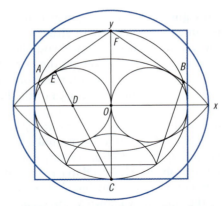

FIGURE 8.9 ■ Lawlor's Construction for Approximately Squaring the Circle

g. Complete the pentagon by stepping the length of the side around the circle.
h. Circumscribe a square around the original circle and circumscribe a circle around the vesica.

Repeat this construction. Then show that the ratio of the radius of the vesica obtained in Step e to that of the original circle is the *golden ratio* Φ. Finally, demonstrate by measurement that the square obtained approximately squares the original circle.

SQUARING THE CIRCLE IN THE GREAT PYRAMID

In Chapter 2, "The Golden Ratio," we examined the claim that the Great Pyramid of Cheops contains the golden ratio. Let's examine a second claim about that pyramid: that it squares the circle.

Perimeter "Squaring" of the Circle

In particular, we want to see if there is any truth to the claim that "the perimeter of the base of the Great Pyramid equals the circumference of a circle with radius equal to the height of the pyramid."[6] The dimensions of the Great Pyramid in terms of Φ, from Figure 8.10, are

$$\text{full base} = 2 \text{ units,}$$

and

$$\text{height} = \sqrt{\Phi}.$$

This makes the perimeter of the base

$$4 \times 2 = 8 \text{ units.}$$

A circle with radius equal to pyramid height ($\sqrt{\Phi}$) has the following circumference:

$$2\pi\sqrt{\Phi} \approx 7.9923 \text{ units.}$$

Therefore, the perimeter of the square and the circumference of the circle agree, to within 0.1%. Within that accuracy, we can state that the Great

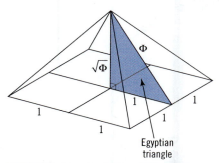

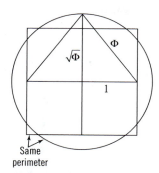

FIGURE 8.10 ■ Approximate Dimensions of the Great Pyramid

FIGURE 8.11 ■ Perimeter Squaring of the Circle in the Great Pyramid

Pyramid's base does in fact approximately square the circle with a radius that is equal to the Great Pyramid's height. See Figure 8.11.

Pi is the ratio of the circumference of a circle to its diameter. You can use the preceding result to get an approximate value for pi. Set the circumference of the circle ($2\pi\sqrt{\Phi}$) equal, approximately, to the base perimeter (8):

$$2\pi\sqrt{\Phi} \approx 8.$$

Divide both sides by $2\sqrt{\Phi}$:

$$\pi = \frac{4}{\sqrt{\Phi}}.$$

The value of pi found from this formula is accurate to within 0.1%.

Area "Squaring" of the Circle

The next claim we want to verify is that "a rectangle whose width equals the base of the Great Pyramid and whose length is twice the height of the pyramid has the same area as a circle whose radius equals the height of the pyramid."[7] The circle is the same as the one we just used for perimeter squaring of the circle. Again, using the Pyramid's dimensions in terms of Φ, and referring to Figure 8.12, we obtain the following area of the indicated rectangle:

$$2(2\sqrt{\Phi}) = 4\sqrt{\Phi} \approx 5.0881 \text{ square units.}$$

The area of a circle of radius $\sqrt{\Phi}$ is

$$\pi r^2 = \pi(\sqrt{\Phi})^2 = \pi\Phi \approx 5.0832 \text{ square units.}$$

Therefore, the rectangle gives the approximate area of the circle, to within 0.1%. This construction does not strictly *square* the circle because it uses a *rectangle*, a square-cornered figure, rather than a square.

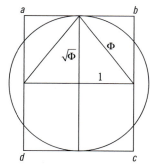

FIGURE 8.12 ■ Area Squaring of the Circle in the Great Pyramid

OTHER EXAMPLES FROM ART AND ARCHITECTURE

In this section, we'll present more examples of squaring the circle, but we won't be as concerned with squaring the circle in a strictly mathematical sense. We'll focus instead on the *idea* of squaring the circle. We'll discuss designs that conspicuously use both the square and the circle, even when they do not have equal perimeters or areas, and we'll even consider examples where a square is not present. As you explore these examples, be sure to keep in mind that these ideas are speculative.

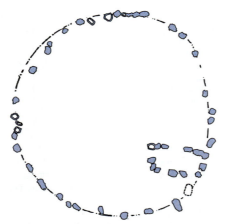

FIGURE 8.13 ■ Thom's Drawing of Castlerigg Stone Circle

Flattened Prehistoric Stone Rings

Although Stonehenge, the most famous stone ring, is circular, not all Neolithic stone rings are circular. Figure 8.13 shows the Castlerigg stone circle in Cumbria, Britain. What is most striking about this ring is that it is *flattened* on one side, as are many of the rings in Britain.

Why go to the trouble of producing a flattened circle when a circular shape is so much easier to produce? Recall that the radius of a circle will fit around the perimeter of the circle six times, plus a bit more—actually 2π times. Some speculate that a stone circle was flattened to make its perimeter *an integral multiple of the radius* drawn to the circular part of its perimeter. For the Castlerigg stone circle, the radius of the primary circle will fit around the perimeter about six times.

Why? According to Keith Critchlow, "Division by six is inherent and fundamental to the circle (its own radius will always mark out six equal parts of its perimeter . . .)." He further states, "The constructions . . . were typical of those which numerically rationalize the perimeter of the primary circle. This balance between 'irrational' geometry and rational numbering is a fundamental reconciliation lying at the roots of sacred geometry. . . . The squaring of the circle is a . . . symbol [of] . . . the establishing of Heaven on Earth. . . ."[8] It was perhaps an attempt to *rationalize the irrational*. You may recall from legend the Pythagoreans' anguish at their discovery of incommensurable magnitudes and irrational numbers, which would have ruined their doctrine based on whole numbers. The story goes that Hippasus was put to death by drowning for making known the discovery of the irrational and incommensurable.

The Sacred Cut

Recall from Chapter 4, "Ad Quadratum and the Sacred Cut," and Chapter 5, "Polygons, Tilings, and Sacred Geometry" that Brunés coined the term *sacred cut*. He believed the "sacredness" of the cut lies in its very nearly solving the riddle of how to square the circle. Figure 8.14 illustrates how the sacred cut can be used to square the circle. According to Brunés, the length of each arc *AOB* equals the length *CD* of the diagonal of half the reference square, to within 0.6%. Therefore, a square with sides equal to such diagonals equals (approximately) the perimeter of a circle composed of four sacred cut arcs.[9]

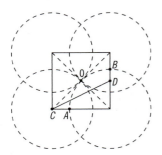

FIGURE 8.14 ■ The Sacred Cut

Vitruvian Man

In his *Ten Books on Architecture*, Vitruvius wrote the following famous passage:

> "... in the human body the central point is naturally the navel.
>> For if a man be placed flat on his back, with his hands and feet extended,
>>> and a pair of compasses centered at his navel,
>> the fingers and toes of his two hands and feet
>>> will touch the circumference of a circle described therefrom.
>> And just as the human body yields a circular outline,
>>> so too a square figure may be found from it.
>> For if we measure the distance from the soles of the feet to the top of the head,
>>> and then apply that measure to the outstretched arms,
>>>> the breadth will be found to be the same as the height. ..."[10]

The Vitruvian Man has been the subject of many illustrations. Figures 8.15 through 8.18 depict just a few. Note the three concentric squares in Figure 8.17. One is inscribed in the circle, one is circumscribed about the circle, and a third is between the other two. This implies that the square of equal area must lie between the inscribed and circumscribed squares.

Vitruvius was introduced in Chapter 1, "Music of the Spheres."

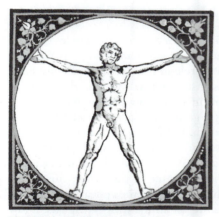

FIGURE 8.15 ■ Vitruvian Man from Fra Giocondo's Edition of Vitruvius

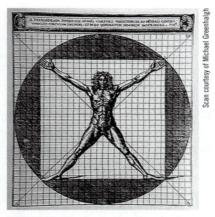

Scan courtesy of Michael Greenhalgh

FIGURE 8.16 ■ Vitruvian Man from Cesariano's Edition of Vitruvius

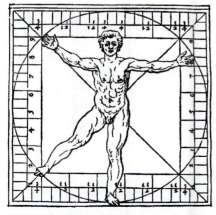

FIGURE 8.17 ■ Vitruvian Man from Scamozzi's *Idea dell'architettura universale*

From Ernst Lehner, *Symbols, Signs, and Signets* (Dover Publications, Inc.)

FIGURE 8.18 ■ An Asian Version of "Vitruvian" Man

FIGURE 8.19 ■ Part of a Page from Leonardo's Notebooks, *Codex Atlanticus*

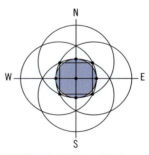

FIGURE 8.20 ■ Orthogonal Vesicas

Of Vitruvius' statement about the man with outstretched arms and legs, Kenneth Clark writes, "It is impossible to exaggerate what this simple-looking proposition meant to men of the Renaissance. To them it was far more than a convenient rule: It was the foundation of a whole philosophy. Taken together with the musical scale of Pythagoras, it seemed to offer exactly that link between sensation and order, between an organic and a geometric basis of beauty, and which was (and perhaps still remains) the philosopher's stone of aesthetics. Hence the many diagrams of figures standing in squares or circles that illustrate the treatises on architecture or aesthetics from the fifteenth to the seventeenth century."[11]

Perhaps the most famous illustration of Vitruvian Man—of squaring the circle—is the one by Leonardo that you saw in Figure 8.1. Augusto Marinoni wrote, "The problem in geometry that engrossed Leonardo interminably was the squaring of the circle. From 1504 on, he devoted hundreds of pages in his notebooks [see, for example, Figure 8.19] to this question of quadrature . . . that so fascinated his mentor Pacioli. In his later years he indulged in a kind of intellectual game . . . [which] produced no appreciable gain for the science of mathematics, but . . . did create a multiplicity of complex and pleasing designs."[12]

Chapter 12, "Brunelleschi's Peepshow and the Origins of Perspective," will have much more information about Leonardo.

Orthogonal Vesicas

Recall from Chapter 7 that a vesica is the figure formed when two circles overlap. *Orthogonal* vesicas are perpendicular to each other. Note the two perpendicular vesica shapes that immediately contain the "circle in the square" in Figure 8.20. Critchlow claims that this circle-squaring construction was used in the layout of Hindu temples, which have their own roots in astronomy and sacred geometry. For example, a Hindu temple is essentially a grand mandala, laid out in relation to the four cardinal directions of North, South, East, and West. To complete the circle-squaring construction that is representative of such temples, perform the following steps. First, draw an East–West axis, and draw two circles with centers on this axis, each circle passing through the center of the other; this forms a vesica. Draw a North–South axis through the center of the vesica. Then draw a second vesica on the North–South axis, centered on the East–West axis. Draw a circle through the centers of the four circles. Finally, connect the four intersection points of the vesicas to get a square. The square is approximately equal in perimeter to the circle, a *squaring of that circle*.

The Mandala[13, 14]

Chapter 6 introduced the *mandala*, which literally means "circle and center," or "holy circle." Recall that the mandala often contains a square as well as a circle. (Figure 6.39 shows exactly this.) The square central pattern may indicate the four cardinal points, or the four directions. They also symbolize the four elements (earth, air, fire, water) and the four seasons. Even if the square does not "square" the circle mathematically, the symbolism is there.

The Dome

The dome may be taken as another example of squaring the circle. Titus Burckhardt, a scholar of Middle Eastern studies and history, writes about sacred buildings or sanctuaries, "The most general meaning of a sanctuary is the reconciliation of earth and Heaven. . . . In the architectural form of the sanctuary, this can be expressed outwardly in several ways; however, the linking of the two . . . poles 'heaven' and 'earth' is expressed with particular eloquence when the sanctuary consists of a square building surmounted by a cupola: The cupola represents heaven, whereas the earth . . . subject to the four elements and the four seasons, is 'square.'"[15] We've already mentioned the Dome of the Rock in Jerusalem. Other domes of major architectural significance are the Hagia Sophia in Istanbul and the Pantheon in Rome. Chapter 11, "The Sphere and Celestial Themes in Art and Architecture," will discuss the dome in greater detail.

Other Designs Combining Square and Circle

Many designs incorporate both square and circle. Figures 8.21 through 8.25 offer a good sampling. Although such designs usually do not show the squaring of the circle in the strict sense of the term, the inclusion of both geometric

FIGURE 8.21 ■ Quatrefoil Outline on a Cathedral

FIGURE 8.22 ■ Inlaid Design

FIGURE 8.23 ■ Rose Window at Lausanne

FIGURE 8.24 ■ *An Allegory of Passion*, Attributed to Hans Holbein the Younger, 1530

FIGURE 8.25 ■ A Celtic Cross

figures in a single design may indicate that the artist had the underlying ideas in mind. In fact, the mason's symbol, which we saw in Figure 7.23, shows the square and circle *indirectly* by means of the instruments used to draw them.

Some paintings from the Renaissance and later show a rectangular frame surmounted by a semicircle, as does the painting by Piero della Francesca shown in Figure 8.26. Figure 8.27 shows how John's arm and Jesus' loincloth continue the line of the circle. The square and circle overlap, as do the two circles in a vesica, but a square replaces one circle. Jesus appears in the area common to both, mediating between heaven and earth. Piero has combined the idea of squaring the circle and the idea of the vesica into one painting.

A further note about this painting: In *Della pittura* (*On painting*), one of Piero's contemporaries, Leone Battista Alberti, made some suggestions for *istoria*, or history paintings. He wrote, "I like there to be someone in the 'istoria' who tells the spectators what is going on, and either beckons them with his hand to look, or with ferocious expressions and forbidding glance challenges them not to come near, as if he wished their business to be secret, or points to some danger or remarkable thing in the picture, or by his gestures invites you to laugh or weep with them."[16] In Piero's *Baptism*, this role is filled by the three angels, one of whom makes eye contact with the viewer. Note the recurring motif of the three graces. The figures are simplified and geometric, earning Piero the nickname "the first Cubist."

From Square to Circle, a 1968 painting by Ad Dekkers (not shown), is an example from twentieth-century abstract geometric art. Apparently, the fascination with squaring the circle continues in later centuries.

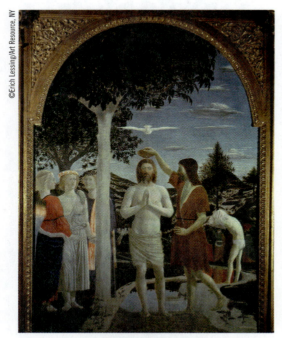

FIGURE 8.26 ■ *The Baptism of Christ*, Piero della Francesca, 1450

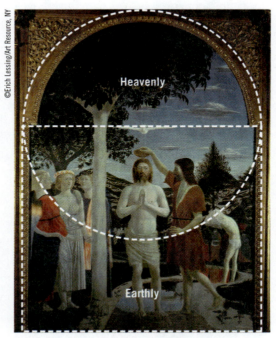

FIGURE 8.27 ■ Overlay of *The Baptism of Christ*

SUMMARY

Precisely squaring the circle with a compass and straightedge construction is impossible. Dante alludes to this at the end of the final canto of *The Divine Comedy*:

> "As the geometrician, who endeavors
> To square the circle,[17] and discovers not,
> By taking thought, the principle he wants . . ."

Their inability to exactly square the circle did not deter artists from using the ideas behind squaring the circle in works of art and architecture. But what are these ideas? Earth, with its square corners of dwellings and four cardinal directions, is often represented by the square. Heaven, perfect and without end, is often represented by the circle. As George Ferguson puts it, "The circle, or ring, has been universally accepted as the symbol of eternity and never-ending existence . . . the square, in contrast with the circle, is the emblem of the earth, and of earthly existence."[18] Carl Jung believed that the circle symbolizes the processes of nature or of the cosmos as a whole, while the square refers to the universe as conceived and projected by man. The circle represents the superconscious aspects of nature, and the square represents the conscious and rational aspects.[19]

So the significance of squaring the circle, either in a strict mathematical sense or in the uniting of both figures in a single design, might be an attempt to reconcile the heavenly and infinite with the earthly and man-made. It may also be seen as an attempt to mediate between the rational and irrational, or, as Kenneth Clark said of squaring the circle, it "seemed to offer exactly that link between sensation and order, between an organic and a geometric basis of beauty. . . ."

Taken as the title of this book, "squaring the circle" may represent the reconciliation of geometry and mathematics on one hand with art and architecture on the other.

The following saying is attributed to late-Gothic stonemasons:

> "A point that goes into the circle,
> Inscribed in the square and the triangle;
> If you would find this point, you possess it,
> And are freed from care and danger;
> Herein you have the whole of art,
> If you do not understand this, all is in vain."[20]

EXERCISES AND PROJECTS

1. Define or describe the following terms:

sacred cut	Vitruvian Man
mandala	orthogonal vesicas

2. Name, describe, and sketch four ways to approximately square a circle.

3. The union of opposites is a major theme of sacred geometry. What geometric symbols or constructions express this idea? Explain each.

4. Test Vitruvius' claims about the proportions of the human body by obtaining actual measurements of a person lying flat on the floor.

5. Why is the sacred cut called *sacred?*

6. Check the accuracy of the value of pi found from the perimeter squaring of the circle in the Great Pyramid.

7. Draw and measure as accurately you can the square and the circle in Leonardo's *Vitruvian Man*. Then calculate how accurately this gives a squaring of the circle, either for perimeter or for area.

8. Do the sacred cut construction. Then determine by calculation how accurately the sacred cut squares the circle.

9. Using an enlarged copy of the layout of the stone circle at Castlerigg, measure the perimeter of the circle. Then measure the radius of the primary circle and divide it into the perimeter. What do you find?

10. Create a personal mandala using any of the elements from this book. Accompany it with a short paper on the subject of the mandala.

11. Make a design based on the squaring of the circle. Use it to make a graphic design, a pillow, a quilt, a screensaver, an inlaid design using cut-out pieces, a tee shirt, or some use of your own.

12. Search an art book or a gallery for paintings that seem to incorporate the theme of squaring the circle. Select a painting, research it further, and write a paper on how and why the artist used this theme.

13. Search an art book for pictures (such as Piero's *The Baptism of Christ*) that contain both the rectangle and the circle. Has the artist used the shapes in a similar way? Write a few paragraphs on your findings.

14. An example of a circle with a perimeter that is an integral multiple of the diameter is found in the Old Testament, I Kings 7:23. Solomon, who was building a house, fetched Hiram, who "made a molten sea, ten cubits from one brim to the other: it was round all about . . . and a line of thirty cubits did compass it round about." Write a short paper commenting on these dimensions.

15. Write a book report on any of the sources listed at the end of this chapter.

16. Write a tongue-in-cheek short paper on why the fast-food chain restaurant Wendy's chose to serve square hamburgers on round buns. Let your imagination run wild! Have a competition for the best paper.

17. Here are some suggestions for short papers or term papers. Come up with your own topic, using these to jog your imagination:

    ■ Attempts at squaring the circle throughout history

    ■ The history of π (pi)

    ■ The influence of Vitruvian Man in art and architecture

- Why an Egyptian king would build circle squaring into a pyramid
- The history of the Rhind Papyrus and its influence on later mathematics
- The mandala
- Neolithic stone rings and speculations about why some were flattened
- Hippocrates' squarable lunes, other than the one shown in this chapter
- Leonardo's squarable curvilinear figures: pendulum, cat's eye, falcate, shark's fin, Cleopatra's headdress, and claw. See *Leonardo's Dessert* on the reading list.
- Leonardo's attempts to square the circle. See Reti, p. 74.
- Tom Stoppard's play, *Squaring the Circle*, and why he chose this title
- Heath's history of squaring the circle in *A History of Greek Mathematics*, pp. 220–235

18. Make a round cake or pizza, and a square one of the same area, as an edible demonstration of squaring of the circle.

Mathematical Challenges

19. The circumference and area of a circle must be greater than that of the inscribed square, but less than that of the circumscribed square, as shown in Figure 8.28. Given a circle of diameter 1, perform the following:
    a. Find the side of the inscribed and the circumscribed square.
    b. Compute the arithmetic mean, the geometric mean, and the harmonic mean of the sides of the two squares.
    c. Using each mean as the side of a square, compute the perimeter and area of the three new squares.
    d. Compare the perimeter and area of each of the three new squares with the circumference and area of the original circle. Which gives the best approximation for circumference? For area? Use a calculator or a spreadsheet.

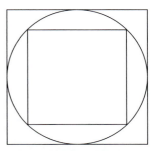

FIGURE 8.28 ■ A Circle with an Inscribed and a Circumscribed Square

20. Do the construction for a circle intersecting a square at the one-quarter points (see Figure 8.5). Show algebraically that the circle and square have approximately the same area. Determine the accuracy of the approximation.

21. Show that the area of a circle is given by a square with a side that equals the mean proportional between its radius and half its circumference.

22. Demonstrate Proposition 1 in Archimedes' *Measurement of a Circle*, which states that the area of a circle is equal to that of a right triangle with legs that equal the radius and the circumference of the circle.

23. *The Quadratrix of Hippias.* Hippias of Elis discovered a curve called the *quadratrix* that has been used to trisect an angle and to square the circle. Research the quadratrix and write a paper on your findings. You may find pages 226–230 of Sir Thomas Heath's *A History of Greek Mathematics* a good starting point.

24. Make an oral presentation on any project in this section you've decided to pursue.

SOURCES

Argüelles, José, and Miriam. *Mandala*. Boston, MA: Shambhala, 1985.

Burckhardt, Titus. *Chartres and the Birth of the Cathedral*. Ipswitch, UK: Golgonooza, 1995.

Campbell, Joseph, with Bill Moyers. *The Power of Myth*. New York: Doubleday, 1988.

Cowen, Painton. *Rose Windows*. London: Thames and Hudson, 1979.

Critchlow, Keith. *Time Stands Still*. New York: St. Martin's, 1982.

Eco, Umberto. *Art and Beauty in the Middle Ages*. New Haven, CT: Yale University Press, 1986.

Ferguson, George. *Signs & Symbols in Christian Art*. London: Oxford University Press, 1954.

Heath, Sir Thomas. *A History of Greek Mathematics*. New York: Dover, 1981. First published in 1921.

Jung, Carl G., et al. *Man and His Symbols*. New York: Dell, 1964.

Lawlor, Robert. *Sacred Geometry*. New York: Thames & Hudson, 1982.

March, Lionel. *Architectonics of Humanism*. Chichester, UK: Academy, 1998.

Newman, James R., ed. *The World of Mathematics*. New York: Simon and Schuster, 1956.

Reti, Ladislao, ed. *The Unknown Leonardo*. New York: McGraw-Hill, 1974.

Robins, Gay, et al. *The Rhind Mathematical Papyrus*. New York: Dover, 1987.

Schwaller de Lubicz, R. A. *The Egyptian Miracle*. Rochester, VT: Inner Traditions, 1985.

Thom, Alexander. *Megalithic Sites in Britain*. Oxford: Clarendon, 1967.

Tompkins, Peter. *Secrets of the Great Pyramid*. New York: Harper, 1971.

Wills, Herbert, III. *Leonardo's Dessert, No Pi*. Reston, VA: National Council of Teachers of Mathematics, 1985.

NOTES

1. Heath, p. 220.
2. Eves, p. 81.
3. Robins, p. 44; Newman, p. 12.
4. Robins, p. 45.
5. Lawlor, p. 21.
6. Tompkins, p. 198.
7. Tompkins, p. 198.
8. Critchlow, p. 85.
9. Kappraff, *Connections*, p. 28.
10. Vitruvius, Chapter 1, Para. 3.
11. Clark, *The Nude*, p. 36.
12. Reti, p. 76.
13. Argüelles, José and Miriam, *Mandala*.
14. Joseph Campbell, *The Power of Myth*.
15. *Chartres and the Birth of the Cathedral*, p. 17.
16. Alberti, *On Painting*, p. 77.
17. This is the Longfellow translation. Other translations say to *measure* the circle. We will give a longer excerpt of this passage in Chapter 11.
18. *Signs & Symbols in Christian Art*, p. 153.
19. Argüelles, p. 38.
20. C. Alhard von Drach, *Das Hüttengeheimnis vom gerechten Steinmetzen-Grund*. Marburg, 1897. From Burckhardt, p. 104.

"In addition to the straight lines, circles, planes, and spheres with which

every student of Euclid is familiar, the Greeks knew the properties of

the . . . ellipse, parabola, and hyperbola. Kepler discovered . . . that

the planets move in ellipses."

J. L. SYNGE

FIGURE 9.1 ■ **Portrait of Baroness de Schlichting, Christina Robertson, 1848**
Réunion des Musées Nationaux/Art Resource, NY

9

The Ellipse and the Spiral

We've studied a great deal of Euclidean geometry up to this point. This type of geometry was named after Euclid, although others such as Archimedes have had major influences in its development. In the seventeenth century, a new kind of geometry was born when the mathematicians René Descartes and Pierre de Fermat merged two branches of mathematics: algebra and geometry. This new *analytic geometry* proved to be so effective that it was used for most of the new developments in geometry. In analytic geometry, the position of a point is represented by *coordinates* on a set of *axes*. As such, it is also called *coordinate geometry*.

A collection of points can define a geometric figure, such as the circle, and algebraic methods can then be used to study that geometric figure in ways that are difficult or impossible in Euclidean geometry. Conversely, an algebraic equation can be described by a *graph* of that equation, that is, a *picture* of the equation. For example, after the circle in Figure 9.2 is drawn on *coordinate axes*, it can be described by the equation $x^2 + y^2 = 25$. These features are especially appealing for those in the visual arts because they allow us to make geometric pictures

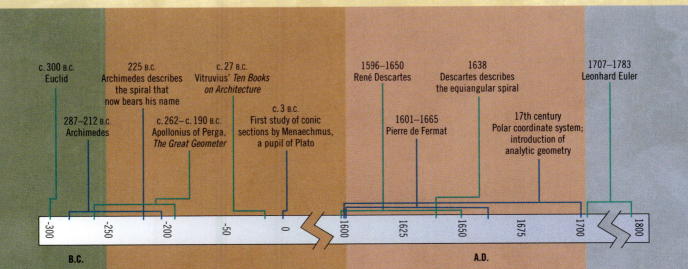

c. 300 B.C.
Euclid

287–212 B.C.
Archimedes

225 B.C.
Archimedes describes
the spiral that
now bears his name

c. 262–c. 190 B.C.
Apollonius of Perga,
The Great Geometer

c. 27 B.C.
Vitruvius' *Ten Books
on Architecture*

c. 3 B.C.
First study of conic
sections by Menaechmus,
a pupil of Plato

1596–1650
René Descartes

1601–1665
Pierre de Fermat

1638
Descartes describes
the equiangular spiral

17th century
Polar coordinate system;
introduction of
analytic geometry

1707–1783
Leonhard Euler

-300   -250   -200   -50   0   1600   1625   1650   1675   1700   1800

B.C.   A.D.

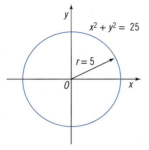

FIGURE 9.2 ■ A Circle and Its Equation

of equations and conversely to write equations for pictures of geometric figures. Up to now, we've made geometric figures by *construction*. Now we have another tool. We can make geometric figures by *calculation*.

In this chapter, you'll see that our old friend the circle is one of the *conic sections*—the set of curves obtained by a plane cutting a cone—and you'll be introduced to another conic section, the ellipse, and its construction. You'll also see some examples of its appearance in art and architecture. After learning how to plot points and curves in *rectangular coordinates*, you'll learn how to construct an ellipse from its equation. After that, we'll explore various spirals. For these, you'll see that another kind of coordinate system, *polar coordinates*, makes the work easier. Finally, we'll discuss the use of the *spiral* in art.

THE ELLIPSE

The ellipse is defined in two different ways: first, as one of the conic sections, and later, as a point moving in a plane under certain restrictions.

The Conic Sections

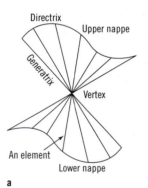

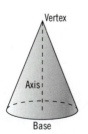

FIGURE 9.3 ■ A Conical Surface and a Cone

A *conical surface* is the surface generated by a straight line that passes through a given point, called the *vertex*, and follows a given curve, as shown in Figure 9.3a. The straight line is called the *generatrix*, or generating line, and the curve is called the *directrix*, or directing line. The vertex separates the conical surface into two *nappes*. Each position of the generatrix is called an *element* of the conical surface.

Now, let the directrix be a *circle*. Draw a perpendicular to the plane of that circle through its center. If the vertex is on that perpendicular, the conical surface is a *circular conical surface*. The line through the vertex and the center of the circle is called the *axis* of the surface. When a plane cuts the circular conical surface, the solid bounded by one nappe of the surface and the plane is called a *cone*. The curve of intersection is called the *base* of the cone. If the base is perpendicular to the axis, the cone is a *right circular cone*, as shown in Figure 9.3b.

When a plane cuts (or sections) a circular conical surface, the curve is called a *conic section*. As you can see in Figure 9.4, the particular conic section obtained depends upon the angle of the sectioning plane. When the cutting plane is perpendicular to the conical surfaces' axis, the section is a circle. When the plane is tilted a bit, but not so much as to be parallel to an element, the section is an

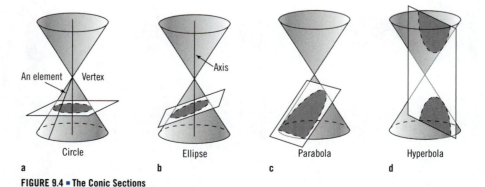

FIGURE 9.4 ■ The Conic Sections

ellipse. When the plane is parallel to an element, the section is a *parabola*. When the plane is steep enough to cut both nappes of the cone, the section is a two-branched *hyperbola*.

Conic sections were first studied in the fourth century B.C. by Menaechmus, a pupil of Plato. He began studying them while trying to solve the problem of constructing a cube twice as large by volume as a given cube, the so-called *Delian* problem, which will be mentioned again in Chapter 10, "The Solids." Euclid wrote four books, now lost, on conic sections. Archimedes devised a way to find the area of the ellipse and of a sector of the parabola. Apollonius of Perga, however, is credited with making the most important contributions to the geometry of the conic sections.

The Ellipse Defined as the Path of a Moving Point

We already have a definition of an ellipse as one of the conic sections. A second definition of the ellipse is given in Statement 88.

88 DEFINITION OF AN ELLIPSE

An ellipse is the set of all points in a plane such that the sum of the distances from each point on the ellipse to two fixed points (called the *foci*) is constant. ●

Statement 89 defines the different parts of an ellipse, and Figure 9.5 shows the typical shape of an ellipse, with the different parts labeled. The figure clearly illustrates that an ellipse has mirror symmetry about both its major and minor axes and has twofold rotational symmetry about its center.

89 PARTS OF AN ELLIPSE

The ellipse has two axes of symmetry: the *major axis* and the *minor axis*. They intersect at the center of the ellipse. A *vertex* is a point where the ellipse crosses the major axis. Half the lengths of the axes are called the *semi-major* and *semi-minor* axes. Each *focus* lies on the major axis. The foci are equidistant from the center. ●

Drawing an Ellipse

A good workshop method for drawing an exact ellipse is to use a loop of string stretched taut around two nails located at the focal points, as in Figure 9.6. This method gives an ellipse that is theoretically exact. This way of drawing an ellipse is sometimes called the *gardener's method*.

APOLLONIUS OF PERGA
(c. 262–c. 190 B.C.)

Apollonius was known by his contemporaries as *The Great Geometer*, and his treatise on conics is considered one of the greatest mathematical works from the ancient world. Of his eight books on conics, seven have survived. His works include: *On the Burning Mirror*, a book about optics in which he discusses the focal properties of the spherical and parabolic mirrors; *On the Cylindrical Helix*; and *Comparison of the Dodecahedron and the Icosahedron*. Apollonius also computed a value of π that was more accurate than what was previously known.

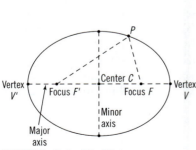

FIGURE 9.5 ■ Parts of the Ellipse

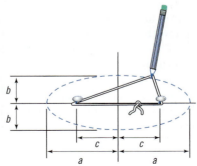

FIGURE 9.6 ■ Drawing an Ellipse with a Loop of String

To use this method, you must be able to find the distance c from the center to each focal point, for a particular ellipse with semi-axes a and b. The three dimensions are related by Equation 90. (See the exercises at the end of this section for a derivation of this formula.)

90 DISTANCE FROM CENTER TO FOCUS OF AN ELLIPSE

$$c = \sqrt{a^2 - b^2}$$

● **EXAMPLE:** Lay out an elliptical mat for a portrait, 12.4 in. long and 8.8 in. wide.

● **SOLUTION:** First, find the distance c from the center to each focus. Here, $a = 6.2$ and $b = 4.4$. From Equation 90, you get

$$c = \sqrt{(6.2)^2 - (4.4)^2} = 4.4 \text{ in.}$$

Draw the two axes of the ellipse, and drive a nail or put a pushpin at each focal point. Next, loop a piece of string around the nails and adjust its length so that the pencil point passes through the ends of each axis. Then, keeping the string taut, draw the ellipse. ◾

The Approximate Ellipse or Oval

"The Renaissance cherished the circle as the shape of cosmic perfection, whereas the Mannerist phase of the Baroque took to the high-strung ellipse, which plays on the ambivalence of roundness vs. extension."

RUDOLPH ARNHEIM

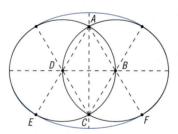

FIGURE 9.7 ◾ Ovato Tondo

Renaissance craftsmen had an approximate method for drawing an ellipse, getting what was called the *ovato tondo*.[1] The construction in Figure 9.7 is taken from Serlio's *The Five Books of Architecture*, Chapter 1, Folio 11. To perform this construction, draw two circles, each passing through the center of the other, intersecting at A and C. (Note that the overlapping circles in the ellipse form a vesica shape.) Then draw line AB and extend it to F. Similarly, draw line ADE. Next, from A swing an arc from E to F. Similarly, swing an arc with C as center, completing the *ovato tondo*.

In later chapters of *The Five Books of Architecture*, Serlio goes on to show the use of the oval in structures. Figure 9.8 is just one example. Also see his oval temple, in Book 5, Chapter 14, Folio 4. Three more of Serlio's oval constructions are presented in the exercises at the end of this section.

Figure 9.9 illustrates another useful workshop method for drawing an approximate ellipse using tangent lines. This method also works for inscribing an approximate ellipse into a parallelogram. To use tangent lines to draw an oval, follow these steps: Draw a rectangle into which one quadrant of the ellipse is to fit. Then subdivide the edge PQ by a number of equally spaced points (seven are shown in Figure 9.9), and subdivide edge QR by the *same number* of points. Number the points as shown in Figure 9.9. Connect point 1 at the top to point 1 at the side, connect 2 to 2, connect 3 to 3, and so on. Draw the ellipse, tangent to the lines drawn in the previous steps. Finally, draw the ellipse in the remaining quadrants by symmetry.

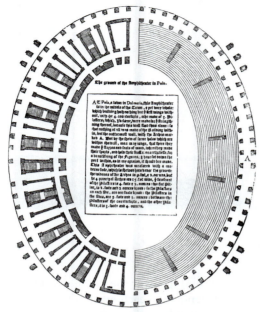

FIGURE 9.8 ■ An Oval Amphitheater from Serlio

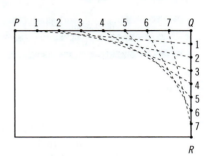

FIGURE 9.9 ■ Oval by Tangent Lines

The Ellipse in Art and Architecture

In art and architecture, elliptical shapes are used prevalently in frames, windows, and arches. For example, the upright ellipse is especially good for framing portraits, as shown earlier in Figure 9.1. Again quoting Arnheim, "For the portrait, the oval lends welcome assistance in the painter's struggle with the human figure, which carries the head high above its center. The upper focal point of the ellipse offers the head . . . a compositional resting place not available in either the tondo or the rectangle. . . . The [horizontal] ellipse can be perceived as the result of the interaction of two spheres of forces. Thus it is a good choice for duets, dialogs, partners, antagonists . . . two centers of energy coping with each other."[2]

The ellipse is popular on United States paper currency, as shown in Figure 9.10. The presidents are placed in elliptical frames, as are depictions of the White House, the Lincoln Memorial, the U.S. Treasury, and the numerals on some denominations. The ellipse is also the most popular shape for the cameo, a type of jewelry that has been popular since the sixth century B.C. in ancient Greece. Figure 9.11 provides a more recent example.

FIGURE 9.10 ■ United States Paper Money

FIGURE 9.11 ■ A Cameo

a   b

FIGURE 9.12 ■ (a) An Elliptical Window; (b) A Semi-Elliptical Arch

The ellipse has the same advantage over the circle that the pointed arch has: Its height and width are independent of each other. This makes it extremely useful as an architectural frame for windows and arches, as seen in Figure 9.12.

■ **EXAMPLE:** The outer frame of the elliptical window in Figure 9.12a is to fit into a rectangle 50 in. high and 40 in. wide. Make a template using nails and a loop of string.

● **SOLUTION:** Find the distance between the nails by using Equation 90, with $a = 25$ and $b = 20$.

$$c = \sqrt{25^2 - 20^2} = 15 \text{ in.}$$

Lay out a 50×40 in. rectangle, draw its centerlines, and drive nails 15 in. from its center. Adjust the string so that the pencil line will pass through the ends of the major and minor axes, and swing the curve. See Figure 9.13.   ■

■ EXERCISES ● THE ELLIPSE

1. Sketch an ellipse. Then draw and define the following terms:

   major axis   minor axis   semi-major axis
   focus   vertex   center
   semi-minor axis

2. Describe at least two methods for drawing an oval.

3. *Distance from Center to Focus of an Ellipse.* We used Equation 90 to find the location for the tacks when drawing an ellipse with a loop of string. See if you can follow and reproduce the following derivation, and then present it to your class.

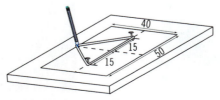

FIGURE 9.13 ■ Drawing an Elliptical Window

Because the sum of the distances from each point P on the ellipse to the foci is constant, no matter where on the ellipse you choose point P, you can say that

$$PF + PF' \text{ is constant.}$$

If P is taken at a vertex V, then

$$VF + VF' = \text{constant.}$$

However, $VF + VF'$ is the length $2a$ of the major axis, so

$$VF + VF' = 2a.$$

Now, let's move P to the end of the semi-minor axis, as in Figure 9.14. Here, PF and PF' are equal. Because their sum is $2a$, each of them must equal a. Using the Pythagorean theorem in triangle PCF gives $a^2 - b^2 = c^2$, or $\sqrt{a^2 - b^2} = c$.

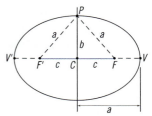

FIGURE 9.14 ■ Figure for Exercise 3

4. An ellipse is 28 cm long and 12 cm wide. Find the distance from its center to each focal point.

5. Draw an ellipse using two tacks and a loop of string. Let the length of the ellipse be 12 in. and the width be 8 in.

6. *Folding an Ellipse.* Fold an ellipse by using the following method, which is illustrated in Figure 9.15. When you're finished, explain why point P is on the ellipse.

   Draw a circle with center at F and cut it out. Fold a diameter (it will be the major axis of the ellipse). On it, locate a point F'. (F and F' will be the foci of the ellipse.) Choose any point Q on the circle. Fold F' onto Q, make crease T, and unfold. (Crease T is a tangent line to the ellipse.) Fold a diameter through Q. Label as P the point where this diameter intersects the preceding crease. (P is a point on the ellipse.) Choose other points on the circle and repeat the construction, getting another point on the ellipse and another tangent to the ellipse with each repetition.

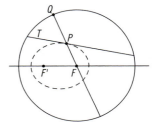

FIGURE 9.15 ■ Folding an Ellipse

7. Repeat the construction in Exercise 6; however, instead of paper, use a computer drawing program such as The Geometer's Sketchpad.

8. In Figure 9.16, the author used a loop of chain stretched around two small pulleys to guide an oxyacetylene cutting torch to cut an ellipse for a steel sculpture. The dimensions of the ellipse are 6 ft by 8 ft. Find the distance between the pulleys.

9. Many computer drawing systems, such as The Geometer's Sketchpad, have built-in tools for drawing ellipses. If you have this feature, learn how to use it.

FIGURE 9.16 ■ Cutting a Steel Ellipse

10. Do the calculations and make a sketch for a template for the outer frame of the semi-elliptical fanlight of Figure 9.12b. Take the width as 60 in. and the height as 24 in.

11. *Serlio's Oval Constructions.* You have already seen one oval construction from Serlio's *The Five Books of Architecture.* Here are three more. Try to reproduce each of them, manually or by CAD.

    a. The first construction method, shown in Figure 9.17a, has an advantage over the constructions shown in Figure 9.17b and Figure 9.17c. With this construction method, you can change the ratio of ellipse length to width. Follow these steps: Set two equilateral triangles, one above the other. Extend the sides of the triangles and choose a point 1 on line BC. From B, swing an arc from 1 to 2, and from C, swing an arc from 1 to 3. From A, swing an arc from 3 to 4, and from D, swing an arc from 2 to 4.

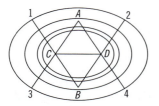
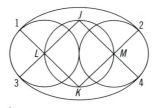
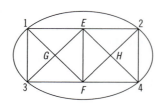

a   b   c

FIGURE 9.17 ■ Serlio's Oval Constructions

b. A second method is laid out in Figure 9.17b. Follow these steps: Draw two tangent circles and draw a third circle passing through their centers, with all three circles of equal size. Then draw a vertical through the center circle, cutting it at *J* and *K*. Next, draw lines from *J* and *K* through the centers of the outer circles, cutting them at 1, 2, 3, and 4. Finally, from *K*, draw an arc from 1 to 2, and from *J*, draw an arc from 3 to 4.

c. A third method is laid out in Figure 9.17c. Follow these steps: Draw a double-square rectangle 1, 2, 3, 4 and draw the diagonals of each square. From *G*, draw an arc from 1 to 3; from *H*, draw an arc from 2 to 4; from *F*, draw an arc from 1 to 2; and from *E*, draw an arc from 3 to 4.

GRAPHING THE ELLIPSE

This section explains how to graph an exact ellipse. However, before you tackle graphing ellipses, you'll need to understand graphing in general. If necessary, you may skip this section without losing continuity, but realize that the methods shown here will enable you to graph any equation, not just that of the ellipse.

Graphing in Rectangular Coordinates

Rectangular coordinates are also called *Cartesian coordinates,* after the Latinized term *Cartesius,* referring to the philosopher and mathematician René Descartes.

You can plot a single number by placing a point on a line, as in Figure 9.18. Select a *zero* point on the line, and choose one direction (usually to the right) for increasing values, marked with an arrowhead. The opposite direction is for decreasing values. Then choose a *scale*—that is, the distance one unit will span on the line—and place and number *tic marks* on the line. To plot a point, simply locate its position on the line. You can plot fractions and irrational numbers by using their approximate decimal values. For example, the numbers 4, π, Φ, 9/2, -1.75, $-\sqrt{30}$, and $-7/2$, and 5.63 are shown on the number line in Figure 9.18.

Let's take a *second* number line and place it at a right angle to the first one, so that each intersects the other at the zero mark, as you see in Figure 9.19. The horizontal number line is called the *x-axis,* and the vertical line is called the *y-axis.* Their point of intersection is called the *origin.* These two axes divide the plane into four *quadrants,* which are numbered counterclockwise from the upper right.

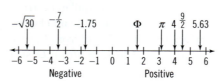

FIGURE 9.18 ■ The Number Line

Figure 9.20 shows a point P in the first quadrant. Its horizontal distance from the origin (three units) is called the *x-coordinate* or *abscissa* of the point P. Its vertical distance from the origin (two units) is called the *y-coordinate* or *ordinate* of the point. The two dimensions taken together are called an *ordered pair* and are written (3, 2), with the x value always given first. They are also called the *rectangular coordinates* of the point, or simply *the coordinates* of the point. It should be clear from Figure 9.21 why we call these coordinates *rectangular* coordinates.

To plot any point (h, k), simply place a point at a distance h from the y-axis and at a distance k from the x-axis. As with the number line, *negative* values of x are located to the left of the origin. Similarly, negative values of y are located below the origin. For example, Figure 9.21 shows the following points:

$$P(4, 1), Q(-2, 3), R(-1, -2), S(2, -3) \quad \text{and} \quad T(1.3, 2.7).$$

Notice that the abscissa is negative in the second and third quadrants and that the ordinate is negative in the third and fourth quadrants. Therefore, the signs of the coordinates of a point indicate the quadrant in which the point lies.

Data obtained by experiment or observation is called *empirical data*. It is usually presented in the form of a set of point pairs—for example, the temperature at various times of the day. Graph each point just as in the preceding section and connect the points. For example, the table in Figure 9.22a shows the temperature in a certain pottery kiln every hour as it heats. In Figure 9.22b, you see a presentation of the kiln temperature versus time. Each axis uses a different scale.

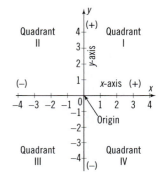

FIGURE 9.19 ■ The Rectangular Coordinate System

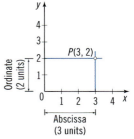

FIGURE 9.20 ■ Graphing a Point

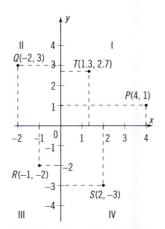

FIGURE 9.21 ■ Some Points Plotted in Rectangular Coordinates

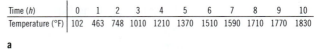

Time (h)	0	1	2	3	4	5	6	7	8	9	10
Temperature (°F)	102	463	748	1010	1210	1370	1510	1590	1710	1770	1830

a

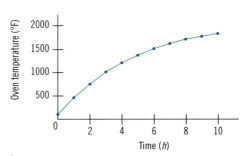

b

FIGURE 9.22 ■ Pottery Kiln Temperature versus Time

First, plot each data point. Because the temperature in a kiln *should* rise smoothly and continuously, you need to connect the data points with a *smooth* curve. Otherwise, you would connect them with straight line segments.

To manually graph an equation that relates two variables x and y, simply select values for x and compute the corresponding values for y. Then plot the resulting set of point pairs and connect them with a smooth curve. When you have an equation, you can use a computer graphics program or a graphing calculator to make the graph, which is a big advantage.

■ **EXAMPLE:** Graph the equation $y = 2x - 1$ for values of x from -2 to 2.

● **SOLUTION:** *Manually.* You can choose the x values, so pick "easy" integer values.

For $x = -2$ $y = 2(-2) - 1 = -5$
For $x = -1$ $y = 2(-1) - 1 = -3$
For $x = 0$ $y = 2(0) - 1 = -1$
For $x = 1$ $y = 2(1) - 1 = 1$
For $x = 2$ $y = 2(2) - 1 = 3$

The set of point pairs $(-2, -5), (-1, -3) \ldots (2, 3)$ is plotted in Figure 9.23. The resulting graph is a straight line. In fact, an equation such as this, in which neither x nor y is raised to a power (called a *first-degree equation*), will always plot as a straight line—hence the name *linear* equation. To plot a linear equation you need to plot only two points, with perhaps a third point as a check. Any point where the curve crosses the x-axis is called an *x-intercept*. Similarly, any point where the curve intersects the y-axis is called a *y-intercept*. Here, the x-intercept is $(\frac{1}{2}, 0)$ and the y-intercept is $(0, -1)$.

By Calculator or Computer. Enter the equation, choose appropriate scales on the x- and y-axes, choose rectangular (not polar) coordinates, and then give the instruction to graph. See your owner's manual for more detailed instructions on how to use your particular calculator or computer graphics program. ■

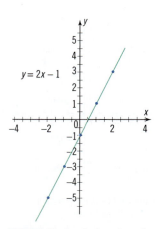

FIGURE 9.23 ■ Graph of $y = 2x - 1$

Equation of an Ellipse

Let's place an ellipse with its axes along coordinate axes and place its center at the origin, as in Figure 9.24. Let $2a$ be the horizontal dimension of the ellipse, and let $2b$ be its vertical dimension, regardless of whether its major axis is horizontal or vertical.

Deriving the equation of an ellipse is not as easy as deriving the equation for a circle. Equation 91a states the formula. If you have the mathematical background, you might want to work it on your own (see Project 38). As with the equation for the circle, you'll get a more convenient form of the equation by solving for y, getting Equation 91b.

91   **ELLIPSE WITH CENTER AT THE ORIGIN**

a. $$\frac{x^2}{a^2} + \frac{y^2}{b^2} = 1$$

b. $$y = \pm b \sqrt{1 - \frac{x^2}{a^2}}$$ ●

As with other curves, you can make a graph manually, by graphing calculator, or by computer. The manual method is explained in the following example. Simply substitute values of x and compute the corresponding values for y. Then plot the resulting table of point pairs.

a

b

FIGURE 9.24 ■ An Ellipse on Coordinate Axes

■ **EXAMPLE:** Graph the first-quadrant portion of the ellipse having a horizontal axis of 12 and a vertical axis of 8.

● **SOLUTION:** Substituting $a = 6$ and $b = 4$ gives this equation:

$$y = \pm 4\sqrt{1 - \frac{x^2}{36}}.$$

Substitute values of x from 0 to 6 and compute y. For first-quadrant values, drop the $\pm$ sign and take positive values only.

x	0.00	1.00	2.00	3.00	4.00	5.00	6.00
y	4.00	3.94	3.77	3.46	2.98	2.21	0.00

These values are graphed in Figure 9.25. If the curve is needed in any other quadrant, simply utilize symmetry about its axes.   ■

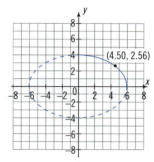

FIGURE 9.25 ■ Graph of an Ellipse

Having the equation of a curve enables you to find the value of y for any value of x.

■ **EXAMPLE:** For the ellipse in the preceding example, find the value of y in the first quadrant, when $x = 4.50$.

● **SOLUTION:** Substituting 4.50 for x in the equation of the ellipse gives the value

$$y = 4\sqrt{1 - \frac{(4.50)^2}{36}} = 2.65.$$ ■

The following example shows how the ability to find a point on the ellipse might be used in an architectural application.

■ **EXAMPLE:** The window in Figure 9.26 is half an ellipse, and it has a width of 5.80 ft and a height of 1.20 ft. Write its equation and find the height of the window at a horizontal distance of 1.50 ft from its center.

● **SOLUTION:** Substitute $a = 2.90$ and $b = 1.20$ into Equation 91b:

$$y = 1.20\sqrt{1 - \frac{x^2}{(2.90)^2}}.$$

FIGURE 9.26 ■ An Elliptical Window

When $x = 1.50$ ft, then

$$y = 1.20\sqrt{1 - \frac{(1.50)^2}{(2.90)^2}} = 1.03 \text{ ft.}$$ ■

■ **EXERCISES ● GRAPHING THE ELLIPSE**

1. Define or describe the following terms:

number line	origin	ordered pair
linear equation	abscissa	ordinate
coordinates	intercept	quadrant
axes	rectangular coordinate system	

2. Draw a number line with a suitable scale, and plot the following numbers: 6, −3, 5/3, $\sqrt{5}$, $-\pi$, and 3.75.

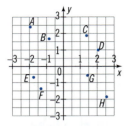

FIGURE 9.27 ■ Some Points in Rectangular Coordinates

3. Write the coordinates of the points in Figure 9.27. Estimate the value of points that don't fall on a grid line.

4. Draw coordinate axes with a suitable scale, and graph each point: $A(3, 5)$, $B(4, -2)$, $C(-2.4, -3.8)$, $D(-3.75, 1.42)$, $E(-4, 3)$, and $F(-1, -3)$.

5. Graph each set of points, connect them, and identify the geometric figure formed.
    a. $(0.7, 2.1)$, $(2.3, 2.1)$, $(2.3, 0.5)$, and $(0.7, 0.5)$
    b. $(2, -\frac{1}{2})$, $(3, -1\frac{1}{2})$, $(1\frac{1}{2}, -3)$, and $(\frac{1}{2}, -2)$
    c. $(-1\frac{1}{2}, 3)$, $(-2\frac{1}{2}, \frac{1}{2})$, and $(-\frac{1}{2}, \frac{1}{2})$
    d. $(-3, -1)$, $(-1, -\frac{1}{2})$, $(-2, -3)$, and $(-4, -3\frac{1}{2})$

6. Three corners of a rectangle have the coordinates $(-4, 9)$, $(8, 3)$, and $(-8, 1)$. Graphically, find the coordinates of the fourth corner.

7. Graph the following points: $(-3, -2)$, $(9, 6)$, $(3, 2)$, and $(-6, -4)$. Connect them with a curve that appears to best fit the data.

8. Graph the following points: $(-10, 9)$, $(-8, 6)$, $(-6, 5)$, $(-4, 4.5)$, $(-2, 4)$, $(0, 5)$, $(2, 6)$, and $(4, 8)$. Connect them with a curve that appears to best fit the data.

9. The following table gives the strength of concrete in lb/sq. in. versus the number of days after pouring. Graph this data.

Days	0	2	4	6	8	10	12	14	16	18	20	22	24	26	28	30
Strength	0	520	972	1365	1708	2006	2265	2490	2687	2857	3006	3135	3247	3345	3430	3504

10. A foundry for casting bronze sculptures has observed that the melting point of a certain bronze alloy varies with the percent of tin, as stated in the following table.

Percent Tin	2	4	6	8	10	12	14	16	18	20
Melting Point °F	1905	1920	1945	1975	2005	2035	2070	2110	2150	2195

    a. Graph the melting point of the bronze versus the percent tin.
    b. From your graph, estimate the melting point for a 15% tin alloy.
    c. From your graph, estimate the percent tin that will give a melting point of 2175°F.

11. For each equation, do the following: Make a table of ordered pairs, taking integer values of x from -3 to 3. Plot the points and connect them with a smooth curve. Find the value of y at the given value of x.
    a. $y = 3x + 1$ at $x = 2.5$
    b. $y = 2x - 2$ at $x = 1.75$
    c. $y = x^2 - 3$ at $x = 1.33$
    d. $y = 4 - x^2$ at $x = 2.34$

12. Given that the lengths of the semi-axes are $a = 4.5$ and $b = 2.3$, complete the following:
    a. Write the equation of the ellipse, taking the origin at the center of the ellipse.
    b. Graph the equation.
    c. Find the height of the ellipse at a horizontal distance of 1.5 units from its center.

FIGURE 9.28 ▪ Window in Baker Library Cupola, Dartmouth

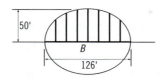

FIGURE 9.29 ▪ Elliptical Arch over an Auditorium

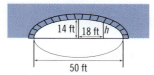

FIGURE 9.30 ▪ Elliptical Bridge Arch

13. *Elliptical Window.* Assume that the height of the elliptical window in Figure 9.28 is 1.25 times its width. Make a drawing of the window. Make suitable assumptions about the internal framing of the window and include them in your drawing.

14. Figure 9.29 shows an elliptical arch over an auditorium. Seven equally spaced vertical members connect the arch to the horizontal beam *B*. Find the height of each vertical member.

15. Figure 9.30 shows an elliptical bridge arch spanning a pool. (The reflection in the water "completes" the ellipse.) Find the headroom *h* 18 ft from the center of the arch.

16. Calculations for other conic sections, such as the circle and the parabola, are completed the same way as for the ellipse. Figure 9.31 shows a circular street whose equation is $y = \sqrt{(650)^2 - x^2}$. Use this equation to find distance *d*.

17. As noted in Chapter 7, "Circular Designs in Architecture," each side of a pointed arch is a circular arc. In Figure 9.32 the equation of the right side of the arch, taking the axes as shown, is $y = \sqrt{100 - x^2}$. Use this equation to find (a) the height of the arch at a distance of 2.5 ft from the centerline and (b) the height of the arch at its peak.

18. Figure 9.33 shows a rectangular opening in a circular arch. The equation of the circular portion is $y = \sqrt{(3.38)^2 - x^2}$. Find the height of the rectangle if it touches the arch as shown.

19. The equation of the round portion of the left arch in Figure 9.34 is $y = \sqrt{(3.60)^2 - x^2}$. Find the height *h* of the column.

20. The equation of the parabolic arch is $y = -x^2/18$, with axes chosen as shown in Figure 9.35. Find the horizontal distance *x* from the center at which the headroom is 10 ft.

21. The suspension bridge cable in Figure 9.36 has the equation $y = x^2/1250$. Find the height of each tower above the roadway.

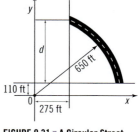

FIGURE 9.31 ▪ A Circular Street

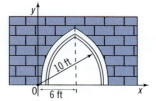

FIGURE 9.32 ▪ A Pointed Arch

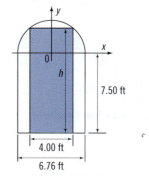

FIGURE 9.33 ▪ A Rectangular Opening in a Circular Arch

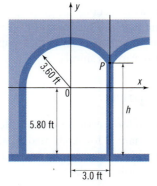

FIGURE 9.34 ▪ Round Arches

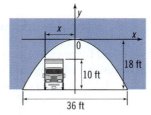

FIGURE 9.35 ▪ Parabolic Arch

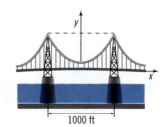

FIGURE 9.36 ▪ Suspension Bridge Cable

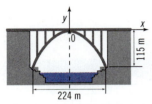

FIGURE 9.37 ■ Parabolic Bridge Arch

22. The parabolic bridge arch supporting a level roadway, shown in Figure 9.37, has the equation $y = -x^2/109$ when the axes are chosen as shown. Find the lengths of the six equally spaced vertical supports.

23. *Computer and Calculator.* Many computer programs (such as Derive and other computer algebra systems, The Geometer's Sketchpad, AutoCAD, and Microsoft Excel) have graphing capabilities. If you have such a program, learn how to use it and try it on any of the problems in these exercises. A graphics calculator, as the name implies, can be used as well.

THE SPIRAL

"From an artistic standpoint, spiral equations provide a vast and deep reservoir from which artists can draw."

CLIFFORD PICKOVER

A *spiral*, in general, is the curve that is generated when a point rotates in a circular path while, at the same time, its distance from the center of the circle increases. There are many kinds of spirals; this section will cover *Archimedean spirals* and *equiangular spirals*, and it will mention others briefly. Spirals are not easy to describe in rectangular coordinates, so you'll be introduced to a new coordinate system: *polar coordinates.*

Graphing in Polar Coordinates

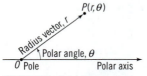

FIGURE 9.38 ■ Comparing Rectangular Coordinates and Polar Coordinates

Suppose you are at point *O* on the path depicted in Figure 9.38, and you want to reach point *P*. You could walk 400 paces along the straight path, make a quarter-turn to the left, and walk another 300 paces. That would be an example of rectangular coordinates. If, however, you walk 500 paces at an angle of approximately 37° to the path, that would be an example of *polar coordinates.*

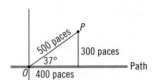

FIGURE 9.39 ■ Polar Coordinates

The polar coordinate system, presented very simply in Figure 9.39, consists of a polar axis passing through a point *O*, which is called the *pole.* The location of a point *P* is given by two numbers: (1) its distance *r* from the pole, called the *radius vector*; and (2) the angle *θ*, called the *polar angle* (sometimes called the *vectorial angle* or *reference angle*). The polar angle is taken as positive when measured counterclockwise from the polar axis and taken as negative when measured clockwise. Therefore, the *polar coordinates* of a point *P* are *r* and *θ*, which are usually written in the form $P(r, \theta)$. For example, a point *P* at a distance of 5 units from the pole with a polar angle of 27° can be written as $P(5, 27°)$.

Using polar coordinate graph paper to plot polar coordinates is convenient, although not necessary. This paper has equally spaced concentric circles, and it has an angular scale in degrees or radians. As with rectangular graph paper, a suitable scale must be chosen and marked on the paper. The points $P(3, 45°)$, $R(1.5, 210°)$, and $S(2.5, 300°)$ are shown plotted on polar coordinate graph paper in Figure 9.40. Now you have the tools to graph the various spirals.

The Archimedean Spiral

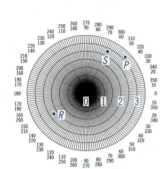

FIGURE 9.40 ■ Points in Polar Coordinates

An Archimedean spiral is the curve you see when you look at the end of a rolled-up rug or the grooves in a phonograph record. It is named for Archimedes, who discovered it around 225 B.C. To construct an Archimedean spiral, utilize the fact that the radii from the center of the spiral to a point on the spiral increase the same amount for equal increments of the angle. In other words, for equal increments of the angle, the radii form an arithmetic progression.

■ EXAMPLE: Construct an Archimedean spiral with a radius that increases by 1 inch for each revolution of the radius vector. Start with a radius of zero at 0°, and plot a point every quarter revolution, up to three revolutions.

● SOLUTION: First, draw the concentric circles and radii of a polar coordinate system (or use polar coordinate paper), as in Figure 9.41.

Because the radius increases by 1 inch per revolution, it will increase by $\frac{1}{8}$ inch for each eighth of a revolution (45°). Therefore, for radii spaced by 45°, we get the following arithmetic progression:

$$0, \tfrac{1}{8}, \tfrac{1}{4}, \tfrac{3}{8}, \tfrac{1}{2}, \tfrac{5}{8}, \tfrac{3}{4}, \tfrac{7}{8}, 1, 1\tfrac{1}{8}, 1\tfrac{1}{4}, \ldots$$

Plot these values for each quarter-revolution, and continue until you have completed three revolutions. Then connect the points, using a French curve or other drafting aid. Note that the spiral passes through the origin. ■

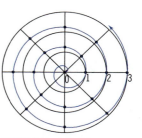

FIGURE 9.41 ■ An Archimedean Spiral

Another way to obtain an Archimedean spiral is from its equation. This method is especially valuable for calculator graphing or computer graphing of the spiral. Refer back to Figure 9.39 and imagine the radius vector rotating so that θ increases at a constant rate, while at the same time the length r of the radius vector increases at a constant rate. You would then produce an Archimedean spiral, where the radius r is directly proportional to the central angle θ. As a polar equation, this is written as shown in Equation 99, where k is a constant of proportionality.

99 POLAR EQUATION OF AN ARCHIMEDEAN SPIRAL

$$r = k\theta$$ ●

As with the ellipse, having an equation allows you to make a graph manually, by graphics calculator, or by computer. To graph the Archimedean spiral manually, select values for θ, either in degrees or radians, and multiply each value by k to get the radius r. This results in a table of point pairs that you can then plot.

■ EXAMPLE: Graph an Archimedean spiral with $k = 1$, with θ in degrees.

● SOLUTION: *Manually.* With $k = 1$, the radius is simply equal to the angle. Therefore, when $\theta = 90°$, $r = 90$, and so forth.

θ (degrees)	0	90	180	270	360	. . .
r	0	90	180	270	360	. . .

Plot these points and connect them with a smooth curve, as you see in Figure 9.42.

By Calculator or Computer. Enter the equation $r = \theta$ into your graphing program, choose polar coordinates rather than rectangular ones, choose a scale, and then graph. ■

If you had used a different value for k, you would have gotten a spiral of the same shape but with different values on the axes. Therefore, to make this spiral exactly match the spiral in the preceding example, all you would have to do is relabel the axes. With a computer-assisted design (CAD) program, to change this spiral into one that is tighter or looser (more or fewer rotations per inch of paper), simply drag the scale to the left or to the right. To obtain more or fewer rotations, drag the end of the spiral counterclockwise or clockwise.

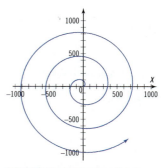

FIGURE 9.42 ■ Archimedean Spiral with $k = 1$

FIGURE 9.43 ■ Chambered Nautilus

The Equiangular Spiral

The *equiangular spiral* was discovered and named in 1638 by Descartes, whom we encountered earlier in this chapter in connection with analytic geometry. This spiral is called *equiangular* because, as you'll soon see, the angle between the tangent to the spiral and the radius drawn to any point on the spiral *is always the same*. The equiangular spiral is also called the *logarithmic spiral*.

Equiangular spirals are frequently found in nature. The chambered nautilus shell in Figure 9.43 is an excellent example of a naturally occurring equiangular spiral. Other examples are sunflower and daisy heads, evergreen cones, ferns, pineapples, cacti, animal horns, tusks and claws, vortices in air and water, and in distant galaxies such as the spiral nebula.

You can construct an equiangular spiral by using the fact that the radii to the curve, measured at equal increments of θ, form a geometric progression.

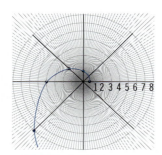

FIGURE 9.44 ■ Graphing the Equiangular Spiral

■ **EXAMPLE:** Construct an equiangular spiral with a radius that doubles every 45°. Start with a radius of $\frac{1}{2}$ and plot a point every 45°.

● **SOLUTION:** For points spaced by 45°, the radii form this geometric progression:

$$\tfrac{1}{2}, 1, 2, 4, 8, 16, \ldots$$

Plot these points and connect them with a smooth curve, as in Figure 9.44. Unlike the Archimedean spiral, this spiral does *not* pass through the origin; however, it keeps getting closer and closer. ■

The construction by whirling rectangles is also based on the fact that the radii to the curve, measured at equal increments of θ, form a geometric progression. Therefore, for three equally spaced radii, the middle radius is the geometric mean (or mean proportional) of the outer two. To construct an equiangular spiral by using whirling rectangles, as in Figure 9.45, follow these steps:

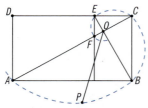

FIGURE 9.45 ■ Constructing an Equiangular Spiral using Whirling Rectangles

1. Draw rectangle *ABCD*, choosing any ratio of width to length. Let *A* and *C* be points on the spiral.

2. Draw the main diagonal *AC*.

3. From *B*, draw a perpendicular *BO* to *AC*. Chapter 3, "The Triangle," demonstrated that an altitude drawn to the hypotenuse of a right triangle is the geometric mean of the segments of that hypotenuse. Therefore, *B* is a point on the spiral. By the same argument, *OB*, *OC*, and *OE* form a geometric progression. Because *B* and *C* lie on the spiral, so must point *E*.

4. Draw *EF* parallel to *BC*. Then *OC*, *OE*, and *OF* form a geometric progression, so point *F* is on the spiral.

5. Connect the corners *A*, *B*, *C*, *E*, *F*, . . . with a smooth curve.

This construction can be continued both inward and outward. Obviously, you are drawing the same whirling rectangles as in Chapter 4, "Ad Quadratum and the Sacred Cut," and the equiangular spiral is the curve connecting the corners of successive rectangles.

The points obtained by this construction, however, may be too far apart to draw a reasonably accurate curve, but you can obtain intermediate points in the following way. To obtain a point on the spiral between, say, *A* and *B*, bisect the angle *AOB* (either by construction or with a protractor). In a separate drawing, lay off *AO* and *OB* end-to-end on the same line, as in Figure 9.46. Construct a semicircle on *AB* and erect a perpendicular *OP* from *O*. Chapter 3 showed that *OP* is the geometric mean between *AO* and *OB*. Therefore, *P* is the distance to the spiral along the bisector of angle *AOB*, and you transfer it to the spiral drawing. Repeat this construction to obtain as many intermediate points as are needed.

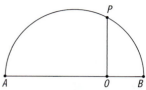

FIGURE 9.46 ■ Plotting Intermediate Points

In the construction in Figure 9.45, you started with a rectangle of any dimensions. If you had used a golden rectangle, you would have gotten what is called a *golden spiral*. In the golden rectangle, the gnomons are squares. This makes it suitable for an approximate construction that uses circular arcs for successive portions of the spiral. To approximately construct a golden spiral, follow these steps: In golden rectangle *ABCD*, construct whirling squares *AEFD*, *EBHG*, *HCJI*, and so forth, as shown in Chapter 4. From *F*, draw circular arc *DE*. From *G*, draw circular arc *EH*. From *I*, draw circular arc *HJ*. Repeat as far as needed. This construction is illustrated in Figure 9.47.

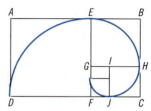

FIGURE 9.47 ■ Approximate Construction of a Golden Spiral

Let's draw tangents at several points on the spiral in Figure 9.48. The angle between tangent and radius vector from *O* is the same at any point on the spiral, hence the name *equiangular spiral*. This property of keeping its shape might have been what prompted the mathematician Jacob Bernoulli (1654–1705) to call it the *miraculous spiral*, or *spira mirabilis*. He was apparently so impressed with its property of self-similarity that he tried to have one engraved on his tombstone, along with the Latin words *Eadem Mutata Resurgo* (In spite of changes, resurrection of the same). However, an Archimedean spiral was carved by mistake.

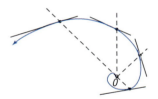

FIGURE 9.48 ■ Tangents to the Equiangular Spiral

As with the Archimedean spiral, you can use the equiangular spiral's equation to make a graph, especially by computer. For the equiangular spiral, the radius *r* is proportional to *e* raised to the angle *θ*. Its polar equation is stated in Equation 100.

100 POLAR EQUATION OF AN EQUIANGULAR SPIRAL

$$r = ke^{a\theta},$$

where *k*, *a*, and *e* are constants, and *e* ≈ 2.71828.   ●

The number *e*, like *π*, is a fundamental number in mathematics. One of its definitions is that it is the limit of $(1 + 1/n)n$ as *n* grows without limit, and this relates it to the continuous growth of various quantities (see "Exercises and Projects").

The number *e* is the base of natural logarithms, and it was introduced by the Swiss mathematician Leonhard Euler (1707–1783) in the mid-eighteenth century. Its numerical value is approximately 2.71828, and like *π* it is irrational. Also like *π*, this number is on your scientific calculator, and it can be treated like any other number.

■ **EXAMPLE:** Plot the equiangular spiral with *k* = 1, *a* = 0.01, and *θ* in degrees from 0 to 510°.

● **SOLUTION:** *Manually.* Make a table of point pairs of θ versus r. Use the e^x or EXP key on a calculator to raise e to a power.

When $\theta = 0°$, $r = e^0 = 1$.
When $\theta = 30°$, $r = e^{(0.01 \times 30)} = e^{0.3} = 1.3$.
When $\theta = 60°$, $r = e^{(0.01 \times 60)} = e^{0.6} = 1.8$.

The remaining values are listed in the following table:

θ	0	30	60	90	120	150	180	210	240	270	300	330	360	390	420	450	480	510
r	1.0	1.3	1.8	2.5	3.3	4.5	6.0	8.2	11.0	14.9	20.1	27.1	36.6	49.4	66.7	90.0	121.5	164.0

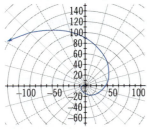

FIGURE 9.49 ■ Equiangular Spiral, $a = 0.01$ and $k = 1$

Plot each point on polar coordinate paper, and connect all the points with a smooth curve, as in Figure 9.49.

By Computer. The method is the same as for the Archimedean spiral, except here you enter the equation $r = e^{0.01\theta}$.

If you had used a different value for k, you would have gotten a spiral of the same shape but with different values on the axes. Therefore, you can make this same spiral exactly match another equiangular spiral simply by relabeling the axes. On the computer, you can change this spiral into one that is tighter or looser (more or fewer rotations per inch of paper), by simply dragging the scale left or right. To obtain more or fewer rotations, drag the end of the spiral counterclockwise or clockwise. ●

■ EXERCISES ● THE SPIRAL

1. Define or describe the following terms:

polar coordinates	polar axis	pole
radius vector	polar angle	equiangular spiral
spira mirabilis	golden spiral	Archimedean spiral

2. Plot each of the following points in polar coordinates: $A(4, 35°)$, $B(3, 120°)$, $C(2.5, 215°)$, and $D(3.8, 345°)$.

3. Graph the spirals in Figure 9.50 by hand, by calculator, or by computer, given their polar equations. Take $a = 1$ for each.
   a. Archimedean spiral
   $$r = a\theta$$
   b. Equiangular spiral, with $k = 1$
   $$r = ke^{a\theta}$$
   c. Hyperbolic spiral
   $$r = \frac{a}{\theta}$$
   d. Parabolic (or Fermat's)
   $$r = a\sqrt{\theta}, \quad \theta = \arctan\frac{y}{x}$$
   e. Lituus
   $$r = \sqrt{\frac{a}{\theta}}$$

4. Construct an Archimedean spiral with a radius that increases by 10 cm for each revolution of the radius vector. Start with a radius of zero at $0°$, and plot a point every quarter revolution, up to two revolutions.

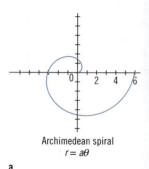

Archimedean spiral
$r = a\theta$

a

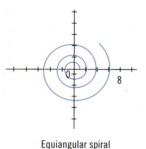

Equiangular spiral
$r = ke^{a\theta}$

b

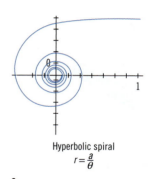

Hyperbolic spiral
$r = \dfrac{a}{\theta}$

c

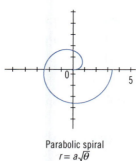

Parabolic spiral
$r = a\sqrt{\theta}$

d

Lituus
$r = \sqrt{\dfrac{a}{\theta}}$

e

FIGURE 9.50 ▪ Some Spirals

5. Construct an equiangular spiral with a radius that doubles every 45°. Start with a radius of 1 in. and plot a point every 45° for one revolution.

6. *Equiangular Spiral by Whirling Rectangles.* Use the *whirling rectangles method of construction* to construct an equiangular spiral.

7. *Golden Spiral by Whirling Squares.* Use the *whirling squares method of construction* to construct a golden spiral. Use the same construction that was used in the preceding exercise, but start with a golden rectangle.

8. *Golden Spiral by Whirling Golden Triangles.* Use the *whirling rectangles method of construction* to construct a golden spiral, but start with a golden triangle. See if you can reproduce the construction in Figure 9.51 without further instruction.

9. *Square-Root Spiral.* Construct a square-root spiral by the following method: Start with an isosceles right triangle with a hypotenuse of $\sqrt{2}$. On its hypotenuse, construct another right triangle where the other leg is 1. The hypotenuse of this triangle is $\sqrt{3}$. Continue adding triangles as shown in Figure 9.52.

10. Draw tangents to the equiangular spiral you created in Exercise 3b at several points. Using a protractor, measure the angle between each tangent and the radius to that point. Summarize your findings in a paragraph or two.

11. Transfer the spiral from Exercise 3b to cardboard or thin rigid plastic. Cut it out to make a template for use in future designs.

12. *Equiangular Spiral by Template.* Make a template from a strip of cardboard about 3/4 in. wide and 6 in. long. Cut one end at an angle, as in Figure 9.53. Follow these steps to draw an equiangular spiral as shown

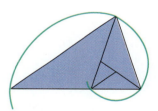

FIGURE 9.51 ▪ Golden Spiral from Golden Triangles

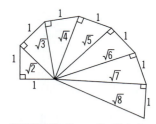

FIGURE 9.52 ▪ Square-Root Spiral

FIGURE 9.53 ▪ Spiral Template

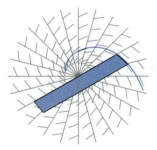

FIGURE 9.54 ■ Equiangular Spiral by Template

in Figure 9.54: Choose a pole for your spiral, and draw radii every 15 degrees. Using your template, draw a number of tangents to each radius. This creates what is called a *tangent field*, a tool sometimes used to solve differential equations. Select a starting point, and sketch a curve whose slope at any point best matches the slope of the tangents in its neighborhood. (You don't need to draw tangents along an entire radius, but just in the vicinity in which the curve seems to be heading.)

THE SPIRAL IN ART AND ARCHITECTURE

"The basic principles underlying the greatest art so far produced in the world may be found in the proportions of the human figure and in the growing plant."[3]

JAY HAMBIDGE

Hambidge was speaking of the golden ratio and the spiral in the quote above, and in this section we'll give a few examples of the spiral in art and architecture.

The Ionic Volute

In his *Ten Books on Architecture*, Vitruvius was the first to describe the three classical orders of architecture: the Doric, Ionic, and Corinthian. Figure 9.55 displays examples of each.

Volutes, the spirals adorning the capitals of the columns, are distinctive features of Ionic columns. The Ionic volute, shown in Figure 9.56, is an approximation to the equiangular spiral, and it is drawn with a series of circular arcs. Edward Edwards writes that actual compass marks have been found on fragments of Ionic capitals.[4]

Doric   Ionic   Corinthian

FIGURE 9.55 ■ The Three Classical Orders of Architecture: Doric, Ionic, and Corinthian

Image from *Illustrated Dictionary of Historic Architecture*, edited by Cyril M. Harris. Used with permission from Dover Publications.

FIGURE 9.56 ■ A Volute at the Olympic Theater, Vicenza

Rochelle Newman and Martha Boles give the following construction for drawing an Ionic volute with compass and straightedge:[5]

1. Draw a rectangle $8 \times 9\frac{1}{2}$ in. This gives a ratio ≈ 1.19, which is said to have been used by the Greeks.

2. Draw the main diagonal and the diagonal of the reciprocal, which is perpendicular to it, as shown in Figure 9.57a. Bisect their angles of intersection, getting AF and GH. Note that the angle bisectors are not parallel to the sides.

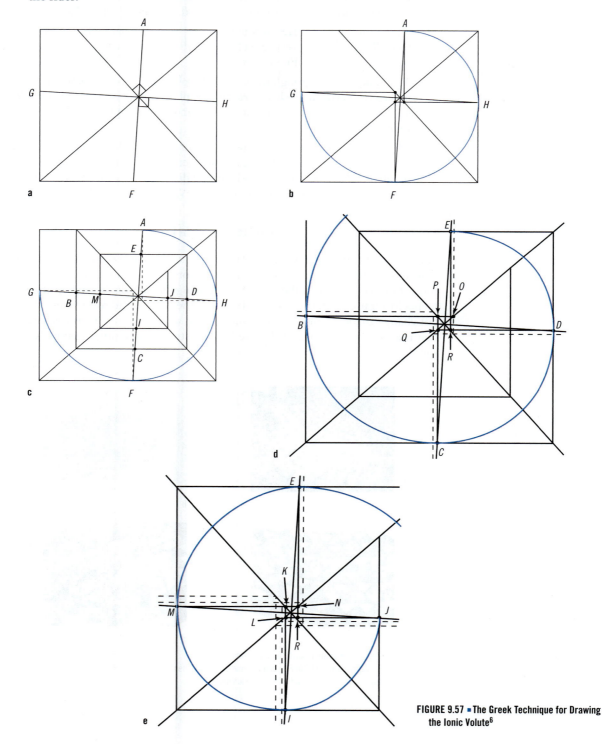

FIGURE 9.57 ■ The Greek Technique for Drawing the Ionic Volute[6]

3. Draw segments parallel to the sides of the rectangle, from *G*, *F*, *H*, and *A*, respectively, to the diagonals, as shown in Figure 9.57b. This forms three squares that provide the framework for the first part of the spiral.

4. Draw arcs through each square using the innermost vertex of each as the point upon which the tip of the compass is placed, as shown in Figure 9.57b.

5. Draw the straight line spiral, as shown in Figure 9.57c. The spiral curve will be tangent to the sides of the straight-line spiral at points *B*, *C*, *D*, and *E*, where it intersects the angle bisectors drawn in Figure 9.57a.

6. To obtain the next series of squares, draw segments parallel to the sides, from *B*, *C*, *D*, and *E*, respectively, to the diagonals, getting *OB*, *PC*, *QD*, and *RE*, as shown in Figure 9.57d.

7. Again, draw arcs through the squares by placing the compass tip on the innermost vertex of each square (that is, place the tip on *O* to cut arc *AB*, on *P* to cut arc *BC*, on *Q* to cut arc *CD*, and on *R* to cut arc *DE*). See Figure 9.57d.

8. The next series of squares, shown in Figure 9.57e, is formed precisely the same way as in Figure 9.57d. Draw segments parallel to the sides from points *E*, *M*, *I*, and *J*, respectively, to the diagonals. To cut the quadrant arcs, the compass tip is placed on points *N*, *K*, and *L*.

Other Spirals

You studied rosettes in Chapter 7, "Circular Designs in Architecture," and saw that most were based on the circle and did not contain spirals at all. There are some, however, in which the curves are equiangular spirals. In each of the pavements in Figure 9.58, the tiles get larger farther out in the rosette, but they do not change shape. In other words, they are all *similar figures*. That is a sure sign that the spiral is equiangular. Figure 9.59 illustrates some other examples of the spiral.

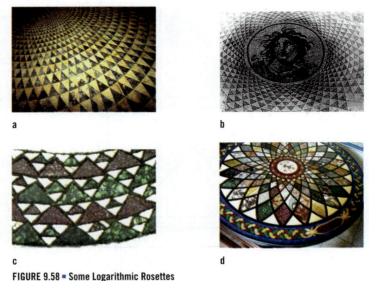

a

b

c

d

FIGURE 9.58 ■ Some Logarithmic Rosettes
(a) Getty pavement; (b) Pompeii pavement; (c) Pavement design in the Sistine Chapel; and (d) Pavement design in the Vatican Museum

FIGURE 9.59 ■ Some Spirals in Art and Architecture

(a) Celtic spirals; (b) Spiral, Xavier del Bac Mission, Arizona; (c) Santa Maria della Stechatta, Parma; (d) Ionic volutes, Pompeii; (e) Crozier with snake and lion in the Louvre Museum; (f) Lampposts near S. Croce, Florence; (g) Spiral bracket, Dartmouth College; and (h) Spiral design in Campo Santo, Pisa

SUMMARY

This completes our brief introduction to analytic geometry. Here we learned how to graph rectangular coordinates, and used these skills as we took a new look at the circle, a geometric figure we studied in detail in Chapters 6, 7, and 8. We saw that the circle belonged to a family of curves called the conic sections, and we studied a closely related curve, the ellipse. Then we learned graphing in polar coordinates and applied it to spirals of various sorts. However, we haven't finished with spirals. We'll see exotic ones when we cover fractals. In the next chapter, we'll leave the world of the plane figures and take a bold leap into the third dimension.

EXERCISES AND PROJECTS

1. *A Definition of the Conics.* Each conic may be defined as the locus of a point in a plane whose distance from a fixed point (the *focus*) is ε times its distance from a fixed line (the *directrix*). The type of conic, ellipse, parabola, or other such shape is determined by the value of ε (called the *eccentricity*). Research this way of defining conics and summarize your findings in a paper, complete with illustrations of how the various conics are formed.

2. *Model of the Conic Sections.* Using clay, damp sand, wood, or Styrofoam, make a model that shows a cone being cut by planes to produce the various conic sections.

3. *Focus and Directrix.* An ellipse may be defined in terms of a *focus* and a *directrix.* Learn how this is done and write a short explanation.

4. *Dürer's Eierlinie.* Repeat the construction of Dürer's *Eierlinie* (egg line)[7] using Figure 9.60, but try to avoid the mistakes that he made. This

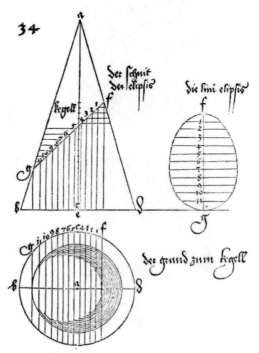

FIGURE 9.60 ■ Dürer's *Eierlinie*

construction should produce a perfect ellipse, rather than the egg-shaped curve Dürer obtained. You can consult a book on *descriptive geometry* that shows how to draw the intersections of two surfaces. Compare your results with your other constructions of the ellipse.

5. *Ovato Tondo.* Construct the ovato tondo. Compare your constructed figure with a mathematically plotted ellipse with the same outer dimensions. One drawback of the ovato tondo is that it always produces an oval with the same ratio of length to width. Calculate that ratio.

6. *Leonardo's Method for Drawing the Ellipse.* Leonardo devised a method for tracing an ellipse using a triangular cut-out, as illustrated in Figure 9.61. Try this method yourself. Make a triangular cut-out *ABC*, and draw two axes, as shown. Place the triangle so that *A* lies on one axis and *B* lies on the other axis. Mark the position of *C*. Move the triangle to other positions and repeat. Make sure that *A* is always on the same axis and *B* is always on the other axis.

7. *Leonardo's Ellipsograph.* Make a model of Leonardo's ellipsograph. This device, shown in Figure 9.62, works the same way as the triangle cut-out, except that it is adjustable. Two points of the compass move along the axes, while the brush traces the curve. Use your model to draw an ellipse that you can compare with the ones you drew by other methods.

8. *Leonardo's Method on the Computer.* Instead of using a triangular cut-out of Leonardo's triangle, draw it in The Geometer's Sketchpad or a similar program. Then cause it to move, as you did by hand, and trace an ellipse. Then change the shape of the triangle and repeat. Change the angle between the two axes and repeat. Summarize your findings in a report.

9. *Ellipsograph.* Figure 9.63 shows a student's ellipsograph, also called a *trammel*, in which two pins slide in two perpendicular slots. Make a model of this ellipsograph. Explain why it works and use it to draw an ellipse. Compare your ellipse with those you drew by other methods.

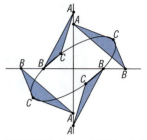

FIGURE 9.61 ■ Leonardo's Triangle for Tracing an Ellipse

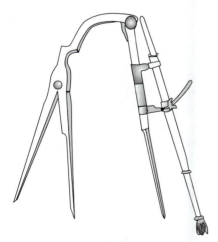

FIGURE 9.62 ■ Leonardo's Ellipsograph

FIGURE 9.63 ■ Victor Ellipse-O-Graph

a

b

c

©Steven Lunetta Photography

d

©Aaron Kohr/Stockphoto.com

FIGURE 9.64 ■ Some Elliptical Architectural Features
(a) Elliptical fanlight; (b) Elliptical trim; (c) Oval windows in the glockenspiel of Salzburg, Austria; (d) Elliptical window

10. *Elliptical Windows and Trim.* For either some or all of the architectural features you see in Figure 9.64, scale the photograph to get an approximate ratio of length to width, and draw an ellipse of those proportions. Draw the internal details, making suitable assumptions as to the dimensions. Then, using these for inspiration, make your own design.

11. Search your campus or neighborhood for elliptical windows, elliptical arches, porch trims, and so forth. Make sketches and take photographs. Try to reproduce some by construction. Write a few paragraphs on your findings.

12. Using heavy cardboard, make a set of ellipse templates that you can later use to trace elliptical openings in matting boards for pictures.

13. *Team Project.* Construct a large ellipse on a ball field or lawn using two driven stakes and a long loop of rope, as if you were laying out an elliptical flower bed or garden. Use pegs to mark points on the ellipse and a hose or rope to indicate the ellipse.

14. *Team Project.* Using a cylindrical trash can or similar object, construct a large Archimedean spiral by winding or unwinding a rope from the can. Use pegs and hose to indicate the spiral.

FIGURE 9.65 ■ A Coaster

15. Identify the two spirals in the coaster in Figure 9.65.

16. *Dürer's Methods for a Spiral.* In *The Painter's Manual*, Dürer shows several ways to construct a spiral. Explore and duplicate at least one of his methods, and write a paragraph of explanation.

17. *Bishop's Crozier.* On p. 55 of *The Painter's Manual*, Dürer shows how to design a Bishop's crozier, such as in Figure 9.59e. Reproduce the construction and write a short explanation.

18. *Serlio's Ionic Volute.* In *The Five Books of Architecture*, Book 4, Chapter 7, Folio 35, Serlio shows how to draw an Ionic volute. Refer to Figure 9.66 as you reproduce his construction.

19. *Alberti's Ionic Volute.* In *The Ten Books of Architecture*, Book 7, Chapter 8, Alberti shows how to draw an Ionic volute. As you study and reproduce his construction, refer to Figure 9.67.

20. *Palladio's Ionic Volute.* In *The Four Books of Architecture*, Book 1, Chapter 16, Palladio shows how to draw an Ionic volute. Refer to Figure 9.68 as you reproduce his construction.

21. Construct an Ionic volute by hand or by using a computer drafting program, and carve it from wood, clay, stone, or another material.

22. *Logarithmic Rosette.* Make a logarithmic rosette by rotating a logarithmic spiral about its center. Use a template or a computer drafting program.

23. *Logarithmic Rosette.* Using the author's method,[8] construct a rosette in which all the tiles are similar triangles, although they can be of different sizes. Figure 9.69 illustrates the construction.

   For the first band, choose the radius of the inner circle. Then subdivide the circle into n equal intervals. Let w be the width of the chord for each interval. Choose the height h for the dark triangles in the first band. Draw n dark isosceles triangles of base w and height h, with their base vertices on the inner circle and heights in the radial direction. Connect the upper vertices of the dark triangles. This defines the white triangles in the first band.

   For the second band, draw n dark triangles, each with a base equal to the base of the first-band white triangles, and geometrically similar to the first-band dark triangles. Connect the vertices of the second-band dark triangles to give the second-band white triangles. These will be geometrically similar to the white triangles in the first band. Repeat the procedure

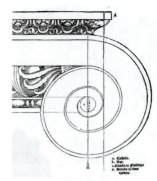

FIGURE 9.66 ■ Serlio's Ionic Volute Construction

From Sebastiano Serlio, *The Five Books of Architecture*

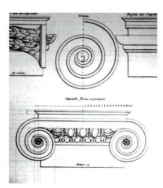

FIGURE 9.67 ■ Alberti's Ionic Volute Construction

From Leon Battista Alberti, *The Ten Books of Architecture*

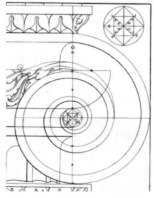

FIGURE 9.68 ■ Palladio's Ionic Volute Construction

From Andrea Palladio, *The Four Books of Architecture*

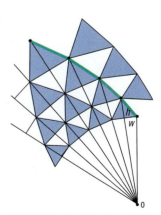

FIGURE 9.69 ■ Drawing a Logarithmic Spiral without Even Knowing It

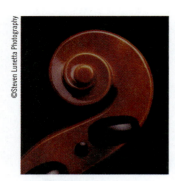

FIGURE 9.70 ■ A Violin Scroll

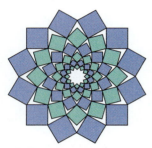

FIGURE 9.71 ■ Spiraling Squares Quilting Pattern

for the subsequent bands. Connect corresponding vertices with a smooth curve. Prove that this leads to a logarithmic spiral.

24. Locate and photograph any spiral architectural features on your campus or in your neighborhood. As you did with the elliptical features, make sketches, take photographs, and try to reproduce some by construction. Write a few paragraphs on your findings.

25. Locate and photograph any spiral plant features on your campus or in your neighborhood. If possible, bring samples to class.

26. *The Fiddlehead.* Learn how to construct the spiral head of a violin-family instrument such as the one shown in Figure 9.70. See Ekwald's "Volutes and Violins" in *Nexus Network Journal*, Vol. 3, No. 4 (Autumn 2001). Carve a violin head.

27. Design a spiral pattern that can be duplicated end-to-end to form a decorative border or frieze pattern.

28. Design and make an article of jewelry based on the spiral.

29. Make a "stained glass" window based on the material in this chapter. You can use actual glass or rigid plastic cut-outs, or simply paint on a sheet of plastic. Install the window temporarily over a window in your classroom.

30. Make a graphic design based on the spiral, and use it for a quilt, tee shirt design, or on fabric for a piece of clothing.

31. *Spiraling Squares Quilting Pattern.* The quilting pattern shown in Figure 9.71 is from Venters, *Mathematical Quilts.* Make this pattern or a similar one of your own design.

32. Make a large, colorful logarithmic rosette.

33. Search an art book or a museum for examples of elliptical frames. Can you see any compositional reasons for why this shape was chosen in the items you found?

34. Here are some suggestions for short papers or term papers:

   ■ Apollonius and the conic sections

   ■ The history of the conic sections since Apollonius

   ■ Descartes and Fermat and the origins of analytic geometry

   ■ Elliptical orbits of celestial bodies

   ■ Why Roman amphitheaters were often elliptical (see http://www.nexusjournal.com/Query07-Ellipses.html)

   ■ Pascal's *Mystic Hexagram*

   ■ Bernoulli's *spira mirabilis*

   ■ The spiral in plants, shells, tusks, and other natural growths

   ■ Spiral galaxies

   ■ Celtic spirals

   ■ The spiral of Archimedes and quadrature of the circle, using Heath, p. 230, as a starting point

   ■ The *maze*, a relative of the spiral

   ■ Bernini's elliptical piazza in front of St. Peter's in Rome, including how the focal points are marked and how the surrounding colonnades appear to someone standing at a focal point

35. *The Helix.* The helix and the spiral are often confused. For example, the term *spiral staircase* refers to a staircase that is actually helical in shape. Research the helix and write a short paper on your findings. Locate some examples of the helix in art and architecture.

36. Write a book report on *Spiral Symmetry*, which is listed in the Sources at the end of the chapter.

37. Bake a loaf of bread or a pastry in the shape of an Archimedean spiral.

Mathematical Challenges

38. *Equation of an Ellipse.* Try to derive the equation for the ellipse shown in Figure 9.72. Follow these steps: Start with the definition of an ellipse: $PF + PF' = 2a$. Drop a perpendicular PQ from P to the x-axis, forming two right triangles. Use the Pythagorean theorem to express PF and PF' in terms of c, x, and y. Simplify, and at a suitable point make the substitution $\sqrt{a^2 - b^2} = c$. Simplify further to get Equation 91a.

39. *Another Definition of the Ellipse.* An *ellipse* is defined as the set of all points in a plane such that the sum of the distances from two fixed points is constant. It can also be defined as the curve obtained by intersecting a plane with a cone. Referring to Figure 9.73, show that these two definitions are consistent.

      Let E be the curve in which a plane intersects a cone, and let P be any point on that curve. Insert a sphere S_1 into the cone so that it touches the cone along circle C_1 and is tangent to the cutting plane at point F_1. Similarly, insert sphere S_2 touching the cone along circle C_2 and the plane at point F_2. Show that $PF_1 + PF_2$ is a constant, regardless of where on E the point P is chosen. (*Hint:* Draw element VP and extend it to where it intersects circle C_2 at L_2. Then use the fact that two tangents drawn to a sphere from a common point are equal.)

40. Figure 9.74 shows a horseshoe arch in a wall. The equation of the circular portion is $y = \sqrt{(2.78)^2 - x^2}$. Use this equation to find dimension A.

41. Prove that the distances from the pole to an equiangular spiral, for equal increments of angle, form a geometric progression. (*Hint:* Write expressions for the radius at θ, 2θ, and 3θ, using Equation 100. Divide the second

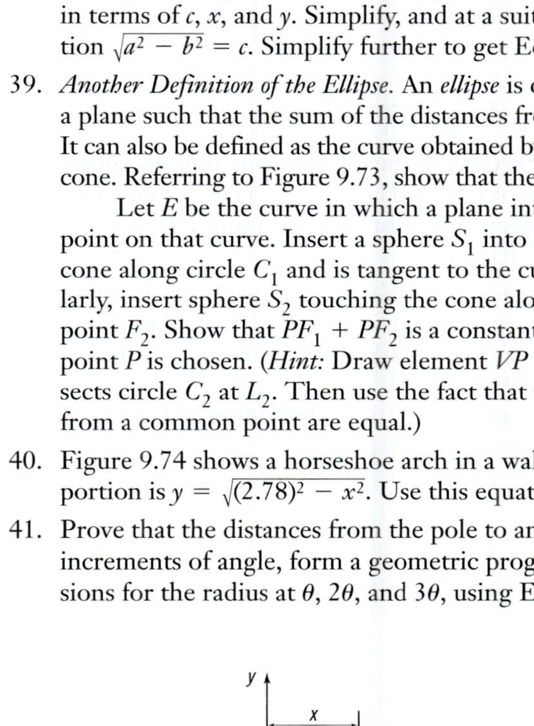

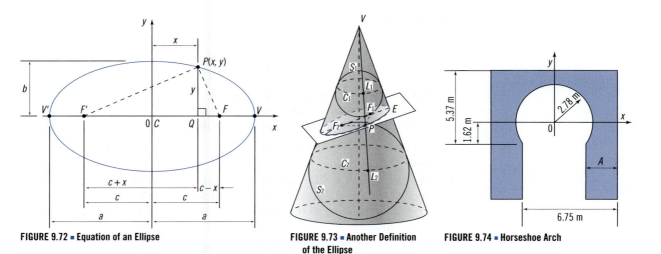

FIGURE 9.72 ■ Equation of an Ellipse

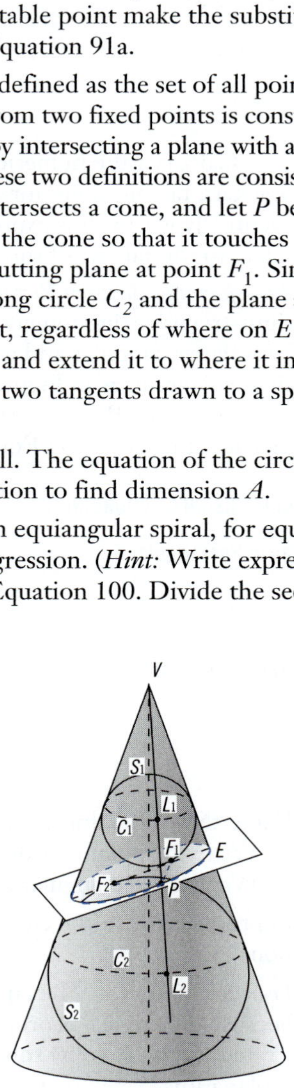

FIGURE 9.73 ■ Another Definition of the Ellipse

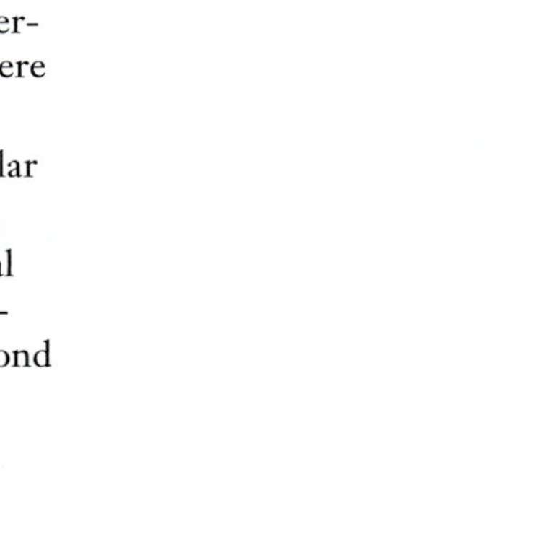

FIGURE 9.74 ■ Horseshoe Arch

radius by the first, and then divide the third by the second. Do you get the same quantity?)

42. *Derivation of e.* Try to derive an expression for *e*, the number in the equation of the equiangular spiral. Suppose a quantity *P* (for example, the amount of money in your bank account) is growing at a rate of *i* percent per year. The amount *S* you would have after *t* years would be $S = P(1 + i)^t$. If the interest were applied not just once per year but *m* times per year, the equation would be

$$S = P\left(1 + \frac{1}{m}\right)^{mt}.$$

Here we have multiplied the number of time periods by *m* and divided the interest rate by *m*. If you make the substitution $n = m/i$, you get

$$S = P\left(1 + \frac{i}{n}\right)^{nit},$$

which can be written

$$S = P\left[\left(1 + \frac{1}{n}\right)^n\right]^{it}.$$

Calculate the numerical value of the expression within the square brackets, taking values for *n* of 1, 10, 100, 1000. . . . What value do you eventually get for this expression? What is its significance? How does it relate to the subject of this chapter?

43. *Transforming Between Rectangular and Polar Coordinates.* If you know some right-triangle trigonometry, you should be able to see the relationship between rectangular and polar coordinates when both are shown on a single diagram, as in Figure 9.75.

Using the Pythagorean theorem and the definitions of the trigonometric ratios gives you Equations 93a and 93b.

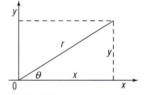

FIGURE 9.75 ■ Rectangular and Polar Coordinates of a Point

93 CONVERSION OF COORDINATES

a. Polar to Rectangular

$$x = r \cos \theta$$
$$y = r \sin \theta$$

b. Rectangular to Polar

$$r = \sqrt{x^2 + y^2}$$

$$\theta = \arctan \frac{y}{x}$$

Using these equations, complete the following:
a. Write the polar coordinates of the point (3, 4).
b. Write the rectangular coordinates of the point (8, 125°).

44. Reproduce any of the curves in Figure 9.76 by graphing the polar equations given.

45. *Logarithmic Spiral.* The equiangular spiral is also called a *logarithmic spiral* because the angle θ is proportional to logarithm of the radius *r*. Demonstrate that this is so by solving Equation 100 for θ. Write a few lines of explanation.

46. *"Squaring" the Circle.* The Archimedean spiral can be used to "square" the circle, except that you arrive at a rectangle rather than a square, as you can see in Figure 9.77. This construction also violates the usual restriction to straightedge and compass. It does, however, give the exact area of the

(*Continues on p. 278.*)

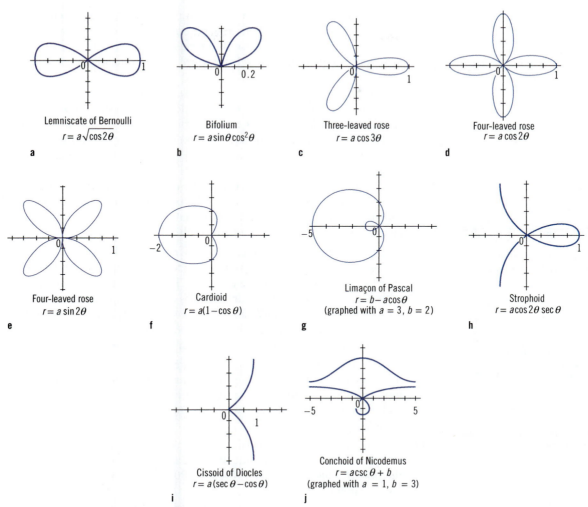

a Lemniscate of Bernoulli
$r = a\sqrt{\cos 2\theta}$

b Bifolium
$r = a\sin\theta\cos^2\theta$

c Three-leaved rose
$r = a\cos 3\theta$

d Four-leaved rose
$r = a\cos 2\theta$

e Four-leaved rose
$r = a\sin 2\theta$

f Cardioid
$r = a(1 - \cos\theta)$

g Limaçon of Pascal
$r = b - a\cos\theta$
(graphed with $a = 3$, $b = 2$)

h Strophoid
$r = a\cos 2\theta\sec\theta$

i Cissoid of Diocles
$r = a(\sec\theta - \cos\theta)$

j Conchoid of Nicodemus
$r = a\csc\theta + b$
(graphed with $a = 1$, $b = 3$)

FIGURE 9.76 ■ Some Well-known Curves in Polar Coordinates

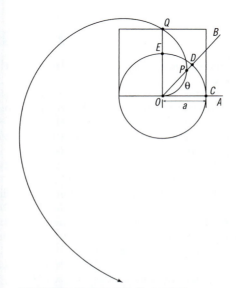

FIGURE 9.77 ■ "Squaring" the Circle with the
Archimedean Spiral

circle. Reproduce this construction: On polar axis OA, draw a circle of radius a, centered at O. Then draw the Archimedean spiral $r = a\theta$ centered at O. Next, locate a point Q on the spiral directly above the pole O. Finally, draw a rectangle of height OQ and width equal to the diameter of the circle.

Explain why the construction works. Show that the area of this rectangle exactly equals the area of the circle. (*Hint:* Choose a point P on the spiral and draw line OPB. Then line segment OP is equal to $a\theta$, and arc CD is also equal to $a\theta$ [if θ is in radians]. Can you say why? In other words, *the radius vector to the spiral is equal to the arc subtended by that radius vector, extended.*)

Now the radius vector OQ cuts the circle at E. Therefore, line OQ equals arc CE, which is one-quarter the circumference of the circle. Use the fact that the area of a circle equals half the product of its radius and its circumference to arrive at the final result.

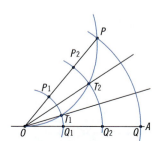

FIGURE 9.78 ■ Trisecting an Angle with the Archimedean Spiral

47. *Angle Trisection.* The Archimedean spiral can be used to trisect an angle, as in Figure 9.78. Reproduce this construction: First, draw an Archimedean spiral and an angle AOP to be trisected. Then, trisect line segment OP with points P_1 and P_2. Next, draw circles centered at O, with radii OP_1 and OP_2, cutting the spiral at T_1 and T_2. Draw lines OT_1 and OT_2 to trisect angle AOP. Explain why this construction works. Modify the construction to subdivide an angle into any number of equal angles.

48. Draw a logarithmic spiral on cardboard, cut it out, and tack it to a board. Tack one end of a string to the spiral near the pole. Insert a pencil in a loop at the other end of the string, as in Figure 9.79. Keep the string taut and wrap it around the spiral while tracing a curve C with the pencil. What is the curve generated?

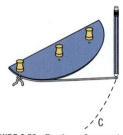

FIGURE 9.79 ■ Tracing a Curve using a Logarithmic Spiral Cut-Out

49. Tack a straightedge to a board. Roll your spiral cut-out from Exercise 48 along the straightedge without slipping, and mark successive positions of the pole P. This is depicted in Figure 9.80. What curve is generated by the moving pole?

50. Make an oral presentation to your class on any of the previous projects.

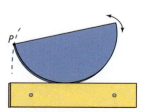

FIGURE 9.80 ■ Spiral Cut-Out for Exercise 49

SOURCES

Arnheim, Rudolph. *The Power of the Center.* Berkeley, CA: University of California Press, 1988.

Boles, Martha, and Newman, Rochelle. *The Golden Relationship: Art, Math & Nature* (4 Volumes). Bradford, MA: Pythagorean Press, 1992.

Bühlmann, Josef. *Classical and Renaissance Architecture.* New York: Helburn, 1916. English translation of Bühlmann's *Die Architektur des Klassischen Altertums und der Renaissance.* Eszlingen: Neff, 1913–19.

Calter, Paul. "How to Construct a Logarithmic Rosette (Without Even Knowing It)." *Nexus Network Journal,* April 2007.

Calter, Paul. *Technical Mathematics.* New York: Wiley, 2007.

Cook, Theodore, *The Curves of Life.* New York: Dover, 1979. First published in 1914.

Coxeter, H. S. M. *Introduction to Geometry,* 2nd ed. New York: Wiley, 1989.

Coxeter, H. S. M., et al. *Geometry Revisited.* Washington, D.C.: Mathematical Association of America, 1967.

Edwards, Edward. *Pattern and Design with Dynamic Symmetry.* New York: Dover, 1967. Reprint of *Dynamarhythmic Design,* 1932.

Ekwall, Åke. "Violins and Volutes: Visual Parallels between Music and Architecture." *Nexus Network Journal,* Autumn 2001.

Hargittai, István, and Pickover, C. A., eds. *Spiral Symmetry*. New York: World
    Scientific, 1991.

Heath, Sir Thomas. *A History of Greek Mathematics*. New York: Dover, 1981. First
    published in 1921.

Huntley, H. E. *The Divine Proportion*. New York: Dover, 1970.

Livio, Mario. *The Golden Ratio*. New York: Broadway Books, 2002.

Pedoe, Dan. *Geometry and the Visual Arts*. New York: Dover, 1976.

Pickover, Clifford A. "Mathematics and Beauty: A Sampling of Spirals and 'Strange'
    Spirals in Science, Nature, and Art." *Leonardo*, Vol. 21, No. 2, pp. 173–181, 1988.

Sharp, John. "Spirals and the Golden Section." *Nexus Network Journal*, Vol. 4, No. 1
    (Winter 2002).

Thompson, Darcy. *On Growth and Form*. New York: Dover, 1992. First published
    in 1942.

Williams, Kim. *Italian Pavements, Patterns in Space*. Houston: Anchorage, 1997.

Williams, Kim. "Spirals and the Rosette in Architectural Ornament." *Nexus Network
    Journal*, Vol. 1 (1999), pp. 129–138.

NOTES

1. Arnheim, p. 91.
2. Arnheim, p. 91.
3. Hambidge, *Elements*, p. xi.
4. Edwards, p. 13.
5. Boles and Newman, p. 211.
6. Newman and Boles, *Universal Patterns*, pp. 212–213. Constructions of the volute
   by Alberti, Serlio, and Palladio are shown in the Exercises and Projects section of
   this chapter. Another construction can be found in Josef Bühlmann, *Classical and
   Renaissance Architecture*, p. 17.
7. Dürer, *The Painter's Manual*, pp. 95–103.
8. "How to Construct a Logarithmic Rosette (Without Even Knowing It)," *Nexus
   Network Journal*, April 2000.

"Let no one destitute of geometry enter my doors."

PLATO

FIGURE 10.1 ■ *Luca Pacioli* (detail), Jacopo de' Barbari, c. 1499
©Erich Lessing/Art Resource, NY

10

The Solids

Let's move now from the two-dimensional world of the plane figures into the three-dimensional world of the solids. We'll start with definitions of surfaces, which define the boundaries of solids, and learn how to compute volumes and areas of familiar solids such as the cube, cylinder, cone, and so forth. *Similar solids*, which are covered in this chapter, are important to artists and architects, who often must translate dimensions from scale drawings or models to murals, sculptures, or buildings.

Next, we'll revisit Plato's *Timaeus* and see his description of how the four elements and heaven itself are made up of the five *Platonic solids*. We will see why only five solids are possible and why Plato thought that those solids, and the triangles that compose them, were the building blocks of the universe.

After we explore these regular polyhedra, we'll examine the semiregular polyhedra (the Archimedean solids and the Kepler-Poinsot solids). We'll show how to construct them and suggest some interesting projects for their use. We'll see how three major Renaissance mathematicians and artists, Piero della

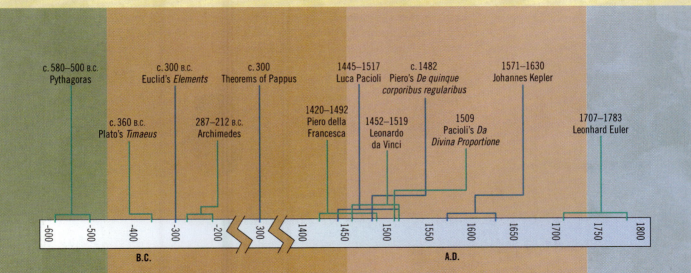

c. 580–500 B.C.
Pythagoras

c. 300 B.C.
Euclid's *Elements*

c. 300
Theorems of Pappus

1445–1517
Luca Pacioli

c. 1482
Piero's *De quinque corporibus regularibus*

1571–1630
Johannes Kepler

c. 360 B.C.
Plato's *Timaeus*

287–212 B.C.
Archimedes

1420–1492
Piero della Francesca

1452–1519
Leonardo da Vinci

1509
Pacioli's *Da Divina Proportione*

1707–1783
Leonhard Euler

-600 -500 -400 -300 -200 300 1400 1450 1500 1550 1600 1650 1700 1750 1800

B.C.   A.D.

Francesca, Luca Pacioli, and Leonardo da Vinci, are connected through their study of polyhedra. Finally, we'll view some examples of the Platonic solids and other polyhedra as they've been used in art.

SURFACES AND SOLIDS

In this section, we'll define some of the basic solids and present some of their properties. We'll start with a discussion of *surfaces*, those geometric entities by which a solid is bounded, and then move on to the various solids. We introduced some of these ideas when we spoke about the conic sections in Chapter 9, "The Ellipse and the Spiral."

Surfaces

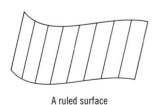

A ruled surface

a

A double-curved surface

b

FIGURE 10.2 ■ Ruled and Double-Curved Surfaces

Just as a *line* may be thought of as being generated by a moving point, a *surface* may be considered to be generated by a moving line, called a *generatrix*. Surfaces may be divided into two major classes: *ruled* and *double-curved*. A moving *straight* line produces a *ruled* surface, and a moving *curved* line gives a *double-curved* surface. In this chapter, we will cover only ruled surfaces. At any point on a ruled surface, you can always place a straightedge in full contact with a ruled surface; however, this is not true of a double-curved surface. See Figure 10.2 for illustrations of both types of surfaces.

If all the points on a straight-line generatrix move in a straight line, a *plane* will be generated. A generatrix, with successive positions that are parallel, generates a *cylindrical* surface. A straight-line generatrix with successive positions that always pass through a fixed point generates a *conical* surface. Both cylindrical and conical surfaces are examples of *single-curved* surfaces. Each position of the generatrix of a ruled surface is called an *element* of that surface. Figure 10.3 illustrates three ruled surfaces.

A straight-line generatrix with successive positions that are not parallel and do not pass through a fixed point generates what is called a *warped surface*, as in Figure 10.4a. Further, a line, straight or curved, when rotated about an axis, generates a *surface of revolution*. See Figure 10.4b. A potter's wheel or lathe produces a surface of revolution.

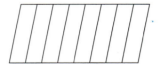

Plane surface

a

Cylindrical surface

b

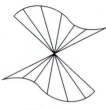

Conical surface

c

FIGURE 10.3 ■ Some Ruled Surfaces

Geometric Solids

The term *geometric solid* means a closed surface in space. Sometimes the word *solid* is taken to mean the surface, and sometimes the word is used to refer to the surface together with its interior. Keep in mind that here we are talking about

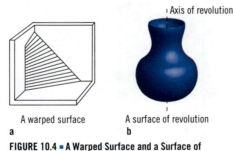

A warped surface

a

Axis of revolution

A surface of revolution

b

FIGURE 10.4 ■ A Warped Surface and a Surface of Revolution

geometric figures and not real objects, so the word *solid* should not be taken in its usual sense to imply *rigidity*, as in "solid as a rock." In this context, we could refer to a soap bubble as a spherical solid even though the soap film and the air enclosed in it are far from rigid.

The *volume* of a solid is a measure of the space it occupies or encloses. A cube with a side of 1 unit has a volume of 1 cubic unit, and you can think of the volume of any solid as the number of such cubes it contains. This may not be a whole number or even a rational number, as you will see with the volumes of cylinders and spheres.

We will discuss three different kinds of *areas* in connection with solids. The *surface area* will mean the total area of the solid, including any ends or bases. The *lateral area*, which we will define for each solid, does not include the area of the base(s). The *cross-sectional area* is the area obtained when a solid is sliced in a certain way.

THE POLYHEDRA

In earlier chapters, you saw that a polygon is a plane figure bounded by line segments. Now, we'll define a *polyhedron* as a solid bounded by polygons. The name *polyhedron* translates from the Latin *poly*, which means "many" and *hedron*, which means "face." Two faces meet in an *edge*, and the point where three or more edges meet is called a *vertex*. The volumes and areas of some common polyhedra are given, without proof, in Figure 10.5. In the following sections, we'll explore a number of basic but very important polyhedra.

Prisms

A *prism* is a polyhedron with two parallel, congruent faces called *bases*. Its remaining faces, called *lateral faces*, are parallelograms. See Figure 10.6 for examples. The lateral faces are formed by joining corresponding vertices of the bases.

104 Cube	a. Volume = a^3 b. Surface area = $6a^2$	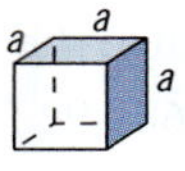	105 Rectangular parallelepiped	a. Volume = lwh b. Surface area = $2(lw + hw + lh)$	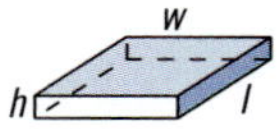
106 Any prism	Volume = (area of base)(altitude)	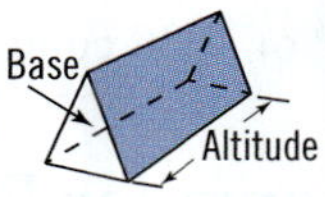	107 Right prism	Lateral area = (perimeter of base)(altitude) *[not including bases]*	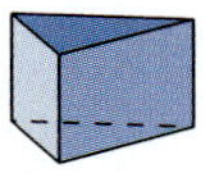
108 Any pyramid	Volume = $\frac{h}{3}$(area of base)	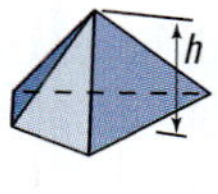	109 Regular pyramid	Lateral area = $\frac{s}{2}$(perimeter of base)	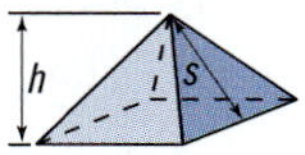
110 Any pyramid	Volume = $\frac{h}{3}(A_1 + A_2 + \sqrt{A_1 A_2})$	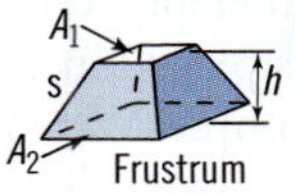	111 Regular pyramid	Lateral area = $\frac{s}{2}$(sum of base perimeters) $= \frac{s}{2}(P_1 + P_2)$	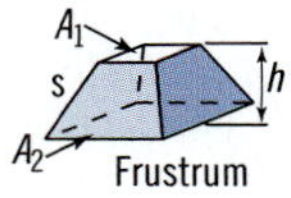

FIGURE 10.5 ■ Volumes and Areas of Some Polyhedra

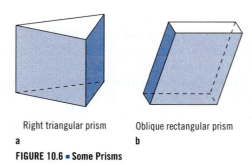

Right triangular prism   Oblique rectangular prism

a   b

FIGURE 10.6 ■ Some Prisms

The *altitude* of a prism is the perpendicular distance between the bases. A prism is named according to the shape of its bases (triangular prism, quadrangular prism, etc.). Also, a prism is called *right* if its bases are perpendicular to its lateral edges and the lateral faces are all rectangles; otherwise, it is called *oblique*. Most buildings are prisms.

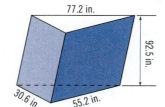

77.2 in.

92.5 in.

30.6 in.   55.2 in.

FIGURE 10.7 ■ An Oblique Triangular Prism

■ **EXAMPLE:** A structural support has the form of an oblique triangular prism, as seen in Figure 10.7. Find its volume.

● **SOLUTION:** First, find the area of the base of the prism by Hero's formula (Equation 30) with $a = 30.6$, $b = 55.2$, and $c = 77.2$.

$$s = \frac{a + b + c}{2}$$

$$= \frac{30.6 + 55.2 + 77.2}{2}$$

$$= 81.5 \text{ in.}$$

$$A = \sqrt{s(s - a)(s - b)(s - c)}$$
$$= \sqrt{81.5(81.5 - 30.6)(81.5 - 55.2)(81.5 - 77.2)}$$
$$= 685 \text{ sq. in.}$$

The volume of the prism is then the area of the base times the altitude:

$$V = 685(92.5) = 63,360 \text{ cubic in.}$$

$$= 36.7 \text{ cubic ft.}$$

(one cubic foot equals 12^3 or 1728 cubic in.)   ■

Rectangular Parallelepipeds

A *parallelepiped* is a prism with bases that are parallelograms. A *right parallelepiped* is a right prism with bases that are parallelograms. A *rectangular parallelepiped* is a right parallelepiped with bases that are rectangles. Frequently called a *rectangular solid*, a rectangular parallelepiped is simply the familiar box shape.

Herms, vertical pillars placed in doorways, on roadsides, or in vineyards for protection, are ancient examples of rectangular solids. They were prisms (either square or rectangular in cross section), and they were surmounted by a carved god—thought to be Hermes, hence the name. The carved god's face was often bearded and the herm was sometimes armless, and had a phallus. Such a pillar is depicted in Figure 10.8.

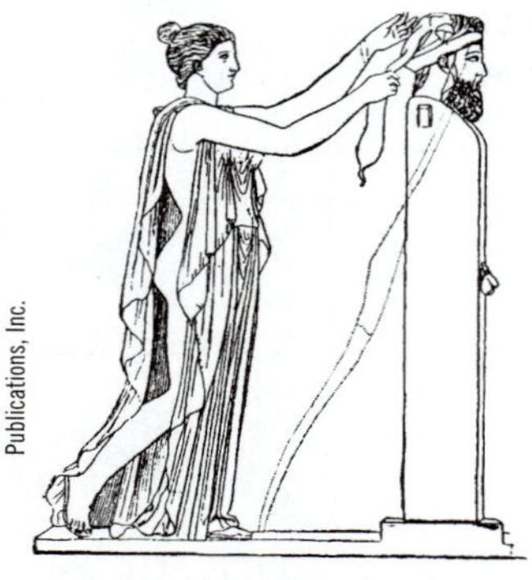

FIGURE 10.8 ■ *Maiden Decorating a Herm,* from an Attic Relief Found Near Naples

■ **EXAMPLE:** The herm shown in Figure 10.8 was carved from a block of stone having a square cross section 20 cm on a side and a height of 125 cm. Find the volume of the original stone.

● **SOLUTION:** To find the volume of the stone, apply Equation 105a:

Volume = length × width × height = (20)(20)(125) = 50,000 cm³.   ■

Cubes

The familiar *cube* is a rectangular parallelepiped having all sides equal. It has six square faces, twelve equal edges, and eight vertices. The cube is also called the *hexahedron*, meaning "six-sided." Unlike the other solids we have covered so far, it is a *regular* polyhedron, so-called because all its faces are congruent regular polygons. Later in this chapter we will discuss the cube as one of the five *Platonic solids*.

■ **EXAMPLE:** Find the volume and surface area of a cubical room that is 10.5 ft on a side.

● **SOLUTION:** To find the volume of the room, apply Equation 104a, with $a = 10.5$ ft:

Volume = a^3 = $(10.5)^3$ = 1158 ft³.

To find the area of the room, apply Equation 104b:

Area = $6a^2$ = $6(10.5)^2$ = 662 ft².   ■

Pyramids

A pyramid is named for the shape of its base: *triangular pyramid, quadrangular pyramid*, and so forth. The pyramids in Figure 10.9 are pentagonal.

The *pyramid* shown in Figure 10.9a is a polyhedron with a *base* that is a polygon and with other faces (called *lateral faces*) that are triangles formed by connecting vertices of the base to a common point V (the *vertex* of the pyramid). The lines connecting the vertices of the base to the vertex of the pyramid are called the *lateral edges* of the pyramid. The *altitude h* of a pyramid is the perpendicular distance from base to vertex.

A *regular* pyramid, as shown in Figure 10.9b, has a base that is a regular polygon and an altitude that passes through the center of that polygon. Its lateral faces are all congruent isosceles triangles, and they form equal angles with the base.

A *frustum* of a pyramid, as shown in Figure 10.9c, is the portion of the pyramid between its base and a plane section parallel to its base. The altitude of the frustum is the perpendicular distance between its base and that plane section.

The *slant height s* of a regular pyramid is the altitude of each triangular face. The slant height of the frustum of a regular pyramid is the perpendicular distance between one edge of the base and one edge of the plane section.

■ **EXAMPLE:** Use the dimensions given for the Great Pyramid of Cheops in Figure 10.10 to determine (a) how many tons of stone, at 3.150 tons/m³, it contains and (b) how many square meters of marble were needed to sheath the exterior surface.

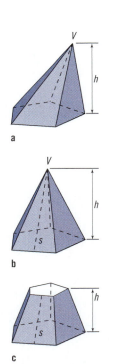

FIGURE 10.9 ■ (a) A Pyramid; (b) A Regular Pyramid; (c) Frustum of a Regular Pyramid

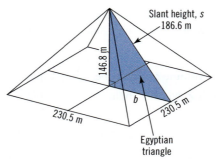

FIGURE 10.10 ■ Approximate Dimensions of the Great Pyramid

● **SOLUTION:** Use these measurements for the Great Pyramid:

$$\text{Base dimension} = 230.5 \text{ m}$$
$$\text{Height} = 146.8 \text{ m}$$
$$\text{Slant height} = 186.6 \text{ m}$$

a. The square base has an area of $(230.5)^2$ m^2. To find the volume, apply Equation 108, with $h = 146.8$ m:

$$\text{Volume} = \frac{h}{3}(\text{area of base})$$
$$= \left(\frac{146.8}{3}\right)(230.5)^2$$
$$= 2{,}599{,}840 \text{ m}^3.$$

You can compute the weight by multiplying the volume by the weight per unit volume:

$$\text{Weight} = 2{,}599{,}840 \text{ m}^3 \ (3.150 \text{ tons/m}^3)$$
$$= 8{,}189{,}497 \text{ tons.}$$

b. The perimeter of the base is $4(230.5)$ m, so by Equation 109 you should obtain the following:

$$\text{Lateral area} = \tfrac{1}{2}(\text{slant height})(\text{perimeter of base})$$
$$= \tfrac{1}{2}(186.6)(4)(230.5)$$
$$= 86{,}023 \text{ m}^2. \qquad ■$$

Symmetry in Three Dimensions

Two points in space have *mirror symmetry* or *bilateral symmetry* about a plane when one point is the reflection of the other about that plane (called the *mirror plane*). Figure 10.11 shows two points (P and Q) that are symmetric about the mirror plane R. Here, R is the perpendicular bisector of line PQ. Further, P and Q are also symmetric about the point O in the plane, which bisects PQ. In addition, P and Q are also symmetric about any line L in the mirror plane passing through O. L is a perpendicular bisector of PQ.

Chapter 3, "The Triangle," introduced some basic ideas about symmetry of plane figures. In this chapter, we'll extend those ideas to solids. (You may want to review that earlier material before proceeding.) A surface or solid is said to have mirror symmetry about a plane if for every point P on that figure, there is another point Q that is symmetric about that plane. In simple terms, half the

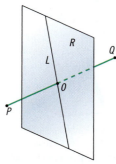

FIGURE 10.11 ■ Mirror Symmetry in Space

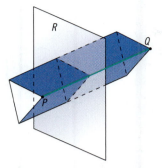

FIGURE 10.12 ■ A Right Triangular
Prism

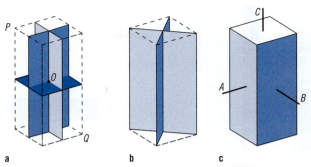

a   b   c

FIGURE 10.13 ■ Symmetries of a Rectangular Parallelepiped with Square
Cross Section

figure is the mirror image of the other half. For example, the right triangular prism shown in Figure 10.12 is symmetric about the mirror plane R. Every point such as P has a symmetrical point Q on the prism.

Let's take this a step further: A rectangular parallelepiped with a square cross section has the three mirror planes shown in Figure 10.13a and two more shown in Figure 10.13b. This solid is also symmetric about its center O because for every point P, there is a mirror image Q.

A surface or solid has *rotational symmetry* about an axis if its appearance is unchanged by a rotation about that axis (called the *axis of rotation*). It has *n-fold* rotational symmetry if n rotations (each less than one revolution) are needed for the figure to return to its original orientation. The rectangular parallelepiped shown in Figure 10.13c has twofold rotational symmetry about each axis A or B, and it has fourfold rotational symmetry about axis C. The regular triangular pyramid shown in Figure 10.14 has *threefold* rotational symmetry about its vertical axis, because it takes three rotations, of one-third revolution each, to return the triangle to its original position. Any *surface or solid of revolution* has rotational symmetry about its axis of revolution. It also has mirror symmetry with respect to any plane containing this axis.

FIGURE 10.14 ■ A Regular Triangular
Pyramid

■ EXERCISES ● THE POLYHEDRA

1. Define or describe the following:

surface	ruled surface	double-curved surface
generatrix	warped surface	surface of revolution
element of a surface	geometric solid	volume
cross-sectional area	lateral area	polyhedron
prism	rectangular parallelepiped	pyramid
mirror symmetry	frustum	surface area
bilateral symmetry	mirror plane	rotational symmetry
axis of rotation	*n*-fold rotational symmetry	

2. Find the lateral area, total area, and volume of each prism in Figure 10.15. Assume that each base is a regular polygon.

3. The support strut shown in Figure 10.16 has a rectangular base that is 12 in. wide and 18 in. deep. Find the lateral area and volume of the strut.

4. Find the lateral area, total area, and volume of each rectangular parallelepiped in Figure 10.17.

5. *Golden Cuboid.* A rectangular parallelepiped with sides that are in the ratio $1 : \Phi : 1/\Phi$ is sometimes called a *golden cuboid*.[1] Find the volume and surface area of a golden cuboid whose shortest side is 1 unit.

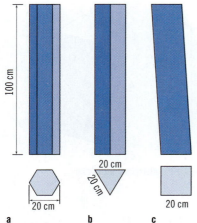

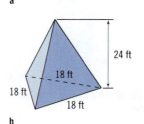

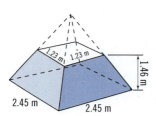

FIGURE 10.15 ■ Some Prisms

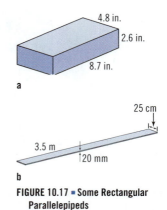

FIGURE 10.16 ■ A Support Strut

FIGURE 10.17 ■ Some Rectangular Parallelepipeds

6. Find the surface area and volume of a cube with the following sides:
   a. 3.75 in.   b. 26.3 cm   c. 2.24 ft

7. Find the volume and lateral area of each pyramid or frustum in Figure 10.18.

8. a. Find the volume enclosed by the pyramidal roof on a square tower. Take the base as 22.0 ft on a side and the height as 24.5 ft, and ignore the overhang.
   b. Find the lateral area of the roof.

9. The pyramidal roof shown in Figure 10.19 has an octagonal base of 4.5 ft on a side and a slant of 14.0 ft. How many square feet of shingles are needed to cover the roof, not counting any waste?

10. A roof, shown simplified in Figure 10.20, is in the shape of a triangular prism. Find the area of the shingled surface and the volume enclosed by the roof.

FIGURE 10.18 ■ Some Pyramids

FIGURE 10.19 ■ A Pyramidal Roof

11. A simplified hip roof is shown in Figure 10.21. Find the area of the shingled surface and the volume enclosed by the roof.

12. Two intersecting roofs are simplified in Figure 10.22. Find the area of the shingled surface and the volume enclosed by the roofs. Make sure you count the portion contained under both roofs only once and not twice.

13. Identify all the mirror planes in the following solids:
    a. Square-based right pyramid
    b. Right prism with a hexagonal base
    c. Rectangular parallelepiped

14. Identify all the axes of rotational symmetry for the solids in the preceding exercise. For those that have *n*-fold rotational symmetry, give the value of *n*.

15. The cube has nine mirror planes, four threefold axes of rotational symmetry, and six twofold axes of rotational symmetry. Identify them all in a model or a sketch. *Hint:* Finding all the mirror planes and axes of symmetry of a cube isn't easy. Use a wooden block or cardboard model to help you to visualize them.

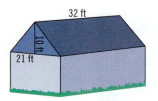

FIGURE 10.20 ■ A Prismatic Roof

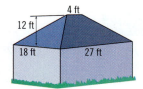

FIGURE 10.21 ■ A Hip Roof

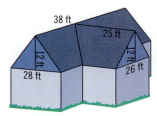

FIGURE 10.22 ■ Intersecting Roofs

CYLINDERS AND CONES

Now let's turn from the polyhedra to other familiar solids, the cylinder and the cone. These solids are, of course, bounded not only by planes, but by curved surfaces as well. We will define each solid in turn.

Cylinders

Earlier we defined a cylindrical surface as the surface produced by a generatrix with successive positions that are parallel to each other. The path followed by the generatrix is a curve called the *directrix*. The directrix may be an open curve or a closed curve. When the directrix is a closed curve, the portion of the cylindrical surface between two parallel planes is called a *cylinder*, as shown in Figure 10.23. Therefore, a cylinder is a solid with two parallel, congruent faces, called *bases*, and with a lateral surface that is formed by joining corresponding points on the bases. The *altitude* of a cylinder is the perpendicular distance between the bases.

This definition of the cylinder is similar to the definition of the prism, except the prism bases are polygons; with the cylinder, the bases can be any closed curve. A cylinder is named according to the shape of its bases (circular cylinder, elliptical cylinder, etc.). The *axis* of a cylinder is the line connecting the centers of its bases. A cylinder is called *right* if its bases are perpendicular to its axis, as in Figure 10.23b; otherwise, it is called *oblique*, as in Figure 10.23a.

The areas and volumes of cylinders are given, without proof, in Figure 10.24. As you can see, some of the formulas for cylinders are the same as for the prism.

a

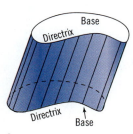

b

FIGURE 10.23 ■ (a) An Oblique Cylinder; (b) A Right Circular Cylinder

■ **EXAMPLE:** A cylindrical mold for a right circular concrete column is 22.5 in. in diameter and 18.5 ft high. How many cubic yards of concrete are needed to fill this mold?

● **SOLUTION:** The column diameter, in feet, is *d*:

$$d = \frac{22.5}{12} = 1.875 \text{ ft.}$$

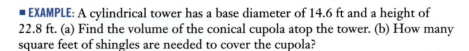

106			107		
Any cylinder	Volume = (area of base)(altitude)		Right cylinder	Lateral area = (perimeter of base)(altitude) *[not including bases]*	
108			109		
Any cone	Volume = $\frac{h}{3}$(area of base)		Right circular cone	Lateral area = $\frac{s}{2}$(circumference of base)	
110			111		
Any cone	Volume = $\frac{h}{3}(A_1 + A_2 + \sqrt{A_1 A_2})$		Right circular cone	Lateral area = $\frac{s}{2}$(sum of base circumferences) = $\frac{s}{2}(P_1 + P_2)$	

FIGURE 10.24 ■ Cylinders and Cones

The area of the base is then $\pi d^2/4$ or 2.76 ft². To find the volume, apply Equation 106:

$$\text{Volume} = \text{area of base} \times \text{altitude}$$
$$= (2.76)(18.5) = 51.1 \text{ ft}^3.$$

Divide this by the 27 cubic feet in a cubic yard:

$$\text{Volume} = \frac{51.1}{27} = 1.89 \text{ yd}^3. \qquad ■$$

Cones

A thorough definition of the cone appears in Chapter 9; take a moment to review the definition. The *frustum* of a cone is just like the frustum of a pyramid, which was explained earlier in this chapter. The formulas for the volume and area of a cone and the frustum of a cone were given in Figure 10.24. Figure 10.25 illustrates a general cone, a right-circular cone, and the frustum of a right-circular cone.

■ **EXAMPLE:** A cylindrical tower has a base diameter of 14.6 ft and a height of 22.8 ft. (a) Find the volume of the conical cupola atop the tower. (b) How many square feet of shingles are needed to cover the cupola?

● **SOLUTION:**

a. The area of the cone's base is $\pi(7.3)^2 = 167$ ft². To find the volume, apply Equation 108:

$$\text{Volume} = \tfrac{1}{3}(\text{height})(\text{area of base})$$
$$= \frac{22.8(167)}{3} = 1270 \text{ ft}^2.$$

b. To find the slant height s, use the Pythagorean theorem:

$$s^2 = (7.3)^2 + (22.8)^2 = 573.$$

Therefore,

$$s = 23.9 \text{ ft.}$$

a

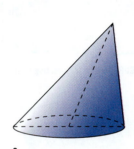

b

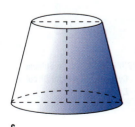

c

FIGURE 10.25 ■ (a) A Cone; (b) A Right Circular Cone; (c) Frustum of a Right Circular Cone

The perimeter of the base is $14.6\pi = 45.9$ ft. To find the lateral area, apply Equation 109:

$$\text{Lateral area} = \tfrac{1}{2}(\text{slant height})(\text{perimeter of base})$$
$$= \frac{23.9(45.9)}{2} = 549 \text{ ft}^2.$$

This is the square footage of shingles needed (not accounting for waste). ■

Solids of Revolution

As you learned earlier, a moving line or curve generates a surface of revolution. Similarly, a polygon or closed curve rotating about an axis will generate a *solid of revolution*, as shown in Figure 10.26. The closed plane figure *C* rotates about an axis *A* to produce a solid of revolution. In this example, the plane figure is a circle, and the solid of revolution is called the *torus* or *anchor ring*. All cross sections perpendicular to the axis of revolution of a solid of revolution are circles. The right circular cylinder is, of course, a solid of revolution.

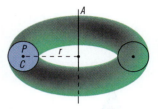

FIGURE 10.26 ■ A Solid of Revolution

You can find the surface area and volume of a solid of revolution by means of the *theorems of Pappus*, named for the Greek geometer Pappus of Alexandria (c. 300 A.D.). Recall the introduction to Pappus in Chapter 3. In Figure 10.26, the center *P* of the figure *C* follows a circular path with length *L* equal to $2\pi r$, where *r* is the radius of the circular path followed by the centroid. Pappus gives us the theorems cited in Statement 112.

112 THEOREMS OF PAPPUS

a. The surface area of the solid of revolution is equal to the perimeter of *C* times the distance *L* traveled by the center of *C*.
b. The volume of the solid of revolution is equal to the area of *C* times *L*. ●

■ **EXAMPLE:** Use the theorems of Pappus to derive the formulas for (a) the volume and (b) the surface area of a right circular cylinder of radius *r* and height *h*, as shown in Figure 10.27.

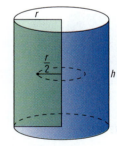

FIGURE 10.27 ■ A Right Circular Cylinder

● **SOLUTION:** Let the cylinder be generated by a rectangle of width *r* and height *h* rotated about an axis. The perimeter of the rectangle is $2h + 2r$, and its area is rh. Its centroid is at a distance $r/2$ from the axis, so it follows a circular path with length that is $2\pi(r/2)$, or πr. Then by using the theorems of Pappus, you can determine the following:

a. Volume of the cylinder = (area of the rectangle)(distance traveled by its center)
$$= (rh)(\pi r) = \pi r^2 h$$

The volume is equal to the area of the cylinder's base times its altitude, as in Equation 106.

b. Area of the cylinder = (perimeter of the rectangle)(distance traveled by its center)
$$= (2h + 2r)(\pi r)$$
$$= 2\pi rh + 2\pi r^2$$

Here, $2\pi rh$ is the lateral area (the perimeter of the cylinder's base times its altitude), and $2\pi r^2$ is the area of the two circular bases. If you exclude the areas of the bases, you'll get Equation 107. ■

■ EXERCISES ● CYLINDERS AND CONES

1. Define or describe the following:

   cylinder   cylindrical surface   conical surface
   frustum   slant height   cone

2. How many cubic feet are in a tapered timber column 30 ft long if one end has an area of 225 sq. in. and the other end has an area of 144 sq. in.?

3. Find the lateral area, total area, and volume of each of the following right circular cylinders:
   a. Base diameter = 34.5 in., height = 26.4 in.
   b. Base diameter = 134 cm, height = 226 cm
   c. Base diameter = 3.65 m, height = 2.76 m

4. Find the lateral area and the volume enclosed by a cylindrical tower having a round base 18.0 ft in diameter and a height of 31.5 ft.

5. A cylindrical chimney has a round base with a 2.55 m outside diameter, a 2.00 m inside diameter, and a height of 7.54 m. Find the volume of masonry in the chimney.

6. *Archimedes on the Cylinder.* Archimedes stated that the lateral area of a cylinder is equal to the area of a circle with a radius that is the mean proportional between the cylinder's height and its base diameter. Verify this using modern notation.

7. Find the lateral area, total area, and volume of each right circular cone or frustum in Figure 10.28.

8. Estimate the volume of a conical pile of sand that is 12.5 ft high and has a base diameter of 14.2 ft.

9. Tapered wooden columns are used to support a tent-like structure at an exposition. Each column is 50 ft high and has a circumference of 5.0 ft at one end and a circumference of 3.0 ft at the other. Find the volume of each column.

10. Identify all the mirror planes in the following solids:
    a. Square-based right pyramid
    b. Right prism with a hexagonal base
    c. Rectangular parallelepiped
    d. Right circular cylinder
    e. Right circular cone

11. Identify all the axes of rotational symmetry for the solids in the preceding exercise. For the axes that have *n*-fold rotational symmetry, give the value of *n*.

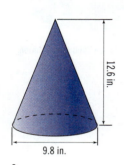

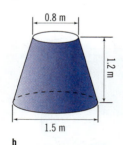

FIGURE 10.28 ■ Some Cones

SIMILAR SOLIDS

We considered similar triangles in Chapter 3, "The Triangle," and similar plane figures in Chapter 5, "Polygons, Tilings, and Sacred Geometry." In this section, we'll expand the idea to cover plane figures of any shape, including those with curved boundaries, and *solids*. We'll again see that the idea of similar figures gives us a rapid and powerful way to compute certain dimensions and areas and in the case of solids, to compute volumes.

Dimensions of Similar Figures

We have already shown that corresponding dimensions of similar plane figures are in proportion. Here, we'll simply add solids to our definition.

94 DIMENSIONS OF PLANE OR SOLID SIMILAR FIGURES

Corresponding dimensions of plane or solid similar figures are in proportion. The ratio of their lengths is the scale factor.   ●

■ **EXAMPLE:** Two similar solids are shown in Figure 10.29. Find the diameter D of the hole.

● **SOLUTION:** The scale factor is $\frac{5.26}{3.15} = 1.670$. To find the diameter, apply Statement 94:

$$D = 22.5(1.670) = 37.6 \text{ mm}$$

Because we are dealing with *ratios* of corresponding dimensions, we don't need to convert all of the dimensions to the same units.   ■

Areas of Similar Figures

The area of a square of side s is equal to the square of one side—that is, $s \times s$ *or* s^2. Therefore, if a side is multiplied by a scale factor k, the area of the larger square is $(ks) \times (ks)$ or $k^2 s^2$. You can see that the area has increased by a factor of k^2, the *square of the scale factor*. An area more complicated than a square can be thought of as being made up of many small squares. Then, if a dimension of that area is multiplied by a scale factor of k, the area of each small square increases by a factor of k^2, and hence the entire area of the figure increases by a factor of k^2. We can state this formally as in Statement 95.

95 AREAS OF PLANE OR SOLID SIMILAR FIGURES

Corresponding areas of similar figures are in proportion, with the constant of proportionality equal to the square of the scale factor.   ●

Therefore, if a figure's sides are doubled, the new area will be *four times* the original area. This relationship is valid not only for plane areas and for surface areas of solids, but also for cross-sectional areas of solids.

■ **EXAMPLE:** The triangular top surface of the smaller solid shown in Figure 10.29 has an area of 4.84 sq. in. Find the area of the corresponding surface on the larger solid.

● **SOLUTION:** We have already found the scale factor, 1.670. To find the area on the larger solid, apply Statement 95:

$$A = 4.84 \, (1.670)^2 = 13.5 \text{ in.}^2$$ ■

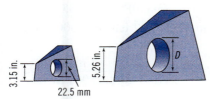

FIGURE 10.29 ■ Similar Solids

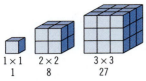

FIGURE 10.30 ■ Volumes of Similar Solids

Volumes of Similar Solids

A cube of side s has a volume s^3. We can think of any solid as being made up of many tiny cubes, each of which has a volume equal to the cube of its side. Therefore, if the dimensions of the solid are multiplied by a scale factor k, the volume of each cube (and hence the entire solid) will increase by a factor of k^3. This relationship is illustrated in Figure 10.30 and described in Statement 96.

96 VOLUMES OF SIMILAR SOLIDS

Corresponding volumes of similar solids are in proportion, with the constant of proportionality equal to the cube of the scale factor.   ●

Forgetting to *square* corresponding dimensions when finding areas and forgetting to *cube* corresponding dimensions when finding volumes are both common mistakes.

■ **EXAMPLE:** Refer again to Figure 10.29. If the volume of the smaller solid is 15.6 in.^{3}, find the volume V of the larger solid.

● **SOLUTION:** By Statement 96, with a scale factor of 1.670, you obtain this volume:
$$V = 15.6(1.670)^3 = 72.7 \text{ in.}^3 \qquad ■$$

Scale Drawings and Models

An important application of similar figures is in the use of *scale drawings*, such as maps, architectural drawings, surveying layouts, and scale models of proposed projects. The ratio of a distance on a drawing or model to the corresponding distance on the actual object is called the *scale* of the drawing or model, or what we have called the *scale factor*. A scale drawing and the actual object are *similar plane figures*, and a scale model and the actual object are *similar solids*. We will use the ideas in this section to convert areas and/or volumes between actual objects and their corresponding drawings or models.

■ **EXAMPLE:** A map has a scale of 1 : 5000. How many acres of land are represented by 168 sq. in. on the map?

● **SOLUTION:** Let A = the area of the land. Because areas of similar figures are proportional to the square of the scale factor,
$$A = 168(5000)^2 = 4,200,000,000 \text{ in.}^2$$
Next, convert from square inches to acres. Because 12 in. = 1 ft, then 12^2 in.^{2} = 1 ft^2. Also, to convert to acres, we use the conversion factor 1 acre = 43,560 ft^2.
$$A = 4,200,000,000 \text{ in.}^2 \left(\frac{1 \text{ ft}^2}{144 \text{ in.}^2} \right)\left(\frac{1 \text{ acre}}{43,560 \text{ ft}^2} \right) = 670 \text{ acres} \qquad ■$$

■ **EXAMPLE:** A room in a 1 : 50 scale model of a house has a volume of 750 cm^2. Find the volume of the actual room.

● **SOLUTION:** Because volumes of similar solids are proportional to the cube of the scale factor, the volume V of the actual room is
$$V = 750(50)^3 = 93,800,000 \text{ cm}^3.$$
Because 100 cm = 1 m, then 100^3 cm^3 = 1,000,000 cm^3 = 1 m^3. Converting gives us this:
$$V = 93,800,000 \text{ m}^3 \left(\frac{1 \text{ m}^3}{1,000,000 \text{ cm}^3} \right) = 93.8 \text{ m}^3. \qquad ■$$

■ EXERCISES ● SIMILAR SOLIDS

1. A wood stove has a firebox volume of 4.25 ft³. What firebox volume would be expected if all the dimensions of the stove were increased by a factor of 1.25?

2. If the stove in Exercise 1 weighed 327 lb, how much would you expect the larger stove to weigh? *Hint:* Assume that the weight is proportional to the volume.

3. A certain solar-heated house stores heat in 155 metric tons of stones; the stones are in a chamber beneath the house. Another solar-heated house is to have a chamber of similar shape but with all dimensions increased by 15%. How many metric tons of stone will it hold?

4. Each side of a square is increased by 15.0 mm, and the area is seen to increase by 2450 mm². What were the dimensions of the original square?

5. The floor plan of a building has a scale of $\frac{1}{4}$ in. = 1 ft. One of the rooms has an area of 40 sq. in. What is the actual room area in square feet?

6. On a drawing having a scale of 1 : 4, the area of a window of a building is 98.2 sq. in. What is the actual window area in square feet?

7. A pipe with a diameter of 3.00 in. discharges 500 gallons (gal) of water in a specified amount of time. What is the diameter of a pipe that will discharge 750 gal in that same amount of time? *Hint:* Assume that the amount of flow through a pipe is proportional to its cross-sectional area.

8. A room in a 1 : 40 scale model of a building has a volume of 46.5 cubic in. Find the volume of the actual room in cubic feet.

9. *Doubling the Volume of a Solid.* According to myth, King Minos of Crete ordered each dimension of his son Glaucus' tomb to be doubled so that the volume would be doubled.[2] Will doubling the dimensions double the volume? Explain your answer.

THE PLATONIC SOLIDS

Earlier in this chapter, we defined a polyhedron as a solid bounded by polygons, and a regular polyhedron, or *regular solid*, as one with faces that are identical regular polygons—that is, all equilateral triangles or all squares, and so on. Only five regular solids are possible: the tetrahedron, octahedron, icosahedron, hexahedron (cube), and dodecahedron.

The prefix *tetra* means four, *hexa* means six, *octa* means eight, *dodeca* means twelve, and *icosa* means twenty. The suffix *hedron* means "face." These polygons, presented in Figure 10.31, have come to be known as the *Platonic solids*. Each Platonic solid can be circumscribed by a sphere, with every vertex touching the sphere. Also, a sphere can be inscribed in each Platonic solid, touching each face at its center.

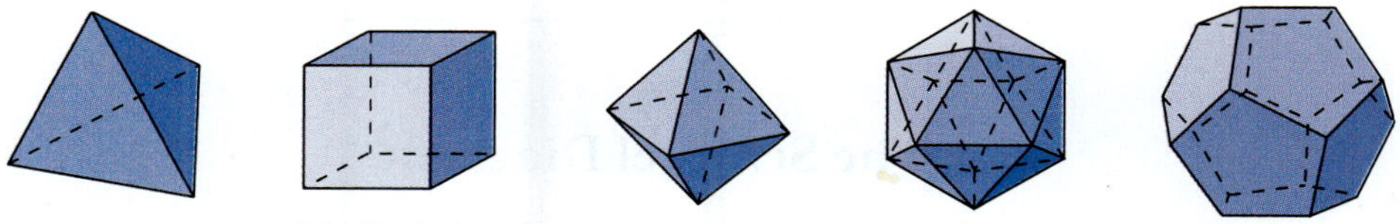

Tetrahedron
a

Cube (Hexahedron)
b

Octahedron
c

Icosahedron
d

Dodecahedron
e

FIGURE 10.31 ■ The Platonic Solids

Why Five and Only Five Regular Solids?[3]

For a *regular* polyhedron, all of the faces must be the same regular polygon (equilateral triangle, square, pentagon, hexagon, etc.). Further, three or more faces must be able to meet to form a convex vertex or corner. That is, the sum of the angles of the polygons meeting at a vertex must be less than 360°. If the faces are *equilateral triangles* with vertex angles that are 60°, it is possible for three, four, or five triangles to form a corner. Six equilateral triangles will form a plane (thus having no corner). Further, seven or more equilateral triangles cannot form a corner. This is because the sum of seven angles that would have to meet at that vertex would be $7 \times 60° = 420°$, which is greater than the allowable 360°. Therefore, only *three* regular polyhedra with faces of equilateral triangles are possible: the tetrahedron, octahedron, and icosahedron.

If the faces are *squares*, with vertex angles of 90°, three squares can form a corner. Four squares form a plane, and more than four are not possible. Therefore, only *one* regular polyhedron with square faces is possible. It is, of course, the cube.

If the faces are *regular pentagons*, with vertex angles of 108°, three pentagons can form a corner because $3 \times 108° = 324°$. Four or more pentagons are not possible. Therefore, only *one* regular polyhedron with pentagons for faces is possible: the dodecahedron.

If the faces are *regular hexagons* with vertex angles of 120°, three form a plane, while four or more are not possible. Therefore, a regular polyhedron with hexagonal sides is not possible.

If the faces are *regular polygons of more than six sides*, each would have a vertex angle greater than 120°; thus three or more are not possible. Therefore, no regular polyhedron can be formed from regular polygons of more than six sides.

The Vertex Net

In Chapter 5, we used a vertex net to describe the polygons about each vertex in a tiling; a tiling having two hexagons alternating with two equilateral triangles about each vertex had a vertex net of 6363. We can use the same idea to describe the Platonic solids. The cube would have a vertex net of 444, and so forth. The vertex net for each Platonic solid, along with additional information about each solid, is given in the following table. (We'll discuss the items in the Element column shortly.)

Table of the Platonic Solids					
Solid	Element	Faces	Vertices	Edges	Vertex Net
Tetrahedron	Fire	4 equilateral triangles	4	6	333
Hexahedron (cube)	Earth	6 squares	8	12	444
Octahedron	Air	8 equilateral triangles	6	12	3333
Icosahedron	Water	20 equilateral triangles	12	30	33333
Dodecahedron	Cosmos	12 regular pentagons	20	30	555

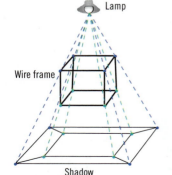

Lamp

Wire frame

Shadow

FIGURE 10.32 ■ Projection of a Polyhedron

The Schlegel Diagram

Now imagine a polyhedron made from a wire frame suspended over a sheet of paper. If you place a lamp directly over one face, the frame will cast a shadow on the paper, as in Figure 10.32. (We will expand on the idea of projection in

Chapter 11, "The Sphere and Celestial Themes in Art and Architecture," and in Chapter 12, "Brunelleschi's Peepshow and the Origins of Perspective.")

The image that is produced is called a *Schlegel diagram*. The upper face appears as a large polygon, with the other faces inside. You have *mapped* the polygon onto a plane, and we say that the Schlegel diagram is a plane map of the polygon. Figure 10.33 shows the Schlegel diagrams for the five Platonic solids. At first glance, each Schlegel diagram seems to have one less face than it should, but you must include the uppermost face of the polyhedron, which appears as the outermost polygon in the Schlegel diagram.

Euler's Theorem for Polyhedrons

Any polyhedron that can be represented by a Schlegel diagram is said to be *simply connected*. Any simply connected polyhedron with V vertices, E edges, and F faces satisfies this condition:

$$V - E + F = 2.$$

This relationship is sometimes called the *Euler Characteristic, Euler's formula,* or the *Euler Number*. This property may have been known to Archimedes. It was stated by Descartes around 1635, and it was independently announced by Euler in 1752. This theorem was one of the early results of what was later to become the field of topology. To demonstrate Euler's formula, let's build a *plane map*, one vertex and edge at a time, as in Figure 10.34, counting vertices, edges, and regions as we go.

You can see that in Figures 10.34a through 10.34d the Euler characteristic has remained at 2. Suppose you now add a vertex Q and an edge PQ, as in Figure 10.34e. This will not add a region, so the Euler characteristic will remain at 2. However, if you instead connect P to R, you add an edge and a region—but not a vertex; therefore, the Euler characteristic remains at 2. You can begin to see that no matter how you extend the figure, the Euler characteristic will remain at 2. What does this have to do with polyhedra? Because Euler's formula holds true for any plane map, and because the Schlegel diagram is a plane map of a simple polyhedron, Euler's formula holds for simple polyhedrons as well.

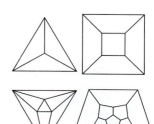

FIGURE 10.33 ■ Schlegel Diagrams for the Platonic Solids

Vertices and Edges	Values of *V, E,* and *F*	Euler Characteristic
One vertex	$V = 1, E = 0, F = 1$ (the plane surface itself)	$1 - 0 + 1 = 2$
Two vertices and an edge	$V = 2, E = 1, F = 1$	$2 - 1 + 1 = 2$
Three vertices and two edges	$V = 3, E = 2, F = 1$	$3 - 2 + 1 = 2$
Three vertices and three edges	$V = 3, E = 3, F = 2$	$3 - 3 + 2 = 2$
Four vertices and four edges	$V = 4, E = 4, F = 2$	$4 - 4 + 2 = 2$

FIGURE 10.34 ■ Demonstration of Euler's Theorem

Duality

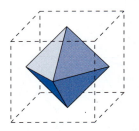

FIGURE 10.35 ■ Duality of Cube and Octahedron

The cube and the octahedron each have 12 edges. The cube has six faces and eight vertices, and the octahedron has eight faces and six vertices. If you connect the centers of the faces of a cube, as in Figure 10.35, you'll get an octahedron. If you connect the centers of the faces of an octahedron, you'll get a cube. Therefore, the cube and the octahedron are said to be *dual* to each other.

The icosahedron and dodecahedron, shown in Figure 10.36, are also duals. Both have 30 edges. Each has as many faces as its dual has vertices, and each has as many vertices as its dual has faces. Connecting the midpoints of the faces of one solid gives the other solid. Interestingly, the tetrahedron is its own dual. Connecting the centers of its faces gives another tetrahedron.

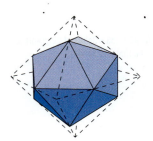

FIGURE 10.36 ■ Duality of Icosahedron and Dodecahedron

Making the Platonic Solids

According to Sir Thomas Heath, the construction of the cube, the tetrahedron (which is simply a triangular pyramid), and the octahedron (which is only a double square pyramid) must have been known to the Egyptians.[4] He also cites evidence attributing the construction of the five regular solids to the Pythagoreans. In fact, Euclid gives the construction of the five regular solids in Book XIII, Propositions 13 to 18, of his *Elements*.[5] You can also construct a solid by cutting the faces shown in Figure 10.37 out of some sheet material, such as cardboard, and gluing them together. Other methods include using sticks or drinking straws for the edges and joining them with glue or clay; carving them from wood, clay, or Styrofoam; and assembling them from a kit.

The Platonic Solids and the Golden Ratio

The Platonic solids have some interesting relationships with the golden ratio. For example, take any edge of an icosahedron and connect its ends with the ends of the opposite edge. The resulting rectangle is a *golden rectangle*. There are three such rectangles, which are mutually perpendicular and intersect at the center of the icosahedron. They are shown in Figure 10.38.

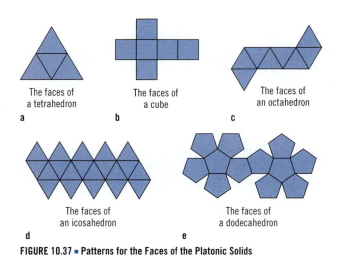

The faces of
a tetrahedron

a

The faces of
a cube

b

The faces of
an octahedron

c

The faces of
an icosahedron

d

The faces of
a dodecahedron

e

FIGURE 10.37 ■ Patterns for the Faces of the Platonic Solids

FIGURE 10.38 ■ Golden Rectangles in an Icosahedron

Similarly, you can form golden rectangles by connecting the centers of the faces of a dodecahedron. Further, the dodecahedron has pentagonal faces, and in Chapter 5 you saw how the pentagon can be subdivided into golden triangles.

The Archimedean Solids

As mentioned earlier, the Platonic solids belong to the group of geometric figures called *regular polyhedra*. We covered the regular polyhedra in the preceding section of this chapter, and here we will discuss some that are called *semiregular*. You can get two sets of semiregular polyhedra from the Platonic solids. The first set is obtained by cutting off the corners of the Platonic solids to get *truncated* polyhedra, which are called the *Archimedean solids*. A second set is obtained by extending the faces of the Platonic solids to form the *star polyhedra*.

For both the Archimedean solids and the Platonic solids, each polygon meeting at each vertex of the polyhedron must be regular, the sum of the angles of the polygons must be less than 360°, at least three polygons are needed to form a vertex of the polyhedron, and all vertices of the polyhedron are identical. The Archimedean solids have an additional condition: There must be at least two kinds of polyhedra at each vertex. With these constraints, each vertex of an Archimedean solid can have a total of only three, four, or five polygons and only two, three, or four *different kinds* of polygons.

Therefore, a vertex net of 3434 is possible, because it has four polygons about a vertex and two different kinds of polygons (equilateral triangles and squares). It gives the solid called the *cuboctahedron*. On the other hand, a vertex net of, say, 334466 is not possible because it would have more than five polygons about a vertex. Further, a vertex net of 34568 is not possible because it has more than four different kinds of polygons about a vertex. Exhausting all the possibilities, given these limits, we arrive at the 13 Archimedean solids in Figure 10.39.

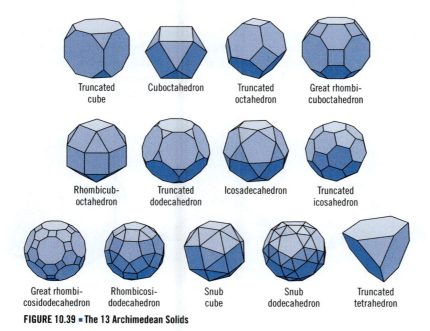

Truncated cube	Cuboctahedron	Truncated octahedron	Great rhombi-cuboctahedron
Rhombicub-octahedron	Truncated dodecahedron	Icosadecahedron	Truncated icosahedron
Great rhombi-cosidodecahedron	Rhombicosi-dodecahedron	Snub cube	Snub dodecahedron	Truncated tetrahedron

FIGURE 10.39 ■ The 13 Archimedean Solids

ARCHIMEDES (c. 287–212 B.C.)

Archimedes was perhaps the most famous ancient Greek mathematician and inventor. Like Pythagoras, he spent time in Egypt early in his career, but for most of his life he lived in Syracuse, his birthplace in the southeast corner of Sicily. Much is known of his life, but like the life of Pythagoras, a great deal is legend.

Archimedes is credited with inventing the *Archimedean screw,* a device for raising water, and he is supposed to have made a star globe and an *orrery,* a device for mechanically representing the motions of the sun, moon, and planets. His name is also given to the *Archimedean spiral* that we studied in Chapter 9.

He is credited with determining the proportion of gold and silver in a crown by weighing it in water. However, the story that he leapt from the bath and

©Erich Lessing/Art Resource, NY

FIGURE 10.40 ■ *Death of Archimedes*, Mosaic (detail), Seventeenth Century

ran naked through the streets shouting "Eureka!" (I have found it!) is probably myth, as is his supposed statement, "Give me a place to stand and I will move the Earth."

He played an important role in the defense of Syracuse against the Romans in 213 B.C. By constructing war machines so effective that they long delayed the capture of the city, he helped delay his own death in battle. However, the stories that Archimedes used a huge array of mirrors to burn the Roman ships besieging Syracuse and that a Roman soldier killed him because he refused to leave his mathematical diagrams are probably just legend. In the mosaic in Figure 10.40, we see Archimedes at work, with a Roman soldier approaching to slay him.

Star Polyhedra

The second obvious way to get another set of solids is to extend the faces of each Platonic solid to form a star, forming the so-called *star polyhedra*. The process of extending the faces to form a star is called *stellation*, and the resulting star polyhedron is sometimes called a *stellated polyhedron* or a *stella polyhedron*, as shown in Figure 10.41. Two star polyhedra, the small stellated dodecahedron and great stellated dodecahedron, were discovered by Louis Poinsot in 1809. When he stumbled across the other two, the great icosahedron and great dodecahedron, he didn't know that Johannes Kepler had discovered them about 200 years earlier. Thus the star polyhedra are also known as the Kepler-Poinsot solids.

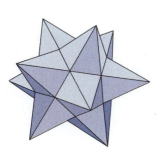
Small stellated dodecahedron
a

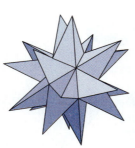
Great stellated dodecahedron
b

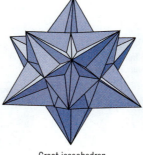
Great icosahedron
c

Great dodecahedron
d

FIGURE 10.41 ■ **The Four Star Polyhedra**

■ EXERCISES ● THE PLATONIC SOLIDS

1. Define or describe the following:

regular solid	duality
icosahedron	Kepler solid
Platonic solid	octahedron
Archimedean solid	dodecahedron
Poinsot solid	semiregular solid
tetrahedron	star polyhedron
hexahedron	

2. Prove or demonstrate that there are five and only five regular polyhedra.

3. Make models of the five Platonic solids using the patterns given in Figure 10.37.

4. Use a computer drawing program to create the templates needed for the preceding exercise, including gluing tabs. Then print the templates and make the models.

5. Construct models of the five Platonic solids using sticks, skewers, or drinking straws.

6. Carve models of the five Platonic solids using a solid material, such as clay, wood, or Styrofoam.

7. Using 6 sticks or straws of equal length, make a tetrahedron. Then use 12 sticks, each half the length of the originals, to connect the midpoints of the edges of the tetrahedron.[6] What new polyhedron have you created?

8. Using 12 sticks or straws of equal length, make an octahedron. Then use 24 sticks, each half the length of the originals, to connect the midpoints of the edges of the octahedron. What new polyhedron have you created?

9. Identify the golden rectangles in your models or drawings of the polyhedra you created in Exercises 7 and 8.

10. For any Platonic solid, identify all mirror planes and axes of rotational symmetry. For each axis of n-fold symmetry, give the value of n.

11. Deduce all possible vertex nets for the Archimedean solids.

12. Make a model of an Archimedean solid. Make your own pattern or use an existing one, such as those found in Dürer, *The Painter's Manual*, pp. 341–345.

13. Make a model of a star polyhedron.

14. Verify Euler's formula for each Platonic solid.

15. A *cuboctahedron* is a solid with edges that are obtained by joining the midpoints of adjacent edges of a cube. Evaluate the Euler characteristic for the cuboctahedron.

16. Make a plane map, as shown in Figure 10.34, with any number of vertices and edges. Verify that Euler's theorem holds true for your map.

17. Find the Euler characteristic for a Platonic solid and its dual. How do they compare?

18. Verify the Euler characteristic of a soccer ball.

19. Obtain a crystal, such as quartz, or a faceted gemstone. Count the vertices, edges, and faces, and verify that Euler's formula holds.

THE CREATION—BY *GEOMETRY*?

". . . the safest general characterization of the European philosophical tradition
is that it consists of a series of footnotes to Plato."

ALFRED NORTH WHITEHEAD

Chapter 1, "Music of the Spheres," introduced Plato and his *Timaeus* in connection with the musical ratios. Now let's revisit the *Timaeus* and show how Plato described the creation of the world using certain polyhedra as building blocks.

The Elements Linked to the Platonic Solids[7]

Chapter 4, "Ad Quadratum and the Sacred Cut," briefly mentioned the Pythagorean *Ten Sets of Four Things*. Here we'll discuss the correspondences between the regular solids and the elements. In Plato's time, people believed that there were *four elements*: earth, fire, air, and water. Plato associated four of the *Platonic solids* with these four elements: "We must proceed to distribute the figures [the solids] we have just described between fire, earth, water, and air. . . ."[8] He does this on the basis of mobility, size, and sharpness:

> "Let us assign the cube to earth, for it is the most immobile of the four
>   bodies and most retentive of shape,
> the least mobile of the remaining figures (icosahedron) to water
> the most mobile (tetrahedron) to fire
> the intermediate in mobility (the octahedron) to air
> the smallest (tetrahedron) to fire
> and the largest (icosahedron) to water
> the sharpest and most penetrating (tetrahedron) to fire
> the least sharp (icosahedron) to water"[9]

The Elements as a Geometric Progression

In the *Timaeus*, paragraphs 31B through 32C, Plato *deduces* the need for the four elements.

"*It is necessary that nature should be visible and tangible . . . and nothing can be visible without fire or tangible without earth. . . .*"

In other words, first we must have fire, to make the world visible, and earth to make it solid.

"*But it is impossible for two things to cohere without the intervention of a third. . . .*"

It seems that we need a kind of glue to hold fire and earth together.

"*. . . and the most beautiful analogy is when in three numbers, the middle is to the last as the first to the middle, . . . they become the same as to relation to each other.*"

The glue Plato chooses is the geometric progression.

"*But if the universe were to have no depth, one medium would suffice to bind all the natures it contains. But . . . the world should be a solid, and solids are never harmonized by one, but always by two mediums.*"

Plato says that the primary bodies, being solids, must be represented by solid numbers (cubes). He says that to connect two plane numbers (squares), one mean is enough, but to connect two *solid* numbers, two means are needed.

In modern terminology, Plato seems to be saying that between two squares (say, 4 and 9) the single geometric mean (in this case, 6) is *rational*:

4, 6, 9.

However, the *single* geometric mean between two *cubes* (say, 8 and 27) is irrational:

$$8, \sqrt{216}, 27.$$

However, *two* geometric means placed between two cubes (again, 8 and 27) *are* rational:

$$8, 12, 18, 27.$$

This is why Plato placed water and air between fire and earth, fabricating them in the same ratio to each other:

$$\frac{\text{fire}}{\text{air}} = \frac{\text{air}}{\text{water}} = \frac{\text{water}}{\text{earth}}.$$

Therefore, the ratio is constant between successive elements, giving a *geometric progression*.

The Cosmos

However, there are *five* regular polyhedra and only four elements. Plato wrote, "There still remained one construction, the fifth; and the god used it for the whole, making a pattern of animal figures thereon."[10] Plato's statement is vague, and he gives no further explanation.

Later Greek philosophers speculated on the existence of a fifth element, while others assigned the fifth polyhedron (the dodecahedron) to *heaven* or the *cosmos*. Aristotle took it as the *aether* that permeated the entire universe, an idea not disproved until as late as 1887.[11] Later the dodecahedron became the *quintessence* (the basic substance of heaven) of medieval alchemists.[12] The dodecahedron has 12 faces, and 12 is associated with the zodiac. This might be what Plato is talking about when he writes of the "*pattern of animal figures*" on the dodecahedron.[13] Also note that the 12 faces of the dodecahedron are pentagons. In Chapter 5, you saw that the pentagon contains the golden ratio, and perhaps this had something to do with equating this figure with the cosmos.

The Three Basic Triangles

Just as molecules are composed of atoms, Plato's solids are, according to the *Timaeus*, composed of right triangles. Plato proclaimed,

> "In the first place it is clear to everybody that fire, earth, air, and water are bodies . . .
> . . . and all bodies are solids . . .
> . . . and all solids are bounded by surfaces . . .
> . . . and all rectilinear surfaces are composed of *triangles*."[14]

The faces of each of the four solids are squares, equilateral triangles, or pentagons. Plato chose to subdivide each square into four isosceles right triangles, and each equilateral triangle into six smaller 30-60-90 right triangles, sometimes called *Timaeus triangles*. (The 30-60-90 right triangle is the same as the familiar drafting triangles that you can buy in any college bookstore.) Plato subdivided each pentagon into ten right triangles, sometimes called *harmonic triangles*. These right triangles are easy to spot in Figure 10.42.

Why did Plato subdivide the faces into *right* triangles? According to Francis Cornford, "Since Plato intends to build his solids out of plane faces, we might expect him to take the equilateral triangle and the square as his elementary plane figures and proceed at once to construct the solids out of the proper numbers

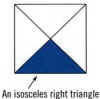

An isosceles right triangle

a

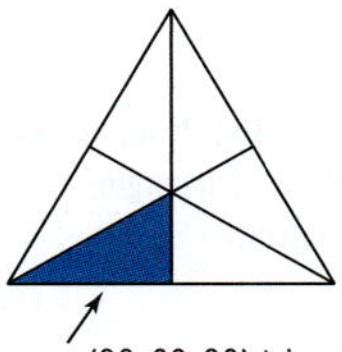

A Timaeus (30-60-90) triangle

b

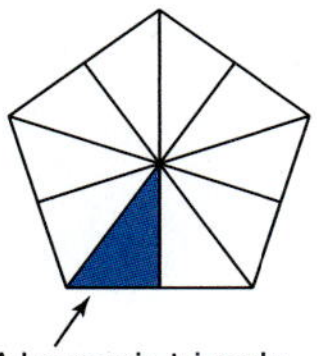

A harmonic triangle

c

FIGURE 10.42 ■ The Three Basic Triangles

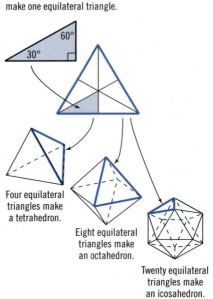

FIGURE 10.44 ■ The Progression from Timaeus
Triangles to Tetrahedron, Octahedron, and
Icosahedron

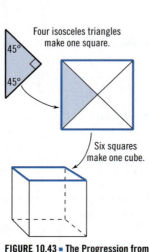

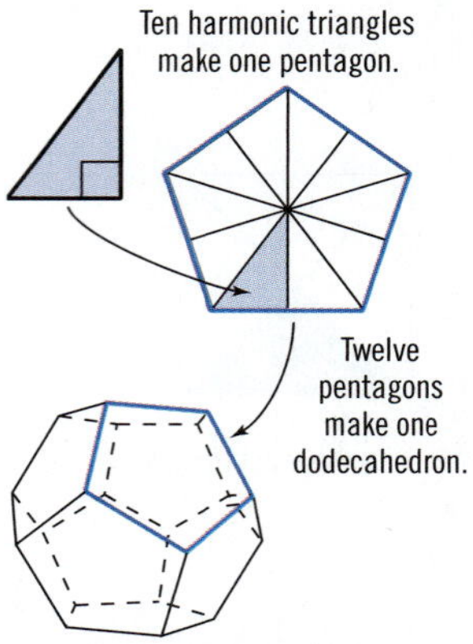

FIGURE 10.43 ■ The Progression from
Isosceles Right Triangles to Cube

FIGURE 10.45 ■ The Progression
from Harmonic Triangles to
Dodecahedron

of such elements. It is by no means obvious why he does not take this simple course. . . ."[15] This breakdown into right triangles might suggest a Pythagorean influence, as his theorem applies only to the right triangle. Figures 10.43, 10.44, and 10.45 show, step by step, the progression from the three basic triangles to the five Platonic solids.

POLYHEDRA IN THE RENAISSANCE

"For he, by Geometrick scale,
Could take the size of Pots of Ale."
SAMUEL BUTLER (1600–1680)

The three famous Renaissance men, Piero della Francesca, Luca Pacioli, and Leonardo da Vinci, are linked by their study of the polyhedra as teachers and pupils. Piero taught Luca, and Luca taught Leonardo. All were somehow involved in the study of Platonic solids. As we will see, one was accused of plagiarizing another.

Polyhedra and Piero

The painter and mathematician Piero della Francesca wrote three books that relate to the three-dimensional material in this chapter. We'll discuss two of the books, *Del abaco* and *De quinque corporibus regularibus*, here, and the third book, *De prospective pingendi*, when we discuss perspective in Chapter 12.

According to Michael Baxandall, a boy in the private or municipal lay schools in Renaissance Florence was educated in two stages. From the age of six or seven, he studied reading, writing, and business correspondence for about four

PIERO DELLA FRANCESCA (1420–1492)

Piero was born in Borgo San Sepolcro (now Sansepolcro), Italy. (Figure 10.46 shows the statue that was erected there in his honor.) In 1439 he worked for Domenico Veneziano in Florence, where he saw the works of Donatello, Luca Della Robbia, Brunelleschi, Masaccio, Fra Angelico, and Alberti. Around 1448 Piero probably worked in Ferrara, where he may have been influenced by northern Italian art. In Rimini, in 1451, he painted a fresco of *Sigismondo Malatesta Before St. Sigismund* in the Tempio Malatestiano. The *Baptism of Christ*, which you saw in Figure 8.27 in Chapter 8, "Squaring the Circle," also belongs to this period.

One of Piero's most famous works is the fresco cycle *The Legend of the True Cross* in the church of S. Francesco at Arezzo. Although another artist started the fresco in 1447, Piero finished it by 1466. In the same period, he created frescoes of the *Magdalen* in Arezzo, the *Resurrection* in San Sepolcro, and the *Madonna del Parto* at

FIGURE 10.46 ■ Statue of Piero in Sansepolcro, Piero's Birthplace

Monterchi. In the late 1450s, Piero painted *The Flagellation of Christ* for the cathedral of Urbino. We will examine its intricate geometry in Chapter 12.

Piero maintained a long association with Federico da Montefeltro, whose court in Urbino was considered *the light of Italy*. Count Federico and his consort, Battista Sforza, appear in a diptych portrait created by Piero. Federico also appears as a kneeling donor in Piero's last known painting, an altarpiece from S. Bernardino, Urbino.

Piero had two passions—art and mathematics. He spent the last two decades of his life in Sansepolcro, where he seems to have abandoned his passion for art in favor of mathematics. There he produced *De prospectiva pingendi, Del abaco,* and *De quinque corporibus regularibus.* Giorgio Vasari says that Piero may have been blind in his last years. He died on nearly the same day that Columbus reached the New World.

years. Then most boys went on to a secondary school, the *abbaco* (which means "arithmetic") for another four years. There they studied mostly mathematics, especially commercial mathematics adapted to the needs of merchants. Piero's *Del abaco* was one of the textbooks used in these schools.

Recall that the *Del abaco* was mentioned earlier in Chapter 1 in connection with the *Rule of Three*, a Renaissance method for solving proportions. Standard-sized containers date only from the nineteenth century, and before then, gauging (the ability to quickly calculate volumes of barrels, sacks, piles of grain, and so on) was a necessary business skill. In the fifteenth century, German merchants used complex prepared rulers, but Italian merchants used geometry and π. Piero's *Del abaco* gives methods and exercises for these computations. The geometry in *Del abaco* is less concerned with proofs, as in Euclid, and more concerned with finding numerical lengths, areas, and volumes—what is sometimes called *mensuration*.

Geometry and Art Appreciation

Baxanhall goes on to say that as with the knowledge of proportion, the skill of gauging affected art. The ability of a merchant to break down complex shapes into simpler geometrical units is similar to the way a painter or sculptor analyzes shapes. Because painters and merchants received the same mathematical education, this was the geometry they knew and used. As a result, the literate public had the same geometrical skills with which to view paintings—and the painters knew this. An obvious way for a painter to invoke a gauger's response was to use the repertory of stock objects from gauging exercises—cisterns, columns, towers, paved floors, and so on.

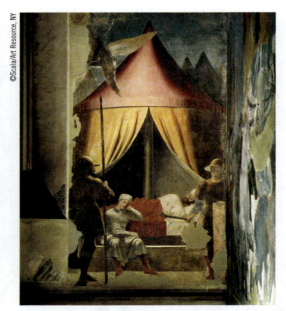

FIGURE 10.47 ■ *Dream of the Emperor Constantine*, Piero della Francesca, c. 1450

For example, consider the scene in Figure 10.47. Almost every mathematics handbook of the time contained an exercise on computing the amount of cloth needed to make a conical tent like the one shown in this painting. A painter who left traces of such analyses was leaving clues that his public was well equipped to find. The geometric figure functions as a *mediator* between artist and merchant, performing a function similar to the one being performed by the seated man in this painting. Notice how the tent pole and the horizontal band on the tent form a *Latin cross*. The highly foreshortened angel, who enters the scene from the viewer's space, upper left, and the light which bursts from him to bathe the scene also act as mediators. Night scenes in painting were very rare at this time.

Polyhedra and Pacioli

Classical Roman lettering was the second main topic in Pacioli's *De divina proportione*. The topic was addressed to stone cutters and builders, and it probably influenced later similar works by Dürer and Geofroy Tory. Pacioli's classical Roman alphabet can be seen on the opening page of each chapter in this book.

Piero wrote *De quinque corporibus regularibus* ("On the Five Regular Bodies") some time after 1482, less than 10 years before his death. The manuscript, illustrated by Piero himself, is in the Vatican Library. Then in 1509 Luca Pacioli wrote a book called *De Divina Proportione*, the entire second part of which was devoted to the Platonic solids and other solids. In it he related the Platonic solids to the golden ratio like this: "As God brought into being the celestial virtue, the fifth essence, and through it created the four solids . . . earth, air, water, and fire . . . so our sacred proportion gave shape to heaven itself, in assigning to it the dodecahedron . . . the solid of twelve pentagons, which cannot be constructed without our sacred proportion. As the aged Plato described in his *Timaeus*."[16] Here Pacioli equated God with Plato's divinity, and wrapped up the Platonic solids, the golden ratio, the creation of the universe, and God, in one neat package.

Da Divina Proportione contained 60 beautifully drawn plates of the various solids, and Luca credits his student Leonardo with these illustrations. He wrote, "The most excellent painter in perspective, architect, musician, and man de tutte vertu doctato, Leonardo da Vinci, who deduced and elaborated a series of diagrams of regular solids. . . ." One of these drawings is shown in Figure 10.48.

FIGURE 10.48 ■ Leonardo's Drawing of the Polyhedra

The section of Luca's *Da Divina Proportione* concerning polyhedra appears to have little relation to the earlier part. Apparently it was stolen from Piero and tacked on, without credit.[18] Vasari had some harsh words for the good friar Pacioli. "The man who should have tried his best to increase Piero's glory and reputation (since he learned everything he knew from him), instead wickedly and maliciously sought to remove his teacher Piero's name and to usurp for himself the honour due to Piero alone by publishing under his own name—that is, Fra Luca Pacioli, all the efforts of that good old man . . . ." And of Piero, Vasari went on to say, "And it often happens that when such a person leaves behind him works which are not quite finished or that are at a good stage of development, they are usurped by the presumption of those who seek to cover their own ass's hide with the noble skin of a lion."[19]

LUCA PACIOLI (1445–1517)

Luca Pacioli was a renowned mathematician, captivating lecturer, teacher, prolific author, religious mystic, and acknowledged scholar in numerous fields. He was a link between the Early Renaissance of Piero and the High Renaissance of Leonardo.

Luca and Piero were close; both were born in Sansepolcro and Luca was tutored in mathematics by Piero. Piero and Luca sometimes walked across the Apennines to the library of Duke Federico of Urbino. Piero introduced Luca to Alberti, a champion of the Italian vernacular, who urged Pacioli to write about mathematics in that language.

Pacioli took vows in 1472, and is usually shown in paintings in a Franciscan habit. He lectured in math in Perugia in 1475. Then he took to the road between 1475 and 1497 to become a traveling teacher of mathematics, while also giving sermons and writing.

In 1494, he published the *Summa de arithmetica, geometrica, proportioni et proportionalita*, a summary of arithmetic, geometry, and algebra, perhaps the sort of book that would now be used in a remedial math course. It was one of the first books to be printed in Venice by the new Gutenberg method. It contained the first mention of double-entry bookkeeping, for which Luca is now known as the "Father of Accounting."

Leonardo da Vinci, impressed by the *Summa*, apparently encouraged Sforza in 1497 to bring Pacioli to Milan to tutor Sforza in mathematics, geometry, and proportion. Leonardo and Pacioli worked together for 10 years, in Milan and Florence. Pacioli is mentioned several times in Leonardo's notebooks, and Leonardo got his knowledge of perspective from Piero through Pacioli. Leonardo made drawings of the polyhedra (Figure 10.48) to illustrate Pacioli's *Da Divina Proportione*.

Luca went to Pisa in 1500 where he lectured on Euclid, and in 1509 he produced a manuscript based on the thirteenth book of *Elements*. The painting at the beginning of this chapter shows Luca and his student Guidobaldo, Duke of Urbino. You can see that Luca is dressed in his monk's robes. Pacioli's hand rests on the thirteenth book of Euclid, which contains the constructions for the polyhedra. The figure on the slate appears to be an isosceles triangle inscribed in a circle, the start of Euclid's construction of the regular pentagon.[17] In the upper left is a rhombi-cuboctahedron. On the table is a dodecahedron on top of a copy of Euclid's *Elements*.

POLYHEDRA AS ART MOTIFS

*"A pyramid, an octahedron, an icosahedron, and a dodecahedron, the
primary figures which Plato predicates, are all beautiful because of the
symmetries and equalities in their relations. . . ."*

PLUTARCH (C. 46–120 B.C.)

In this section, we'll examine a short collection of examples of polyhedra in art.
Polyhedra have served as art motifs from prehistoric times right up to the present. We'll also study some symbolism attached to the cube, one of the Platonic
solids, and review the crystal, a polyhedron that is commonly used in jewelry.

Polyhedra in Art

**FIGURE 10.49 ■ Etruscan
Dodecahedron**

**FIGURE 10.50 ■ Mosaic from San
Marco Cathedral, Venice, Uccello,
1425–1430**

Neolithic solids have been found in Scotland.[20] The British Museum has icosa-
hedral dice from the Ptolemaic dynasty and dodecahedral objects from Etruria.
Excavations near Padua have unearthed the Etruscan dodecahedron (c. 500 B.C.)
that is shown in Figure 10.49. It was probably used as a toy.[21]

The mosaic in Figure 10.50 has been attributed to Paolo Ucello. It appears
to depict a star dodecahedron. Ucello was deeply involved in the newly developed technique of linear perspective, which we will study in Chapter 12.

The famous engraving by Albrecht Dürer that is presented in Figure 10.51
shows an irregular polyhedron, as well as a sphere, a magic square, and compasses. (In Chapter 11, you'll see how the sphere was often used to represent the
whims of fate, a fitting element for this somber scene.) Just a few of the other
objects you'll notice here are tools, scales, a rainbow, and an hourglass with time
running out. People who have analyzed the polyhedron in this engraving have
decided that it is actually a cube with opposite corners cut off.[22]

The painting presented in Figure 10.52 shows German mathematician
Johannes Neudorfer giving his son a math lesson. Neudorfer is pointing out

**FIGURE 10.51 ■ *Melencolia I*, Dürer,
1514**

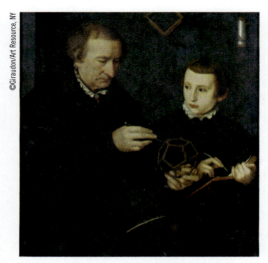

**FIGURE 10.52 ■ Picture of Johannes Neudorfer and His Son,
Neufchatel, 1561**

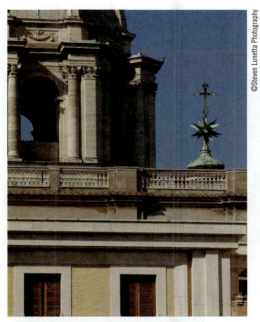

FIGURE 10.53 ■ Star Polyhedron atop the Sacristy
of St. Peter's, Rome

FIGURE 10.54 ■ *Stars,* Escher, 1948

something about the dodecahedron he holds in his hand. Other details of the painting include a cube and a cylinder hanging overhead.

A polyhedron is sometimes found atop the cupola of a church or other structure. The sacristy of St. Peter's in Rome has a star polyhedron topped by a cross, as you can see in Figure 10.53, and Michelangelo's Medici Chapel of San Lorenzo in Florence is surmounted by a polyhedron. Sant'lvo alla Sapienza and San Andrea delle Valle, also in Rome, show star polyhedra. English structures topped by polyhedra include Salisbury Cathedral, Wimbourne St. Giles Parish Church, and Merton College Chapel, Oxford. Protestant churches in Hungary are commonly decorated with polyhedra, symbolizing the star of Bethlehem.

Moving forward in time in our examples of polyhedra, we have Figure 10.54, M.C. Escher's *Stars*. This engraving is composed of all kinds of intersecting polyhedra, octahedra, tetrahedra, cubes, and chameleons. Note the similarity between the most prominent polyhedron and Leonardo's illustrations for Pacioli's book, Figure 10.48.

The artists of the surrealist movement attempted to express the workings of the subconscious. This might include combining objects that to the conscious mind seem unrelated. The Swiss artist Alberto Giacometti (1901–1966) often included polyhedra in his earlier surrealist works. Consider Figure 10.55.

The shape of Salvador Dali's *Sacrament of the Last Supper* in Figure 10.56 is a golden rectangle, and the huge dodecahedron engulfing the table contains the golden ratio, as we saw earlier.

The geodesic dome is yet another polyhedron. Specifically, it utilizes tetrahedrons, octahedrons, and closely packed spheres. Figure 10.57 presents a geodesic dome that was designed by Buckminster Fuller and his student Shoji Sadao. This particular geodesic dome is 250 ft in diameter and 200 ft high; it housed the U.S. Pavilion at Expo 67 in Montreal. It is now known as the Biosphere.

FIGURE 10.55 ■ *The Surrealist Table*, Alberto Giacometti, 1933

FIGURE 10.56 ■ *Sacrament of the Last Supper*, Salvador Dali, 1955

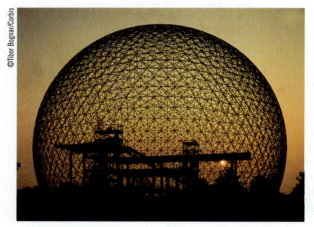

FIGURE 10.57 ■ A Geodesic Dome

FIGURE 10.58 ■ *Truncated Close-Packing Octahedra, Rhombidodecahedra, and Cubes*, Harriet Brisson, 1976

The sculpture by Harriet Brisson, shown in Figure 10.58, is another twentieth-century work that uses polyhedra. She wrote the following about her sculpture.

"I construct many close packing pieces. I begin with one of cubes and octahedra that fill space completely.

I make a space filling tensegrity structure of close-packing octahedra, cubes and rhombic dodecahedra. As it rotates its form seems to change – elusive – planes overlap – appear solid – separate – open up – turn to close again.

Twenty-seven truncated octahedral close-pack in a tensegrity structure endlessly forming larger and larger cubes—expanding in all directions.

The close-packings are sample units of space filling polyhedra."[23]

FIGURE 10.59 ■ *Platonic Forms*, Computer Graphic, Lucio Saffaro, 1989

The Italian artist Lucio Saffaro has made a study of the polyhedra. He has created new polyhedral shapes that he has depicted by traditional means, such as oil on canvas, and by computer graphics, as shown in Figure 10.59.

Symbolism of the Cube

The square, as you have seen, is often associated with the earthly, the mundane, and the terrestrial. The cube as well, with its six square faces, is sometimes associated with the earth. Recall the Platonic solids we examined earlier in this chapter. In art, the cube has two further, conflicting, attributes. On one hand, it is the symbol of *stability*, in contrast to the unstable sphere upon which Lady Luck is often portrayed standing. On the other hand, the cube as a *die* is a symbol of fate, as in the phrase "the die is cast," or games of chance or gambling. At the crucifixion, soldiers cast lots for Jesus' cloak.

The Crystal

The *crystal* is a type of polyhedron that is especially fascinating. In the early nineteenth century Edgar Allen Poe (1809–1849) wrote,

> "Let us examine a crystal. We are at once interested by an equality between sides and between the angles of one of its faces; the equality of the sides pleases us; that of the angles doubles the pleasure.
>
> On bringing to view a second face in all respects similar to the first, this pleasure seems to be squared; on bringing to view a third it appears to be cubed, and so on.
>
> I have no doubt, indeed, that the delight experienced, if measurable, would be found to have exact mathematical relations as I suggest; that is to say, as far as a certain point, beyond which there would be a decrease in similar relations."[24]

For Poe, *degrees of delight* seem to form a geometric progression.

In the "Symbolism in the Visual Arts" section of *Man and His Symbols*, Aniela Jaffé wrote about the enormous appeal of stones, particularly the crystal. "In many dreams of the nuclear center, the Self also appears as a crystal. The mathematically precise arrangement of a crystal evokes in us the intuitive feeling that even in so-called 'dead' matter there is a spiritual ordering principle at work. Thus the crystal often symbolically stands for the union of extreme opposites—of matter and spirit."[25] Here we have the recurring theme of *union of opposites*. In art, one well-known portrayal of a crystal is by M. C. Escher, as shown in Figure 10.60.

FIGURE 10.60 ■ *Crystal,* Escher, 1947

SUMMARY

In this chapter—the first of several dealing with the third dimension—we studied the basic solids and then the regular and semiregular solids. In Chapter 1, we learned how Plato, in his *Timaeus*, described the creation of the world in terms of the musical ratios. In this chapter, we saw how the regular polyhedra played a role in this process of creation. Another main idea you found here was how the elements were composed of the regular polyhedra. This may seem naive to us now, but quoting Mario Livio, "While [Plato's] description does not conform with our modern understanding of the structure of matter, the central idea—that the most fundamental particles in our universe and their interaction can be described by a mathematical theory that possesses certain symmetries—is one of the cornerstones of today's research in particle physics."[26] Plato wrote that the universe was composed of tiny triangles, and now we think it is made up of tiny balls, which is not such a great difference. Also in this chapter, we profiled Piero della Francesca, whose main interests in art and mathematics make him one of the most important figures in the interdisciplinary subject matter of this book. We also gave profiles of Archimedes and Luca Pacioli, both important figures in the history of mathematics.

The sphere, which is considered the most perfect solid, has been conspicuously absent here. We will cover it in the following chapter.

EXERCISES AND PROJECTS

1. Name and describe the five Platonic solids. Which four solids correspond to the four elements? With what is the fifth solid associated?

2. Euclid gives constructions for the five regular solids in Book XIII, Propositons 13–17. Repeat any of these constructions. You may do them manually, or you may use The Geometer's Sketchpad or another CAD program.

3. Manually or by CAD, repeat Euclid's construction of the Platonic solids inscribed in a sphere from Book XIII, Proposition 18.

4. Find the dimensions of I. M. Pei's pyramid that is located in front of the Louvre in Paris. Use those dimensions to calculate the approximate square footage of glass used to make the pyramid and to calculate the volume enclosed by the pyramid.

5. The conical tank shown in Figure 10.61 has a base diameter equal to its height. It is being filled with water at the rate of 2.25 cubic meters per hour. Write a program or spreadsheet that will compute and display the volume of water in the tank and its depth D, for every 5 minutes.

6. On your campus or in your community, locate and photograph buildings and roof styles that are made up of the basic solids covered in this chapter. Take measurements and compute the areas and volumes of at least one.

7. Using a CAD program that can draw three-dimensional objects, make a computer image of a polyhedron. Shade or color the faces to make the image look solid.

8. Make a screensaver using the computer-drawn polyhedron you created in the preceding exercise.

9. Use your computer-drawn polyhedron from Exercise 7 as the basis for a painting or a graphic. Use it to design wallpaper, wrapping paper, greeting cards, a tiling, a mosaic, a pillow, a tee shirt, a quilt, or a dress.

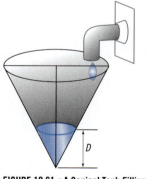

FIGURE 10.61 ■ **A Conical Tank Filling with Water**

10. Make a soft sculpture or pillow based on a polyhedron.

11. Make a box kite with a polyhedral design. Go fly the kite.

12. Make a polyhedral lampshade, using translucent plastic, fiberglass, glass, or stained glass.

13. Design and make a table-sized sculpture based on a polyhedron. You can make it open by using thin members to define the edges; you can make it closed by using sheet copper, brass, or thin wood sheets to define the faces; or you can make it solid by using clay, Styrofoam, wood, or stone.

14. Design and make either a functional or purely decorative polyhedral ceramic.

15. Design and make a mobile using the five Platonic solids, as in the student project shown in Figure 10.62, or you can design and make one using the 13 Archimedean solids.

16. Design and make a model for a monumental outdoor public sculpture based on the polyhedron.

17. Make a polyhedral set of children's blocks.

18. Make a polyhedral house for your pet.

19. Make a model of a polyhedron using a kit, such as the *Zome Advanced Mathematics Creator Kit*, which is available from Key Curriculum Press.

20. Using 30 sticks or straws of equal length, make an icosahedron. Then use 60 sticks, each half the length of the originals, to connect the midpoints of the edges of the icosahedron. What new figure have you created?

21. Use stiff cardboard to create three golden rectangles as large or as small as you like. Slit each so that the cards can be assembled into the star shown in Figure 10.63. Connect the corners of the rectangles with string or thin sticks. What polyhedron have you created?

22. Make a model of the 12-faced dodecahedron. On each face put one of the 12 signs of the zodiac, as in the student's project shown in Figure 10.64.

23. Figure 10.65 shows an exquisitely made model of a complex star polyhedron. Make a star polyhedron such as this one.

24. Design and make a model for a complex of buildings incorporating polyhedral shapes.

25. Make a wire frame model of a polyhedron and project it onto a plane, as in Figure 10.32. Trace the projected image to produce a Schlegel diagram. Try it with other polyhedra.

26. Demonstrate surfaces and solids of revolution by rapidly rotating a bent wire with a hand drill or other device. Darken the room and observe the rotation with a strobe light or under fluorescent lighting.

FIGURE 10.62 ■ Student's Platonic Solid Mobile

FIGURE 10.63 ■ Three Golden Rectangles Assembled to Form a Star for the Model in Exercise 21

FIGURE 10.64 ■ Student's Model of the Dodecahedron

FIGURE 10.65 ■ Student's Star Polyhedron

27. Make an edible polyhedron from bread sticks and cheese.

28. Have a polyhedral bake-off. Bake treats in the different shapes covered in this chapter and bring them to class.

29. Read and report on any of the books listed in the sources at the end of this chapter.

30. Here are some suggestions for term papers or short papers:

   ■ The French architect Etienne-Louis Boullée, who in a famous essay *La Théorie des Corps* investigated the various geometric solids and their effects upon the senses

   ■ The German astronomer Johannes Kepler, who in 1596 published *The Cosmic Mystery*. In this book he speculated that because each of the six known planets orbited within a sphere, each pair of spheres might be separated by one of the five regular solids, each solid circumscribed about a sphere as well as inscribed in the next larger sphere. (You will see a model of Kepler's scheme in Chapter 11, Figure 11.34.)

   ■ The ties between Piero della Francesca, Luca Pacioli, and Leonardo da Vinci

   ■ The Delian problem

   ■ Plato's description of the creation of the world in *Timaeus*

   ■ The history of the elements, starting with earth, air, fire, and water, up to the periodic table we have today

   ■ The life and work of Archimedes

   ■ The *Archimedes Palimpsest*, an ancient book containing calculations by Archimedes that were later overwritten with prayers around 1200 A.D.

   ■ Piero della Francesca and how his work touches on many of the ideas in this book

   ■ The historical problem of the *duplication of the cube*. See Heath, p. 245, as a starting point.

   ■ Buckminster Fuller and the geodesic dome

   ■ The symbolism of the crystal

Mathematical Challenges

31. *Doubling the Cube.* By what factor must the side of a cube be multiplied to give a new cube of twice the volume?

32. *Doubling the Cube: The Delian Problem.* According to one story, in order to get rid of a certain pestilence, the inhabitants of Delos were instructed by their oracle to double the volume of the cubical altar to Apollo. This gave rise to the cube-duplication problem known as the *Delian* problem. This problem was supposedly studied by Plato, who maintained that Apollo did not really care about the size of his altar but had posed this problem to reproach the Greeks for their neglect of mathematics and geometry.[27] There are several geometric constructions for the solution to the Delian problem. Figure 10.66 illustrates one of three given by Dürer.[28] To begin, let s be the side of the original cube. Then perform the following steps:

   a. Draw line *dae*, where $da = ae = 2s$.

   b. Construct a semicircle on *de* with center at *a*.

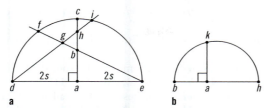

FIGURE 10.66 ■ Dürer's Solution to the Delian Problem

c. Draw perpendicular *ac* and bisect at *b*.

d. Draw line *eb* and extend to *f*.

e. From *d* draw a line *di*, adjusted so that *gh* = *hi*. This is another example of a *neusis* construction, a type that we used earlier.

f. In a *new figure*, lay out *ba* and *ah* from the preceding figure, on the same line.

g. Construct a semicircle on *bh*.

h. Erect a perpendicular *ak* from *a*. Then *ak* is the side of the doubled cube.

  Repeat the construction, and show by measurement that it works. Compare your answer with that obtained in the preceding exercise. Is Dürer's method exact or approximate?

33. Use the theorems of Pappus to derive formulas for the surface area and volume of a cone.

34. Use the theorems of Pappus to prove that for the torus of Figure 10.26, its volume is $2\pi^2 kr^2$ and its surface area is $4\pi^2 kr$. Here r is the radius of the circle and k is the distance from the circle's center to the axis of revolution.

35. Make an oral presentation to your class about any of the previous projects.

SOURCES

Bairati, Eleonaora. *Piero della Francesca*. New York: Crescent, 1991.

Baxandall, Michael. *Painting and Experience in Fifteenth Century Italy*. Oxford: Clarendon Press, 1972.

Calter, Paul. *Technical Mathematics*. New York: Wiley, 2007.

Cornford, Francis. *Plato's Cosmology*. New York: Harcourt, 1937.

Coxeter, H. S. M. *Introduction to Geometry*, 2nd ed. New York: Wiley, 1989.

Coxeter, H. S. M., et al. *Geometry Revisited*. Washington, D.C.: Mathematical Association of America, 1967.

Critchlow, Keith. *Order in Space: A Design Source Book*. London: Thames & Hudson, 1969.

Critchlow, Keith. *Time Stands Still*. New York: St. Martin's, 1982.

Emmer, Michele, ed. *The Visual Mind: Art and Mathematics*. Cambridge, MA: MIT Press, 1993.

Euclid. *The Thirteen Books of the Elements*. New York: Dover, 1956.

Eves, Howard. *An Introduction to the History of Mathematics*. New York: Holt, 1953.

Ghyka, Matila. *The Geometry of Art and Life*. New York: Dover, 1977.

Hargittai, István. *Symmetry, a Unifying Concept*. Bolinas, CA: Shelter, 1994.

Ivins, William. *Art & Geometry: A Study in Space Intuitions*. Cambridge, MA: Harvard University Press, 1946.

Kappraff, Jay. *Connections: The Geometric Bridge Between Art and Science.* New York: McGraw-Hill, 1990.

Kinsey, L. C., et al. *Symmetry, Shape, and Space.* Emeryville, CA: Key College Publishing, 2002.

Lawlor, Robert. *Sacred Geometry.* New York: Thames & Hudson, 1982.

Livio, Mario. *The Golden Ratio.* New York: Broadway Books, 2002.

March, Lionel. *Architectonics of Humanism.* Chichester, UK: Academy, 1998.

Newman, James R., ed. *The World of Mathematics.* New York: Simon and Schuster, 1956.

Pedoe, Dan. *Geometry and the Visual Arts.* New York: Dover, 1976.

Pederson, Mark A. "The Geometry of Piero della Francesca." *The Mathematical Intelligencer,* Vol. 19, No. 3, Summer 1997, pp. 33–40.

Plato. *Timaeus.* Ed. and trans. by John Warrington. London: Dent, 1965. Original c. 360 B.C.

Runion, Garth E. *The Golden Section.* Palo Alto, CA: Seymour, 1990.

Vasari, Giorgio. *The Lives of the Artists.* Oxford: Oxford University Press, 1991. Originally printed in 1550.

Wenninger, Magnus J. *Polyhedron Models for the Classroom.* Reston, VA: National Council of Teachers of Mathematics, 1966.

Wittkower, Rudolf. *Architectural Principles in the Age of Humanism.* New York: Random House, 1965.

NOTES

1. Runion, p. 113.
2. Eves, p. 82.
3. Richter, p. 30.
4. Euclid, Vol. 3, p. 438.
5. Euclid, Vol. 3, p. 438.
6. This problem, and the following two, are from Kappraff, p. 295.
7. *Timaeus,* Para. 56.
8. *Timaeus,* Para. 56.
9. *Timaeus,* Para. 56.
10. *Timaeus,* Para. 55c. Cornford, p. 218.
11. Livio, p. 68.
12. Coxeter, *Introduction to Geometry,* p. 149.
13. Pedoe, p. 262.
14. *Timaeus,* Para. 53.
15. Cornford, p. 211.
16. Richter, p. 42.
17. March, p. 160.
18. Bouleau, p. 92.
19. Vasari, *Lives of the Artists.*
20. Chritchlow, *Time Stands Still,* Chap. 7.
21. Coexter, *Regular Polytopes.* New York: Dover, 1973, p. 13.
22. "The literature on Melancholia is more extensive than that on any other engraving by Dürer: That statement would probably remain true if the last two words were omitted." This quotation is by Campbell Dodgson in his book *Albrecht Dürer* (London: Medici Society, 1926, p. 94). A more recent analysis by the art historian Patrick Doorly explains how Melencolia I is an illustration of Plato's *Greater Hippias* or *On the Beautiful.* See *The Art Bulletin,* Vol. LXXXVI, No. 2, June 2004, p. 255, 261.

23. http://www.number53.com/harrietbrisson
24. Linn, p. 63.
25. Jung, p. 221.
26. Livio, p. 70.
27. March, p. 65.
28. Dürer, *The Painter's Manual*, p. 347.

"The Bible shows us the way to go to heaven,

but not the way the heavens go."

CESARE CARDINAL BARONIO

FIGURE 11.1 ▪ *Creation of the World and the Expulsion from Paradise* (detail), Giovanni di Paolo, 1445

The Metropolitan Museum of Art, Robert Lehman Collection, 1975 (1975.1.31). Image © The Metropolitan Museum of Art.

11

The Sphere and Celestial Themes in Art and Architecture

When Spinosa wrote "the only perfect form is the circle," he must have been thinking in only two dimensions. What form could be more perfect than that of the sphere? This important solid deserves a chapter to itself. We'll begin with its geometry, covering areas, volumes, zones, segments, and sectors.

The dome is an architectural feature related to the sphere. Although seldom a true hemisphere, the dome has so many similar characteristics that we'll discuss it here. Next, we'll discuss the *terrestrial sphere*, the approximately spherical solid on which we live, and the *celestial sphere*, the imaginary sphere within which we live. We'll discuss the problem of locating objects on or within each of these spheres, and we'll explore how each sphere is modeled using globes, maps, orreries, and so forth.

Because our perception of time is due to the rotations of the terrestrial sphere within the celestial sphere, we'll discuss various kinds of time, some instruments used to measure time, and temporal themes in art.

We've searched far and wide for geometric art motifs. In this chapter, we will expand our search further into the heavens. We'll examine an assortment of spheres and celestial themes in art, including the symbolism of the sphere,

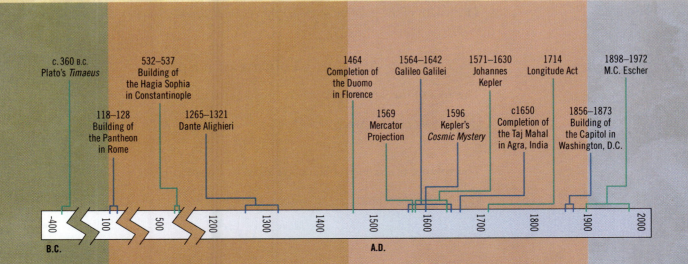

c. 360 B.C.
Plato's *Timaeus*

118–128
Building of
the Pantheon
in Rome

532–537
Building of
the Hagia Sophia
in Constantinople

1265–1321
Dante Alighieri

1464
Completion of
the Duomo
in Florence

1569
Mercator
Projection

1564–1642
Galileo Galilei

1596
Kepler's
Cosmic Mystery

1571–1630
Johannes
Kepler

c1650
Completion of
the Taj Mahal
in Agra, India

1714
Longitude Act

1856–1873
Building of
the Capitol in
Washington, D.C.

1898–1972
M.C. Escher

portrayals of astrologers and astronomers and astronomy itself, and recent uses of the heavens in art.

This chapter is closely related to Chapter 6, "The Circle," where we studied celestial circles and the rainbow. Chapter 1, "Music of the Spheres," referenced the cosmos, the Pythagorean Music of the Spheres, and Plato's *Timaeus*, which described the creation of the world in terms of geometry.

THE SPHERE

"When, the diameter of a semicircle remaining fixed, the semicircle is carried round and restored again to the same position from which it began to be moved, the figure so comprehended is a sphere."[1]

EUCLID

As usual, let's begin the chapter with geometric definitions and relationships, before we move to applications of the sphere in art and architecture. Heath, in his translation of Euclid's *Elements*, pointed out that Euclid's definition is "not properly a definition of a sphere but a description of a mode of generating it." The current and more usual definition is found in Statement 113.

113 DEFINITION OF A SPHERE

A sphere is the set of points in space at a given distance from a fixed point.   ●

The given point is, of course, the center O of the sphere, and the fixed distance is the radius r, as shown in Figure 11.2. The diameter d is twice the radius.

A *circle* is a section of a sphere cut by a plane. If the plane passes through the center of the sphere, the resulting figure is a *great circle*, with a radius equal to that of the sphere. A great circle divides the sphere into two hemispheres. Sections that do not pass through the sphere's center give *small circles*.

The *axis* of a great or small circle is a line through its center perpendicular to the plane of the circle. The axis passes through the center of the sphere. The points where the axis cuts the sphere are called the *poles*. A *meridian* is a great circle passing through the poles.

Any straight line cutting a sphere is a *secant*. The line segment cut out of a secant by the sphere is a *chord*.

Only one great circle can be drawn through any two given points on a sphere. The shorter arc between those points is called the *geodesic* between them. The length of the geodesic is called the *spherical distance* between the two given points.

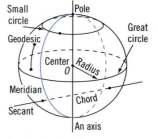

FIGURE 11.2 ■ Geometry of the Sphere

Area and Volume of a Sphere

The **area of the surface** of a sphere of radius r is given by Equation 114.

114 SURFACE AREA OF A SPHERE

$$\text{Area} = 4\pi r^2$$ ●

The **volume** of a sphere of radius r is given by Equation 115.

115 VOLUME OF A SPHERE

$$\text{Volume} = \left(\frac{4}{3}\right)\pi r^3$$ ●

Based on these equations, a sphere of radius 2.55 cm would have a surface area of $4\pi(2.55)^2 = 81.7$ cm^2 and a volume of $(4/3)\pi(2.55)^3 = 69.5$ cm^3.

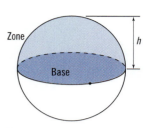

Spherical Zones, Segments, and Sectors

As you can see in Figure 11.3, a *spherical zone* of a sphere is the portion of the surface of the sphere lying between two parallel planes. If one of the planes is tangent to the sphere, you have a zone of one base (Figure 11.3a); otherwise, you have a zone of two bases. The *altitude* of the zone is the perpendicular distance between the two planes (Figure 11.3b). A *zone* can also be thought of as a surface of revolution formed when an arc of a circle is rotated about a diameter.

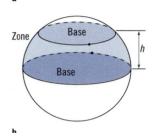

a

116 SURFACE AREA OF A SPHERICAL ZONE

The surface area of a spherical zone (of either one or two bases) is equal to the circumference of a great circle of that sphere times the altitude of the zone.   ●

A *spherical segment* is the solid bounded by a spherical zone and its base(s). A spherical segment of one base may also be thought of as the solid of revolution generated when the segment of a circle is rotated about the diameter perpendicular to the chord of the segment. The volume of a spherical segment of two bases of radii *a* and *b* and having an altitude *h* is given by Equation 118.

b

FIGURE 11.3 ▪ Zones of a Sphere

118 VOLUME OF A SPHERICAL SEGMENT

$$V = \frac{\pi h}{6}(3a^2 + 3b^2 + h^2)$$

For a segment of one base, either *a* or *b* is zero.   ●

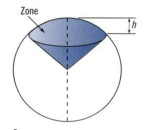

A *spherical sector* is a solid of revolution generated by rotating the sector of a circle about a diameter. A spherical sector can have one base, as shown in Figure 11.4a, or two bases, as shown in Figure 11.4b. The arc of the segment generates a zone.

a

   The two formulas for the volume of a spherical sector are given in Equation 119.

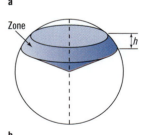

119 VOLUME OF A SPHERICAL SECTOR

a. The volume of a spherical sector is one-third the product of the area of its zone and the radius of the sphere.

b. The volume of a spherical sector of altitude *h* in a sphere of radius *r* is

$$V = \left(\frac{2}{3}\right)\pi r^2 h$$
                     ●

b

FIGURE 11.4 ▪ Spherical Sectors

■ **EXAMPLE:** Figure 11.5 shows a zone, a segment, and a sector in a sphere of radius 10 cm. Find (a) the surface area S of the spherical zone, (b) the volume V_1 of the segment, and (c) the volume V_2 of the sector.

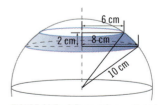

FIGURE 11.5 ▪ A Zone, a Segment, and a Sector of a Sphere

● **SOLUTION:**

a. The circumference of a great circle in the given sphere is as follows:

$$2\pi(10 \text{ cm}) = 62.8 \text{ cm}.$$

   Apply Equation 116 to find the area of the zone:

$$S = (62.8 \text{ cm})(2 \text{ cm}) = 125.6 \text{ cm}^2.$$

b. Substitute into Equation 118 using $a = 8$, $b = 6$, and $h = 2$ to find the volume of the segment:

$$V_1 = \left(\frac{2\pi}{6}\right)[3(8)^2 + 3(6)^2 + 2^2] = 318 \text{ cm}^3.$$

c. Apply Equation 119 to find the volume of the sector:

$$V_2 = \left(\frac{2}{3}\right)\pi(10^2)(2) = 419 \text{ cm}^3. \qquad ■$$

■ EXERCISES ● THE SPHERE

1. Sketch a sphere. Then sketch, label, and define the following terms:

sphere	great circle	small circle
hemisphere	pole	axis
meridian	secant	chord
geodesic	spherical distance	spherical zone
spherical sector	spherical segment	

2. Find the volume and surface area of a sphere having each given radius:
   a. 744 in.   b. 1.55 m

3. Find the volume and radius of a sphere having a surface area of 46 cm^2.

4. Find the surface area and radius of a sphere that has a volume of 462 ft^3.

5. Find the total volume of 100 steel balls, each 2.50 in. in diameter.

6. Describe the axes and planes of symmetry of the sphere.

7. *Archimedes on the Area of a Sphere.* Archimedes stated that the area of a sphere was equal to the area of four of its great circles. Verify this using modern notation.

8. *Archimedes' Hat-Box Theorem.* Figure 11.6 shows a cylinder and a sphere, each with the same radius. Make two slices through each of them. The slices should be the same distance apart and perpendicular to the axis of the cylinder. The theorem then states that the surface area of the slice of the sphere is equal to the surface area of the slice of the cylinder.
   a. Using the formulas in this chapter and in Chapter 10, "The Solids," find these areas and verify the theorem.
   b. How would you extend the theorem to a complete sphere and the circumscribed cylinder?

9. *Archimedes on the Volume of a Sphere.* Archimedes stated that the volume of a sphere was equal to four times the volume of a cone with a base that is a great circle of the sphere and with a height that equals the radius of the sphere. Verify this using modern notation.

10. *Archimedes' Comparison of the Sphere and the Cylinder.* Archimedes stated that a circumscribed cylinder (bases included) is one and a half times as large as the sphere in both surface area and volume. Verify this using modern notation.

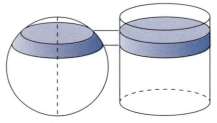

FIGURE 11.6 ■ Archimedes' Hat-Box Theorem

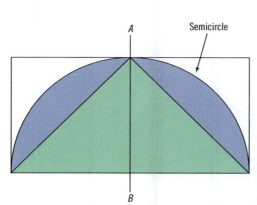

FIGURE 11.7 ■ A Cone Inscribed in a Hemisphere Inscribed in a Cylinder

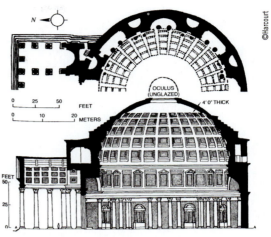

FIGURE 11.8 ■ The Pantheon, Rome, c. 118–128 A.D.

11. *Archimedes' Tomb.* When rotated about axis *AB*, the object in Figure 11.7 generates three solids of revolution: (1) a cone inscribed in (2) a hemisphere, which is itself inscribed in (3) a cylinder. Archimedes, the legend goes, was so pleased with this discovery that he ordered this figure to be engraved on his tomb. It also commemorates what is said to be his favorite work, *On the Sphere and the Cylinder.* Show that the volumes of these three solids are in the ratio 1 : 2 : 3.

12. Verify that the surface area of a sphere is $\frac{2}{3}$ the surface area of the circumscribed cylinder, bases included.

13. Find the area of a spherical zone having a height of 3.5 in. in a sphere of radius 14 in.

14. Find the volume of a spherical segment of height 4.5 cm, with base radii that are 6.3 cm and 8.6 cm.

15. Find the volume of a spherical sector of altitude 34 ft in a sphere of radius 124 ft.

16. *The Pantheon.* Refer to the drawing of the Pantheon in Figure 11.8. It is possible to inscribe a sphere that lies along the inner surface of the dome and that is tangent to the pavement. Find the volume of that sphere, taking its diameter as 142.5 ft.

17. *Cubing the Sphere in the Great Pyramid.*[2] Using the proportions of the triangular cross-section of the Great Pyramid from Chapter 1, enclose that triangle in a rectangle, as shown in Figure 11.9. That rectangle has an area of $2\sqrt{\Phi}$ (where Φ is the golden ratio). Show that this area is approximately equal to the area of one-eighth of a sphere of radius $\sqrt{\Phi}$. Use the approximate value $\pi \approx 4/\sqrt{\Phi}$ that we obtained in Chapter 8. (We have not really drawn a cube with an area equal to that of a sphere; instead, we have drawn a plane area approximately equal to the area of an octant of a sphere. This is somewhat analogous to squaring the circle.)

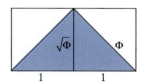

FIGURE 11.9 ■ Cubing the Sphere

THE DOME

A *dome* is a convex curved vault that is circular or polygonal in cross-section. The term is often interchangeable with cupola. The shape of the dome can be a hemisphere, but it can also be elongated, flattened, or bulbous. Many domes are

FIGURE 11.10 ■ Dome, Drum, Pendentives, and Lantern

solids of revolution, but they can also have a polygonal cross-section, often octagonal. Each has rotational symmetry about a vertical axis through its center.

Domes can be built from a variety of materials: wood, stone, metal, etc. They can be a single thickness or have two or more shells. A dome can spring directly from the structure below, or it can be separated from that structure by a drum. A *drum* is a cylindrical or polygonal wall just below the dome. A *pendentive* is a curved wall surface in the upper corner of a room that provides a transition between a polygonal structure below and the circular dome or drum above. It is a roughly triangular portion of a spherical surface. A dome can be topped by a *lantern*, a structure with windows that is often used to crown a dome or other kind of roof. Figure 11.10 shows how each of these pieces might appear in an overall structure.

A Short History of the Architectural Dome

The history of the dome falls into two major periods: the periods before and after domes were placed on pendentives. Early tribes used domes over houses and other buildings. The Romans were apparently the first to build very large hemispherical domes resting on circular bases. The Pantheon in Rome, which is shown in Figures 11.8 and 11.11, is a good example. Built by Hadrian, the Pantheon, at 142 feet in diameter, remains the largest masonry dome in the world. The *oculus*, or opening, at its summit, is an unusual feature of this dome and is the only window in the building. The oculus is unglazed and is open to the weather.

The Byzantine architects solved the problem of supporting a dome over a square or polygonal space with their invention of pendentives. Their solution ushered the second and more important period of dome construction. One of

FIGURE 11.11 ■ The Pantheon, Rome,
c. 118–128 A.D.

FIGURE 11.12 ■ Hagia Sophia, Constantinople (Istanbul),
Completed 537 A.D.

the earliest examples of this type of construction—the Hagia Sophia, which is shown in Figure 11.12—is also one of the largest. The vast scale of Hagia Sophia, with a dome that is 107 feet in diameter and rises to a height of 180 feet, was never duplicated.

Another great innovation at Hagia Sophia was a crown of 40 windows penetrating the base of the dome, which was strengthened by buttresses between the windows. This method of lighting was used in St. Mark's in Venice as much as 500 years later. Later Byzantine churches had smaller domes, so any such windows would have been inadequate. A solution was to raise the dome on a cylindrical drum, which would accommodate windows of a reasonable size.

During the Middle Ages, the Gothic architecture in the West generally used no domes. However, domes were used at the crossings in many churches in Italy. The largest of these crossing domes, elliptical in plan, is in the Duomo in Pisa. The dome regained popularity during the Renaissance and Baroque periods. Generally, baptisteries built in Italy at that time were vaulted. During the Renaissance, the dome often was raised on a high drum and crowned with a lantern. The Duomo in Florence, which is shown in Figure 11.13, was built from Brunelleschi's designs and partly under his direction between 1420 and 1446. After his death in 1464, builders completed it from his models. This dome is octagonal and built of two masonry shells carried by 8 colossal ribs and 16 intermediate ribs that meet in a ring at the summit. The dome is 139 feet in diameter and 380 feet high to the top of the lantern. To see some spectacular views of Florence, visitors today can climb a staircase between the two shells to the lantern.

As you saw in the section on cruciform churches in Chapter 4, "Ad Quadratum and the Sacred Cut," a number of large Italian churches that were built after the fifteenth century had a square or polygonal plan. These centrally symmetrical churches usually had a dome on pendentives over the crossing. In 1506 Bramante began building a church for Julius II on the site of the older Basilica of St. Peter's at Rome, with a central dome 137 feet in diameter on four arches with pendentives. Under the direction of later architects, this grew into the colossal church of St. Peter's. Michelangelo's final design for the dome, unfinished at the time of his death in 1565, was constructed with slight changes by

FIGURE 11.13 ■ The Florence Duomo

other architects. Shown in Figure 11.14, the dome is 405 feet in total height and built in two shells resting on a magnificent drum adorned with 16 coupled column buttresses.

In the Muslim world, the octagonal wooden Dome of the Rock (Mosque of Omar) in Jerusalem (see Figure 5.68) is the largest and perhaps the earliest Arab dome. The swelling or bulbous dome was invented in Persia. These domes were built of brick, on pendentives, with ingeniously interlaced ribs and vaults of great beauty. This type of dome spread to India, where it was reproduced in stone and marble. For example, the exquisite alabaster dome of the Taj Mahal is shown in Figure 11.15.

The fantastic bulbous lanterns and towers of some Russian churches, such as St. Basil's in Moscow, display the late Byzantine drum elongated into a high narrow turret. The dome is replaced by a bulbous termination, dubbed an *onion head*. Saint Basil's, which is shown in Figure 11.16, was built by Tsar Ivan IV, who is better known as Ivan the Terrible.

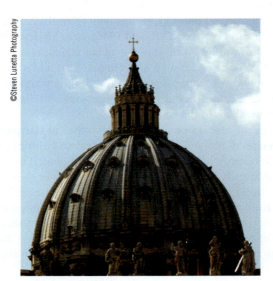

FIGURE 11.14 ■ St. Peter's, Rome

FIGURE 11.15 ■ The Taj Mahal, Agra, India, 1660

FIGURE 11.16 ■ Saint Basil's, Moscow, c. 1554–1560

In the United States, the dome was first used for the state capitol building in Boston in 1795, and it has been a prominent feature of state capitals since then. The dome of the United States Capitol in Washington, D.C., is made of cast iron; its inner shell covers a rotunda 90 feet in diameter. Its cornerstone was laid in 1793 by George Washington. Since it was first built, the U.S. Capitol has undergone a series of building, rebuilding, and extensions which continue to the present day. A new visitor's center is due to open in 2008.

The *geodesic dome*, developed in the twentieth century by R. Buckminster Fuller, is a semiregular polyhedron constructed of lightweight struts forming a skeleton or constructed of flat plates. A geodesic dome can be supported by walls, but it can also sit directly on the ground as a complete structure. A large geodesic dome was used to house the U.S. exhibit at Expo 67 in Montreal in 1967. See Figure 10.57 in Chapter 10.

The Dome of Heaven

What could be more natural than associating the ceiling of a building with the sky? Many ancient civilizations thought of their sacred buildings as symbols of the universe and associated their ceilings with the heavens. Blue ceilings with stars painted on them appear in Egyptian tombs and Babylonian palaces. The association is even stronger when the ceiling is not a flat surface but is instead a dome. When discussing the name *Pantheon*, the Roman historian Dio Cassius (c. A.D. 150–235) wrote, "It is called thus possibly because it included the images of many gods in its statues . . . but I believe that the reason is the similarity of its cupola-form to the heavens."[3]

Wittkower has written, "The dome as a symbol of the sky has a long pedigree. . . . The material for the celestial character of domes in antiquity has been collected in an exemplary manner by Karl Lehmann . . . who followed the conception up into the western, Islamic and Byzantine worlds. . . ."[4] Wittkower later says, "A cosmic interpretation for the dome remained common well into the eighteenth century."[5]

Lehmann himself provides this observation: "One of the fundamental artistic expressions of Christian thought and emotion is the vision of heaven depicted in painting and mosaic on domes. . . ."[6] Lehmann points out that most of the older decorations are on flat ceilings, but later the heavenly visions are more often found in cupolas. He speculates that the symbolism may have actually fostered greater use of the dome. "At least one driving force [for the development of domed construction for churches] may have been the firmly established symbolism of the dome as a celestial hemisphere."[7]

In his history of the University of Virginia, P. A. Bruce describes an odd scheme, conceived by Thomas Jefferson, for teaching astronomy. "The concave ceiling of the Rotunda is proposed to be painted a sky-blue, and spangled with gilt starts in their position and magnitude copied exactly from any selected hemisphere of our latitude." Jefferson's notebook gave elaborate instructions for building a movable seat in which the operator, seated on a saddle at the end of an oak sapling, could be transported to any point of the concave ceiling in order to adjust the stars to their appropriate positions.[8]

THE TERRESTRIAL SPHERE

Let's apply some sphere definitions to Earth, which is approximately spherical, with a radius of 3960 miles. Actually, the distance from pole to pole is about 27 miles less than the diameter at the equator. We'll start with the Earth's *axis of rotation*, which cuts the planet at the north and south *poles*. A plane perpendicular to that axis at its midpoint cuts the Earth's surface in a great circle called the *equator*.

Any great circle passing through the poles is called a *meridian*. The particular meridian passing through the Royal Observatory at Greenwich, England, is called the *prime meridian*. See Figure 11.17.

Latitude and Longitude

The *latitude* of a point *P* on the Earth is the angle, measured at the Earth's center, between that point and the equator. A small circle in a plane parallel to that of the equator is called a *parallel of latitude*. The *longitude* of a point *P* on the Earth is the angle, measured at the Earth's center and in the plane of the equator, between the meridian containing that point and the prime meridian.

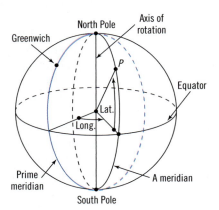

FIGURE 11.17 ■ The Terrestrial Sphere

THE LONGITUDE PROBLEM

A series of maritime disasters caused by faulty navigation prompted the British government to pass the Longitude Act of 1714. It offered a prize to anyone who could determine the longitude at sea to within one-half a degree of arc. The longitude prize attracted serious scientists but also many crackpots with strange proposals.

Look closely at Figure 11.18. This engraving by the English painter/engraver William Hogarth shows lunatics in the notorious insane asylum of Bedlam. Note the telescope and the globe with its lines of latitude and longitude. John Harrison eventually solved the longitude problem using a series of chronometers that were not affected by temperature.

FIGURE 11.18 ■ "The Longitude Lunatic" in *A Rake's Progress*," Hogarth, 1735

If you assume the Earth to be a sphere of radius 3960 mi and you know the angle that the points subtend at the Earth's center, you can find the distance between two points on the Earth's surface using Equation 62. This is easy if the two points are on the same meridian or on the equator.

Finding the distance between any two points on the Earth's surface requires a little trigonometry. We explain how in the Exercises and Projects at the end of this chapter.

■ **EXAMPLE:** How many miles north of the equator is a town of latitude 43.6° N?

● **SOLUTION:** Equation 62 requires the angle to be in radians, so we make the conversion as follows:

$$\theta = 43.6° \left(\frac{\pi \text{ rad}}{180°} \right) = 0.761 \text{ rad}.$$

Let s represent the distance from the town to the equator. By Equation 62, you have the following:

$$s = r\theta = 3960 \text{ mi } (0.761) = 3013 \text{ mi}. \qquad ■$$

Temple and Church Orientation

The four cardinal directions—north, south, east, and west—had a great deal to do with the placement of churches and temples. We mentioned these directions in Chapter 4, "Ad Quadratum and the Sacred Cut," when we discussed the Cosmic cross. This cross is not difficult to lay out using the shadow of an upright stick cast by the sun. Given the importance of the sun and its movements to life on Earth, it is not surprising that sacred buildings have often been aligned to the cardinal directions. For example, the Great Pyramid of Cheops has sides aligned with the Cosmic cross to within several minutes of arc.[9] Vitruvius, in his *Ten Books on Architecture*, explains how a temple should be approached from the west. "If the choice is free . . . the temple and the statue . . . should face the western quarter of the sky. This will enable those who approach the altar . . . to face the direction of the sunrise . . . and those who are undertaking vows [will] look towards the quarter from which the sun comes forth, and likewise the statues themselves appear to be coming forth out of the east. . . ."[10]

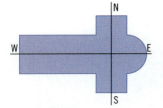

FIGURE 11.19 ■ **Typical Church Alignment**

Interestingly, the verb *orient* means to face the east.

This idea was apparently not lost on church builders. According to Burckhardt, "God arises like the sun, in the east, and on this symbolism almost all later churches are founded. . . ." That is, churches were oriented east–west, with the altar toward the east. He continues, "Early Christians saw in the sun, which on Easter morning rises [almost] precisely in the east, the natural image of the savior who had risen from death to life." He further states, "Since the nave runs from west to east, it follows that the transept lies in a north–south direction; the axial cross of the building thus corresponds to the axial cross of the heavens. . . ."[11] Figure 11.19 shows this alignment. Janet Fisher writes, "The full afternoon sun . . . lights the front of most churches, because they were built with the same orientation. Inside, the congregation facing the main altar faces the east, the direction of the rising sun and the rising Christ. Every evening across Europe the last light illumines the rose windows above the front doors of countless churches."[12]

Globes and Maps

The Earth's features can be portrayed using a *terrestrial globe* or a map. A globe is simply a small replica of the Earth that uses a spherical surface to represent a larger spherical surface. Globes come in all sizes; they can be small enough to fit in a pocket or large enough to walk inside. As well as being functional objects, they can be works of art. The photograph in Figure 11.20 shows a terrestrial globe and a celestial globe.

Representing a spherical surface on a flat surface (that is, by mapping) is more complicated than representing it on another sphere. The representation is usually created by *projecting* points from the sphere to corresponding points on a map. (*Projective geometry* is the branch of mathematics concerned with projections.) Before we study map projections, let's discuss projection in general.

Central projection is the projection of one configuration, such as line *ABC* on the given plane *P* in Figure 11.21, onto another plane *P'* by means of rays drawn from a fixed point *O*, the *center of projection*. In Figure 11.21, the points *A'*, *B'*, and *C'* are the projections of points *A*, *B*, and *C*. The two figures made to correspond with each other by such a projection are said to be *in perspective*, and the image is called a *projection* of the original figure.

FIGURE 11.20 ■ **A Celestial and a Terrestrial Globe**

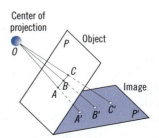

FIGURE 11.21 ■ **Central Projection**

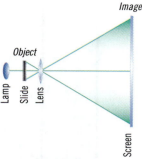

FIGURE 11.22 ■ A Slide Projector

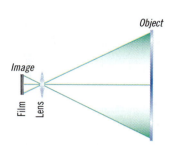

FIGURE 11.23 ■ A Camera

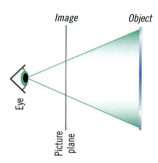

FIGURE 11.24 ■ Perspective

In the slide projector shown in Figure 11.22, the lens (considered as a point) is the center of projection. In optics, the words *object* and *image* are often used to name the planes P and P', with light traveling from object to image. In this example, the slide is the plane of the object and the screen is the image plane. If the screen were the object and the slide were the image, the geometry would be exactly the same.

In the camera shown in Figure 11.23, the image on the film is a projection of the object being photographed, with the lens being the center of projection. In this example, all points on the object (such as a tree) will not usually be in a single plane, and only the points that are at a specific distance will be "in focus"; points not at that specific distance will be projected either behind or in front of the film plane. While there is only one distance, for a given lens, at which an object is in perfect focus, there is a range of distances at which an object appears acceptably in focus. This range is called *depth of field*.

When the center of projection is the artist's eye, as portrayed in Figure 11.24, the image plane is the picture plane, and the object is the scene being painted. We will cover perspective in detail in Chapter 12, "Brunelleschi's Peepshow and the Origins of Perspective."

Rays that are parallel produce what is called *parallel projection*. This occurs when the center of projection is at a great distance, as it might be when a shadow is cast by the sun. See Figure 11.25. If, in addition, the rays are perpendicular to the plane of projection, they produce what is called *orthogonal projection*. If the two planes are parallel, the object and the image will be identical.

If both P and P' are flat planes, as in the preceding examples, we have a *plane projection*. A second common type of projection, of more interest to us for mapping, is created when we project points from a sphere to a plane. When the center of projection is a point on the surface of the sphere, we have what is called *stereographic projection*, as seen in Figure 11.26. In this example, the center of projection O is the north pole of the sphere. The plane P is usually located in one of two places: (a) tangent to the sphere at the south pole or (b) in the plane of the equator. A ray from O cuts the sphere at A and cuts the plane at A'. We say that A is *mapped* onto A'.

If the center of projection is at the center of the sphere, we have *gnomonic projection*, which is depicted in Figure 11.27. In this example, the plane P is tangent to the sphere. A ray from O cuts the sphere at A and cuts the plane at A'.

Only on a globe can terrestrial areas and shapes be drawn with reasonable accuracy, even though the Earth is a somewhat flattened sphere and not geometrically perfect. On maps, especially maps of large areas, distortions are inevitable; however, the distortions can be minimized by selecting the projection most suitable for the use of the map. A great variety of map projections have been invented to meet specific needs.

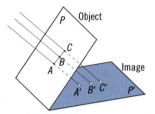

FIGURE 11.25 ■ Parallel Projection: A Shadow Cast by the Sun

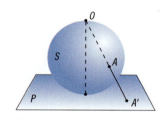

a

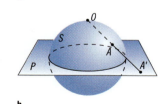

b

FIGURE 11.26 ■ Stereographic Projection

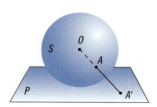

FIGURE 11.27 ■ Gnomonic Projection

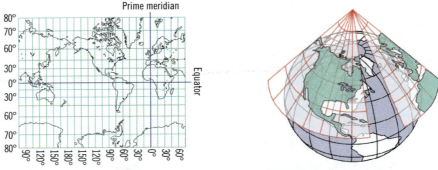

FIGURE 11.28 ■ Mercator Projection

FIGURE 11.29 ■ Conic Projection

There are three classes of maps: (1) those made by projection onto a plane; (2) those made by projection onto a cylinder or cone, and then unrolled; and (3) those made by arbitrary constructions. All are called *projections*, although only the first is a true projection. The term *projection* is well established; therefore, we will use it here. The following are descriptions of some common types of maps.

A *Mercator projection* is shown in Figure 11.28. In 1569 Gerardus Mercator solved the projection problem by producing his famous world map with the meridians vertical and parallels of latitude horizontal. Their spacing is in proportion to the *secant* (a trigonometric function) of the latitude. The Mercator projection has many advantages in spite of the great distortions it causes in the higher latitudes. Compass bearings can be plotted as straight segments on these maps, which have been traditionally used as nautical charts.

Conic projections, as seen in Figure 11.29, are derived from a projection of the globe on a cone drawn with the point above either the north pole or the south pole and tangent to the Earth at some selected parallel.

Polyconic projections are often used for a series of maps. Each cone is placed tangent to the globe at the particular latitude of each map. The familiar topographic maps issued by the U.S. Geological Survey are polyconic projections.

A *cylindrical projection* results from projection onto a cylinder touching the Earth at the equator. Parallels of latitude appear as horizontal lines, and meridians appear vertical.

Great-circle charts are maps of large areas, such as the entire Pacific Ocean. They are ordinarily drawn to very small scales and use gnomonic projection.

Azimuthal projections (also called *zenithal projections*) picture a portion of the Earth as a flattened disk, tangent to the Earth at a specified point. The viewpoint may be the center of the Earth (gnomonic projection) or on the opposite side of the Earth's surface (stereographic projection). A *polar projection* is an azimuthal projection drawn to show Arctic and Antarctic areas.

■ EXERCISES ● THE TERRESTRIAL SPHERE

1. Sketch a terrestrial sphere. Indicate and label the following:

Earth's axis of rotation	a meridian
the equator	the angle of longitude of a city
the angle of latitude of a city	a parallel of latitude
north and south poles	the prime meridian

2. Define or describe the following terms:

projection	central projection	plane projection
stereographic projection	center of projection	parallel projection

3. Briefly describe each kind of map:

   Mercator projection   cylindrical projection   great-circle chart
   conic projection   polyconic projection   azimuthal projection
   polar projection

4. *Distance Between Two Points on Earth on the Equator or on the Same Meridian.*
   Assume a town is at latitude 35.2° N. Find the distance from the town to
   the north pole. *Hint:* For this exercise and the one to follow, assume the
   Earth is spherical with a radius of 3960 mi.

5. Two points on the Earth's equator have longitude of 74.5° W and 91.3° W.
   Find the distance between them.

THE CELESTIAL SPHERE

When you're looking at the night sky, it is impossible to tell how far away celestial objects are, and it is easy to imagine that all the stars and planets are at the
same distance on the surface of a huge sphere. This *celestial sphere* is an imaginary sphere of infinite radius. Its center is the center of the Earth. However, for
practical observation, the center of the celestial sphere can be thought of as a
point *on* the Earth's surface, because the Earth's diameter is insignificant compared to the distances to the heavenly bodies. We visualize the stars as moving
across the surface of the celestial sphere; the stars' apparent movement across
the sphere is an illusion, because it is actually the Earth that is moving.

Celestial Coordinates

Four systems of celestial coordinates are used today: the *horizon system*, which
locates a body by its distance up from the horizon and at a particular compass
direction; the *ecliptic system*, which gives the location of a planet in the ecliptic
(the apparent path of the sun); the *galactic system*, which uses the plane of
our galaxy as a reference; and the *equatorial system*, which is illustrated in
Figure 11.30. Only half the celestial sphere is shown with a finite radius.

   The equatorial or *equinoctial* system is the one most commonly used in
astronomy. The Earth's axis, extended, cuts the celestial sphere at the north and
south *celestial poles*. The heavens seem to turn around these poles. *Polaris*, the
north star, is located near the north celestial pole.

   If the plane of the Earth's equator is extended, it will intersect the celestial
sphere at the *celestial equator*, or *equinoctial*. Further, if the plane of each meridian

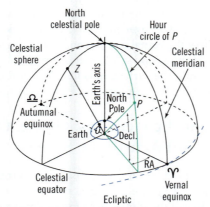

**FIGURE 11.30 ■ The Equatorial System of Celestial
Coordinates**

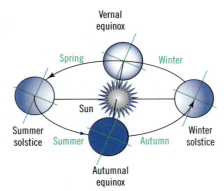

FIGURE 11.31 ■ The Solstices and Equinoxes

on Earth is extended, it will intersect the celestial sphere at a *celestial meridian*. Therefore, the celestial meridians and the celestial equator exactly correspond to the Earth's meridians and equator; any point on Earth can be projected onto the celestial sphere, and vice versa. The zenith Z of an observer O on Earth is a point directly over the observer's head. The great circle passing through the celestial poles and a particular celestial body P is called the *hour circle* of that body.

Geometrically, the Earth rotates around the sun once a year in a plane that is inclined to the Earth's equator by about 23.5 degrees, as shown in Figure 11.31. (Of course, the distances in this figure are not to scale.) The intersection of this plane with the celestial sphere is called the *ecliptic*, a small section of which was shown in Figure 11.30. The sun appears to move along the ecliptic.

If you observe the elevation of the noontime sun, you can see that its elevation is greatest in summer and least in winter. Its greatest elevation, which is called the *summer solstice*, occurs on about June 21. Its least elevation, which is called the *winter solstice*, occurs on about December 21. Midway between these dates, the sun appears to be over the Earth's equator, on about March 21 and September 21. These points are called the spring or *vernal equinox* and the fall or *autumnal equinox*. The spring equinox is also called the *first point of Aries*, and the autumnal equinox is also called the *first point of Libra*. The equinoxes are the two points on the celestial sphere in which the ecliptic and the celestial equator intersect. The two solstices and two equinoxes divide the ecliptic into four quadrants, and each of these is further subdivided into three arcs, giving the twelve signs of the zodiac, a common subject in art that will come up again later in this chapter.

To locate a point on Earth, we use its latitude and longitude. Similarly, to locate a celestial body on the celestial sphere, we use the declination and right ascension. The *declination* (DEC) of a celestial body is the angle between the body and the celestial equator, measured along the hour circle of that body, in degrees. It is similar to latitude on Earth. The *right ascension* (RA) of a celestial body is the angle between the hour circle of the vernal equinox and the hour circle of the body, usually measured in hours, minutes, and seconds. It is similar to longitude on Earth.

We are just scratching the surface of celestial coordinates. For further exploration on your own, see the Exercises and Projects at the end of the chapter.

Models of the Heavens: Globe, Orrery, and Armillary Sphere

There are several kinds of models of the heavens. When they're well made, they easily qualify as art objects. Like the terrestrial sphere, the celestial sphere can be portrayed on a globe. Recall the globes in the photograph in Figure 11.20. The

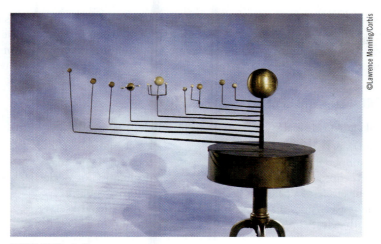

FIGURE 11.32 ■ An Orrery

orrery and the armillary sphere also portray the heavens. For example, the orrery, or planetaria, that we see in Figure 11.32 is a little model of the solar system.

In the *Timaeus*, Plato described how the Creator formed the circular paths for the stars. After marking off intervals of the musical ratios on strips of "soul-stuff," the Creator "cut the whole fabric into two strips which he placed crosswise at their middle points to form a shape like the letter X; he then bent the ends round in a circle and fastened them to each other . . . to make two circles, one inner and one outer."[13] According to translator Desmond Lee, "In writing about the strips or rings that are to carry the planets, Plato probably had in mind . . . an armillary sphere."[14] The word *armillary* is from the Latin *armilla*, meaning "bracelet." In the examples shown in Figure 11.33, you can see that the armillary sphere is a skeleton globe. In art, the armillary, like the celestial globe, is an attribute of astronomy, one of the Seven Liberal Arts. It is a symbol of the universe, and of *Urania*, the muse of astronomy. In a painting, an armillary sphere or globe, accompanied by books and other scholarly items, may signify education.

Johannes Kepler, the Renaissance astronomer and astrologer, theorized that because there were six planets orbiting within an invisible sphere and there were five regular solids, each of which could be inscribed in a sphere, it was possible that the cosmos was constructed so that one of the five regular solids was located between each pair of the spheres that carried the six planets. To help visualize this arrangement, see Figure 11.34. This nest of alternating planets and

FIGURE 11.33 ■ Armillary Spheres

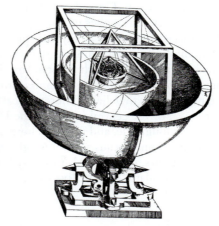

FIGURE 11.34 ■ Kepler's Model of the Universe

JOHANNES KEPLER
(1571–1630)

Johannes Kepler is best known for his discovery of the three principles of planetary motion, which are now known as Kepler's laws. Kepler observed that celestial bodies have elliptical, not circular orbits. Instead of a purely geometric description of the motions of heavenly bodies, Kepler introduced the concept of *force*, which led to the concept of these motions in terms of dynamics.

Kepler was also the first to correctly explain how humans see. He designed telescopes and accurately explained how they worked, thereby becoming one of the founders of modern optics.

solids, which we studied in Chapter 10, was the main theme of Kepler's *Cosmic Mystery*, which was published in 1596. Of course, his observations did not fit this incorrect scheme, but they may have helped lead him to the principles of planetary motion that now bear his name.

■ EXERCISES ● THE CELESTIAL SPHERE

1. Sketch a celestial sphere and indicate and label the following terms:

north celestial pole   a celestial meridian
equinoctial   the vernal equinox
the ecliptic   declination of a celestial body
hour circle of a celestial body   celestial equator
right ascension of a celestial body   zenith of an observer
south celestial pole   the autumnal equinox

TIME

We measure the passage of time by the motions of the celestial spheres. The Earth's rotation about its axis gives us day and night, the moon's motion determines the month, the Earth's rotation about the sun defines the year, and the tilt of the Earth's axis as it makes this rotation gives us the seasons. We will look at each of these time periods in turn and point out the common art themes associated with each.

Different Kinds of Time

When someone asks what time it is, we usually look at a watch or a clock and report what we see there. Time, however, is a bit more complicated than that. There are several kinds of time.

Standard Time: When you read a clock, you get the standard time at that location. The standard time is the same everywhere within a particular time zone. For example, let's say it is 3:00 P.M. in Topeka, Kansas, which uses Central Standard Time (CST). It will also be 3:00 P.M. in Peoria, Illinois, which is in the same time zone.

Daylight Saving Time: In the summer months in the United States, we set our clocks ahead by one hour, giving us Daylight Saving Time (DST). For example, if the time is 3:00 P.M. in Topeka in July, the Central Daylight Saving Time is 3:00 P.M. and the Central Standard Time is 2:00 P.M.

Local Civil Time: Local Civil Time (LCT) is based on the motion of a fictional sun that moves uniformly through the heavens so that the length of a day is exactly the same at any time of year. The word *local* is used to indicate that it applies only to locations on the observer's meridian. Within each time zone, the local time agrees with the standard time at only one meridian, which is called its *standard meridian*. For example, although Topeka and Peoria have the same standard time, they have different local times, because they are at different longitudes. For the central region of the United States, the standard meridian is at 90° west longitude, and Topeka's longitude is 95° 42′, or 5° 42′ west of its standard meridian. Because 15° of longitude equals 1 hour of time, 5° 42′ converts to 0.38 hours, or about 23 minutes. Therefore, when the clock reads 3:00 P.M. Central Standard Time in Topeka, its local civil time is 3:00 P.M. minus 23 minutes, or 2:37 P.M.

Local Apparent Time: The *apparent* time is not based on a fictional sun moving uniformly, but on the *actual* sun. The Local Apparent Time (LAT) is the hour angle of the sun, measured from the observer's meridian, plus 12 hours. The apparent time is that indicated by a sundial. For example, when the sun is on the observer's meridian, the local hour angle of the sun is zero degrees, so the local apparent time is 0 + 12, or 12 noon. In Topeka, you would expect this to occur when the Local Civil Time is 12:00 P.M. (or when the standard time is 12 minus 23 minutes, or at 11:37 A.M.).

The Equation of Time

The sun will be on the observer's meridian when LCT is 12 noon, which is only four times during the year. The rest of the year it will be either earlier or later, by as much as 15 minutes. The difference between the apparent time and civil time is called the *equation of time:*

Equation of Time = Local Apparent Time − Local Civil Time.

You can obtain the equation of time from tables or from a graph such as the one shown in Figure 11.35.

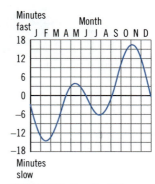

FIGURE 11.35 ■ Equation of Time

■ **EXAMPLE:** A sundial in Topeka reads 3:00 P.M. on July 31. What would a clock read at that moment?

● **SOLUTION:** The sundial gives the Local Apparent Time: 3:00 P.M. From Figure 11.35, at July 31, the equation of time is −6 minutes:

Local Apparent Time − Local Civil Time = −6 minutes.

Therefore, local time would be as follows:

LCT = LAT + 6 minutes = 3:00 P.M. + 6 = 3:06 P.M.

Earlier we noted that the LCT at Topeka was 23 minutes earlier than its standard time:

Central Standard Time at Topeka = LCT at Topeka + 23 minutes

= 3:29 P.M.

Finally, in July, we add an hour for Daylight Saving Time:

Central Daylight Time at Topeka = 3:29 P.M. + 1:00 P.M. = 4:29 P.M.

Therefore, when a sundial reads 3:00 P.M. on July 31 in Topeka, a clock would read 4:29 P.M. ■

Instruments to Measure Time

Sundials, sun calendars, analemmas, and meridian lines have all provided humans, both in ancient and modern times, with different ways to mark the passing of time.

Over the centuries the sundial, the instrument perhaps best known for measuring time, has appeared everywhere. Sundials come in all sizes, from ones you can fit into your pocket to large dials mounted on buildings. Sundials, indicating the passage of time and eventual death, are a symbol of *vanitas.* We'll explore the concepts of vanitas and melancholia in greater depth in the section of this chapter titled "The Sphere and the Heavens as Art Motifs."

References to sundials go back as far as the Old Testament. Both Isaiah 38:8 and 2 Kings 20:11 mention the *Dial of Ahaz,* the King of Judah. Herodotus

FIGURE 11.36 ■ **Sundial at Chartres**

FIGURE 11.37 ■ **Sundial at Longwood Gardens**

mentions the sundial as well, ". . . and the sun-dial and the division of the day into twelve parts, the Greeks learnt from the Babylonians."[15] Vitruvius also mentions sundials; he devoted a whole chapter to "Sundials and Water Clocks."[16] Sundials are present in the octagonal Tower of the Winds. Some experts believe that at one time the tower contained as many as eight or nine sundials. This edifice was shown earlier in Figure 5.69 in Chapter 5. In Figure 11.36, you see a sundial still marking the time at Chartres Cathedral. Figure 11.37 shows a sundial at the Longwood Gardens botanical gardens in Pennsylvania. The Longwood sundial displays the equation of time we examined earlier. The equation of time is often engraved on or near a sundial to give the number of minutes to be added or subtracted from the dial reading.

Sun clocks are not the only instrument we can build to measure the passage of time; we can also construct *sun calendars*. These constructions usually involve the observation of a shadow or projection of the sun at a particular time of day, often noon. Some calendars mark only particular times of the year, such as solstices and equinoxes. Using the same idea as a gun sight, calendar circles, such as Stonehenge, may use the alignments of stones to determine a solstice or equinox based on when the sun sets in a particular direction. For example, a sun dagger in Chaco Canon, New Mexico, marks the summer solstice. At Casa Grande (Great House), Arizona, the four walls are aligned to the cardinal points of the compass. Openings in the walls align with the sun and moon to provide times for planting, harvest, and celebrations.

Figure 11.38 shows an upright stick, or *gnomon*, with the sun causing it to cast a shadow on the ground. The tip of the shadow at noon is called the *noon mark*. When pegs are driven into the ground at the noon mark throughout the year, the pegs form a figure-8 pattern, the *analemma*.

At the vernal equinox, when the sun is over the equator, the noon mark will be at point *E*. As the sun orbits ever higher above the equator, the noon mark moves closer to the gnomon. After the summer solstice, the sun is lower in the sky and the noon mark begins to move away from the gnomon. At the autumnal equinox, the noon mark is again at *E*. As the sun moves below the equator, the noon mark moves further away from the gnomon, toward the winter solstice. In addition to the north–south movement of the noon mark, we have the east–west movement of the mark according to the Equation of Time, as described earlier. The two movements combine to create the analemma's figure-8 shape, which is detailed in Figure 11.39. Older globes frequently contain an analemma.

FIGURE 11.38 ■ **The Noon Mark Tracing an Analemma**

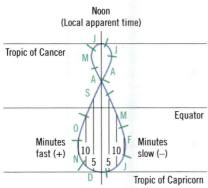

Noon
(Local apparent time)

Tropic of Cancer

Equator

Minutes
fast (+)

Minutes
slow (−)

Tropic of Capricorn

FIGURE 11.39 ■ An Analemma

FIGURE 11.40 ■ Parma Town Hall

Sun calendars are often found on the walls of public buildings, as you can see in the photograph of the Town Hall in Parma, Italy, in Figure 11.40. The analemma is to the left of the windows; two sundials, one for morning and one for afternoon, are to the right.

Meridian lines, or *meridiana*, offer yet another means of tracking time. J. L. Heilbron points out that for the six centuries from the late Middle Ages into the Enlightenment, the church supported astronomy. Also at that time, a career as a meridian line maker was largely underwritten by the church. The main purpose of the meridian makers' astronomical investigations was to fix the date of Easter, based on the following specifics. Jesus ate the last supper on Passover, and was crucified the same day. He was buried the following evening and rose on the eighth day. Therefore, Easter, the celebration of the resurrection, is linked to Passover. However, the date of Passover was set by the rabbis. Therefore, early Christians would have had to apply to them to find the date of Easter—they found this intolerable. Eventually, the Council of Trent decreed that priests were to learn how to compute the date of Easter. The best tool they found to help determine this critical date was the *meridian line*, a north–south line in a dark building with an aperture in its roof.

Even the great Victorian-era author Charles Dickens wrote about the meridian line, with regard to the Italian city of Bologna. "The colleges, and churches too, and palaces: and above all the academy of Fine Arts . . . give it a place of its own in the memory. Even though these were not, and there were nothing else to remember it by, the great Meridian on the pavement of the church of San Petronio, where the sunbeams mark the time among the kneeling people, would give it a fanciful and pleasant interest."[17] In Figure 11.41, you can see the meridian line that so enraptured Dickens. An aperture in the ceiling of the cathedral projects a shaft of sunlight onto a bronze strip on the pavement below, which is engraved with the days of the year and the signs of the zodiac. Figure 11.42 shows a cross-section of the cathedral and the geometry of the meridian line and the aperture.

FIGURE 11.41 ■ Meridian Line, S. Petronio, Bologna

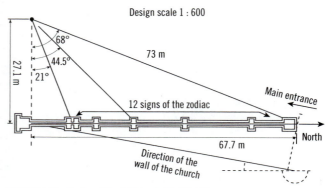

FIGURE 11.42 ■ Design of the S. Petronio Meridian Line

Other meridian lines are found in the Duomo in Florence, the Torre dei Venti in Rome, the Duomo in Palermo, and S. Maria della Angeli, Rome. The light through the aperture at the Torre dei Venti meridian line is especially striking.

Time as an Art Motif

The passage of time, which affects our lives in such important ways, is understandably a frequent artistic subject. We will examine a few examples here: hours of the day, times of the day, days of the week, months of the year, and seasons of the year.

The *canonical hours* were the times of day allocated to specific prayers. The Roman church had eight such hours: Matins, Lauds, Prime, Terce, Sext, Nones, Vespers, and Compline. A *book of hours* contained prayers for each of the canonical hours. Such books were popular in the later Middle Ages. Small and lavishly decorated, such as the example in Figure 11.43, they contributed to the development of Gothic illumination.

The day was divided into four periods: dawn, day, evening, and night. All four periods are portrayed by Michelangelo in the Medici chapel. Dusk and Dawn are depicted in Figure 11.44; Dusk (left) and Dawn (right) are below the central figure, called *Il Pensiero*, of Lorenzo de Medici. Lorenzo is sometimes taken as a representation of melancholia, which we will discuss later in this chapter. Note the spirals, which we studied in Chapter 9, "The Ellipse and the Spiral." Also note the translational symmetry in the decorative band running behind the reclining figures—we encountered them in Chapter 3, "The Triangle."

For another view of a day, look for William Hogarth's *Four Times of the Day*, where the irreverent creator of *A Rake's Progress*, shown earlier, gives us his moral commentary on how the four times of day—morning, noon, afternoon, and night—differ for the rich and the poor in seventeenth-century England.

The hours add up to days, and the days add up to weeks. Because the four phases of the moon add up to a complete lunar cycle of 28 days, each phase is seven days, which defines the week. Not surprisingly, seven is one of the main astrological numbers. Seven is also identified with the seven planets known in

FIGURE 11.43 ■ A Page from a Book of Hours, Bourges or
Loire Valley, Early Fifteenth Century

FIGURE 11.44 ■ Medici Chapel Figures of Dusk and Dawn,
and Lorenzo de Medici, Michelangelo

early times. In contrast to what we know today, at one time the sun and the moon were grouped with the five known planets. Like the number four, the number seven has its *correspondences*. Refer to Appendix E under the heading "Seven," where each planet is identified with a specific day of the week and a particular metal from alchemy, and each has its own symbol.

The next increment of time is the month. Using pictures to depict the individual months is not a modern concept. Sets of pictures, with an appropriate task pictured each month, can be found in cathedrals, Books of Hours, tapestries, and paintings. The tasks that are portrayed vary. However, the spring months are usually represented by the preparation of fields, orchards, and vineyards; the summer months are usually illustrated with scenes of people working those fields, orchards, and vineyards; the autumn months are depicted by the harvesting of corn and wheat, and the gathering of grapes for wine making; and the winter months are illustrated with people cutting wood and feasting around a fire.

A Book of Hours, mentioned earlier, was often made for a particular patron, such as the famous one made for the Duc de Berry. In addition to daily prayers, these books usually contained calendar pages, showing feast days, holidays, and so forth. Take a look at a page from the Duc de Berry's Book of Hours in Figure 11.45. This page for September shows grapes being harvested. *Zodiacal Man*, another illustration from this book, appears in Figure 4.63 in Chapter 4. For another view, look for *Signs of the Zodiac and the Labors of the Months* on the West Façade of Amiens Cathedral.

The final measurement of earthly time—one which is obviously closely tied to the months—is the seasons. In Greek mythology, the seasons came about when Demeter's daughter, Persephone, was abducted to the underworld by Hades. Demeter, after a great deal of fuss, eventually arranged with Zeus for Persephone to spend half the year in the underworld, which became autumn and winter, and half the year above ground, which became spring and summer.

FIGURE 11.45 ■ A Page from the *Très Riches Heures du Duc de Berry,* **Pol de Limbourg and His Brothers, c. 1414–1418**

Four is another important astrological number. The month is subdivided into the four phases of the moon, and the year is divided into four seasons. The four seasons are a common theme in art. Spring is sometimes depicted as a young woman, as flowers, as lovers, or as birds; summer is commonly depicted with ears of corn, fruit, reapers, swimmers; autumn is depicted with grapes and grape vines; and winter is presented by ice skaters or by an old man, heavily wrapped. The four seasons are often associated with the *four ages of man.* For yet another view, look for Paul Cezanne's lovely depiction of the four seasons.

■ EXERCISES • TIME

1. Define or describe the following terms:

standard time	daylight saving time	local civil time
local apparent time	equation of time	analemma
meridiana		

2. What is the Local Civil Time in Pittsburgh on November 1 when the standard time is 2.00 P.M.? Take the longitude of Pittsburg as 80° west.

3. What is the Local Apparent Time in Pittsburgh when the standard time is 2.00 P.M.? Take the longitude of Pittsburg as 80° west.

4. A sundial in Pittsburgh reads 2.00 P.M. on October 1. What is the standard time at that moment? Take the longitude of Pittsburg as 80° west.

THE SPHERE AND THE HEAVENS AS ART MOTIFS

In this section, we'll examine a varied collection of artworks that use the sphere, celestial and otherwise, and we'll look at other art involving the heavens. These works of art are from different times in history and from different traditions; their common thread is that they are all related to the sphere or to the cosmos. Central themes to watch for include: (1) fortune, instability, and inconstancy; (2) vanitas and melancholia; and (3) globes representing power through discovery, expansion, and new territories. We'll also consider the recognition given to astrologers and astronomers, and the thought-provoking portrayals of the sphere given by M. C. Escher and other twentieth-century artists.

Symbolism of the Sphere

Fortuna, or *Lady Luck*, is a common subject in art. We have seen her portrayed with a wheel, but she is frequently shown standing on a sphere, or with a soap bubble, indicating chance or uncertainty. She may be shown blindfolded or with a cornucopia, symbolizing the favors she may capriciously hand out. Dice are sometimes present, as well as billowing drapery, indicating the variable winds of chance. In Dürer's portrayal of Fortune, which is shown in Figure 11.46, the instability of the comparatively small sphere is emphasized by the thinness of the cane. Fortune holds the aphrodisiac plant eryngium, denoting luck in love. See also Albrecht Dürer's *Fortuna (Das grosse Glück)* and Cesare Ripa's *Fortune*.

Erwin Panofsky writes that Michelangelo's *Dream*, which is shown in Figure 11.47, "denotes the human mind called back to Virtue from the Vices. (See Appendix E for further information on the seven virtues and seven vices.) Here the sphere upon which the youth reclines may indicate the instability of the task of abandoning vice. But the sphere here is a terrestrial globe, so we may have a depiction of man placed between the unreal life on Earth and the celestial realm represented by the angel."[18]

Image scanned from *The Complete Engravings, Etchings & Drypoints of Albrecht Dürer*, Walter Strauss, ed. New York: Dover 1972, Plate 7, page 17.

FIGURE 11.46 ▪ *Fortune (Das kleine Gluck)*, Dürer, 1495

The Samuel Courtauld Trust, Courtauld Institute of Art Gallery, London

FIGURE 11.47 ▪ *Dream (Il sogno)*, Michelangelo, c. 1533

©Cameraphoto/Art Resource, NY

FIGURE 11.48 ■ *Inconstantia,* from the Scrovegni Chapel, Giotto, c. 1305

In Figure 11.48, Inconstancy (*Inconstantia*) is depicted as a young woman on a sphere, who is about to fall backward even though she tries to keep her balance by spreading her arms.

Vanitas is a related artistic theme that often uses the sphere in the form of soap bubbles. The concept is not vanity in the sense of vainness, conceit, or self-importance. Rather, it refers to the emptiness of worldly possessions and the transience of life. Consider Figure 11.49. Aside from soap bubbles, a vanitas still life may contain a skull, a clock or an hourglass indicating the passage of time, a candle that can easily be snuffed out, an overturned cup or other vessel, and transient drops of dew.

Spheres often appear in depictions of *melancholia*, such as in Dürer's *Melencolia I* in Figure 10.51 in Chapter 10. In that painting, Melencolia is surrounded with the objects of geometry, one of the Seven Liberal Arts over

©Erich Lessing/Art Resource, NY

FIGURE 11.49 ■ *Vanitasstilleven (Vanitas Still Life),* D. Bailey, 1651 which Saturn presided. The neo-Platonist Marsilio Ficino wrote, "all men who excel in the arts are melancholics." Dürer knew of Ficino's works, and his *Melencolia* is said to be based on Ficino's statement. Other famous melancholics in art include Michelangelo's *Il Pensiero* (the statue of Lorenzo de Medici shown in Figure 11.44) in the Medici chapel, and Rodin's *Thinker*.

A melancholic person is also called *saturnine*, for a daughter of Saturn. Of the four humours, mentioned in Chapter 4, it is the most frequently portrayed, probably because many artists are melancholic. Italo Calvino writes, "Ever since antiquity it has been thought that the saturnine temperament is the one proper to artists, poets, and thinkers, and that seems true enough. Certainly literature would never have existed if some human beings had not been strongly inclined to introversion, discontented with the world as it is, inclined to forget themselves for hours on end and to fix their gaze on the immobility of silent words . . . I have always been saturnine. . . . I am a Saturn who dreams of being a Mercury, and everything I write reflects these two impulses."[19]

A globe may signify an interest in discovery, expansion, and new territories. It is sometimes seen in portrayals of the liberal arts and may signify education. Lastly, a sphere or orb in the hands of a monarch symbolizes his or her rule over the world, and it was first used as such by Roman emperors. When topped by a cross, it was a symbol of the Holy Roman emperors, as in Figure 11.50. Sometimes Jesus is shown holding an orb to proclaim him as the savior of the world.

FIGURE 11.50 ■ *Emperor Sigismund,* Dürer

Portrayals of Astronomers and Astronomy

Astrologers and astronomers have always been popular subjects for painters, such as those in the following illustrations and paintings. First, consider the work of Jan Vermeer. While Vermeer painted some landscapes, what he seemed to like best were interior paintings containing one or two people, at most three. His subjects were usually placed in a room into which light streamed from a window, such as the portrayal of an astronomer in Figure 11.51.

FIGURE 11.51 ■ *The Astronomer,* Vermeer, 1668

FIGURE 11.52 ■ *Astronomy*, Raphael

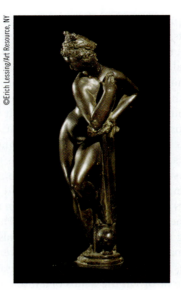

FIGURE 11.53 ■ *Allegory to Astronomy*, Giamobologna

Astronomy is one of the Seven Liberal Arts, part of the *quadrivium* (see Appendix E under the heading "Seven"). As with the other liberal arts, artists were fond of depicting allegories to astronomy.

The small rectangular panel shown in Figure 11.52 is on the ceiling of the same Vatican room as Raphael's *School of Athens.* Samuel Edgerton wrote that the panel shows "Astronomy, personified perhaps as *Urania* (the traditional muse of that science) looking from outside at the geocentric cosmos enclosed in a crystalline sphere."[20]

The Flemish artist Jean de Boulogne, known as Giambologna, sculpted many statues of Venus. The one of Venus Urania in Figure 11.53 is also known as an allegory to astronomy. Just above the base are an armillary sphere and a plummet, and near Venus' hands are a ruler, compasses, and a framing square. Giambologna also created a sculpture called *Architecture*, containing many of the same instruments. In Figure 11.54, notice Urania's crown of stars, which we'll soon see is similar to that found in some depictions of the Virgin.

Astronomical Instruments

Instruments used by astronomers, from small astrolabes to large installations, are often beautifully made and ornamented. They easily qualify as art objects. For example, the Maharajah Jai Singh II of Jaipur constructed five astronomical observatories in India, such as the one shown in Figure 11.55. They were located at Jaipur, Delhi, Ujjain, and Benares. These large outdoor structures, built between 1727 and 1734, were capable of taking many kinds of astronomical measurements. They are commonly known as *Jantar Mantars.* The astrolabe in Figure 11.56 is another example of an astronomical instrument.

The Sphere and Astronomy in Twentieth-Century Art

M. C. Escher's art has appeared in a number of chapters in this book, and here is some more. Escher used the sphere in a great many of his works. He wrote, "The most beautiful and simple form of the unbounded and still enclosed plane is the surface of the sphere. Only after various other compositions on a flat plane did I come

FIGURE 11.54 ■ *Urania, Muse of Astronomy,* Eighteenth-Century Painting

FIGURE 11.55 ■ Astronomical Sites in India

to use the sphere as a plane for representation." In this section, we'll examine a few of those sphere-related works here, with commentary in the artist's own words.[21]

Escher's *Hand with Reflecting Sphere* is shown in Figure 11.57. According to Escher himself, "Such a globe reflection collects almost one's whole surroundings in one disk-shaped image. The whole room . . . everything, albeit distorted, is compressed into that one small circle. Your own head, or more exactly the point between your eyes, is in the center. No matter how you turn or twist yourself, you can't get out of that central point. You are immovably the focus of your world."

When discussing *Three Spheres II,* which is shown in Figure 11.58, Escher said, ". . . three balls are placed side by side: a translucent one, a reflecting

FIGURE 11.56 ■ An Astrolabe

FIGURE 11.57 ■ *Hand with Reflecting Sphere,* Escher, 1935

FIGURE 11.58 ■ *Three Spheres II,* Escher, 1946

MAURITS CORNELIUS (M. C.) ESCHER (1898–1972)

The Dutch graphic artist M. C. Escher (Figure 11.59) studied at the School of Architecture and Decorative Arts in Haarlem in the Netherlands. He traveled for several years throughout Europe, sketching as he went. His drawings sometimes showed an unusual use of perspective. He married and raised three sons. He eventually returned to the Netherlands, where he lived and continued to draw until the end of his life.

©2005 The M. C. Escher Company–Holland

FIGURE 11.59 ■ *M. C. Escher*

His works remain very interesting to mathematicians. Some of the ideas present in his works are infinity, portrayal of mathematical objects such as the Mobius strip, the polyhedra, the sphere, recursion and feedback, symmetry, mirrors and reflections, impossible structures, tilings of the plane, duality, metamorphosis (topological change), and relativity. Many of the images showing these ideas are reproduced in this book.

one, and an ordinary one. The mirroring globe in the middle connects all three, for it reflects the other two as well as myself, sketching the three balls before me."

Escher created even more works that feature the sphere, including: *St. Bavo's, Haarlem, Still Life with Reflecting Sphere, Sphere with Angels and Devils, Sphere with Human Figures, Balcony, Three Spheres I, Dewdrop, Order and Chaos, Concentric Rinds, Sphere Surface with Fish, Sphere Spiral, Carved Beechwood Ball with Fish*, and *Eye*.

The twentieth century has seen a great many sculptures on celestial themes, often featuring some sort of alignment with the cosmos and relying on sight lines or cast shadows. For example, the concrete and earth sculpture shown in Figure 11.60, which is nearly 300 ft in diameter. It has slits cut in the directions of the equinoxes, solstices, and moonrise.

The armillary sculpture you see in Figure 11.61 is also a functioning sundial.

In Figure 11.62, the noon shadow of the gnomon falling on the carved marble analemma gives the date and the sign of the zodiac.

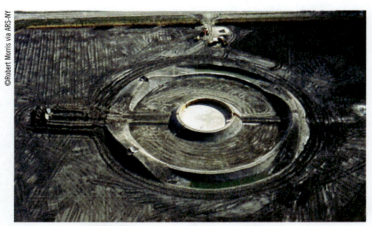

©Robert Morris via ARS-NY

FIGURE 11.60 ■ Observatory, Robert Morris, 1970–1977

FIGURE 11.61 ■ *Armillary VII*, Calter, 1987

FIGURE 11.62 ■ Sun Disk, Moon Disk, Calter, 1994

The Medieval Spheres of Heaven

In medieval times, the universe was subdivided into four concentric realms or hierarchies. In ascending order, they were (1) *the realm of matter*—formless, lifeless dirt, rocks, and so forth; (2) *the realm of nature*—terrestrial, corruptible, the realm of man, animals, and plants; (3) *the cosmic soul*—celestial, incorruptible, containing the nine spheres of heaven (empyrium, fixed stars, sun, moon, and the five known planets); and (4) *the cosmic mind*—supercelestial, incorruptible, containing prototypes for everything below. Realms one and two, where we live, were called *sublunary*. Realms three and four, containing the moon and all the celestial bodies, were called *translunary*. The translunary world was further subdivided into the *nine spheres of the heavens*: the sun and moon, the five known planets, the fixed stars, and the *primum mobile*. The opening picture for this chapter, Figure 11.1, is a wonderful painting by Giovanni di Paolo that depicts the geocentric universe.

We might think that Earth, being at the center of this medieval universe, was special, but just the opposite was believed. According to E. M. W. Tillyard, "far from being dignified . . . the earth in the Ptolemaic system was the cesspool of the universe, the repository of its grossest dregs."[22] In Giovanni di Paolo's painting, the Earth is shown as a *mappamondo*, or world map, surrounded by the three other elements: water, air, and fire. The Earth at the center is brown, and its bright red border clearly marks the boundary between the sublunary realm and the translunary realm. Then come the moon and the planets, all blue—except for the sun, shown yellow-white with a gilded sunburst, and Mars, shown pink for the red planet.

After that, we find the fixed stars with signs of the zodiac and the *primum mobile* (the first moved), which regulates the motion of all the spheres beneath it. Beyond the nine spheres is the *Empyrean heaven*, the home of God. Giovanni di Paolo shows no ring for the Empyrean, just a region beyond the last ring, implying that it cannot be contained by a circular boundary. The spheres of heaven are represented on a flat surface by circles. The number of rings used in pictures of this sort varies. For one thing, theologians couldn't decide whether the Empyrean occupied a definite sphere or whether it was infinite and unknowable, which posed a problem for artists.

For additional discussion of the medieval spheres of heaven, see Appendix E.

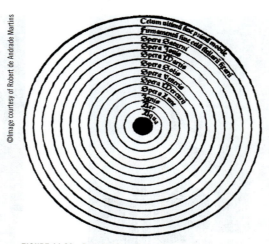

FIGURE 11.63 ■ Drawing from Peter Apian's *Cosmographia*, Antwerp, 1539

DANTE ALIGHIERI (1265–1321)

Dante was born in Florence, but political turmoil there led to his banishment from the city in 1302. He spent his years of exile in Verona and other northern Italian cities, and he went to Paris around 1308. During his exile, Dante wrote *De vulgari eloquentia (Concerning the Common Speech)*, a treatise on the literary use of the Italian language. He started writing *La Divina Commedia (The Divine Comedy)* in about 1307 and completed it shortly before his death. The statue of Dante that you see in Figure 11.64 stands outside the Uffizi Gallery in Florence.

FIGURE 11.64 ■ Statue of Dante in Florence

The probable source for Giovanni di Paolo's picture and others like it was Johannes de Sacrobosco's *Sphaera mundi* (c. 1230). It presented an elementary view of the universe, giving Greek cosmology with a Christian spin. It became very popular and was used as a textbook even into the seventeenth century. *Sphaera mundi* is an illustrated account of the geometry behind the astronomy of Ptolemy. In Figure 11.63, you can see a later illustration of the Ptolemaic system.

The result of all this is what Edgerton calls the *geometrization of heavenly space*. This is the counterpart of geometrization of terrestrial space, achieved with linear perspective where all receding lines traveled obediently to a neat vanishing point at infinity, the subject of the following chapter on perspective. This is part of a world view ruled by a general conception of an orderly universe that God created out of chaos where, according to the *Book of Wisdom* (also known as the *Wisdom of Solomon*, from the apocrypha of the Old Testament), God had arranged all things according to number, weight, and measure.[23]

Each sphere of heaven has artistic associations, and many, such as Venus and Mars, bear the names of Greek gods who have well-known representations in art. Many of these motifs can be found in Appendix E under the heading "Nine."

Dante's *Divine Comedy*

Sacrobosco's descriptions may have influenced the portrayal of the spheres of heaven and hell in the *Divine Comedy* of Dante Alighieri. Dante's greatest work, the *Divine Comedy*, is an allegorical tale of the poet's journey through hell, purgatory, and heaven. In each realm, Dante meets mythological, historical, and contemporary figures, each symbolizing a particular fault or virtue, and each getting in return a particular punishment or reward. Dante is guided through hell by Virgil, who is the symbol of Reason to Dante. He is escorted through heaven by Beatrice, whom he regards as an instrument of divine will. Dante met Beatrice, his great love, in 1274. Not long after Beatrice's death, he wrote *La Vita Nuova* (The New Life), a work that narrates the course of his love for her and his premonition of her death.

The pre-Raphaelite artist and poet Dante Gabriel Rossetti (1828–1882) created a series of drawings and paintings based on his own translation of Dante's

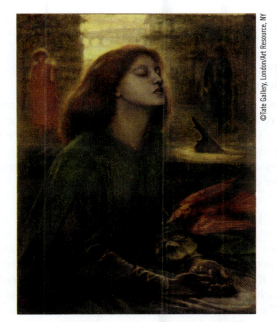

FIGURE 11.65 ■ *Beata Beatrix,* Dante Gabriel Rossetti, 1872

FIGURE 11.66 ■ Map of the Spheres of Heaven

Vita Nuova. He used the likeness of his own wife, Elizabeth Siddal, for Beatrice. The painting by Rossetti that you see in Figure 11.65 depicts Beatrice's death as a trance-like state, with the possibility of a future awakening.

Paradiso, the third book of the *Divine Comedy,* tells of the journey that Dante and Beatrice take from Earth to the Empyrean heaven, visiting each of the nine spheres of heaven along the way. Figure 11.66 shows the Earth and the nine spheres, each with its own kind of angel and keyed to the particular canto in *Paradiso.*

Dante's *Paradiso* has been illustrated by different artists over the years, including paintings by Giovanni di Paolo, created about 1445, and etchings by Gustav Doré, drawn in the 1900s. Another famous set of illustrations for the *Divine Comedy,* mostly unfinished sketches for paintings, was created by Botticelli in the 1400s. In Figure 11.67, Dante and Beatrice approach Sphere 8 of the fixed

FIGURE 11.67 ■ *Seven Planets,* Giovanni di Paolo, 1445

stars and look back at the planets they have just traversed. Their journey is described in *Paradiso*, Canto XXII,

> "And all the seven were shown to me, and I thought:
> How swift they are in moving and how great,
> And each one from the other how remote!"

Pope-Hennessey points out some of the significant details in Giovanni di Paolo's painting: "The figure in the center is Leda holding her twins, Castor and Pollux, the Constellation of Gemini. Below left is the sun, a scarlet figure in a flaming chariot, and on the right the moon, Mars, Mercury, Jupiter, Venus, and Saturn. The figures are not arranged in the sequence in which they appear in the *Paradiso*, but in the order of the days of the week."[24]

The *Immacolata*

A popular celestial art theme in medieval times was the Virgin standing on the moon, usually wearing a crown of stars, as was Urania in Figure 11.54. Called the *Immacolata*, it comes from Revelation 12:1. "And a great portent appeared in heaven, a woman clothed with the sun, with the moon under her feet and on her head a crown of stars." So, like the sun, she is centered in the cosmos, three planets above, three below, head in the stars, and feet on the moon. The moon, in depictions of the Madonna, symbolizes *chastity*. It was popular with Tiepolo, and especially with Dürer. See Figure 11.68. See also Tiepolo's *Immaculate Conception*, painted in 1767.

In Figure 11.69, we have another *Immacolata*. This one was painted in 1612 by Cigoli.[25] Look carefully at the moon. It is pockmarked. This may not seem important now; however, at the time, it defied convention and church doctrine that showed the moon either as a crescent or as a smooth orb, believed to be as perfect and flawless as the person standing upon it.

Why was Cigoli's moon pockmarked? Panofsky explains: "The painter, as a good and loyal friend [to Galileo] paid tribute to the great scientist by representing the moon under the virgin's feet exactly as it had revealed itself to Galileo's telescope—complete with . . . those little . . . craters which did so much to prove that the celestial bodies did not essentially differ . . . from our earth."[26] Galileo's drawings of the moon are shown in Figure 11.70.

FIGURE 11.68 ■ *Virgin on the Crescent,* Dürer, 1499

FIGURE 11.69 ■ *Pauline Chapel Dome* in S. Maria Maggiore, Rome, Cigoli, 1612

©Fotographia Vasari

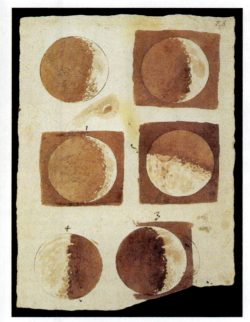

FIGURE 11.70 ■ Galileo's Moon Drawings

GALILEO GALILEI (1564–1642)

Galileo Galilei, Italian mathematician, astronomer, and physicist, was born in Pisa. He was the son of Vincenzo Galilei, a musician. He was educated at a monastery and, in 1581, entered the University of Pisa to study medicine, but he later withdrew because of lack of funds.

While in the Pisa cathedral, Galileo supposedly noted that a swinging lamp always took the same amount of time to complete an oscillation—no matter how large the swing. He verified this observation later in life and suggested that the pendulum might be used to regulate clocks. Galileo's papers on the hydrostatic balance (1586) and the center of gravity in solids (1589) brought him recognition and the position of mathematics lecturer at the University of Pisa. Later he became chair of mathematics at the University of Padua, where he stayed for 18 years and performed most of his best work. There he proved that the path of an object thrown into the air is parabolic, and that falling bodies accelerate uniformly. That he dropped weights from the leaning tower of Pisa is probably a legend.

FIGURE 11.71 ■ Statue of Galileo

In 1609 Galileo learned of the recent invention of the telescope, and soon he built his own. He was the first to use a telescope to study the skies and soon made several important discoveries: The surface of the Moon is not smooth, as had been supposed; the Milky Way system is composed of a collection of stars; and Jupiter has satellites. He saw spots on the sun and observed the phases of Venus.

Galileo left Padua in 1610 to become "philosopher and mathematician" to the Grand Duke of Tuscany. In 1613, he wrote that Copernicus was right: The planets revolve around the sun, rather than the Earth being the center of the universe. This started years of conflict with the Catholic Church, and it eventually led to a trial in 1633 in which Galileo was convicted of having held and taught Copernicus' doctrine. Ordered to recant, Galileo said that he "abjured, cursed, and detested" his past errors. He was sentenced to house arrest at his estate at Arcetri near Florence, where he remained for the last eight years of his life. Figure 11.71 shows the statue of Galileo that stands outside the Uffizi Gallery in Florence.

FIGURE 11.72 ■ Fountain of Neptune, P. della Signoria, Florence

FIGURE 11.73 ■ Zodiac Pavement from the Floor of the Synagogue Bet Alfa, Third to Fourth Century B.C.

Stars and the Zodiac

We've seen something of the roles that the planets and Earth's moon have played in the art of the past. Now we'll examine a few pieces of art in which stars figure prominently, especially the signs of the zodiac. Also see Appendix E, under the heading "Nine."

The paddle wheels in the Fountain of Neptune, shown in Figure 11.72, are carved with the signs of the zodiac. The mosaic pavement in Figure 11.73 shows the Greek sun-god Helios surrounded by the twelve signs of the zodiac, each with its Hebrew name. Four winged women, one in each corner, represent the four seasons.

Chartres Cathedral, which we discussed in Chapter 7, "Circular Designs in Architecture," shows the strong influence of astrology. In addition to zodiac signs over a doorway and in the zodiac window shown in Figure 11.74, it has the towers of the sun and moon and a sundial that we saw in Figure 11.36. A gap in Chartres' south-facing rose window allows a shaft of white sunlight to shine on a brass marker on the floor on St. John's day in midsummer.

The painting by Tintoretto that you see in Figure 11.75 shows a rather literal explanation for the *Origin of the Milky Way*. Hercules, son of Jupiter,

FIGURE 11.74 ■ Zodiac Window, Chartres

FIGURE 11.75 ■ *Origin of the Milky Way*, Tintoretto, c. 1577

©Digital image, The Museum of Modern Art/Licensed by Scala/Art Resource, NY

FIGURE 11.76 ■ *Starry Night,* Van Gogh, 1889

cannot be immortal because his birth was the result of Jupiter's affair with a mortal. To fix this, Jupiter has his incredibly tolerant wife Juno nurse Hercules. Some milk spurts upward from her breast to form the stars of the Milky Way.

The star-filled sky apparently caught the imagination of Edvard Munch, who painted two works he titled *Starry Night.* So, of course, did Vincent Van Gogh, who made perhaps the most famous painting with the same title. It is shown here in Figure 11.76. Note the spirals. Stars are also prominent in Van Gogh's *Cafe Terrace at Night.*

SUMMARY

Building on the earlier chapters about the circle, we began this chapter with the geometry of the sphere. We then surveyed domes and the terrestrial and celestial spheres. We examined some ways in which these spheres have been portrayed, in globes, armillaries, orreries, and maps. Because the motions of the celestial spheres give us our concept of time, we briefly explored the passage of time, some of the instruments that are used to measure it, and the passage of time in art. Finally, we surveyed the sphere and the cosmos as used by artists—from the symbolism of the sphere, to the medieval spheres of heaven, to portrayals of astronomy and astronomers, and to celestial art in the twentieth century.

Here we studied the geometrization of celestial space; in the next chapter, on perspective, we will see the geometrization of terrestrial space.

At the end of the final Canto of the *Divine Comedy,* Dante saw a "lofty light appearing like three circles of three colors . . . and one seemed reflected by the other, as rainbow by rainbow." In the very last paragraph, at the end of his fantastic journey down to hell and back and through purgatory and up through the spheres of heaven, Dante doesn't talk about Beatrice or God. He talks about *geometry.*

> "As the geometrician, who endeavors
> To *square the circle,*[27] and discovers not,
> By taking thought, the principle he wants,

Even such was I at that new apparition;
I wished to see how the image to the circle
Conformed itself, and how it there finds place;

But my own wings were not enough for this,
Had it not been that then my mind there smote
A flash of lightning, wherein came its wish.

Here vigor failed the lofty fantasy:
But now was turning my desire and will,
Even as a wheel that equally is moved,

The Love which moves the sun and the other stars."

So we end our journey to the heavens with love and with geometry. What more could anyone who is fond of art and mathematics want?

EXERCISES AND PROJECTS

1. Define or describe the following terms:

dome	cupola	pendentive
drum (of a dome)	lantern (of a dome)	empyrean
zodiac	Fortuna	melancholia
vanitas	primum mobile	

2. *Lénárt Sphere*. The Lénárt sphere is a plastic globe with accessories used for exploring geometry on a spherical surface. If a Lénárt sphere and study guide are available, use them to do some simple explorations. Summarize your findings in a paper.

3. *Team Project, the Analemma*. Find a spot that will not be disturbed for a few months, erect a gnomon (an upright stick), and plot a noon mark each day. Connect the marks to trace a portion of an analemma. You may have to remove the gnomon to prevent it from being disturbed, but arrange to replace it each day in the same spot, using a carpenter's level to set it vertical. A mark on a south-facing window can also serve as a "gnomon," casting a shadow on the floor of your room. Take photos as the work progresses and write a report when you are finished.

4. *Analemma*. Find a table giving the equation of time and another giving the declination of the sun.[28] Graph the equation of time against the sun's declination, giving each day of the year a point on the graph. What is the shape of your graph? Write a few paragraphs describing and explaining your work.

5. *Map Projections*. Research and write a paper on map projections. Find samples of each type of map, and make a display for your class. Explain the advantages and shortcomings of each.

6. *Team Project, an East–West Line*. On a flat spot on campus, use a gnomon, string, and the sun's shadow to lay out an east–west line. Then use string to erect a perpendicular to that line to obtain a true north–south line. Compare this direction to a compass reading. Do they differ? If they do, why?

7. *East–West Line*. In Ken Follett's *The Pillars of the Earth*, a builder named Tom lays an east–west line. "In the ground at the east end he had planted an iron spike with a small loop in its top like the eye of a needle. . . ."

Then at sunrise, "the red disk of the sun showed over the top of the wall. Tom shifted his position so that he could see the edge of the sun through the small loop in the spike at the far end . . . [and] held the second spike in front of him so that it blocked his view of the sun . . . [and] pressed its pointed end into the damp earth, always keeping it precisely between his eye and the sun. . . . The sun shone into his eye through the two loops. The two spikes lay on a perfect east-west line . . . [that] would provide orientation for the new cathedral."[29] Write a short paper in which you evaluate the validity of this method.

8. *Models.* Make a model of one or more of the following instruments:

   - An equatorial sundial[30]
   - A horizontal sundial
   - An orrery
   - An astrolabe[31]
   - A perpetual calendar
   - A nocturnal
   - A tower of the winds
   - A sun calendar
   - A geodesic dome
   - An armillary sphere

   Write a paper explaining how you constructed your model and how it operates. An example of a student project, an armillary sphere, appears in Figure 11.77.

FIGURE 11.77 ■ Student's Armillary Sphere

9. Reproduce Kepler's model of the universe.

10. Create a painting with a spherical or celestial theme.

11. Make a graphic design with a spherical or celestial motif. Use it for wallpaper, wrapping paper, a greeting card, a screensaver, a tiling, or a mosaic.

12. Make a fabric design with celestial subject matter. Use it for a pillow, a tee shirt, a quilt, a dress, or a scarf.

13. Make an illustrated "Book of Hours" for your own life. Outline your own "Labors of the Twelve Months" and include anything you feel is appropriate. Study the *Très Riches Heures* for ideas. See http://humanities.uchicago.edu/images/heures/heures.html.

14. Make a sculpture using elements that are spherical, or make one that is based on the armillary sphere.

15. Find a website that gives the current appearance of the night sky. Find out what is happening celestially at the moment and verify by actual observation of the night sky. Summarize your findings in a short paper and/or an oral presentation to your class. If possible, point out your findings to classmates at night.

16. Research a method for locating a star in the night sky, at a particular hour of the night, and at your own latitude and longitude. Make a sighting device from wood and cardboard using two drafting protractors for angular scales. For an example, see Figure 11.78. Summarize your findings in a paper, an oral presentation, or a live demonstration.

FIGURE 11.78 ■ A Home-Made Sighting Device

17. Here are some suggestions for short papers or term papers:
    - Dava Sobel's book *Galileo's Daughter*
    - The longitude problem (see Dava Sobel's *Longitude* in this chapter's "Sources" section for a possible reference)
    - The Royal Observatory at Greenwich, England
    - The Map room at Villa Farnesina
    - The *Kalendar and Kompost of Shepherds* (1493), which tells how each of the planets supposedly affects our behavior
    - Eratosthenes' measurement of the diameter of the Earth about 2000 years ago
    - The art and music of Hildegarde von Bingen
    - The medieval structuring of the heavens
    - Brunelleschi's amazing feat of erecting a dome over the Florence Duomo
    - The history of the globe
    - The life of Gerhardt Mercator
    - The celestial and geographical works of Ptolemy
    - Albrecht Dürer's extensive instructions for constructing sundials[32]
    - The history of the dome
    - Dante's *Divine Comedy*
    - The concept of *vanitas* in Ecclesiastes, a book from the Old Testament
    - The Maharajah Jai Singh's interest in astronomy and the several observatories he built in India
    - The Maharishi Vedic Observatory on the campus of the Maharishi University of Management in Fairfield, Iowa
    - The nine medieval spheres of heaven (See Appendix E under the heading "Nine.")

18. Some of the themes in this chapter have also been used in music, such as Vivaldi's *Four Seasons* and Gustav Holst's *The Planets*. Can you find other musical works pertaining to any of this chapter's themes?

19. The *Quattro Stagioni*, or Four Seasons, is a popular pizza in Italy. Each quadrant of the pizza has a different topping. Make a Quattro Stagioni and share it with your classmates.

Mathematical Challenges

20. *Angle Between Any Two Points on the Earth.* The angle D (measured at the Earth's center) between two points on the Earth's surface can be found from this equation:

$$\cos D = \sin L_1 \sin L_2 + \cos L_1 \cos L_2 \cos(M_1 - M_2),$$

where L_1 and M_1 are the latitude and longitude, respectively, of one point, and L_2 and M_2 are the latitude and longitude of the other point. Find the angle between Pittsburgh (latitude 41.0° N, longitude 80.0° W) and Houston (latitude 29.8° N, longitude 95.3° W).

21. *Distance Between any Two Points on the Earth's Surface.* Using the results of Exercise 20, apply Equation 62 to find the distance in miles between Pittsburgh and Houston. Check your answer by taking measurements on a terrestrial globe.

22. Find the distance between two ships in the Pacific Ocean if one has a latitude of 15.8° N and a longitude of 153.6° E, and the other is at latitude of 32.7° N and at longitude of 175.3° E. Compare your results with measurements on a terrestrial globe.

23. If you have access to a global positioning system (GPS), mark a waypoint at some starting position. Take a trip and mark a second waypoint at your destination. Using the latitude and longitude of each waypoint given by the GPS, calculate the distance between the two waypoints. Compare your results with the distance given by the GPS and with that given by a road map.

24. *Archimedes' Method of Equilibrium for the Volume of a Sphere.*[33] Refer to Figure 11.79 as you work through this challenge. On a horizontal line L, draw a hemisphere of radius r. Draw PE perpendicular to L. Circumscribe a rectangle $OABC$ about the hemisphere. Extend OE and BC to D to form right triangle ODC.

    By rotating the hemisphere, rectangle, and triangle about L, form three solids of rotation: the sphere, cylinder, and cone. At a distance x from O, take a vertical slice through the three solids. Make the thickness δ of the slice so small that each slice can be considered a cylinder.

    a. Show that the volumes of the slices are as follows: The sphere is $\pi x(2r - x)\delta$; the cylinder is $\pi x^2\delta$; and the cone is $\pi x^2\delta$. Place a point T on line L at a distance $2r$ from O. Remove the slice of sphere and the slice of cone and hang them from T.

    b. Given that the *moment of a volume* about a point O is the product of the volume and the distance from its centroid to O, show that the moment about O of the slices of sphere and cone combined, after being hung from T, is equal to $4\pi r^2 x\delta$.

    c. Show that the moment about O of the slice of cylinder, unmoved from its original position, is equal to $\pi r^2 x\delta$.

    Note that the moment of sphere and cone combined is four times the moment of the cylinder. Continue to make thin slices. Hang each slice of sphere and cone from T; leave each slice of cylinder where it is. When you're finished, the entire volume of the sphere and cone will be hanging from T, and the entire cylinder will be just where it originally was. The moment of (sphere + cone) is then four times that of the cylinder:

    Moment about O of sphere and cone = 4 × Moment about O of cylinder

    (Volume of sphere + Volume of cone)$2r$ = 4(Volume of cylinder)r **(1)**

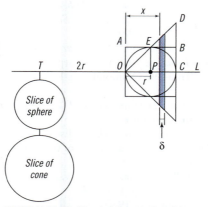

FIGURE 11.79 ■ Archimedes' Computation of the Volume of a Sphere

      d. Use Equation 108 (see Appendix F) for the volume of a cone, to show that its volume here is $8\pi r^3/3$.

      e. Substitute the volume of the cone into (1) and solve for the volume of the sphere.

25. Using Statement 116 for the surface area of a spherical zone, derive the formula for the area of the entire sphere.

26. Using Statement 116 for the surface area of a spherical zone, show that Statement 117 is true.

117 SURFACE AREA OF SPHERICAL ZONE OF ONE BASE

The surface area of a spherical zone of one base is equal to that of a circle with a radius that is the chord of the generating arc.　　　　　　　　　　　　　　　　　●

27. Using Statement 119 for the volume of a spherical sector, derive Equation 118 for the volume of a spherical segment.

28. Show that the volume of the spherical segment of Exercise 27 is equal to the sum of the volumes of a sphere of radius *h*/2 and two cylinders of altitude *h*/2, and having radii of *a* and of *b*.

29. The two formulas for the volume of a sector are given in Equations 119a and 119b. Verify that they are equivalent.

30. Make an oral presentation to your class on the project you have chosen.

SOURCES

Adzema, Robert, et al. *The Great Sundial Cutout Book*. New York: Hawthorn, 1978.

Burckhardt, Titus. *Chartres and the Birth of the Cathedral*. Ipswitch, UK: Golgonooza, 1995.

Calter, Paul. *Technical Mathematics*. New York: Wiley, 2007.

Cornford, Francis. *Plato's Cosmology*. New York: Harcourt, 1937.

Coxeter, H. S. M., et al. *Geometry Revisited*. Washington, D.C.: Mathematical Association of America, 1967.

Coxeter, H. S. M. *Introduction to Geometry*, 2nd ed. New York: Wiley, 1989.

Dante. *The Divine Comedy*. Trans. H. F. Cary, Ill. Gustave Doré. London: Cassell, 1885.

Edgerton, Samuel. *The Heritage of Giotto's Geometry*. Ithaca, New York: Cornell University Press, 1991.

Escher, M. C. *Escher on Escher, Exploring the Infinite*. New York: Abrams, 1989.

Eves, Howard. *An Introduction to the History of Mathematics*. New York: Holt, 1953.

Heilbron, J. L. *The Sun in the Church, Cathedrals as Solar Observatories*. Cambridge, MA: Harvard University Press, 1999.

Lehmann, Karl. "The Dome of Heaven." *Art Bulletin*, Vol. XXVII, 1945, pp. 1–27.

Panofsky, Erwin. *Studies in Iconology*. New York: Harper, 1939.

Plato. *Timaeus and Critias*. Trans. by Desmond Lee. London: Penguin, 1965. Original c. 360 B.C.

Pope-Hennessy, John. *Paradiso. The Illuminations of Dante's Divine Comedy by Giovanni di Paolo*. New York: Random House, 1993.

Pseudo Dionysius. *Complete Works*. New York: Paulist, 1987.

Rosenbusch, Robert. "The Pantheon as an Image of the Universe." *Nexus Network Journal*, Vol. 6, No. 1 (Spring 2004).

Saad-Cook, Janet. "Natural Phenomena, Earth, Sky, and Connections to Astronomy." *Leonardo*, Vol. 21, No. 2, 1988, pp. 123–134.

Smith, Baldwin. *The Dome*. Princeton, 1950.

Sobel, Dava. *Galileo's Daughter*. New York: Walker, 1999.

Sobel, Dava. *Longitude: The True Story of a Lone Genius Who Solved the Greatest Scientific Problem of His Time*. New York: Walker, 1995.

Tillyard, E. M. W. *The Elizabethan World Picture*. New York: Vintage.

Vitruvius. *The Ten Books on Architecture*. New York: Dover, 1960.

Waugh, Albert E. *Sundials, Their Theory and Construction*. New York: Dover, 1973.

Wilkins, Peter. "The Pantheon as a Globe-Shaped Conception." *Nexus Network Journal*, Vol. 6, No. 1 (Spring 2004).

NOTES

1. Book XI, Definition 14.
2. Tompkins, p. 200.
3. Dio's *Roman History*, with an English translation by Earnest Cary, LIII, 27, Hist. LIII, 27.
4. Wittkower, p. 9.
5. Wittkower, p. 10.
6. Lehmann, p. 1.
7. Lehmann, p. 26.
8. Bruce, P. A., *A Short History of the University of Virginia*. New York, 1920.
9. Tompkins, p. 203.
10. Vitruvius, p. 116.
11. Burckhardt, p. 15.
12. Fisher, p. 19.
13. *Timaeus*, Sec. 36.
14. Lee's translation of the *Timaeus*, p. 35.
15. Euterpe, II, para. 109.
16. Book IX, Chapter VIII.
17. From Charles Dickens's description of Bologna in *Pictures from Italy*.
18. Panofsky, p. 225.
19. Calvino, Italo. *Six Memos for the Next Millenium*. Cambridge, MA: Harvard University Press, 1998, p. 52.
20. Edgerton, *The Heritage of Giotto's Geometry*, p. 196.
21. From *Escher on Escher*.
22. Tillyard, p. 39.
23. *Wisdom of Solomon*, 11, 20. In that same book, Chapter 7, verses 18 and 19, Solomon is said to have knowledge of the alternations of the solstices, the changes in the seasons, the cycles of the year, and the constellations of the stars.
24. Pope-Hennessey, p. 144.
25. Ostrow, Steven. "Cigoli's *Immacolata*." *Art Bulletin*, June 1996, p. 218. See also "The Heritage of Giottos Geometry."
26. *Galileo as a Critic of the Arts*. E. Panofsky. Isis, XLVII, 1956, pp. 3–4. Quoted in Ostrow's article, p. 222.
27. This is the Longfellow translation. Other translations say "measure the circle." I mentioned this passage in Chapter 8.
28. Look in a book on sundial construction, such as Waugh's *Sundials: Their Theory and Construction*, pp. 205 and 206.
29. At the end of Chapter 5, p. 339.
30. Robert Adzema gives many patterns for different sundials in *The Great Sundial Cutout Book*.
31. Cardboard kits for the sundial, astrolabe, perpetual calendar, and nocturnal are available from Paul MacAlister and Associates, Box 157, Lake Bluff, IL 60044.
32. Dürer, *The Painter's Manual*, pp. 247–257.
33. Eves, p. 319.

"Perspective is the rein and rudder of painting."

LEONARDO DA VINCI

FIGURE 12.1 ▪ *Putti Engaged in the Study of Geometry and Perspective* (detail), Illustration by Samuel Wale, c. 1761

12

Brunelleschi's Peepshow and the Origins of Perspective

Let's turn our attention to the early Renaissance and a survey of perspective. In Chapter 11, "The Sphere and Celestial Themes in Art and Architecture," we studied the structuring of celestial space; here we'll examine the structuring of *terrestrial space*. First, you'll learn what *perspective* is, including one-point or *linear perspective*, *two-point perspective*, and finally *three-point perspective*. We will then see that the entire theory of perspective can be developed from the fact that the apparent size of an object decreases as its distance from the viewer's eye increases. We explored the concepts of *projection* in Chapter 11, and *section* in Chapter 9, where we intersected a plane with a cone to get the curves called the conic sections. In this chapter, you'll see how a picture drawn in perspective can be an example of projection and section. You will then learn how to draw pictures in one-point and two-point perspective.

The developments that led up to the Renaissance invention of perspective go all the way back to Euclid, and they include geometry, geography, optics, mirrors, surveying, and the influence of commercial developments in Florence. We will explain the development of perspective with Brunelleschi's experiment,

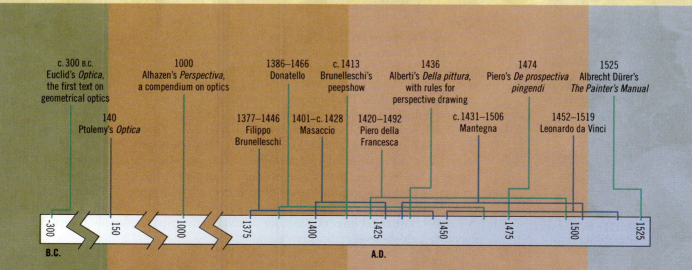

c. 300 B.C.
Euclid's *Optica*,
the first text on
geometrical optics

1000
Alhazen's *Perspectiva*,
a compendium on optics

1386–1466
Donatello

c. 1413
Brunelleschi's
peepshow

1436
Alberti's *Della pittura*,
with rules for
perspective drawing

1474
Piero's *De prospectiva
pingendi*

1525
Albrecht Dürer's
The Painter's Manual

140
Ptolemy's *Optica*

1377–1446
Filippo
Brunelleschi

1401–c. 1428
Masaccio

1420–1492
Piero della
Francesca

c. 1431–1506
Mantegna

1452–1519
Leonardo da Vinci

-300
B.C.

150

1000

1375

1400

1425

1450

1475

1500

1525
A.D.

called his *peepshow*. Closely tied to this is Alberti's *Treatise on Painting*, containing the first written exposition of perspective theory in the early Renaissance. In Chapter 4, "Ad Quadratum and the Sacred Cut," we mentioned four reasons to frame a picture: to separate the world of the picture from the surrounding world, to enable the artist to control the composition, to make the picture portable, and to serve as a *window*. In Alberti's treatise, he describes a painting as a window through which we view the world, expecting to see an accurate portrayal of what we would see from a real window. It's a brief step from this to an examination of how three early Renaissance artists used perspective to enhance the composition of their paintings, and we'll see how the use of perspective enhanced the status of artists. Our coverage would not be complete without a look at the contributions that Piero della Francesca, Leonardo da Vinci, and Albrecht Dürer made to the theory and use of perspective. We'll close with a quick survey of what has happened to perspective since the Renaissance, and we'll present a few examples of perspective in more recent works—until the use of perspective began its decline in the twentieth century.

WHAT IS PERSPECTIVE?

FIGURE 12.2 ■ A Cube, Viewed Straight-On

Linear perspective in art is a system for representing objects located in three-dimensional space on a flat surface. It takes into account how the distance of such objects from the viewer will affect their appearance on the flat surface, called the *picture plane*. We say "linear" perspective to distinguish this geometric system from aerial or atmospheric perspective, which considers change in color and sharpness with distance.

We will use a series of views of a cube to help explain perspective. First, let's look at a cube straight-on. Of course, it looks like a square, as you can see in Figure 12.2.

Now, let's keep the cube facing toward us, but move our viewpoint so that a side is visible, as you see in Figure 12.3. The illustration doesn't look right. It looks like two squares side by side. Intuitively, we know that the side of the cube should appear shorter than the front, which directly faces the viewer. Any object that is not perpendicular to the viewer's line of sight will appear shorter than it actually is. To make the drawing appear more realistic to the viewer, let's shorten the side, as in Figure 12.4. This technique is called *foreshortening*.

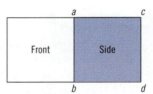

FIGURE 12.3 ■ The Cube from a Different Viewpoint

However, the drawing still doesn't look right. Again, we know from experience that things appear smaller the farther away they are. Therefore, edge *cd*, being farther away, should appear shorter than edge *ab*. Take a look at Figure 12.5. That's better!

Lines *ac* and *bd* are called *orthogonals*. They are the horizontal lines that are perpendicular to the viewing plane, which is also called the *picture plane*. If you extend the orthogonals, they intersect at a point *VP*, which is called the *vanishing point*. Lines *ae* and *bg* are horizontal lines that are parallel to the picture plane. They are called *transversals*. They do *not* meet at a vanishing point. This figure is an example of a drawing in *one-point perspective*.

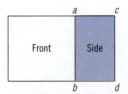

FIGURE 12.4 ■ The Cube with the Side Foreshortened

Now, let's slide the cube vertically so that its lower edge is at eye level and again extend the orthogonals to locate the vanishing point. The *VP* is now at the same height as the lower edge of the cube. Next, lower the cube so that its upper edge is at eye level, and the *VP* is even with the upper edge. Therefore, the vanishing point for orthogonals is at the observer's eye level. Figure 12.6 gives us an example of each. In fact, you will soon see that the vanishing point for any pair of parallel horizontal lines that are not parallel to the picture plane is always at the observer's eye level.

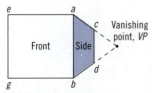

FIGURE 12.5 ■ The Cube with a Vanishing Point

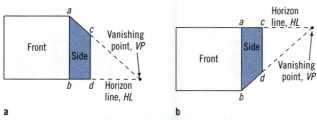

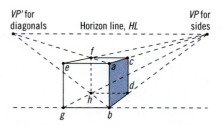

a

FIGURE 12.6 ■ (a) The Cube with the Vanishing Point Even with the Bottom;
(b) The Cube with the Vanishing Point Even with the Top

FIGURE 12.7 ■ The Cube with the Upper and Lower
Edges Receding to the Same Vanishing Point

A horizontal plane drawn through the location of the observer's eye is called the *observer's horizon*. From the previous observations of the cube, you can see that the vanishing point for horizontal lines is always on the observer's horizon. We will denote the horizon line as *HL*. Next, let's *lower* the cube so that its upper edge is *below* the viewer's horizon. The cube's upper face *acfe* now becomes visible, as shown in Figure 12.7. Line *ef* is also an orthogonal, and when it is extended it must pass through the same vanishing point as the other orthogonals.

You have seen that the orthogonals, which are parallel to each other, recede to a vanishing point on the observer's horizon. In fact, any set of horizontal, parallel lines will recede to a vanishing point on the observer's horizon. The diagonals *af* and *bh*, being parallel, will converge to a vanishing point, which is labeled *VP'*. The foreshortened distance *bd* will appear right only if the viewer is at the correct *viewing distance* from the picture plane. A little later, you will see that if *bdhg* is a *square*, the viewing distance is equal to the distance between the vanishing points *VP* and *VP'*.

Let's now turn the cube so that no face is parallel to the picture plane, while keeping edges *ab*, *cd*, and *eg* vertical, as in Figure 12.8. The top and bottom edges, extended, converge to *two vanishing points, VR and VL*, hence the name *two-point perspective*. Because the top and bottom edges are horizontal, the two vanishing points lie on the horizon line *HL*.

Note that the box in Figure 12.7 is *not* drawn in two-point perspective, even though it has two vanishing points. In Figure 12.7, the second vanishing point *VP'* is for the *diagonals* of the boxes; in Figure 12.8, it is for the orthogonals of one of the *faces*.

Finally, let's *tilt* the cube so we have no horizontals or verticals. The edges of the three faces converge to *three* vanishing points, and we now have *three-point perspective*, as shown in Figure 12.9. Because the top and bottom edges of the box are *not* horizontal, their vanishing points are not on the horizon line.

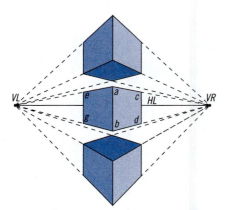

FIGURE 12.8 ■ Three Boxes in Two-Point Perspective

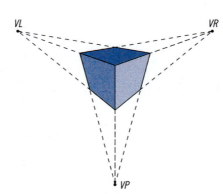

FIGURE 12.9 ■ A Box in Three-Point Perspective

A PERSPECTIVE PRIMER

THE INVENTION OF DRAWING
According to legend, the Corinthian maiden Dibutade invented drawing by tracing the silhouette of her lover projected onto a wall by the light of a single flame. Figure 12.10 portrays this moment.

FIGURE 12.10 ▪ *The Invention of Drawing*, Suvée, c. 1791

As you learned in the introduction to this chapter, the entire theory of perspective can be developed from the fact that the apparent size of an object decreases as its distance from the viewer increases. This is the phenomenon that makes railroad tracks appear to converge in the distance.

Suppose you view an object, say a box, through a window, and trace the outline of the box on the glass without changing the position of your eye. You would get a perspective picture of the box on the glass, as in Figure 12.11. In this illustration, the viewer's eye *E* and the box are separated by a picture plane. Everything beyond the picture plane is called the *object*, even though there may be many objects or no clearly defined object. A line from the eye *E* to a point on the object is called a sight line. The entire bundle of sight lines is called the *visual pyramid* or *visual cone*. The point where a sight line to a point *Q* on the object pierces the picture plane is called a *projection of Q* onto the picture plane. The sight line from *E* perpendicular to the picture plane is called the *centric ray*. It pierces the picture plane at point *VP*, which is called the *principal vanishing point*. The perpendicular distance from eye to picture plane is called the *viewing distance*. Note that the perspective picture, including the vanishing point *VP*, lies entirely on the picture plane, although in this figure the vanishing point may appear to lie behind the picture plane.

Let's see how to construct a perspective picture and develop some rules of perspective. Using the demonstrations in Figures 12.2 through 12.9 of the preceding section, we can immediately derive the following statements regarding two-point perspective.

120

Straight lines on an object appear as straight lines on the picture plane.   ●

121

Vertical lines on an object appear as vertical lines on the picture plane.   ●

122

Horizontal lines on an object that are parallel to the picture plane appear as horizontal lines on the picture plane.   ●

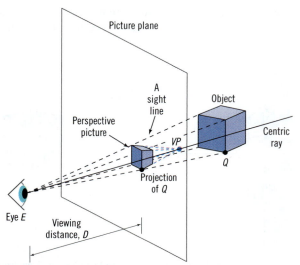

FIGURE 12.11 ▪ Viewer, Picture Plane, and Object

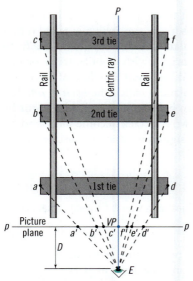

FIGURE 12.12 ■ Plan View of a Railroad Track

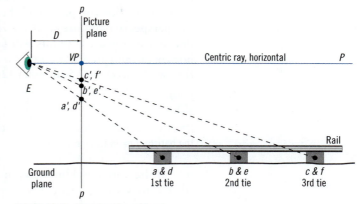

FIGURE 12.13 ■ Elevation View of Track

You can make more observations about one-point perspective by studying a railroad track. Figure 12.12 shows the *plan view* (the view from above) of a railroad track with two horizontal rails, *abc* and *def*, and three horizontal ties, *ad*, *be*, and *cf*. Draw a picture plane, *pp*, shown edgewise in the figure, at right angles to the rails. The rails, being perpendicular to the picture plane, are the orthogonals. Similarly, the ties, and any other lines parallel to the picture plane, are the transversals.

Then place an observer's eye at *E* at a viewing distance *D* from the picture plane, and draw orthogonal *EP*, the centric ray. Label the intersection of the centric ray with the picture plane *VP*. A ray from point *a* to *E* cuts the picture plane at *a'*, so *a'* is a projection of point *a* onto the picture plane. Similarly, *d* has a projection *d'* on the picture plane. Therefore, railroad tie *ad* will have an apparent length *a' d'* at the picture plane. Similarly, ties *be* and *cf* have apparent lengths *b' e'* and *c' f'*.

The farther a tie is from the observer, the smaller its projection on the picture plane. If you were looking only at the picture plane and not at the actual track, the more distant ties would appear to be smaller. As the ties shrink in apparent length, the rails *appear to converge*. This, of course, agrees with ordinary observation. As you continue down the track, the ties *appear to vanish*; that is, the lengths of the ties eventually reduce to zero. Note that the lines of sight to the ends of each subsequent tie become closer to being parallel to the centric ray, and each projection lies closer and closer to point *VP*. The projections from both ends of a tie located at an infinite distance from the observer would be at *VP*. It is no surprise that *VP* is called a *vanishing point*. The vanishing point appears *only* on the picture plane, *not* on the object.

Receding parallel lines on an object will meet at a vanishing point somewhere in the picture plane. ●

Let's look at an *elevation view* of the track, as shown in Figure 12.13. The track is shown lying in a horizontal plane called the *ground plane*. Its intersection with the picture plane is called the *ground line GL*. (The ground line *GL* and the horizon line *HL* are viewed end-on here, so they are not visible.)

The centric ray *EP*, being perpendicular to the picture plane, is horizontal. It lies in the horizontal plane through the observer's eye, which is called the *horizon plane*. (The horizon plane and ground plane are defined *geometrically*; they do not usually correspond with the actual horizon with its hills, trees, or buildings, or the actual ground with its bumps and hollows.) The line of intersection between the horizon plane and the picture plane is called the *horizon line HL* of the perspective picture. It is viewed end-on here so is not visible. Sight lines drawn from the eye to points ever farther down the track become more and more horizontal, and points infinitely far will have sight lines that *are* horizontal. From this, we can conclude Statement 124.

124

Receding horizontal parallel lines on an object will meet at a vanishing point somewhere on the horizon line in the picture plane.   ●

If the horizontal lines are, in addition, perpendicular to the picture plane (that is, they are orthogonals), then we can conclude Statement 125.

125

Receding orthogonals on an object will meet at the principal vanishing point *VP* on the horizon line in the picture plane.   ●

Finally, let's *combine* the plan view and the elevation view, both drawn to the *same scale*, into a single perspective view of the tracks, as shown in Figure 12.14. For this construction, the distance *D* from the eye *E* to picture plane *pp* must be the same in both plan view and elevation view.

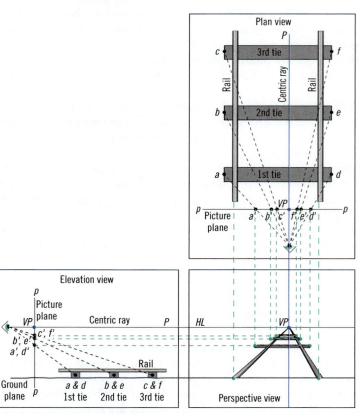

FIGURE 12.14 ■ Perspective Drawing of the Rails and Ties

Place the plan view above the perspective drawing and the elevation view to the side. For clarity, the three views are shown as if they were drawn on separate sheets of paper. In practice, they would be on a single sheet. From the plan, project lines down to the perspective drawing to show the horizontal position of the projection of each point *on the picture plane* (not on the object). From the elevation view, project horizontal lines that give the vertical position of the projection of each point on the picture plane. The intersection of the vertical and horizontal projections of a particular point gives the location of that point on the perspective view. The projections of the vanishing point *VP* from each view intersect at the vanishing point in the perspective view. Through *VP*, draw the horizon line *HL*. Notice that in the perspective view, not only do the ties appear shorter as they recede, but the *horizontal distance* between them appears to diminish as well. As mentioned earlier, this is called foreshortening, and it is difficult for artists to portray convincingly.

You have just seen that orthogonals on an object will appear to meet at the principal vanishing point *VP* on the horizon line in the picture plane. Suppose that the rails are *not* perpendicular to the picture plane—that is, they are *not* orthogonals. Ordinary experience and a construction similar to the previous one show that they also appear to intersect at a vanishing point, but not the same one as for orthogonals.

Figure 12.15 shows a plan view of rails that are not parallel to the centric ray. Lines of sight to points on the rails, at ever greater distances, approach a line that is parallel to the tracks. The rails appear to converge at *VP'* where a line from the eye, drawn parallel to the tracks, cuts the picture plane. Therefore, we have Statement 126.

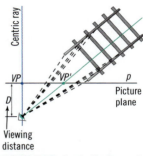

FIGURE 12.15 ■ Tracks Not Perpendicular to Picture Plane

The vanishing point for parallel lines on the object is at the same angle from the eye as the angles that the parallels make with the picture plane.   ●

If the rails are *horizontal*, their vanishing point will be on the horizon line. If the rails are tilted up, their *VP* will be above the horizon; if the rails are tilted down, then the *VP* will be below the horizon.

Figure 12.16 shows a plan view of two squares, with sides parallel and perpendicular to the picture plane, and their diagonals. By Statement 125, the sides that are perpendicular to the picture plane will appear to meet at the vanishing point *VP* on the picture plane. However, by Statement 126, the *diagonals* of the squares will appear to meet at *VP'*, where *EV'* is a line parallel to the diagonals. However, because the diagonals are at 45° to the picture plane, the distance between the vanishing points must equal the viewing distance *D*. Therefore, we have Statement 127. Recall that we made such an assertion in reference to Figure 12.5.

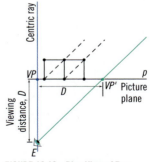

FIGURE 12.16 ■ Plan View of Two Squares

The vanishing point *VP'* for the diagonal of a square with sides that are parallel and/or perpendicular to the picture plane is at a distance from the principal vanishing point *VP* equal to the viewing distance *D*.   ●

We have made the perspective drawing shown in Figure 12.14 using projections from plan and elevation views of the track, and developed some rules of perspective that can help artists make convincing freehand perspective drawings. There are various other methods for making an accurate perspective drawing from plan and elevation views of a building, which is sometimes called *mechanical perspective*. The method we will now show is perhaps one of the clearest.

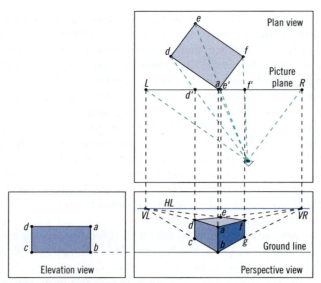

FIGURE 12.17 ▪ A Box Drawn in Perspective

As with our drawing of the railroad tracks, the three views are presented as if they were drawn on separate sheets of paper. Place the elevation view to one side of the space selected for the perspective view, as in Figure 12.17. Extend the base *cb* of the box to form the ground line. Choose a suitable height for the eye above the ground line, and draw the horizon line *HL* parallel to the ground line. A low horizon line—a *bug's-eye perspective*—can produce dramatic effects, as you will see in the paintings of Mantegna. In this example, we will use a high horizon line, from which the top of the object is visible, giving what is called *bird's-eye perspective*.

Place the plan above the perspective view. As before, both views of the object must be drawn to the same scale. Draw the picture plane on edge, parallel to the ground line. The picture plane is usually drawn so that it *touches* the leading edge of the object, so that the leading edge will appear in true length. For a drawing in one-point perspective, place the face of the box along or parallel to the picture plane. For two-point perspective, *tilt* the plan, as shown in Figure 12.17, with an angle of 30° to 45° usually giving good results.

Select an eye point (sometimes called the *station point*) on the plan. When selecting its location, we will get a better perspective view if we (a) try to keep the object within a 30° field of vision, (b) have the centric ray aimed more or less toward the center of the object, (c) avoid lines of sight that coincide with the sides of the box, and (d) place the eye point so that it is not too close to the picture plane. If the viewing distance is too short, the vanishing points *VL* and *VR* will be too close together, and the near corner of the box, angle *cbg*, will look like an acute angle rather than a right angle.

On the plan, draw a sight line through the eye point parallel to side *af* of the box to where it intersects the picture plane at *R*. Drop a perpendicular from *R* to the horizon line to locate the right vanishing point *VR*. A similar construction will locate *VL*.

To draw the leading edge *ab* of the box, drop a perpendicular from this edge in the plan to the ground line. If this edge touches the picture plane, as shown, it will appear in true length in the perspective view. It is sometimes called a *measuring line*. Transfer its length from the elevation view. Draw lines from ends *a* and *b* of the leading edge to each vanishing point.

To locate vertical *cd* in the perspective view, draw a sight line from the eye point to corner *d* in plan. From the projection *d'* where this sight line pierces the picture plane, drop a perpendicular to the perspective view. Its intersection with the lines receding from *a* and *b* to *VL* gives the height *cd* of this vertical. Vertical *gf* is found in a similar way. Point *e* is located at the intersection of lines drawn from *d* to *VR* and from *f* to *VL*.

■ EXERCISES ● A PERSPECTIVE PRIMER

1. Define or describe the following terms:

foreshortening	orthogonal	vanishing point
linear perspective	picture plane	viewing distance
visual cone	projection	centric ray
transversal	horizon plane	bird's-eye perspective
observer's horizon	two-point perspective	three-point perspective
bug's-eye perspective		

2. Figure 12.18 shows a box drawn in one-point perspective. Using the same vanishing point, add at least three more boxes of any size, inside the circle. Choose any depth for each box. Include one box in bird's-eye perspective, one that has bug's-eye perspective, and one box above the horizon. For convenience, use an enlarged photocopy of this figure or download it from our website.

3. Figure 12.19 shows a box drawn in two-point perspective. Using the same vanishing points, add at least three more boxes of any size, inside the circle. Choose any depth for each box. As before, show one box in bird's-eye perspective, one in bug's-eye perspective, and one above the horizon.

4. Figure 12.20 shows a box drawn in three-point perspective. Using the same vanishing points, add at least three more boxes of any size, inside the circle. Choose any depth for each box. Again, show one box in bird's-eye perspective, one in bug's-eye perspective, and one above the horizon.

5. View a scene through a window. Make a perspective drawing by tracing the image of what you see onto the glass using a china marker or felt pen. To save your sketch, draw it on a clear plastic sheet taped to the window. Be careful not to move your eye position while tracing.

6. Make a plan view and an elevation view of a box. Then, using the method shown in Figure 12.17, draw a perspective view of the box in one-point perspective.

7. Repeat Exercise 6, but draw the box in two-point perspective.

8. Perform Exercises 6 and 7 using The Geometer's Sketchpad or other CAD software. Then drag the eye point and note how the perspective view changes. Summarize your findings in a paragraph or two.

9. Using only a straightedge and a pencil, make a perspective drawing of an architectural scene.

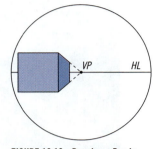

FIGURE 12.18 ■ Drawing a Box in One-Point Perspective

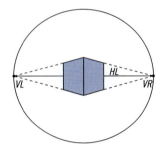

FIGURE 12.19 ■ Drawing a Box in Two-Point Perspective

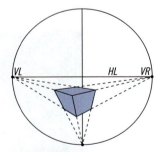

FIGURE 12.20 ■ Drawing a Box in Three-Point Perspective

PAINTINGS BEFORE PERSPECTIVE

The playful drawing you see in Figure 12.21 is by William Hogarth, the English painter and engraver and is the frontispiece for a book on perspective. In it, Hogarth intentionally disregarded the rules for illustrative purposes. A thorough understanding of the principles of perspective will help you avoid making those same mistakes accidentally.

For a picture in one-point perspective, parallel lines, when extended, meet at a vanishing point behind the object and on the observer's horizon. Four mistakes in a perspective drawing are especially common: (1) having parallel lines *that do*

FIGURE 12.21 ■ *Perspectival Absurdities*, Hogarth, 1754

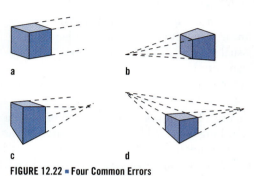

FIGURE 12.22 ■ Four Common Errors
(a) No convergence; (b) Vanishing point in front;
(c) No common vanishing point; (d) Tilted horizon

not converge at all, (2) placing the vanishing point in *front* of the object, (3) having sets of parallel lines converge to *more than one* vanishing point, and (4) failing to place the vanishing point for different sets of horizontal parallels on the same horizontal line. Figure 12.22 illustrates examples of each of these mistakes.

These four deviations from "correct" perspective can be found mostly in paintings made before the development of perspective, and occasionally afterward. Here are some other examples. Feel free to make photocopies of these pictures and draw perspective lines on them to test for the four common errors. Figure 12.23 provides a good starting place. In this example, the parallel lines do not converge. In Figure 12.24, parallel lines converge in front of the picture, rather than behind it. Figure 12.25, a fresco from the upper church of St. Francis of

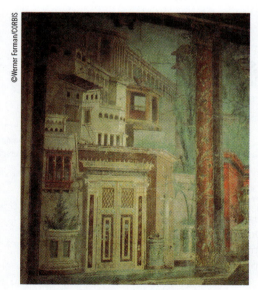

FIGURE 12.23 ■ Architectural Composition, Boscoreale

FIGURE 12.24 ■ Artist's rendering of the Initial Word Panel of Psalm 114, from the Kaufmann Haggadah

FIGURE 12.25 ■ *St. Francis Renouncing the World,* Giotto

FIGURE 12.26 ■ Ten-Dollar Bill, Series 1995

Assisi, is even a little more complex in that there is *no common vanishing point* for all the parallels. The 1995-series United States ten-dollar bill in Figure 12.26 shows two-point perspective, but the orthogonals on the right *do not converge* to a single point. Not all modern pictures show an accurate use of perspective.

SETTING THE STAGE

As with most discoveries, perspective theory did not emerge from a vacuum. The underlying ideas had been accumulating for centuries. Recall Vitruvius' *Ten Books on Architecture*, which appeared about 25 B.C. Vitruvius wrote, "Perspective is the method of sketching a front with the sides withdrawing into the background, the lines all meeting in the center of a circle."[1] Unfortunately, he did not elaborate on that idea. Elsewhere in his book, Vitruvius' reference to Greek and Roman stage design implied an understanding of the vanishing point.

Perspective is an example of the geometric operations of projection and section, where projection lines from an object to the eye are sectioned, or cut, by a picture plane. This has roots in the conic sections, where projection lines from a circle to a point form a cone, which can then be sectioned by a plane to give a circle, an ellipse, a parabola, or a hyperbola, depending on the angle of the cutting plane. We introduced the conic sections in Chapter 9, "The Ellipse and the Spiral" (see Figure 9.4).

Ptolemy apparently knew about perspective, but he applied it only to maps. In his *Geographia*, which was written around A.D. 140, he applied the principles of geometric optics to the projection of the spherical surface of the Earth onto a flat surface to produce a map. Refer back to Figures 11.26 and 11.27. He is said to have made the first known linear perspective construction for drawing a map of the world. Figure 12.27 is taken from *Geographia*.

Although the main application of perspective is in art, it is an *optical* phenomenon and, therefore, has its principal roots in geometrical optics. Refer back to Figures 11.22, 11.23, and 11.24. Many works on optics existed before the development of perspective in the fifteenth century, and there is evidence that those works were in collections available to Brunelleschi and Alberti, the pioneers in perspective.

Euclid's *Optica*, c. 300 B.C., was the first text on geometrical optics. The terms *visual ray* and *visual cone* are defined in it. Ptolemy's *Optica*, c. A.D. 140, was another early text on geometrical optics, and it included theories on refraction. The *centric ray* is defined by Ptolemy as the ray that is not refracted when passing from one material—air, glass, water—to another. Galen's *De usu partium*,

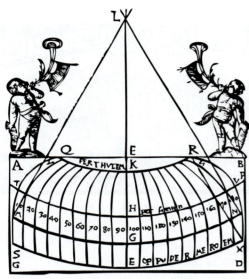

FIGURE 12.27 ■ Figure from Ptolemy's *Geographia*

FIGURE 12.28 ■ Self-Portrait in a Convex Mirror, Parmigianino, 1524

c. A.D. 175, contained an early but erroneous description of how the eye creates images. The book was still important, however, as a stepping stone in the development of the theory of perspective. The Islamic mathematician, astronomer, and physicist Alhazen wrote *Perspectiva*, an important compendium on optics, around A.D. 1000. It integrated the works of Euclid, Ptolemy, and Galen. Roger Bacon's *opus Majus*, c. 1260, also included a section on optics. John Pecham's *Perspectiva communis*, c. 1270, was another treatise on optics that was widely available during the Renaissance. Finally, Blasius of Parma's *Quaestiones perspectivae*, c. 1390, was a popular adaptation of the works of Bacon and Pecham.

An interest in mirrors was another contributing factor in the development of perspective theory. The flat, lead-backed mirror was introduced in the thirteenth century, and apparently it fascinated artists and writers as much as it did those interested in optics. Dante made several references to mirrors. Giotto is said to have painted "with the aid of mirrors." Alberti recommended looking at a painting in a mirror to expose its weaknesses. In his *Della pittura* he wrote, "A good judge for you to know is the mirror. I do not know why painted things have so much grace in the mirror. It is marvelous how every weakness in a painting is so manifestly deformed in the mirror."[2] Parmigianino was another artist entranced by mirrors. He was so pleased with his image in a barber's convex mirror that he had a carpenter turn a wooden sphere and cut off a segment on which he painted the self-portrait shown in Figure 12.28.

The mirror, possibly the easier-to-make convex type, was supposedly a standard piece of furniture in the studios of late medieval painters. The increased use of mirrors by artists generated interest in geometrical optics and provided a way to see a real scene, like an actual landscape, on a flat plane. The convergence of parallel lines to a vanishing point when seen in a real-life physical scene is easily ignored because it is so familiar. However, when seen on the unfamiliar flat surface of a mirror, parallel convergence is less likely to go unnoticed. This may have caused artists to look for the same phenomenon in the real world.

There are obvious similarities between perspective and indirect measurement of objects using surveying techniques. Both involve lines of sight from an object to the eye, usually cutting a plane in which the instrument is located, as in

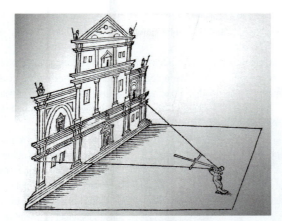

FIGURE 12.29 ■ Measurements using Two Rods and a Stretched Cord, from Francesco di Giorgio Martini, *Trattato di Architettura Ingegneria e Arte Militare,* c. 1478

the *cross-staff* (also called a *radio astronomico* or, as Leonardo referred to it, a *bacolo of Euclid*) shown in Figure 12.29. Here the plane of the instrument would be analogous to the picture plane in the perspective construction.

BRUNELLESCHI AND ALBERTI

Filippo Brunelleschi and Leon Battista Alberti are credited with formalizing and codifying the concept of perspective in the Renaissance, when perspective came into regular use. Many of the following discussions of artists are accompanied by observations made by the Italian art historian Giorgio Vasari, a respected painter himself. In the 1500s, he wrote an encyclopedia of biographies of major artists called *Le Vite delle più eccellenti pittori, scultori, ed architettori*, or simply *Le Vite*.

Brunelleschi's Peepshow

The sculptor, architect, and engineer Brunelleschi is credited with the development of perspective. Giorgio Vasari wrote glowingly of Brunelleschi, "Nature has created many men who are small and insignificant in appearance but who are endowed with spirits so full of greatness and hearts of such boundless courage that they have no peace until they undertake difficult and almost impossible tasks and bring them to completion. . . . This can be clearly seen in Filippo di Ser Brunellesco. . . ."[3]

Let's look more closely at the experiment through which Brunelleschi presented the concept of perspective. Arnheim has dubbed it *Brunelleschi's peepshow*. Imagine that it is sometime during the early fifteenth century in the Piazza del Duomo in Florence, and you see Brunelleschi standing in the west door of the unfinished cathedral. Brunelleschi has recently lost the commission for the north doors and not yet started on the dome. He beckons to you and asks you to face the octagonal baptistery across the piazza. Look again at Figure 5.66. He holds up a picture of the baptistery painted on a panel; its unpainted back is toward you and blocks your view of the actual baptistery. He has you squint through a small hole in the painting. According to Brunelleschi's biographer Antonio Manetti, on the painted side the hole would be about the size of a lentil, and on

FILIPPO BRUNELLESCHI (1377–1446)

After training as a goldsmith and sculptor, Brunelleschi competed with Lorenzo Ghiberti and others for the famous commission to make the bronze reliefs for the door of the Baptistery of Florence. Ghiberti won, and Brunelleschi's extreme disappointment might be the reason he switched from sculpture to architecture.

His first major architectural work was the *Ospedale degli Innocenti,* the Foundling Hospital, which contained the relief sculptures of children shown in Chapter 6, "The Circle." He also designed the Basilica of San Lorenzo and its sacristy, which is known as the "old sacristy" to distinguish it from Michelangelo's "new" sixteenth-century sacristy in the same church (also called the Medici Chapel). He gave the sacristy the shape of a cube topped by a hemispherical dome. Brunelleschi also designed the Pazzi Chapel adjacent to the Church of Santa

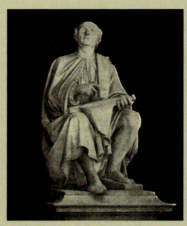

FIGURE 12.30 ■ Statue of Brunelleschi

Croce (for which we saw examples of roundels in Chapter 6); Santa Maria degli Angeli and the Church of Santo Spirito, both in Florence; and military architecture in various cities. He died in Florence and is buried in the cathedral there.

Figure 12.30 shows a statue of Brunelleschi looking up at the famous dome he built for the Duomo in Florence, which is shown in Figure 11.13 and is his major work. The cathedral was begun in 1296. By 1418 construction had reached the stage at which the enormous technical problems of constructing a vault above the huge octagonal drum had to be solved. Brunelleschi devised a successful method to construct the dome without the usual wooden centering, and he invented the machinery necessary to carry it out. In the statue he holds dividers, the symbols of the architect and geometer.

FIGURE 12.31 ■ The Peepshow

the back of the panel the hole would open in somewhat of a pyramid shape and be about the size of a coin at its widest point. Through the hole you see a mirror that reflects the painting itself; in the mirror you see the front of the painting. To help visualize this, take a look at Figure 12.31. Brunelleschi whisks away the mirror so that you see the real baptistery through the peephole. You are amazed because the real baptistery and the painting are so similar. However, this is no ordinary painting. It is said to be the world's first accurate perspective picture.

In reality, the small panel, which was about one-half *braccio* on a side, is now lost. (A braccio is a Florentine unit of measure equal to one-third the height of a person, or about 23 in.) Some writers say it was painted on a mirror, with the sky portion left unpainted, so that it reflected the actual sky and clouds. One writer even speculated that Brunelleschi painted the baptistery directly on top of its mirror image. Others say it was a wood panel with burnished silver to reflect the sky. Either way, a mirror would have enhanced the effect.

Michael Kubovy believes that Brunelleschi, instead of just looking directly at the picture, used the peephole and mirror to create a more dramatic illusion of depth.[4] Further, the reduced aperture increased the depth of field. As you may know, depth of field is the range of distances within which objects are in sharp focus, a phenomenon familiar to photographers. Thus, distinguishing the nearby panel from the actual far building was made more difficult for someone viewing Brunelleschi's panel because of the increased depth of field and reduced depth perception. Also, viewing through a peephole blocked extraneous information that would provide clues that the viewer was looking at a picture and not the real scene.

The hole was drilled in the panel right at the vanishing point of the painting, so that the perspectival illusion had its greatest effect. Later artists found that a good illusion of depth could still be created when the viewer's eye was *not* at the vanishing point. However, this so-called *robustness of perspective* was probably not yet known to Brunelleschi. Finally, the peephole forced the viewer to use just *one* eye, thereby reducing the stereoscopic effect obtained with binocular vision. The corresponding reduction in depth perception made it harder for the viewer to tell the painting from the actual building. Another possible reason for the use of the peephole and mirror was pure showmanship. Brunelleschi's second perspective painting of the Palace of the Signori (the Palazzo Vecchio) in Florence did not involve a peephole and mirror.

Alberti's Many Contributions

Because Brunelleschi left no written record of his perspective findings, it remained for Alberti to be the first to put theory into writing. According to Vasari, "Leon Battista Alberti, having studied the Latin language and having practiced architecture, perspective, and painting, left behind books written so well that although countless modern artisans have proved more excellent than him in practice, they have been unable to equal him in writing about their craft."[5] Alberti made another significant contribution: He is said to have invented the *camera obscura*, such as the one shown in Figure 12.32. The camera obscura is a box with a dark interior and a small hole or lens centered on one face through which light rays enter and form an image on the opposite face of the box. Any camera, including the camera obscura, automatically makes perspective pictures, so it is hard to imagine that this did not influence his development of perspective theory.

FIGURE 12.32 ■ A Camera Obscura

Look again at the profile of Alberti that appears in Chapter 1, "Music of the Spheres." In Florence, Alberti's friendships with Donatello and Brunelleschi led him to write his influential book on painting, *Della pittura*. In it the rules for perspective drawing were put into writing for the first time. In addition, Alberti gave practical information for painters and advice on how to paint *istoria*, or history paintings. He said, for example, that a painting should have an *introducer*, someone who looks out of the painting at the viewer and thus connects the viewer to the scene. This function is played by the Madonna in Masaccio's *Trinity*, as you will soon see. He also gave advice on how to set up a studio, how to live, and so forth. We will discuss only the section in *Della pittura* that gives the first written account of linear perspective.

Alberti believed that you can create a correct image of an object as seen through a window by tracing the outline of the object on the window glass, which represents the picture plane.[6] The construction that Alberti actually describes in his *Della pittura* has come to be known as the *costruzione legittima*, or the "legitimate construction." It is presented in Figure 12.33. This construction results in a horizontal grid, like the paving stones in a plaza, sometimes called an *Albertian grid*, which is then used to proportion other items in the picture. In his description, he has the orthogonals and transversals both spaced by 1 braccio. The construction is performed in two parts: the first, sometimes called the *vanishing point operation*, establishes the orthogonals, and the second, the *distance point operation*, locates the transversals.

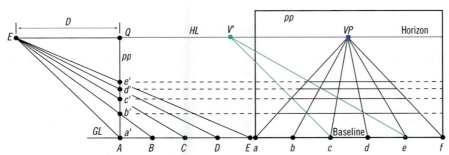

FIGURE 12.33 ■ Alberti's *Costruzione Legittima*

We first perform the vanishing point operation:

1. Draw a rectangular picture plane *pp*.

2. Decide how high a painted person, standing on the base line, is to be. Trisect that height. (Each third is equal to 1 braccio. Note that because dimensions appear to shrink with distance, this scale is true only for objects in the immediate foreground.)

3. Subdivide the base line into braccia, getting as many points *a*, *b*, *c*, *d*, and so on, as will fit.

4. Arbitrarily locate the vanishing point *VP* somewhere on the picture.

5. Draw a horizontal through *VP* to represent the horizon.

6. Connect *VP* to the points *a*, *b*, *c*, and so on, to obtain the orthogonals of the pavement grid.

We next do the distance point operation:

7. Extend the base line to the left and subdivide that portion of the base line into bracci, with points *A*, *B*, *C*, and so on.

8. Through the leftmost point *A*, draw a vertical *pp*. The line *pp* represents the picture plane as seen edgewise

9. Locate eye point *E* on the horizon line, to the left of *pp* by an amount equal to the chosen viewing distance *D*.

10. Draw lines *EA*, *EB*, *EC*, etc., from *E* to the base line. These lines will intersect the vertical at points *a'*, *b'*, *c'*, and so on.

11. Finally, project *a'*, *b'*, *c'*, and so on, to the right to form the transversals.

Therefore, each quadrilateral in the pavement grid is a square measuring 1 braccio on a side, as seen in perspective. Although none of these grid lines may actually appear in the final picture, they are used to give scale to the other items in the picture. It is like having *perspective graph paper*.

To check the accuracy of the construction, Alberti says to draw a number of diagonals through the grid squares. Not only should each diagonal pass exactly through the corners of the grid squares, but if properly drawn they should meet at a common point *V'* on the horizon line, a distance from *VP* equal to the viewing distance. Recall that we earlier established that the vanishing point for the diagonals of a square was at a distance from the principal vanishing point equal to the viewing distance.

THREE RENAISSANCE ARTISTS

Renaissance artists were quick to adopt this concept of perspective in their painting and relief sculpture. Here we examine the works of only three, but these gentlemen were perhaps the most prominent of the time.

Donatello

The earliest surviving use of linear perspective in art is attributed to Donatello, who is considered by many to be the greatest sculptor of the early Renaissance in Italy—and perhaps one of the greatest sculptors of all time. About Donatello, Vasari wrote that ". . . devoting himself to the art of design, he became not only an unusually fine sculptor . . . but also grew experienced in stucco, quite skilled in perspective, and highly esteemed in architecture."[7]

DONATELLO (1386–1466)

Donato di Niccolò di Betto Bardi was born in Florence. Around 1405 he joined the workshop of Lorenzo Ghiberti and worked as an assistant in the modeling of the north doors of the Baptistery. In the competition of 1401, Ghiberti was chosen over Brunelleschi for this commission.

Donatello made a marble statue of David at the Palazzo Vecchio, which was later eclipsed by Michelangelo's huge statue on the same theme. Other early works include a seated marble figure of St. John the Evangelist for the cathedral façade, a wooden crucifix in the church of Santa Croce, statues of the five prophets for the niches of the campanile, and large roundels of the four evangelists in the Pazzi Chapel of Santa Croce. Later, Donatello made a statue of St. Louis of Toulouse for a niche at Orsanmichele; a large-scale free-standing bronze David; sculptural decoration for the Old Sacristy in San Lorenzo; a wooden statue of St. John the Baptist in Santa Maria Gloriosa dei Frari, Venice; and a wooden figure of Mary Magdalen in the Florence Baptistery. He sculpted the bronze group *Judith and Holofernes*, a bronze statue of St. John the Baptist for Siena Cathedral, and twin bronze pulpits for San Lorenzo.

Donatello made two marble statues, *St. Mark* and *St. George*, for exterior niches of Orsanmichele, the church of Florentine guilds. The *St. George* has been replaced by a copy; the original is now in the Bargello. The relief carving below the *St. George* is presented in Figure 12.34. This carving is particularly interesting because Donatello appears to be using linear perspective. Unfortunately, the orthogonals are not defined well enough to let us locate a precise vanishing point. In this panel, which depicts St. George slaying the dragon, Donatello invented a technique called *rilievo schiacciato* (ski-a-chat-o), or *flattened relief*. The technique involved extremely shallow carving throughout, but the forms were not shaped in the usual way. They appear as though they were *painted* with the chisel.

However, there is no doubt about Donatello's use of perspective in his bronze *Feast of Herod*, which is shown in Figure 12.35. This panel, which was made for the octagonal baptismal font of the Siena Cathedral shown in Figure 5.64, is widely recognized as the first use of linear perspective. Although Donatello created several works of bronze for the baptismal font of San Giovanni in Siena, this is his earliest. Donatello also made two statuettes of Virtues and three nude *putti*, or child angels, for this font.

FIGURE 12.34 ■ The Relief Below *St. George*, Donatello, c. 1412

FIGURE 12.35 ■ *Feast of Herod*, Donatello, c. 1425

FIGURE 12.36 ■ *Miracle of the Believing Donkey*

In the *Feast of Herod*, the orthogonals are not very long or prominent, but they appear to intersect at a vanishing point near the elbow of the musician in the central window. Because the orthogonals do not quite meet in a single vanishing point, Donatello's use of linear perspective is imperfect here. However, this imperfection is small and may have occurred during the casting of the bronze rather than in the original design. At any rate, the perspective construction is far more advanced than in his *St. George* relief and is apparently based on theory.

In 1443 Donatello left Florence for Padua, where he made the *Gattamelata*, which was the first large equestrian statue made since antiquity. This statue is the ancestor of all the equestrian monuments erected since. Donatello also created some important works, such as a bronze crucifix and a new high altar, for the Church of Sant'Antonio of Padua. Of most interest to us is the series of relief sculptures he made for the altar; they depict the four miracles of St. Anthony. These sculptures show an even more dramatic use of perspective than his *Feast of Herod*. His bronze panel *Miracle of the Believing Donkey*, for example, uses a vanishing point so low that the floor is not even visible. The effect is dramatic and causes the barrel vaulting to appear to explode forward rather than recede nicely into the background. This relief, which is shown in Figure 12.36, no doubt influenced Mantegna, who was to paint his Ovetari chapel frescos a few years later at the Church of the Eremitani, just a few blocks north of the Church of Sant'Antonio. We will discuss Mantegna a bit later.

Masaccio

MASACCIO
(1401–c. 1428)
Masaccio is considered by many to be one of the big three, along with Brunelleschi and Donatello, who laid the foundations for the art of the Renaissance in Italy. He was born in what is now San Giovanni Valdarno, but there is no record of where he received his artistic training.

Unfortunately, his career lasted only six years; however, in that short time, Masaccio transformed the art of the Renaissance. Although he left no workshop or pupils to carry on his style, he greatly influenced many painters of the Early and High Renaissance.

Vasari observes of a group of great artists, "It is Nature's custom, when she creates a person of great excellence in any profession, to create not just one man alone, but another as well. . . And how true this is we can observe from the fact that Florence produced in the same period Filippo Brunelleschi, Donatello, Lorenzo Ghiberti, Paolo Ucello, and Masaccio, all most excellent artisans. . . ."[8] Of this distinguished group, we now turn to Masaccio, creator of the first known perspective painting.

Masaccio's first known painting—the triptych *Madonna and Child with Saints* (c. 1422), which is now in the Uffizi Gallery in Florence—shows no sign of the use of perspective. However, three other paintings, which were all done in the last four years of his short life, do show perspective and are the works on which Masaccio's fame rests. Of the three, Masaccio's *Trinity*, which was painted for S. Maria Novella in Florence around 1427, is usually considered to be the oldest surviving perspective painting. See Figure 12.37.

Trinity was painted two years later than Donatello's *Feast of Herod*. In it, you can see a pyramid of figures topped by God, who holds the cross, and the Holy Spirit represented by the dove. St. John the Baptist is on the right and the

FIGURE 12.37 ■ *Trinity*, Masaccio, c. 1427

Madonna is on the left. The Madonna looks directly at us. She plays the role of Alberti's introducer. Masaccio is said to have consulted with Brunelleschi on this painting. He constructed a grid framework, tooled right into the surface, which showed rigorous linear perspective. Even the nails are shown in perspective. He placed the vanishing point at the eye level of an average person; he probably believed that this would enhance the illusion of depth. The painting also shows other features of Renaissance art: facial features that are individualized, not idealized, a characteristic of painting in the 1400s; controlled direction of light; strong modeling of light and shade; and great depth and illusion. It exhibits a *classical vocabulary*—that is, it shows architectural details from Greek and Roman structures, such as Corinthian pilasters and barrel vaulting. Note the skeleton below. The inscription reads, "I was once that which you are, and what I am you also will be."[9] The inscription is a *memento mori*, which is a reminder of one's own death. In this case, the inscription is prophetic because Masaccio died within a year of its creation.

In addition to spatial organization, illusion of depth, and a structural focus, perspective can provide *narrative* focus. Because the eye invariably travels to the vanishing point of a picture, Renaissance artists did not hesitate to put something important at or near that point. Let's look at Masaccio's *Tribute Money* (c. 1425), a painting that we first mentioned in Chapter 7, "Circular Designs in Architecture." In this fresco, which is shown in Figure 12.38, Peter is asked by a collector for the tax levied on all Jews.[10] Jesus tells Peter to catch a fish and that he will find the money in its mouth. Note that Peter appears *three times in the same painting:* talking to the Roman tax collector, taking money from the mouth

©Scala/Art Resource, NY

FIGURE 12.38 ◾ *Tribute Money*, Masaccio, c. 1427

FIGURE 12.39 ◾ Bust of Mantegna in
Sant' Andrea, Mantua

of a fish on the shore of Lake Galilee, and finally paying the tax. This painting shows clear and effective use of perspective. It not only provides a visual struc-ture for the painting, but it also provides a *narrative focus* by placing the vanish-ing point at Jesus' head. This painting also shows a believable sky, receding mountains of the Arno valley, consistent lighting, and what is called *atmospheric perspective*, which we will encounter again later.

Mantegna

Regarding Mantegna, Vasari wrote, "Andrea was so kind . . . that his memory will always endure . . . in the entire world. Ariosto lists him among the most illustrious painters of his times."[11] From Mantegna's large body of work, we will look at just two examples of how he used perspective with dramatic effect.

In the 1450s, Mantegna worked on the fresco decoration of the Ovetari Chapel in the Eremitani Church in Padua. In his *St. James before Herod Agrippa*, which is shown in Figure 12.40, the pavement is laid in a strict grid, the *Albert-ian grid* that we have already constructed. The two lowest scenes in the cycle are made extremely dramatic by having a vanishing point *below* the bottom frame of the paintings, as in *St. James Led to His Execution* (see Figure 12.41). This is some-times called *di sotto in su* ("from below to above") perspective, or what we have been calling bug's-eye perspective. Having the viewpoint below the frame exag-gerates the height of the scene and the grandeur of the architecture. This place-ment may have been due to Donatello's influence. You have already seen that Donatello used a similar low viewpoint a few years earlier in his *Miracle of the Believing Donkey*, which is also in Padua. It is also interesting that the vanishing point in *St. James before Herod Agrippa* is located in the frame *between* this panel and the one immediately to its left (not shown) and is shared by that painting. Unfortunately, most of Mantegna's frescoes in the Ovetari Chapel were dam-aged by a bomb during World War II.

Despite his strong independence, Mantegna entered the service of Ludovico Gonzaga in Mantua, in 1459. There he received a fixed income and created what became his best-known surviving work, the so-called *Camera degli Sposi* (Wedding Chamber), or *Camera Picta* (Painted Room), one panel of which is shown in Figure 12.42. In one entire room in the Palazzo Ducale, Mantegna developed a self-consistent illusion of a total environment. That is, the perspectives of all four walls and the ceiling are designed to be perfect when viewed from a single point in the room.

Especially noteworthy is the scene painted on the flat ceiling directly above the center of the room. Looking up, the viewer sees what appears to be an *oculus*,

FIGURE 12.40 ■ *St. James Before Herod Agrippa*, c. 1455

FIGURE 12.41 ■ *St. James Led to His Execution*, Mantegna, c. 1455

a circular opening to the sky. Through this "opening" can be seen *putti* (little angels) and ladies around a balustrade in dramatically foreshortened perspective. The eye point for the oculus is the same point as that for the four walls and the ceiling. Mantegna's oculus was the most influential illusionistic ceiling of the early Renaissance. This illusionism, which depends on the viewer standing at a single point in the room, began a tradition of ceiling decoration that was followed for centuries. The Oculus is shown in Figure 12.43.

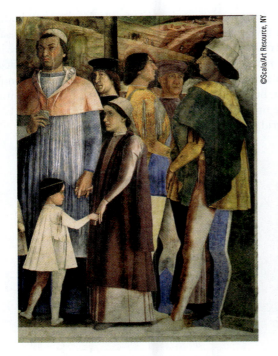

FIGURE 12.42 ■ A Section of the Wall Painting in the Camera degli Sposi (Wedding Chamber), 1474

FIGURE 12.43 ■ The Oculus

LATER DEVELOPMENTS

We have focused on three Early Renaissance artists who used perspective—but of course, there were others, including Lorenzo Ghiberti, Domenico Veneziano, Paolo Ucello, and Jacopo Bellini. Another Italian master working in this period was Piero della Francesca, who wrote about and practiced perspective. Leonardo da Vinci and Albrecht Dürer were also theorists on perspective as well as practitioners of it. This section gives a brief overview of the contributions of these three. We'll follow this overview with some later events in the history of perspective.

Piero, Leonardo, and Dürer

First, let's consider Piero della Francesca, whose profile is given on page 305. According to Vasari, "[Piero] was regarded as an uncommon master of the problems of regular bodies in both arithmetic and geometry . . . who . . . also excelled in painting."[12] In fact, Piero was one of the greatest practitioners of linear perspective, and he even wrote a book on perspective, *De prospectiva pingendi* (On Perspective in Painting), around 1474. This was 39 years after Alberti's *Treatise on Painting* of 1435, and it was the first of Piero's three books. (The others were the *Trattato del abaco* and *De quinque corporibus regularibus*.) The manuscript of *De prospectiva pingendi* was handwritten by Piero himself. He also illuminated it himself with diagrams on various geometric, proportional, and perspectival problems. It is considered an extension of Alberti's work, but it is more explicit. Piero was evidently familiar with Euclid's *Optics*, as well as the *Elements*,[13] to which he often referred.

Did Piero practice what he preached? According to Martin Kemp, "The evidence of his paintings suggests that he did exercise an . . . extraordinary degree of meticulous, time-consuming, geometrical care over the perspectival projection of architectural forms. . . ."[14] Kemp gives Piero's *Flagellation*, which is shown in Figure 4.33, as an example.

Leonardo, in his own distinctive way, also practiced what he preached. He wrote, "Those who are in love with practice without science are like the sailor who boards a ship without rudder or compass, who is never certain of where he is going."[15] Perspective may have been developed by Brunelleschi, and codified by Alberti and Piero, but it was perfected by Leonardo. Leonardo's notes on perspective are apparently lost, but he made great use of perspective in his paintings, such as in the study of the unfinished *Adoration* that is shown in Figure 12.44. Note the strict Albertian grid on the pavement.

©Scala/Art Resource, NY

FIGURE 12.44 ■ *The Adoration of the Magi,* Leonardo, 1481

FIGURE 12.45 ■ *Annunciation,* Leonardo, c. 1472

Leonardo's *Annunciation,* which is presented in Figure 12.45, shows a carefully worked-out perspective framework. Incised lines beneath the paint on this wood panel show his construction. Note, however, that the Virgin's arm appears too long. Studies have shown that Leonardo departed from the correct perspective here for the sake of a more expressive gesture, a common practice in the Renaissance.

The *Last Supper,* which is shown in Figure 12.46, is perhaps Leonardo's most famous perspective painting. Like other paintings depicting Jesus' last meal with his apostles, this one is placed in a refectory or dining hall—actually in the Convent of Santa Maria delle Grazie in Milan. The vanishing point is placed at Jesus' right eye, where he dominates the foreground. Even his arms, which lie along the lines of the visual pyramid, reinforce the perspective.

In his notebooks, Leonardo described three kinds of perspective: size, color, and disappearance (atmospheric perspective). Together, they describe the appearance of distant objects as smaller, less distinct, paler, and bluer. For a good example of this, look for Leonardo's *The Virgin of the Rocks,* which he painted around 1508.

No discussion of perspective is complete without including the work of Albrecht Dürer, whose profile is given on page 136. In contrast to his earlier pictures, Dürer's works showed good perspective after about 1500. According to Panofsky, Dürer was indebted to the Italians for this. He had taken a special trip to Bologna in 1506 to learn the secret theoretical foundation for a process he already knew how to do mechanically. In a letter he wrote in 1506, he said, "I shall ride to Bologna where someone is willing to teach me the secrets of perspective. . . ."[16] The name of Dürer's teacher in Bologna is not known, but Panofsky speculates that it might have been Pacioli or Bramante. An excerpt

FIGURE 12.46 ■ *Last Supper,* Leonardo, c. 1497

LEONARDO DA VINCI (1452–1519)

Leonardo was born in Vinci, and at the age of about 15 was apprenticed to Andrea del Verrocchio, where he learned painting, sculpture, and related skills. During that period he painted details for Verrocchio's *Baptism of Christ*, two Annunciations, the *Madonna with the Carnation*, the *Madonna Benois*, and the *Portrait of Ginevra de' Benci*. He also worked in the workshop of Antonio Pollaiuolo, where he may have been introduced to the study of anatomy. Around 1477 Leonardo left his teacher's workshop to work independently. Surprisingly, in 1482 he left Florence to enter the service of the Duke of Milan, leaving behind unfinished commissions. Some scholars speculate that the sophisticated Neoplatonism that pervaded Florence under the Medici repelled this practical man.

Leonardo spent 17 years in Milan. He maintained an extensive workshop, employing apprentices and students. His paintings during this period include *Lady with an Ermine, The Virgin of the Rocks, Last Supper*, and the ceiling painting of the Sala delle Asse in the Milan castle. He was also consulted in the fields of architecture, fortifications, military matters, and hydraulic and mechanical engineering.

In Milan, Leonardo increased his scientific studies, and he planned treatises on painting, architecture, mechanics, and human anatomy. The studies and sketches he made in notebooks and on individual sheets of paper amounted to thousands of pages. Heavily illustrated with sketches, and all in Leonardo's left-handed mirror writing, only a fraction of these have survived.

After the Duke fell from power in 1499, Leonardo left Milan with Luca Pacioli. In Mantua, the multitalented and never idle Leonardo

FIGURE 12.47 ■ *Self-Portrait, Leonardo, c. 1512*

drew a portrait of Isabella d'Este, which is now in the Louvre. In Venice, he offered advice on how to ward off a threatened Turkish incursion. He then returned to Florence, where he was well received after his long absence. In 1502 Leonardo left Florence again to enter the service of Cesare Borgia as a military architect and engineer. He traveled with Borgia for ten months and surveyed his territories.

Back in Florence in 1503, Leonardo received a prized commission: to create a monumental mural for the Palazzo Vecchio, portraying the *Battle of Anghiari*. During this period, he also painted the *Mona Lisa*, performed dissections, made observations about the flight of birds, and studied the movement of water. In 1506 Leonardo was invited to go for a time to Milan; he stretched that visit to six years. As a result, the *Battle of Anghiari* remained unfinished after three years of work.

Leonardo's duties in Milan were light. He continued to work on the *Virgin and Child with St. Anne* and *Leda*, and he completed a second version of *The Virgin of the Rocks*. He again taught students, including the young nobleman Francesco Melzi, a friend and companion until his death. Leonardo filled notebooks with mechanical, mathematical, anatomical, optical, geological, and botanical studies. During this time, he made the self-portrait shown in Figure 12.47.

In 1513, after the ouster of the French from Milan, Leonardo went to Rome. No commissions came his way, and in 1516 Leonardo, with Francesco Melzi, left Italy forever. He spent his last three years in Cloux, near the King's summer palace, editing his studies and making his descriptions of *The Deluge*.

FIGURE 12.48 ■ *Saint Jerome in His Study*, Dürer, 1514

from Piero's book on perspective appears in Dürer's notebooks, and his development is similar to Piero's, suggesting that his instructor was close to Piero.

Dürer returned from Italy with an understanding of Alberti's *Costruzione legittima*, which he calls *der nähere Weg* (the Shorter Route). Because Dürer had been trained in a late Gothic tradition, the concept of perspective came to him as a revelation, and he seems to have regarded it as a kind of visual magic. *Saint Jerome in His Study*, which is shown in Figure 12.48, is one of his works that displays a strong use of perspective.

In the last section of Book Four of his *Painter's Manual*, Dürer gives instructions for perspective drawing. In it he also includes the two well-known drawings of his perspective apparatus. In Figure 12.49, you can see that Dürer used Alberti's idea of the picture plane as a window in the design of these devices to produce perspective drawings.

Panofsky notes that these devices were known to Alberti, Leonardo, and Bramantino, and that aside from improvements to these devices, Dürer added

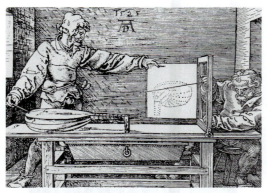

FIGURE 12.49 ■ Perspective Apparatus

FIGURE 12.50 ■ *Paris, A Rainy Day*, Gustave Callebotte

nothing to perspective theory. However, Panofsky goes on to say, Dürer's *Manual* is important because it is the first Northern writing to give a scientific basis to perspective.[17]

Perspective After the Renaissance

Perspective was used extensively in the art of the four centuries following the Renaissance and many treatises were written on the subject. The ideas of projection and section were even expanded into a branch of mathematics called *projective geometry* by Gerard Desargues (1593–1662), a French geometer, architect, and engineer. For a good account of these developments, see Kemp's *The Science of Art*.

Into the nineteenth century, perspective was still widely used in art. Figure 12.50 is an example from that period. In the twentieth century, however, with its cubists, futurists, constructivists, Dadaists, minimalists, and so on, artists no longer considered paintings to be windows. They were pieces of canvas with paint on them. Consider Jackson Pollock's work in Figure 12.51. Then came the notion of *integrity of the picture plane* and the notion of *flatness*. Pictures were so flat that art critics argued over the thickness of paint, and some artists thinned their oils with turpentine to eliminate even that thickness. Perspective was dead, at least for the avant-garde, although illustrators and other artists continued to use it.

FIGURE 12.51 ■ *White Light*, Jackson Pollock

FIGURE 12.52 ▪ *Relativity*, Escher, 1953

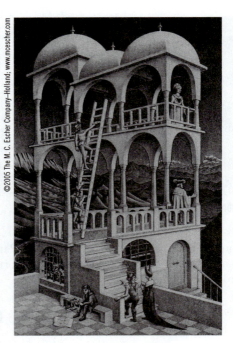

FIGURE 12.53 ▪ *Belvedere,* Escher, 1958

FIGURE 12.54 ▪ Study for *Belvedere*, (detail) Escher, 1958

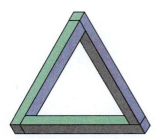

FIGURE 12.55 ▪ The *Penrose Tribar*

M.C. Escher, whose profile appears on page 348, used traditional perspective in his earlier works—but then as his art evolved, he began to manipulate perspective to achieve interesting and often humorous results. Consider Figure 12.52. In this piece, each surface in the structure is simultaneously wall and floor, or wall and ceiling. It is reminiscent of William Hogarth's *Perspectival Absurdities;* look again at Figure 12.21.

In the Escher work *Belvedere*, which is shown in Figure 12.53, the seated man at the bottom of the picture seems to be just one of many subjects with a story to tell. Look closely and you'll see that he holds a cube with an upper surface that is seen from above and with a lower face that is seen from below. In Figure 12.54, you can see that cube up close. A cube such as this is called an *impossible figure*. The *Penrose Tribar*, shown in Figure 12.55, is another impossible figure.

SUMMARY

We began this chapter by describing what perspective is, and then we showed how to make a picture in perspective. Armed with this basic understanding of perspective, we traced its roots in geometry, optics, measurement, and cartography, from the first century to the fifteenth century. Early in the fifteenth century, in Florence, Brunelleschi made the first known perspective picture and Alberti described perspective theory in writing. The perspective construction was then used by Donatello, Masaccio, Mantegna, and others, who produced a flood of perspective paintings and sculptures during the Renaissance. We viewed examples of the use of perspective right up to the twentieth century, and, finally, we learned the reasons for its decline.

The use of perspective enhanced the status of artists, and it was recognized and appreciated by their audiences who understood the mathematics. An artist who used perspective followed a rational, mathematical system and did not place objects within a picture at whim. Many Renaissance Florentines—skilled in business arithmetic; able to compute volumes of barrels, flasks, and piles of grain; and able to perform simple surveying tasks—could recognize and appreciate the appearance of a rational system, even if they did not fully understand its principles.

Much early art was religious; therefore, perspective, as with the other types of geometry we have been studying, was given a religious spin. Some theologians did not hesitate to find relationships between optics and moral doctrine, including attempted correlations between *spiritual seeing* and *visual seeing*. Further, the harmonious organization of visual space provided by perspective was taken by many to symbolize the harmonious organization of heaven.

We have finished our journey through the first, second, and third dimensions; let's now venture between the whole numbers to the *fractional* dimensions.

EXERCISES AND PROJECTS

1. How many perspectival absurdities can you find in Hogarth's drawing in Figure 12.21?

2. What is the difference between *perspective* and *foreshortening*?

3. What is the basic idea upon which all perspective theory is based?

4. Name and discuss two uses for perspective.

5. Who is credited with the invention of perspective? When and where did this occur?

6. Who first described how to make a perspective picture?

7. Describe four common errors in "incorrect" perspective.

8. Make plan and elevation views of a simple house or building. Then use the method shown in Figure 12.17 to draw a two-point perspective view of the house.

9. Reproduce Alberti's *Costruzione legittima* by hand or with a computer drawing program. If you use The Geometer's Sketchpad, drag the eye point, change the viewing distance, and then change the viewing height. Write a few paragraphs describing your findings.

10. Draw a pavement with tiles that are square with the picture plane—that is, the *Albertian grid*. Then add objects to the picture at various distances from the picture plane: cubes, walls, doors and windows, people, etc. Use the grid to estimate heights and widths at different depths.

11. Find a suitable building on campus and reproduce Brunelleschi's peepshow.

12. Make a drawing or painting using perspective.

13. Make a skeleton cube out of wire or sticks. Then use a light source (the smaller the better) to project a shadow of the cube onto a sheet of cardboard, as in Suvée's *Invention of Drawing*. Move the light source and screen, and turn the cube to get different projected shadows. When you find an interesting arrangement, trace the shadow onto the cardboard. Write a few paragraphs describing your findings.

14. Use a camcorder connected to a TV or a digital camera connected to a computer to obtain perspective views of a selected scene. Draw perspective lines on the screen with dry erase markers. Describe your findings.

15. *Camera Obscura.* Make a simple camera obscura. Use it to project a perspective picture onto an interior wall. Describe your camera and your findings in a short paper.

16. *Pinhole Camera.* Find a description of a simple pinhole camera and make one. Use it to photograph or view a scene. Draw perspective lines on your image, and write a summary of your findings.

17. Using glass or clear plastic for the picture plane, make one of Dürer's perspective devices. Use it to make a perspective drawing of some object, marking it on the picture plane with a felt-tipped marker or china marker. Alternatively, use a laser pointer mounted on a tripod to project a laser beam through the plastic to the object, marking the spot where it pierces the picture plane.

18. From an art book, photocopy a few paintings that show the use of one-point perspective. Draw orthogonals and locate the vanishing point. Try to explain why the artist placed it there.

19. Flip through an art history book looking for examples of incorrect perspective, and photocopy the examples. Write a short paper on your findings.

20. Alberti wrote that viewing a painting in a mirror will expose its weaknesses. Try it. Write your assessment of several paintings (your own, paintings in a museum or gallery, from a website or art book). Then view each painting in a mirror and write a new critique. Can you give reasons for any changes of opinion?

21. The British artist David Hockney wrote the book *Secret Knowledge: Rediscovering the Lost Techniques of Old Masters* (London, Thames & Hudson, 2001). In it, he claims that many painters in the past used optical devices such as the lens, mirror, camera obscura, and camera lucida as drawing aids. Research the debate that these controversial theories caused, and summarize it in a short paper. Give your own opinion of Hockney's claims.

22. Find and study a description of Desargues' theorem, Equation 45b, in a book on projective geometry. Write a paper on Desargues' theorem. Relate the theorem to perspective in art.

23. Complete Exercise 22. Then draw the house shown in Figure 12.56 in two-point perspective. Find vanishing points *VR* and *VL* for the sides of the house and vanishing points *P* and *Q* for the slanted edges of the roof. If the house is properly drawn, a vertical line from *Q* will pass through *VR* and *P*. Why is this so? Explain your answer using Desargues' theorem, Equation 45b.

24. Demonstrate Desargue's theorem using The Geometer's Sketchpad.

25. Write a report on any book listed in the sources at the end of this chapter.

26. Using the following suggestions as a starting point, write a short research paper or a term paper relating to the material in this chapter.

  ■ The American artist Thomas Eakins' use of perspective
  ■ The use of mirrors in art
  ■ Illusionistic rooms
  ■ Piero's work on perspective
  ■ Leonardo's contributions to perspective
  ■ Dürer's work on perspective
  ■ Serlio's work on perspective (see Book 2, Chapter 2)
  ■ "Atmospheric" or "aerial" perspective
  ■ The history of geometric optics and its influence on perspective
  ■ The decline of perspective in the twentieth century

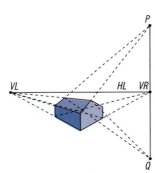

FIGURE 12.56 ■ A House in Two-Point Perspective

Mathematical Challenges

27. Imagine you are drawing a row of telephone poles that are all the same height and are equally spaced. Show that the heights of the poles in the picture plane diminish as a harmonic progression.

28. Make an oral presentation to your class on any of the previous projects.

SOURCES

Alberti, Leon Battista. *On Painting*. A translation of *Della pittura*. New Haven, CT: Yale University Press, 1956.

Cole, Alison. *Perspective*. London: Kindersley, 1992.

Cole, Rex. *Perspective for Artists*. New York: Dover, 1976. First published in 1921.

Dunning, William V. *Changing Images of Pictorial Space*. Syracuse, NY: Syracuse University Press, 1991.

Dürer, Albrecht. *The Painter's Manual*. Trans. by Walter Strauss. New York: Abaris, 1977. Originally printed in 1525.

Edgerton, Samuel. *The Heritage of Giotto's Geometry*. Ithaca, NY: Cornell University Press, 1991.

Edgerton, Samuel. *The Renaissance Rediscovery of Linear Perspective*. New York: Basic Books, 1975.

Forseth, Kevin. *Graphics for Architecture*. New York: Van Nostrand, 1980.

Ivins, William. *Art & Geometry: A Study in Space Intuitions*. Cambridge, MA: Harvard University Press, 1946.

Kemp, Martin. *Leonardo on Painting*. New Haven, CT: Yale University Press, 1989.

Kemp, Martin. *The Science of Art*. New Haven, CT: Yale University Press, 1990.

Kubovy, Michael. *The Psychology of Perspective and Renaissance Art*. Cambridge, UK: Cambridge University Press, 1986.

Newman, James R., ed. *The World of Mathematics*. New York: Simon and Schuster, 1956.

Norling, Ernest. *Perspective Made Easy*. New York: Macmillan, 1939.

Pederson, Mark A. "The Geometry of Piero della Francesca." *The Mathematical Intelligencer*, Vol. 19, No. 3, Summer 1997.

Pozzo, Andrea. *Perspective in Architecture and Painting*. New York: Dover, 1989. Original printed c. 1707.

Sanders, Cathi. *Perspective Drawing with The Geometer's Sketchpad*. Emeryville, CA: Key Press, 1994.

Serlio, Sebastiano. *The Five Books of Architecture*. New York: Dover, 1982. A reprint of the edition of 1611.

Vasari, Giorgio. *The Lives of the Artists*. Oxford, UK: Oxford University Press, 1991. Originally printed in 1550.

Vredman de Vries, Jan. *Perspective*. New York: Dover, 1968. Originally printed c. 1604.

NOTES

1. Vitruvius, p. 14.
2. Alberti, p. 83.
3. Giorgio Vasari (1511–1574) was an Italian painter, architect, and art historian. He is the designer of the famous Uffizi Gallery in Florence. He is best known for his *Lives of the Most Excellent Painters, Sculptors, and Architects*, c. 1550, which has earned him the reputation as the founder of modern art history and criticism.
4. Kubovy, p. 49.
5. Vasari, p. 178.
6. Ivins, p. 78.
7. Vasari, p. 147.
8. Vasari, p. 101.
9. Hartt, p. 211.
10. Matthew 17:24–27.
11. Vasari, p. 249.
12. Vasari, p. 163.
13. Kemp, *The Science of Art*, p. 27.
14. Kemp, *The Science of Art*, p. 30.
15. Kemp, *Leonardo on Painting*, p. 52.
16. Kemp, *The Science of Art*, p. 54.
17. Panofsky, p. 609.

"Why is geometry often described as 'cold' and 'dry'? One reason lies in its inability to describe the shape of a cloud, a mountain, a coastline, or a tree. Clouds are not spheres, mountains are not cones, coastlines are not circles . . . many patterns of Nature are so irregular and fragmented that compared with Euclid . . . Nature exhibits an altogether different level of complexity. . . ."[1]

BENOIT MANDELBROT

FIGURE 13.1 ■ *God the Geometer* (detail), Frontispiece of Bible Moralisée, c. 1220–1250
©Erich Lessing/Art Resource, NY

13

Fractals

This book has been a journey through the dimensions: from zero-dimensional points to one-dimensional lines; to two-dimensional polygons, circular figures, ellipses and spirals; and on to three-dimensional solids and the three-dimensional illusion of perspective pictures. In this, our final chapter, we travel *between* dimensions—say, to the 1.585th dimension, a *fractal* dimension.

Let's begin this final leg of our journey by discussing *dimension* in general and *fractal dimension* in particular; we'll see how that discussion leads us directly to a definition of a *fractal*. We'll see how real landscapes and real objects differ from the geometrically perfect lines, planes, and solids of Euclidean geometry. After we build on our previous work with similarity, we'll explore the concept of *self-similarity*, one of the striking features of a fractal. Next, we'll see how we can create a large number of our own fractals, using several methods, and finally, we'll look at some fractals in art.

Look closely at Figure 13.1. The Earth that God as geometer is creating in this thirteenth-century painting looks startlingly similar to some of the figures

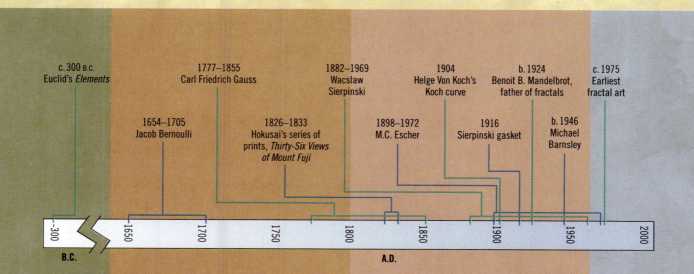

c. 300 B.C.
Euclid's *Elements*

1777–1855
Carl Friedrich Gauss

1882–1969
Wacslaw Sierpinski

1904
Helge Von Koch's Koch curve

b. 1924
Benoit B. Mandelbrot, father of fractals

c. 1975
Earliest fractal art

1654–1705
Jacob Bernoulli

1826–1833
Hokusai's series of prints, *Thirty-Six Views of Mount Fuji*

1898–1972
M.C. Escher

1916
Sierpinski gasket

b. 1946
Michael Barnsley

-300

1650

1700

1750

1800

1850

1900

1950

2000

B.C.

A.D.

developed from mathematical concepts that only came to be recognized in the twentieth century. When translated, the legend (not shown) accompanying this illustration reads, "Here creates God sky, earth, sun, moon, and all elements."

DIMENSION

"Dimension is not easy to understand. At the turn of the century it was one of the major problems in mathematics to determine what dimension means and which properties it has . . . And since then the situation has gotten worse because mathematicians have come up with some ten different notions of dimension: topological dimension, self-similarity dimension, box-counting dimension, capacity dimension, information dimension, Euclidean dimension, and more. . . . The details can be confusing even for a research mathematician."[2]

PEITGEN ET AL.

While exploring fractals, we will confine ourselves to the Euclidean dimension E, the topological dimension D_T, and the fractal dimension D.

Euclidean Dimension E

Euclidean geometry uses the familiar integer dimensions: 0 for a point, 1 for a line, 2 for a surface, and 3 for a solid. In the definitions at the beginning of Book I of the *Elements*, and later in Book XI, Euclid wrote,

"A *point* is that which has no part.
A *line* is a breadthless length.
A *surface* is that which has length and breadth only.
A *solid* is that which has length, breadth, and depth."

In other words, Euclid defines Statement 128.

128 EUCLIDEAN DIMENSION *E*

Points are zero-dimensional. Lines are one-dimensional. Surfaces are two-dimensional. Solids are three-dimensional. ●

Euclidean dimensions have been described in various ways:

- *Boundaries.* The *surfaces* are the boundaries of a solid, *lines* (curved or straight) are the boundaries of a surface, and *points* are the boundaries of lines. Euclid referred to this when he wrote:

   "The extremities of a line are points.
   The extremities of a surface are lines.
   The extremities of a solid are surfaces."

- *The Number of Parameters Needed to Describe the Location of a Point.* One number is needed to give the location of a point on a line, two numbers are needed to locate a point in a plane, and three numbers are needed to locate a point in space.

- *The Degrees of Freedom Available.* On a line, a point can move in one direction; in a plane, it can move in two directions; and in space, it can move in three directions. This refers to mutually perpendicular directions, such as those of coordinate axes. Motions in any direction can be broken down into components in such orthogonal directions. For example, a movement in the northeast direction can be considered as a combination of a northward movement and an eastward movement.

■ *The Figure Obtained When Another Figure Is Cut.* When a line is cut, we get points; when a surface is cut, we get lines; and when a solid is cut, we get surfaces.

Topological Dimension D_T

As we expand our discussion, we will denote the Euclidean dimension by E, the letter used by Mandelbrot.[3]

Identifying the Euclidean dimension of most geometric objects, such as a triangle, is easy. However, identifying the Euclidean dimension of a strange figure such as the *Sierpinski triangle*, a few iterations of which are shown in Figure 13.2, can be tricky. (The circle shown is *not* part of the Sierpinski triangle. We will need it to find the topological dimension in a later example.) To construct this triangle, you just need to follow a few steps. Starting with an equilateral triangle, first bisect each side, forming four smaller equilateral triangles. Next, remove the center triangle, leaving the three outer triangles. Then, simply repeat the steps. This figure is also called the *Sierpinski gasket.* It was introduced in 1916 by the Polish mathematician Wacslaw Sierpinski (1882–1969).

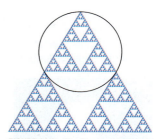

FIGURE 13.2 ■ **The Sierpinski Triangle**

As you can see, this triangle is so riddled with holes that calling it two-dimensional does not seem right. We can use the topological dimensions to describe such figures. The topological dimension D_T is what we usually mean by ordinary dimension or Euclidean dimension. A point has a topological dimension of 0—that is, $D_T = E = 0$. A line has a topological dimension of 1, a surface has a topological dimension of 2, and so forth. Like the Euclidean dimension, the topological dimension is always an integer. Statement 129 offers a more rigorous definition.

129 TOPOLOGICAL DIMENSION D_T

a. A set has topological dimension 0 if every point has an arbitrarily small neighborhood with a boundary that does not intersect the set.

b. A set has topological dimension D_T if each point in the set has arbitrarily small neighborhoods with boundaries that meet the set in a set of dimension $D_T - 1$. ●

For example, Figure 13.3a shows a collection of points in a plane. You can draw a closed curve around any point—if you can make it as small as needed—that does not intersect another point. Therefore, the collection has the topological dimension 0. A collection of points in *space* has a topological dimension 0 because you can draw a hollow sphere around any point that does not intersect another point. Figure 13.3b shows a line in a plane or in space. If you draw a circle (or a sphere, in space) about any point on the line, no matter how small, it intersects the line in points. Therefore, the topological dimension of the line is 1. Figure 13.3c shows a surface. If you draw a circle about any point on the surface, no matter how small, it will intersect the surface in a line. Therefore, the topological dimension of the surface is 2.

■ **EXAMPLE:** Find the topological dimension of the Sierpinski triangle, shown in Figure 13.2.

● **SOLUTION:** You can draw circles, such as the one shown, that intersect the triangle in a set of points, so the topological dimension of the Sierpinski triangle is 1.   ■

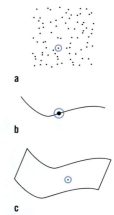

FIGURE 13.3 ■ **Points, Line, and Surface**

BENOIT B. MANDELBROT (b. 1924)
Benoit B. Mandelbrot, shown in Figure 13.4, is a Fellow at IBM's Thomas J. Watson Research Center in New York, a Professor Emeritus of Mathematics at Yale University, and a Battelle Fellow at Pacific Northwest National Laboratory. Because of his work in the field of extreme and unpredictable irregularity in natural phenomena and his landmark book, *The Fractal Geometry of Nature*, Mandelbrot is called the "Father of Fractals."

FIGURE 13.4 ■ Portrait of Professor Mandelbrot

Fractal Dimension *D*

A dimension of 1 for the Sierpinski triangle is certainly not a satisfying concept—what is the dimension of a coastline, a crumpled sheet of paper, or a babbling brook? As Mandelbrot said, "Clouds are not spheres, mountains are not cones, coastlines are not circles . . . many patterns of Nature are . . . irregular and fragmented. . . ." In other words, the world is *very crinkly*, and Euclidean geometry is not adequate to completely describe it. A better way to handle such problems is with Mandelbrot's newly invented *fractal geometry*.

According to Mandelbrot, "Scientists will (I am sure) be surprised and delighted to find that not a few of the shapes they had to call grainy, hydralike, in between, pimply, pocky, ramified, seaweedy, strange, tangled, tortuous, wiggly, wispy, wrinkled, and the like, can henceforth be approached in . . . quantitative fashion."[4] One such quantitative measure is the fractal dimension *D*, a measure of the complexity, crinkliness, or thickness of a fractal. Consider Figure 13.5. As you delve further into this chapter, you will see that the Koch curve construction, which is shown in Figure 13.5a, has a fractal dimension of 1.262 and the Sierpinski triangle, which is shown in Figure 13.5b, has a fractal dimension of 1.585. Intuitively, the Sierpinski triangle *looks* more complex than the Koch snowflake. For one thing, it has more detail inside its boundary than the snowflake has.

Euclidean or topological dimensions are always integers; however, a fractional dimension is not limited to an integer—although it certainly can be an integer. For example, you will see that the fractal dimension of the Sierpinski tetrahedron (shown in Figure 13.11) is exactly 2.

We can derive a formula for the fractal dimension by first examining objects of one, two, and three dimensions: a line, a square, and a cube, as shown in Figure 13.6.

One Dimension (D = 1) (a) Bisecting a line gives two pieces. We say that the scale factor *m* is 2, because the length of the original line is twice that of each smaller segment. If we let *n* be the number of resulting pieces, here *n* = 2. (b) Trisecting a line gives three pieces, each $\frac{1}{3}$ the length of the original. Therefore, the scale factor *m* is 3, and the number of pieces *n* is 3.

Two Dimension (D = 2) (a) Bisecting each side of a square gives four (or 2^2) pieces, each with a side half as long as that of the original. Therefore, *m* = 2 and *n* = 2^2. (b) Trisecting each side of a square gives us nine (or 3^2) pieces, each $\frac{1}{3}$ the length of the original. Therefore, *m* = 3 and *n* = 3^2.

Three Dimension (D = 3) (a) Bisecting each side of a cube gives eight (or 2^3) pieces; therefore, *m* = 2 and *n* = 2^3. (b) Trisecting each side of a cube gives twenty-seven (or 3^3) pieces; therefore, *m* = 3 and *n* = 3^3.

These results are assembled in Table 13.1.

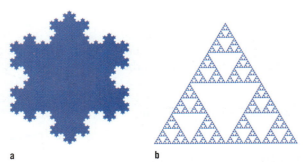

a   b

FIGURE 13.5 ■ (a) The Koch Snowflake; and (b) the Sierpinski Triangle

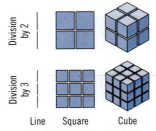

Line   Square   Cube

FIGURE 13.6 ■ Lines, Squares, and Cubes

TABLE 13.1 ■ Finding the Fractal Dimension

Dimension D	Scale Factor m	Number of Pieces n
1 (line)	2	2
1 (line)	3	3
2 (square)	2	$2^2 = 4$
2 (square)	3	$3^2 = 9$
3 (cube)	2	$2^3 = 8$
3 (cube)	3	$3^3 = 27$

The relationship between D, m, and n is

$$n = m^D.$$

Solving this equation for D requires logarithms (see Exercises and Projects at the end of this chapter). The resulting expression is given in Equation 130.

> You might recognize this equation as a power function. It is usually written in the form $y = x^n$, where n is the power to which x is raised to obtain y.

130 FRACTAL DIMENSION D

$$D = \frac{\log n}{\log m}$$ ●

We call D the fractal dimension, but obviously it applies to nonfractals as well.

In Equation 130, log n means "the logarithm of n." Either natural logarithms or common logarithms will work—just be sure to use the same kind of logarithm for both the numerator and the denominator. To find the natural logarithm, simply use the **ln** key on your calculator. For common logarithms, use the **log** key.

■ **EXAMPLE:** The fractal shown in Figure 13.7 is formed by repeatedly trisecting each line segment in the previous stage. Find its fractal dimension.

● **SOLUTION:** Each line is replaced with three others that are one-third the length; therefore, $n = 3$ and $m = 3$. Apply Equation 130:

$$D = \frac{\ln 3}{\ln 3} = 1.$$

Because any quantity divided by itself equals 1, the result is the same when a line is divided into any number of equal parts. ■

As expected for a line segment, we obtained a dimension of 1. Now let's find the fractal dimension of a fractal.

■ **EXAMPLE:** Find the fractal dimension of the Koch curve. The construction is shown in Figure 13.8. To do this, first trisect line segment AB, shown in (a); remove the middle segment, and replace it with two others of the same length, forming an equilateral triangle, as in (b). Then, for each line segment in the new figure, trisect, remove the middle segment, and replace with two others of the same length, as in (c). Repeat. This curve, three iterations of which are shown in (d), was introduced in 1904 by the Swedish mathematician Helge Von Koch.

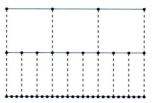

FIGURE 13.7 ■ Repeated Trisections of a Line Segment

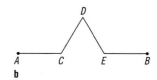

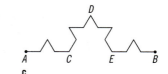

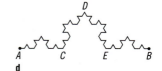

FIGURE 13.8 ■ The Koch Curve

● **SOLUTION:** Each line is replaced with four others that are one-third the length; therefore, $n = 4$ and $m = 3$. Apply Equation 130 using natural logarithms:

$$D = \frac{\ln 4}{\ln 3} \approx \frac{1.386}{1.099} \approx 1.262.$$ ■

Definition of a Fractal

You have just seen that for a straight line, the fractal dimension equals its topological or Euclidean dimension; both equal 1. However, for the Koch curve the fractal dimension (1.262) *exceeds* the topological dimension (1). According to Mandelbrot, that is what makes this figure a fractal.

131 DEFINITION OF A FRACTAL

A fractal is a set for which the fractal dimension exceeds the topological dimension.[5]

$$D > D_T$$ ●

Mandelbrot wrote, "I coined the word *fractal* from the Latin adjective *fractus*. The corresponding Latin verb *frangere* means 'to break': to create irregular fragments. It is therefore sensible . . . that in addition to fragmented . . . fractus should also mean 'irregular.'"[6] The term *fractal* applies both to *natural* fractals, such as ferns, and to *mathematical fractals*, such as those we will generate in this chapter.

■ **EXAMPLE:** Find and compare the topological and fractal dimensions of the Sierpinski triangle that was shown in Figure 13.5b.

● **SOLUTION:** In a previous example, we found the topological dimension of the Sierpinski triangle to be $D_T = 1$. For the fractal dimension, the triangle is composed of three other self-similar pieces, so $n = 3$. The side of each smaller triangle is half the size of the original, so $m = 2$. Apply Equation 130:

$$D = \frac{\log 3}{\log 2} \approx 1.585.$$

The fractal dimension of 1.585 exceeds the topological dimension of 1, which verifies that the Sierpinski triangle is a fractal. You have studied polygons of various sorts, but Mandelbrot would call the Sierpinski triangle a *teragon*. *Tera* is from the Greek—meaning "monster" or "strange creature." ■

The Coastline of Britain

"You know we built planets, do you? . . . Fascinating trade . . . doing coastlines was always my favorite. Used to have endless fun doing the little bits in fjords. . . ."[7]

DOUGLAS ADAMS

Chapter 3, "The Triangle," introduced the Harpedonaptai, or rope stretchers. Recall that in Figure 3.1, the rope stretchers were relocating boundary markers after the flooding of the Nile. As noted in Chapter 3, the very word *geometry* means "to measure the land." Suppose we used Harpedonaptai to measure the coastline of Britain. Imagine them taking their knotted rope, stretching it along the shoreline, and counting knots as they go. How long should their rope be, and how far apart should the knots be spaced? Fifty feet? One foot? One inch? One millimeter? The smaller the unit of measurement used, the larger will be the result. In fact, Mandelbrot caused a stir when he proved that the coastline of England was *infinitely long*.

■ **EXAMPLE:** Find the distance from *A* to *B* in Figure 13.9 by stepping it off with dividers. Assume that the straight-line distance *AB* is 6 in.

● **SOLUTION:** If we open our dividers to a width of 6 in., we get a measurement of 6 in. With an opening of 1 in., we get a measurement of 9.3 in. Smaller openings give the following:

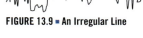

FIGURE 13.9 ■ An Irregular Line

Dividers Opening	Distance *AB*
6	6
1	9.3
$\frac{1}{2}$	10.4
$\frac{1}{4}$	13.8

Smaller openings of the dividers give larger distances, as we pick up more detail in the curve. With a curve as complex as a coastline—where, with smaller dividers setting, we start measuring around stones and grains of sand—the distance between two points will approach infinity. ■

■ EXERCISES ● DIMENSION

1. Define or describe the following:

   Euclidean dimension   topological dimension   fractal dimension

2. Define a fractal.

3. Give the topological dimension for each figure:
   a. The Sierpinski hexagon in Figure 13.10
   b. The Sierpinski tetrahedron in Figure 13.11

4. Find the fractal dimension of each figure:
   a. The Sierpinski triangle in Figure 13.2
   b. The Sierpinski carpet in Figure 13.12
   c. The Koch curve in Figure 13.13
   d. The Sierpinski tetrahedron in Figure 13.11

FIGURE 13.10 ■ The Sierpinski Hexagon

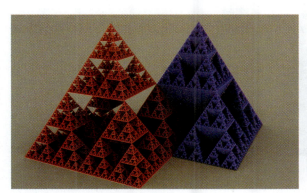

FIGURE 13.11 ■ The Sierpinski Tetrahedron

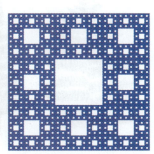

FIGURE 13.12 ■ The Sierpinski Carpet

FIGURE 13.13 ■ The Koch Curve

SELF-SIMILARITY

"So, naturalists observe, a flea
Hath smaller fleas that on him prey;
And these have smaller fleas to bite 'em,
And so proceed ad infinitum . . ."[8]

JONATHAN SWIFT

Jonathan Swift touched on the idea of *self-similarity* over 260 years ago. A figure exhibits self-similarity if the overall shape of the figure is the same as that of a smaller part of that figure, and if the shape of that smaller part is the same as that of an even smaller part, and so on. *Similarity* is one of the most useful ideas in geometry. We already discussed similar triangles and scale factor in Chapter 3, "The Triangle"; similar polygons in Chapter 5, "Polygons, Tilings, and Sacred Geometry"; and similar solids in Chapter 10, "The Solids." We saw that the area of a plane figure increases as the square of the scale factor, and that the volume of a solid increases as the cube of the scale factor.

Self-similarity extends the idea of similarity. It is one characteristic of many natural objects—that is, a piece of the object resembles the whole object. Self-similarity is what causes photographers to put rulers in certain photographs to indicate scale. Without reference points, viewers can not distinguish between a picture of a pebble and a picture of a boulder. To further make the point, consider Figure 13.14. A whole fern frond is shown in (a), one "branch" of the frond is shown in (b), and a branch of the branch is shown in (c). Any one of them looks like an entire fern frond.

Examine the image in Figure 13.15. Scale modelers rely on self-similarity when they use a twig or a weed to represent a tree. In reality, this "tree" is a twig 8-in. high.

Self-similarity in a natural object, such as a fern leaf, breaks down after a while—for example, when we get down to molecules and atoms. However, geometric figures can be self-similar at any magnification. For example, the equiangular or logarithmic spiral we saw in Figure 9.50 was self-similar. Magnifying it did not change its appearance. The Swiss mathematician Jacob Bernoulli was so impressed with the self-similarity of this spiral that he even wrote a treatise on it called the Spira Mirabilis ("wonderful spiral"). For his tombstone, he chose the words *Eadem Mutato Resurgo* ("In spite of changes, resurrection of the same").[9]

a   b   c
FIGURE 13.14 ■ **Fern Frond**   **FIGURE 13.15** ■ **A Tree?**

If you magnify one-quarter of the Koch curve shown in Figure 13.16 by a factor of 3, you'll get an exact duplicate of the entire curve. Therefore, each quarter is a self-similar copy of the entire curve. As you've seen before, the ratio of lengths in the original figure to that in the self-similar piece is called the *scale factor*. As such, from each portion of the curve, you can get the entire curve by applying a scale factor of 3. Conversely, from the whole, you can get each piece by applying a scale factor (or *contraction factor*) of $\frac{1}{3}$.

If you magnify one-quarter of that already magnified curve, you'll again get an exact duplicate of both preceding curves. You can continue magnifying as far as you wish, and you will always get a self-similar copy of the original curve. Keep in mind that we are talking about a *mathematically generated* figure that will show the same structure no matter how many times it is magnified. By contrast, the same figure printed on paper will only show ink dots and paper fibers under high magnification.

■ **EXAMPLE:** Identify a portion of the Koch curve in Figure 13.16 that is similar to the whole and give its scale factor.

● **SOLUTION:** Many pieces of the Koch curve are similar to the whole. Here we show one-quarter of the curve which, magnified by a scale factor of 3, reproduces the entire curve. ■

The Koch curve is an excellent example of *strict self-similarity*. Magnify *any part* of the curve and you will get a self-similar copy of the whole. The Sierpinski triangle also has this property. However, it is possible to have lesser degrees of self-similarity. For example, imagine a book cover that displays a picture of the same book. The book in the picture has, of course, a picture of the same book on its cover, and so forth, as illustrated in Figure 13.17. The book cover has self-similarity, but only at the point where all the images converge. This property is called *self-similarity at a point*. Some other point on the book—for example, a corner—does not contain copies of the whole.

The fractal tree in Figure 13.18 exhibits self-similarity, but not everywhere. The trunk, for example, does not contain similar copies of the whole. Such a fractal is called *self-affine*.

All fractals exhibit some self-similarity, but not all figures that have some self-similarity are fractals. The square in Figure 13.19, for example, is shown divided into smaller, similar copies of itself. Each smaller square can be subdivided into even smaller squares, and so on, with each square similar to all the others. However, this figure is not a fractal.

What then, makes a figure a fractal? As you've seen, a fractal is a figure with a fractal dimension that exceeds its topological dimension.

■ EXERCISES ● SELF-SIMILARITY

1. Define or describe the following terms:
   similarity   self-similarity
   strict self-similarity magnification factor
2. Identify the largest of three smaller triangles in the Sierpinski triangle in Figure 13.2 that are similar to the whole. What is its magnification factor?
3. For the Sierpinski hexagon in Figure 13.10:
   a. Identify a self-similar piece that has a magnification factor of 3.
   b. Find the number of such pieces.
   c. For the next smallest of the self-similar pieces, find the magnification factor and the number of pieces.

FIGURE 13.16 ■ Self-Similarity of the Koch Curve

FIGURE 13.17 ■ A Book Cover

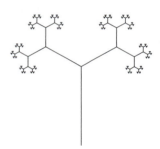
FIGURE 13.18 ■ A Fractal Tree

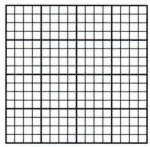
FIGURE 13.19 ■ A Square Subdivided into Self-Similar Pieces

4. For the Sierpinski carpet in Figure 13.12:
   a. Identify a self-similar piece that has a magnification factor of 3.
   b. Find the number of such pieces.
   c. For the next smallest of the self-similar pieces, find the magnification factor and the number of pieces.

5. Using the Sierpinski triangle in Figure 13.2, count the number of triangles at each stage, from 0 to 4. At each stage, give the magnification factor. Make a table showing the stage, magnification factor, and number of triangles.

Stage	0	1	2	3	4
Magnification Factor	1	2			
Number of Triangles	1	3			

6. Using the Sierpinski carpet in Figure 13.12, count the number of squares at each stage, from 0 to 4. At each stage, give the magnification factor. Make a table showing the stage, magnification factor, and number of squares.

Stage	0	1	2	3	4
Magnification Factor	1	3			
Number of Squares	1	8			

7. Use a CAD program to draw a logarithmic spiral. Then repeatedly zoom in and print a new copy at each step. Compare your copies at each magnification, and write a few paragraphs about your findings.

HOW TO CREATE A FRACTAL

"The Backbone of Fractals [is] Feedback and the Iterator."[10]

PEITGEN ET AL.

So far we have defined fractals and described their dimensions and properties. Now you will see how a fractal is created and how to make a variety of fractals of your own.

Iteration and Feedback

The process of *iteration* occurs when the same operation is carried out many times and the output of one operation is fed back and used as input for the next iteration. This *feedback* process can be visualized as a machine, as presented in Figure 13.20. The first input to the machine is called the *seed*. After that, the output of one iteration is the input for the next. A home heating system is a simple, everyday example of feedback. Cool air causes the thermostat to turn the furnace on, which in turn warms the air, which in turn causes the thermostat to turn the furnace off, which allows the air to cool, and so forth. The method of alternation that was used to find the diagonal of a square in Chapter 4, "Ad Quadratum and the Sacred Cut," is another example of iteration and feedback.

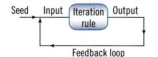

FIGURE 13.20 ■ The Feedback Machine

Iteration is usually performed with numbers computed by equations, but you can also perform iteration on geometric figures.

1. Begin with a geometric figure called the *seed*. This is sometimes called Stage 0.

2. Then perform a geometric operation, called the *iteration rule*, on the seed. This produces a new figure, or Stage 1.

3. Next, *iterate*—that is, go back to step 2 and apply the iteration rule to the new figure.

Continue iterating to generate a sequence of figures (or stages) called the *orbit* of the iteration. The figure that results from a great many iterations is called the *fate* of the orbit.

■ **EXAMPLE:** Use the iteration rule "Decrease each side of the square to half its length" to determine the orbit and its fate for the geometric iteration in which the seed is a square with a side that is 1 in.

● **SOLUTION:** The orbit is a sequence of squares decreasing in size, as in Figure 13.21, and the fate of this orbit is a point. (Applying the same iteration rule to any geometric figure would eventually result in a point.) ■

FIGURE 13.21 ■ Shrinking Squares

A loop with *linear feedback* (in which straight lines remain straight) generates what is called a *deterministic fractal*, such as the Sierpinski triangle. If you magnify any part of such a fractal, you will see a smaller but exactly similar version of the original. The Koch snowflake is a good example of a deterministic fractal.

Mandelbrot discovered that if the loop has *nonlinear* feedback, straight lines will become curves or swirls, producing another class of fractals called *nonlinear fractals*. The Mandelbrot set, shown in Figure 13.22, is the best-known example. As with the classical fractal, you get more copies of the original figure as you zoom in. They are usually not exact copies, so the image is less predictable.

A feedback loop with *random* feedback elements will generate imitations of nature—clouds, plants, mountains. For example, Figure 13.23 is a computer-generated random fractal, not a photograph.

Creating Deterministic Fractals

In this section, we will create only deterministic fractals. The iteration in the shrinking squares example did not produce a very exciting result. However, you will see that applying even a simple iteration rule can produce a final figure of great complexity and beauty—a *fractal*. We will first make a fractal by hand.

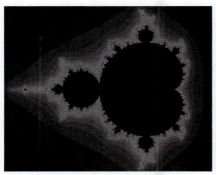

FIGURE 13.22 ■ A Nonlinear Fractal: The Mandelbrot Set

©Richard Dion Wilson/iStockphoto.com

FIGURE 13.23 ■ A Random Fractal: Plant Forms

FIGURE 13.24 ■ Generating the Sierpinski Triangle

■ **EXAMPLE:** *The Sierpinski Triangle.* Using a seed that is a shaded equilateral triangle, apply the iteration rule "From this triangle, remove an equilateral triangle, leaving three congruent equilateral triangles." Sketch two more iterations in this orbit.

● **SOLUTION:** Figure 13.24 shows four iterations in the orbit, obtained by computer. This figure shows the well-known Sierpinski triangle, and it displays strict self-similarity. In the preceding section, we calculated the fractal dimension of the Sierpinski triangle to be 1.585.   ■

Next, we will use a copy machine to create a fractal. Each copy becomes the original for the next iteration; you continuously make *copies of copies.* A "copy machine" can be an actual photocopier, a scanner, a camera, or a computer drafting program. The process is very simple. Start with a seed. Then apply this iteration rule: "Make *n* smaller copies and assemble in a given pattern." Repeat the iteration rule.

■ **EXAMPLE:** Using a copy machine, make a Sierpinski triangle.

● **SOLUTION:** The process is illustrated in Figure 13.25. Let's use the equilateral triangle I as the seed. Make three half-size copies of the original triangle and assemble them into the larger triangle II. Then make three half-size copies of II and assemble as in III. Repeat as many times as desired. This process of cutting and pasting is also called the *collage method.*   ■

■ **EXAMPLE:** *Fractal Tree.* Using a copy machine, make a fractal tree, such as the one shown in Figure 13.26.

● **SOLUTION:** Start with a stem with two branches at 120°, each half the length of the stem. Make two half-size photocopies and paste them onto the branches as shown. Make half-size copies of the new figure and paste them into the figure as well. Continue as far as desired.   ■

You can easily get the first few stages of a fractal by hand or by copy machine, but it soon becomes tedious to develop the full image—and it becomes more difficult to be precise. When the process is tedious, you can turn to the computer. In fact, the study of fractal geometry did not become popular until computers became available. The following sections show several ways in which you can use a computer to create fractals.

The steps for creating a fractal by CAD are simple. First, draw a basic figure. Next, replace sections of the basic figure with smaller replicas of the basic figure, creating a new figure. Then, replace sections of the new figure with smaller replicas of the figure itself. Repeat as many times as desired.

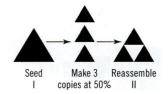

Seed   Make 3   Reassemble
 I   copies at 50%   II

a

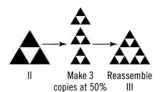

 II   Make 3   Reassemble
          copies at 50%   III

b

FIGURE 13.25 ■ The Sierpinski Triangle Created with a Copy Machine

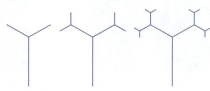

FIGURE 13.26 ■ Fractal Tree

■ EXAMPLE: Generate the Koch curve using The Geometer's Sketchpad. The steps are illustrated in Figure 13.27. The fractal dimension of this curve was introduced in Figure 13.13.

● SOLUTION: The Geometer's Sketchpad uses the **Iterate** command to provide feedback. Consult your CAD manual for specific instructions to complete the following steps:

1. Draw the basic figure, as shown in Figure 13.27a.[11]
2. Choose endpoints A and B, and choose **Iterate** from the **Transform** menu.
3. The figure from A to B is mapped in turn onto lines AC, CD, DE, and EB. The mapped figures will appear lightly on the drawing.
4. Click **Iterate**, and the mapped portions will appear with normal line weight on the drawing, as you can see in Figure 13.27b.
5. Hide the original four line segments.
6. For more iterations, simply **Select All** and press the plus key **(+)** repeatedly. Figure 13.27c shows a few more iterations. Pressing the minus key **(−)** repeatedly will reverse the iterations one at a time. ■

■ EXAMPLE: Construct a Pythagorean fractal tree, such as you see in Figure 13.28.

● SOLUTION: This figure is related to the fractal tree in Figure 13.26; however, in this example, you'll work with squares and triangles rather than lines. The steps are fairly simple.

1. Draw a square.
2. Attach a right triangle with its hypotenuse along one side of the square.
3. Attach a square to a leg of the triangle.
4. To each square attach a triangle that is similar to the first.
5. Go to step 3. Iterate as many times as desired. ■

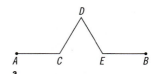

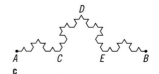

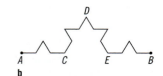

FIGURE 13.27 ■ Generating the Koch Curve

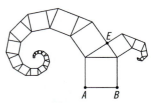

FIGURE 13.28 ■ Pythagorean Fractal Tree

Creating Fractals Using the Chaos Game

The *chaos game*—a name coined by Michael Barnsley—also uses a seed, and again the output of one iteration becomes the input for the next interation. However, instead of there being *one* iteration rule, there can be *several*; in the chaos game, the rule that is used is chosen at random. Although the rule is chosen at random, the results are not random. You still get a deterministic fractal. In Figure 13.29, this concept is presented as a feedback machine, and the random selection of the iteration rule is depicted as a wheel of fortune.

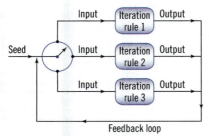

FIGURE 13.29 ■ Feedback Machine for the Chaos Game

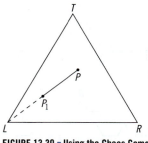

FIGURE 13.30 ▪ Using the Chaos Game to Create a Sierpinski Triangle

■ **EXAMPLE:** Generate the Sierpinski triangle manually using the chaos game.

● **SOLUTION:** The following chaos game for the Sierpinski triangle is one of many possible chaos games. Refer to Figure 13.30 when following these steps:

1. Locate three points, T, R, and L, to define the vertices of a triangle.

2. Select the seed: Choose any point P within the triangle.

3. Use the iteration rule "Roll the die to select a vertex." For a roll of 1 or 2, move halfway to T. For a roll of 2 or 3, move halfway to R. For a roll of 4 or 5, move halfway to L (as shown in the figure).

4. Mark a new point P_1 halfway between the previous point and the chosen vertex. Because the position of each point depends on the position of the preceding point, errors can accumulate. Plot your points very accurately.

5. Go to step 3. Iterate as many times as desired.

6. Erase the first ten points.

What pattern do you get for a large number of rolls of the die after the first ten points are discarded? A random scatter of points? A completely filled triangle? A shapeless blob? No. Amazingly, you'll get the Sierpinski triangle! ■

Chaos Game by Calculator

Now that you have seen how the game works when the iterations are done by hand, you can generate points more quickly using a computer or a calculator. The following table contains a program for building a Sierpinski triangle using a TI graphing calculator. Using this program as a guide, you should be able to rewrite the instructions for your own calculator.

TI Graphing Calculator Program for the Sierpinski Triangle[12]	Comments
:ClDrw	*Clears the display*
:0→Xmin:1→Xmax:0→Ymin:1→Ymax	*Sets the boundaries of the viewing window*
:Disp "X=":Input X	*Prompts user for the X seed value*
:Disp "Y=":Input Y	*Prompts user for the Y seed value*
:0→C	*Initializes counter to zero*
:Lbl Line 1	*Labels this point in the program as "Line 1"*
:C+I→C	*Increments the counter*
:PtOn(X,Y)	*Plots a point at (X,Y)*
:If C>1500:End	*Checks counter and ends program if greater than 1500*
:Rand→N	*Stores a random number (the die toss) in N*
:If N<(1/3):Goto Line2 :If N>(2/3):Goto Line3	*Checks the counter to see which vertex has been chosen and directs control to the proper computation below*
:X/2→X:Y/2→Y:Goto Line1	*Computes a new X and Y and loops back to Line1*
:Lbl Line2	*Labels line 2*
:(X+I)/2→X:Y/2→Y:Goto Line1	*Computes a new X and Y and loops back to Line1*
:Lbl Line3	*Labels line 3*
:(X+.5)/2→X:(Y+I)/2→Y:Goto Line1	*Computes a new X and Y and loops back to Line1*

Chaos Game Using BASIC

Now let's program the chaos game on the computer. The program shown here is one of several BASIC programs that we will use. It is from Peitgen et al., who noted that BASIC is an "outdated, inefficient, and unstructured language which is known to prevent writing good programs."[13] So why do they use it, and why should we use it here? We use BASIC because it is easily available, is usually free or built-in, is so simple that someone not familiar with programming can understand it, and is easily translated into another language for use on a computer or on a graphing calculator.

The following program uses the chaos game to generate the Sierpinski triangle. The line that tells the program which video adapter you are using is missing. You'll need to supply that instruction, such as "Screen 9." It also assumes a screen layout as in Figure 13.31.

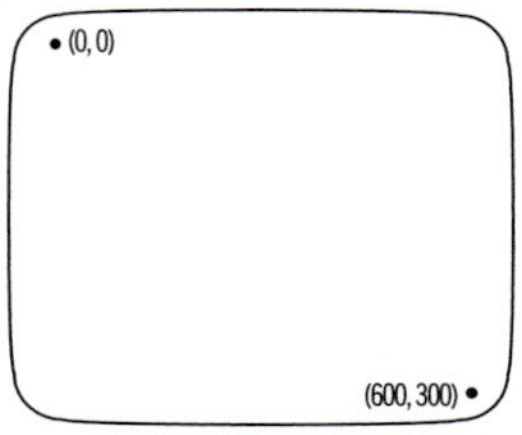

FIGURE 13.31 ■ Screen Layout for the Sierpinski Triangle BASIC Program

BASIC Program for the Sierpinski Triangle	Comments
10 CLS	*Clears the screen*
30 X=300:Y=300	*Gives the coordinates of the seed*
50 FOR N=1 TO 100000	*Counts the number of iterations*
60 DIE=INT(RND*3)	*Randomly rolls the die, from 0 to 3*
70 IF DIE=0 THEN X=300+(X-300)/2:Y=20+(Y-20)/2:GOTO 100	*Lines 70, 80, and 90 compute X and Y for the new point location, depending on the number on the die*
80 IF DIE=1 THEN X=550-(550-X)/2:Y=300-(300-Y)/2:GOTO 100	
90 IF DIE=2 THEN X=50+(X-50)/2:Y=300-(300-Y)/2:GOTO 100	
100 PSET(X,Y)	*Turns on a pixel at the new point*
120 NEXT N	*Iterates by looping back*
130 END	

Chaos Game by Spreadsheet

In the preceding program, each point is plotted immediately after its location is computed, and the program loops back to compute the next point. With a spreadsheet, it is necessary to compute *all the points* and store them, and then to plot them in a separate step.

Here are the first few lines of a spreadsheet, with comments in italics, to create a Sierpinski triangle. It is in Excel, but the formulas have a similar format in other spreadsheets. The formulas in row 2 must be copied down to the next few thousand rows. Each row will give the *x*- and *y*-coordinates of one point on the Sierpinski triangle, in columns D and E. After the spreadsheet is constructed, a scatter plot of columns D and E will show the Sierpinski triangle.

	A Die	B *X* at Vertex	C *Y* at Vertex	D *X*	E *Y*
1				1 X *seed value*	1 Y *seed value*

2	INT(3*RAND())
	
	Gives a random number between 0 and 2 to simulate the toss of a die
		IF(A2=0, 0,
		IF(A2=1, 2, 1))
		
		Selects the X-coordinate of the vertex chosen by the die value
			IF(A2=0, 0,
			IF(A2=1, 0, 1.5))
			
			Selects the Y-coordinate of the vertex chosen by the die value
				(B2+D1)/2
				
				Computes a new X halfway between the last X and X at the chosen vertex
					(C2+E1)/2
					
					Computes a new Y halfway between the last Y and Y at the chosen vertex

Here is a printout of the first 18 lines of the spreadsheet.

Die	X at Vertex	Y at Vertex	X	Y
			1.0000	1.0000
0	0	0	0.5000	0.5000
1	2	0	1.2500	0.2500
1	2	0	1.6250	0.1250
1	2	0	1.8125	0.0625
0	0	0	0.9063	0.0313
2	1	1.5	0.9531	0.7656
1	2	0	1.4766	0.3828
2	1	1.5	1.2383	0.9414
1	2	0	1.6191	0.4707
1	2	0	1.8096	0.2354
1	2	0	1.9048	0.1177
1	2	0	1.9524	0.0588
1	2	0	1.9762	0.0294
2	1	1.5	1.4881	0.7647
1	2	0	1.7440	0.3824
1	2	0	1.8720	0.1912
2	1	1.5	1.4360	0.8456
1	2	0	1.7180	0.4228

Chaos Game by Fractalina

The Fractalina program is available online at http://math.bu.edu/DYSYS/ applets/fractalina_more.html. With it, you can play the chaos game with a variety of different configurations. You can enter up to 15 vertices, move them around, choose various compression ratios and rotations, and then display the resulting image. When you first open Fractalina, you will see three points at the vertices of a triangle: red, green, and blue. This configuration will give the Sierpinski triangle. Click **Start** to see the Sierpinski triangle appear. Click **Stop** to end the computation.

You can magnify portions of this image by clicking on a point and then dragging the mouse to form a window. When you then click **Start**, you see the portion of the image within the window magnified to fill the entire screen. To return to the original image, click **Zoom Out**. You can move a vertex by dragging it to a new position or by entering its new coordinates in the control panel. Both the x- and y-coordinates run from -150 to $+150$, with $(0, 0)$ at the center of the window. You can similarly change both the rotations and the compression ratios at each vertex.

Fractalina has other stored fractals as well. Pull down the lower-right menu to find the Koch curve, the Sierpinski carpet, and others. To play a chaos game other than those stored, you need to select a number of vertices and their properties. The control panel is to the right of the main window. For each vertex, it lists the color, the x- and y-coordinates, the rotation about the vertex, and the compression ratio. To add a new vertex, click **New Point**. A new vertex appears at the origin. You can then move this point to a new location and change the various quantities. The **Kill Point** key removes the last point on the list in the control panel.

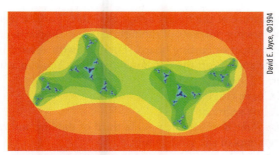

FIGURE 13.32 ■ A Julia Set

Generating Fractals Using Fractint

Fractint is a freeware fractal generator created for PCs. It is very versatile, has many great features, and is constantly being upgraded and improved. You can use Fractint to view fractals that are too complex for us to program. The name *Fractint* was chosen because it uses *integer arithmetic* for calculations rather than the usual floating-point arithmetic. The huge number of iterations needed to produce an image can be computed faster with integer arithmetic. You can download the latest version from http://www.fractint.org.

After you download Fractint, press **Enter** to bring up the main menu. In order to properly display the images, select your particular video mode. Then simply choose a fractal type from the list provided under **Select Fractal Type**. Press the F2 key to obtain a description of the fractal, and press **Enter** to display its parameters. You can then change those parameters or accept the default values. Finally, press **Enter** to display the image.

To bring up a zoom box, press the **Page Up** key. You can reduce the image by repeatedly pressing that key. Enlarge the zoom box by pressing the **Page Down** key, and use the arrow keys to move the image. Press **Enter** to cause the contents of the zoom box to fill the screen. Pressing the plus (**+**) or minus (**−**) key will cause the colors to cycle through various combinations and create fascinating effects. Other operations, too numerous to mention here, are available. The Julia set is a well-known fractal display that can be viewed on Fractint. It is named for the French mathematician Gaston Julia (1893–1978). See Figure 13.32.

■ EXERCISES • HOW TO CREATE A FRACTAL

1. Define or describe the following terms:

iteration	feedback	feedback machine
seed	iteration rule	orbit
fate (of an orbit)	deterministic fractal	nonlinear fractal
random fractal	chaos game	

2. For the shrinking squares that were shown in Figure 13.21, perform the following steps:
   a. Describe the orbit and fate.
   b. Construct several stages in the orbit by hand or using CAD.
   c. Using a spreadsheet, compute the sides of the various stages.

3. Describe the orbit and fate of the ad quadratum figure. See Figure 13.33. We also examined this figure in Chapter 4.

4. Draw several stages in the orbit of each of the following fractals. Draw them by hand or using a copy machine or computer drawing program.

   These figures show only the first two stages of each fractal, so of course they do not look very interesting. You need to construct several stages

a

b

c

FIGURE 13.33 ■ The Ad Quadratum Figure

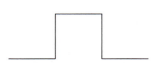

FIGURE 13.34 ■ Hat Curve or Koch Stool

FIGURE 13.35 ■ Variation on the Sierpinski Triangle

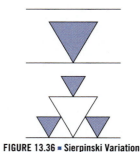

FIGURE 13.36 ■ Sierpinski Variation

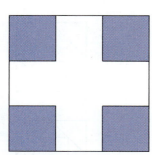

FIGURE 13.37 ■ Cantor Dust

before the nature of the fractal becomes visible. For fractals that you will form by removing portions of the figure, try pasting dark paper onto a rigid plastic sheet or cutting board. Then cut with a sharp blade and straightedge and lift off the unwanted portions. You can also use adhesive shelf liner.

a. **Koch Curve** (Figure 13.13)
   Seed: A line segment
   Iteration Rule: Remove the middle third of the line and replace it with two segments of the same length.

b. **Sierpinski Carpet** (Figure 13.12)
   Seed: A square
   Iteration Rule: Connect the trisection points of each side with the opposite ones. Keep only the eight boundary squares.

c. **Hat Curve or Koch Stool** (Figure 13.34)
   Seed: A line segment
   Iteration Rule: Remove the middle third of the line and replace it with three segments of the same length, as shown.

d. **Variation on the Sierpinski Triangle** (Figure 13.35)
   Seed: An equilateral triangle
   Iteration Rule: Connect the trisection points, as shown. Keep only the six triangles on the border.

e. **Sierpinski Variation** (Figure 13.36)
   Seed: A line segment
   Iteration Rule: To each line segment, add an equilateral triangle as shown.

f. **Cantor Dust** (Figure 13.37)
   Seed: A line segment
   Iteration Rule: Remove the middle third of the line.

g. **Cantor Square** (Figure 13.38)
   Seed: A square
   Iteration Rule: Place a one-third size copy in each corner of the square.

h. **Quadratic Koch Island** (Figure 13.39)
   Seed: A square
   Iteration Rule: Replace each side with the figure shown.

i. **Plus Sign** (Figure 13.40)
   Seed: A square
   Iteration Rule: Trisect each side and remove the corner squares.

FIGURE 13.38 ■ Cantor Square

FIGURE 13.39 ■ Quadratic Koch Island

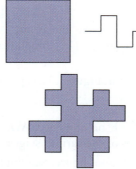

FIGURE 13.40 ■ Plus Sign

j. **The Diagonal** (Figure 13.41)
Seed: A square
Iteration Rule: Trisect each side and keep only the three squares on a diagonal.

k. **Triangle Variation I** (Figure 13.42)
Seed: An equilateral triangle
Iteration Rule: Join the midpoints of the sides to form a new triangle. Remove the new triangle.

l. **Triangle Variation II** (Figure 13.43)
Seed: A right triangle
Iteration Rule: Join the midpoints of the sides to form a new triangle. Remove the new triangle.

m. **Circles in a Circle** (Figure 13.44)
Seed: A circle
Iteration Rule: Replace the circle with two circles, each half the diameter of the original.

n. **Split Rectangles** (Figure 13.45)
Seed: A rectangle
Iteration Rule: Cut vertically into thirds, and remove the middle piece.

o. **Sierpinski Hexagon** (Figure 13.46)
Seed: A regular hexagon
Iteration Rule: Make a copy at one-third reduction and place it vertex-to-vertex around a hexagon.

p. **Fractal Tree** (Figure 13.26)
Seed: A vertical line, with two half-sized lines at 120°
Iteration Rule: From the endpoint of each branch, add two half-sized branches at 120°. You first saw this in Figure 13.18.

q. **Golden Fractal Tree** (Figure 13.47)
Seed: A vertical line, with two lines at 120°
Iteration Rule: From the endpoint of each branch, add two branches at 120°, each having a reduction factor of $1/\Phi$ (the reciprocal of the golden ratio). With this reduction factor, the clusters of branches will, in the limit, just touch.[14]

5. *Koch Snowflake.* Starting with the Koch curve shown in Figure 13.13, draw the Koch snowflake shown in Figure 13.5a. Use any method.

6. *Variation on the Sierpinski Triangle.* Using a copy machine, repeat the Sierpinski triangle construction with a different seed but with the same iteration rule. What do you find?

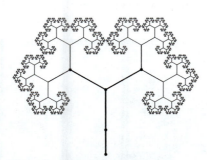

FIGURE 13.47 ■ Golden Fractal Tree

FIGURE 13.41 ■ The Diagonal

FIGURE 13.42 ■ Triangle Variation I

FIGURE 13.43 ■ Triangle Variation II

FIGURE 13.44 ■ Circles in a Circle

FIGURE 13.45 ■ Split Rectangles

FIGURE 13.46 ■ Sierpinski Hexagon

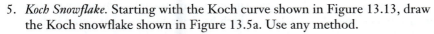

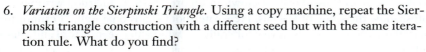

7. *Sierpinski Triangle by Chaos Game.* Using the chaos game, plot several points by hand on the Sierpinski triangle.

8. Using Fractalina, create the Sierpinski triangle. Then repeat with a different overall compression factor, and repeat with a different compression factor for each vertex.

9. Using Fractalina, play the chaos game with the five vertices of a regular pentagon.

10. Using Fractalina, play the chaos game with the six vertices of a regular hexagon and a compression factor of $\frac{2}{3}$.

11. *Chaos Game by Programmable Calculator or Computer.* Write a program for your graphing calculator or computer to create the Sierpinski triangle by playing the chaos game.

12. *Using Fractint.* Download Fractint and use it to display some fractals. Change some parameters for a fractal and observe what happens.

FRACTAL-RELATED ART

"The fractal 'new geometric art' shows surprising kinship to Grand Masters painting or Beaux Arts architecture. An obvious reason is that classical visual arts, like fractals, involve very many scales of length and favor self-similarity . . . and it came in through an effort to imitate nature . . . [so] it well may be that fractal art is readily accepted because it is not truly unfamiliar."[15]

MANDELBROT

Mandelbrot could almost be speaking of our various systems of proportions, those based on the musical ratios, the square, and the golden ratio. The following works of art, old and new, are all somehow related to the material of this chapter. Most of these works are not actually fractals; however, true fractal art can be seen in great abundance in many "galleries" on the Web.

Leonardo appeared to anticipate chaos and fractals with his many drawings of turbulence and of the deluge, which he described in great detail in his notebooks.[16] One of those works is shown in Figure 13.48. Leonardo's writings also foreshadow the study of fractals. He wrote about stains on a wall, "I cannot forbear to mention . . . a new device for study . . . which may seem trivial and almost ludicrous . . . [but] is extremely useful in arousing the mind. . . . Look at a wall spotted with stains, or with a mixture of stones . . . you may discover a resemblance

FIGURE 13.48 ■ *Deluge,* c. 1514

FIGURE 13.49 ■ *The Great Wave,* Katsushika Hokusai, c. 1829–1833

FIGURE 13.50 ■ Escher's Study of Bartolomeo's Twelfth-Century Patterns in the Ravello Cathedral

to landscapes . . . battles with figures in action . . . strange faces and costumes . . . and an endless variety of objects . . . confusedly, like the sound of bells in whose jangle you may find any name or word you choose to imagine."[17] The passage almost sounds like a description of a fractal pattern.

The Japanese artist Hokusai created thousands of prints during his prolific career. From about 1826 to 1833, he created a famous series entitled *Thirty-Six Views of Mount Fuji.* Included in this series is *The Breaking Wave Off Kanagawa*— or, more simply, *The Great Wave*—which is shown in Figure 13.49. It portrays a scene in which a large wave dwarfs Mount Fuji, seen in the background, while it threatens to destroy the boats beneath it. Hokusai's works include some of the finest examples of Japanese landscape printmaking. Note the self-similarity between the large wave and the smaller and smaller wavelets.

Figure 13.50 displays studies by Escher of designs on a pulpit. Note the tilings' resemblance to the Sierpinski triangle. The Escher drawing in Figure 13.51 suggests the idea of iteration and feedback, a notion also present in Escher's *Mobius Strip.* Another Escher design is shown in Figure 13.52. Mandelbrot wrote, "Is some [fractal art] reminiscent of M. C. Escher? It should be, because Escher had the merit of letting himself be inspired by the hyperbolic tilings in Fricke & Kleine, which . . . relate closely to shapes that are being incorporated into the fractal realm."[18]

In Figure 5.67, you saw the octagonal Castel del Monte with its eight octagonal towers. The pattern used in that castle was the basis of the fractal design

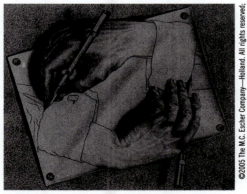

FIGURE 13.51 ■ *Drawing Hands,* Escher, 1948

FIGURE 13.52 ■ *Circle Limit IV,* Escher, 1960

FIGURE 13.53 ■ A Fractal Based on the Castel del Monte, Susanne Krömker

FIGURE 13.54 ■ The Paris Opera House

FIGURE 13.55 ■ *Pythagorean Fractal Tree*, Koos Verhoeff

FIGURE 13.56 ■ *Gaussian Hills That Never Were*, Plate C11 from Mandelbrot's *The Fractal Geometry of Nature*

you see in Figure 13.53. It was prepared at the Interdisciplinary Center for Scientific Computation, University of Heidelberg.

According to Mandelbrot, "A high period Beaux Arts building is rich in fractal aspects."[19] The Paris Opera House, which is shown in Figure 13.54, exemplifies the scaling feature of fractal geometry.

The sculpture presented in Figure 13.55 is a three-dimensional model of the Pythagorean fractal tree that we studied earlier in this chapter. The artificial landscape in Figure 13.56 is named for Carl Friedrich Gauss (1777–1855). The *Gaussian distribution* of probability (the well-known *bell-shaped curve*) was used in the computations that produced this image. In *Mandala II*, which is shown in Figure 13.57, a large marble circle is subdivided into three circles, each of which is subdivided into three, each of which is subdivided into three, continuing as far as possible with hammer and chisel.

FIGURE 13.57 ■ *Mandala II* (detail), Calter, 1992

SUMMARY

We began this book with extremely simple geometric figures—points and lines. We ended our journey with incredibly complex fractal figures. We began with Euclidean geometry, both plane and solid; then we touched upon analytic geometry, the ellipse, and the spiral in Chapter 9; we considered projective geometry when we covered perspective; and we finished with the new field of fractal geometry.

Michael Barnsley wrote, "Fractal geometry will make you see everything differently. There is a danger in reading further. You risk the loss of your childhood vision of clouds, forests, galaxies, leaves. . . and much else besides. Never again will your interpretation of these things be quite the same."[20]

We hope that geometry, fractal or otherwise, will make *you* see the world differently and will enrich your appreciation of art and architecture.

EXERCISES AND PROJECTS

1. *Audio Feedback.* Place a microphone near a speaker that produces sounds that, after amplification, are fed into the microphone. Alternatively, use two walkie-talkies or CBs face to face, one sending and one receiving. Speak into the microphone and listen to the sounds of chaos produced by this closed feedback loop.

2. *Video Feedback.* Place a video camera so that it takes a picture of the same image that is the output of the camera. Watch the patterns of chaos.

3. *Parallel Mirrors.* Place two mirrors parallel or nearly parallel to each other so that you get an infinity of repeated images. Photograph your result.

4. Draw several stages in the orbit of each of the following figures. Create the images by hand, or use a copy machine or computer drawing program.
   a. **The Letter H** (Figure 13.58)
      Seed: A square
      Iteration Rule: Subdivide the square into a nine-square, and remove two squares as shown.

   b. **The Letter L** (Figure 13.59)
      Seed: A square
      Iteration Rule: Subdivide the square into four squares and remove one square as shown.

   c. **The Letter X** (Figure 13.60)
      Seed: A square
      Iteration Rule: Subdivide the square into a nine-square, and remove four squares as shown.

5. Using the preceding projects as guides, make fractals using your own initials. Use them to decorate some personal items: book plates, a tee shirt, writing paper, and so forth.

6. Draw several stages in the orbit of the following figure. Create the images by hand, or use a copy machine or computer drawing program.

   Pythagorean Fractal Tree (Figure 13.61)
      Seed: A square surmounted by an isosceles triangle
      Iteration Rule: Attach a reduced square and isosceles triangle to each side of the triangle.

FIGURE 13.58 ■ The Letter H

FIGURE 13.59 ■ The Letter L

FIGURE 13.60 ■ The Letter X

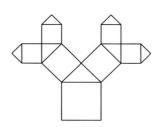

FIGURE 13.61 ■ Pythagorean Fractal Tree with Obtuse Isosceles Triangles

7. *Variations on the Pythagorean Fractal Tree.* Using Figure 13.61 as your guide, draw several stages in the orbit of each of the following figures.

   a. **Pythagorean Tree with Different Shape of Triangle**
   Seed: A square surmounted by a non-isosceles triangle
   Iteration Rule: Attach a reduced square and a non-isosceles triangle to each side of the triangle.

   b. **Pythagorean Tree in Which Orientation of Triangle Is Changed**
   Seed: A square surmounted by a triangle
   Iteration Rule: Attach a reduced square and triangle to each side of the triangle. However, *reverse the orientation* of the triangle every time.

   c. **Pythagorean Tree with Equilateral Triangles**
   Seed: A square surmounted by an equilateral triangle
   Iteration Rule: Attach a reduced square and equilateral triangle to each side of the triangle. You should get a semiregular periodic tiling, like the ones in Chapter 5, "Polygons, Tilings, and Sacred Geometry."

8. Take photographs of a fern and each of its self-similar parts. Enlarge each photo to the same size. Compare them. Do the same with cauliflower, trees, and so forth.

9. *Flip Book.* Print successive images in the orbit of any fractal. Bind them at one edge to make a flip book. For the best results, use stiff paper and keep the size small.

10. Using a fractal design or self-similarity as the basic idea, create a painting, graphic, wallpaper, wrapping paper, greeting card, screensaver, tiling, mosaic, or poster. You can also create a fabric design on a pillow, tee shirt, quilt, or dress.

11. Using any picture (perhaps from a digital camera) as the seed, make three half-sized photocopies and arrange them in a triangle. Create several stages of the orbit and transform them into a graphic.

12. Make a fractal using the photocopy or collage method. For your initial image, use a self-portrait, your pet, or something else that interests you. Don't use very many stages and make sure the images are large enough to be identified. Use your final image as a graphic design for a tee shirt or sweatshirt.

13. Make a "stained glass" window based on the material in this chapter. You can use actual glass, rigid plastic cutouts, or simply paint on a sheet of plastic. Install it temporarily over a window in your classroom.

14. Make a sculpture that is based on a fractal or that exhibits self-similarity in its design.

15. Make a three-dimensional model of the Sierpinski tetrahedron. Use cardboard and a glue gun to make the first few levels of detail. Then indicate the next few levels by drawing or by pasting computer-generated images onto your model.

16. The following BASIC program will generate a Barnsley fern. Modify it to suit your computer or graphing calculator and try it.

Program for Barnsley Fern	Comments
10 SCREEN 9	
20 X=100:Y=100	*Sets the seed location*
30 CLS	*Clears the screen*
40 FOR =1 TO 10000	*Starts the feedback loop and sets the number of iterations*
50 RN=RND	*Selects a random number between 0 and 1*
60 IF RN<.85 THEN 120	*Lines 60, 70, and 80 transfer control to lines below depending on the value of the random number.*
70 IF RN<.92 THEN 140	
80 IF RN<.99 THEN 160	
	One of the next four lines is randomly chosen, each iteration, to compute a new X and Y.
100 X=0:Y=.17*X:GOTO 180	*This line is used about 1% of the time.*
120 X=.85*X+.04*Y:Y=-.04*X+.85*Y+3:GOTO 180	*This line is used about 85% of the time.*
140 X=.2*X-.26*Y:Y=.23*X+.22*Y+1.4:GOTO 180	*This line is used about 7% of the time.*
160 X=-.25*X+.28*Y:Y=.26*X+.24*Y+4	*This line is used about 7% of the time.*
180 PSET (50*(X+4),15*Y)	*Turns on a pixel at the new point (X, Y)*
190 NEXT I	*Closes the iteration loop*

17. Write a BASIC program to generate the Koch curve.

18. The Web has many fractal galleries. Download and print a representative sample of images, and write a short critique for each. Summarize your opinions about fractal art on the Web.

19. Look for hints of self-similarity in art history books, a museum, or a gallery. Select one or two paintings for closer inspection, and write a few paragraphs on your findings.

20. Mandelbrot said that high-period Beaux Arts buildings are rich in fractal aspects. Find some examples and write a paper on your findings, with illustrations. Relate your findings to the various systems of proportions you studied in this book.

21. Write a book report on any book in this chapter's list of sources.

22. Here is another list of suggestions for short papers or term papers. Use them as a starting point to come up with your own topic.

    ■ The history of fractal geometry

    ■ The different kinds of dimension

    ■ The *box-counting dimension*, which is used to find the fractal dimension of coastlines and other natural features where the fractal dimension described in this chapter cannot be used

    ■ The *divider dimension*, which is used for natural features, as is the box-counting dimension

    ■ Why the chaos game produces the Sierpinski triangle, and not just a random scattering of points

- The use of fractals to generate realistic copies of natural features such as clouds, mountains, and so forth
- The relationship between the various systems of architectural proportions covered in this text, and the self-similarity of fractal structures
- The use of fractals to make movies such as *Star Wars* and *Toy Story*
- How studios, such as Pixar, use fractals to generate images for movies
- The fourth dimension and how it influenced art in the early part of the twentieth century. For starters, see *The Fourth Dimension and Non-Euclidean Geometry in Modern Art*, by Linda Dalrymple Henderson.
- How the "new mathematics" and "new physics" at the start of the twentieth century were "akin to the cubist paintings and atonal music that were upsetting established standards of taste in the arts at about the same time."[21]
- Randomness in art

23. Mandelbrot means "almond bread." (As you learned in Chapter 7, "Circular Designs in Architecture," a mandorla is an almond-shaped figure.) Make some mandelbrot and share it with your class.[22]

24. Make a fractal cookie array. Bake 29 triangular cookies and arrange them in a Sierpinski triangle. You can also make them out of mandelbrot.

Mathematical Challenges

25. Using Equation 130, derive the equation for fractal dimension D. Start with the results provided in Table 13.1 (p. 397), which gave $n = m^D$. Solve for D using logarithms.

26. The following BASIC program will create a Mandelbrot set. If you understand complex numbers, research the Mandelbrot set to learn how its points are selected. Then modify it to suit your computer or graphing calculator and try it. Write a short paper on your findings.

Program for Mandelbrot Set	Comments
	Lines 10 and 20 select one point (S, T) at a time to be tested.
10 FOR T=-1 TO 0 STEP .005	*Steps vertically from -1 to 0*
20 FOR S=-2 TO 1 STEP .005	*Steps horizontally from -2 to 1*
30 X=S:Y=T	
40 FOR K=1 TO 80	*Start of loop to test the point (S, T) 80 times*
50 Z=X^2-Y^2+S	*Square of real part added to original S*
60 W=2*X*Y+T	*Square of imaginary part added to T*
70 IF Z^2+W^2>4 THEN 120	*Test: If Z^2+W^2 exceeds 4, the test fails and we move to the next point.*
80 X=Z:Y =W	
90 NEXT K	
100 SP=(2+S)*200:TP=(1+T)*150	*Scales and translates S and T to fit the screen*
110 PSET(SP,TP)	*Turns on a pixel at the chosen point*
120 NEXT S	
130 NEXT T	

27. Make an oral presentation to your class on any of the projects in this section.

SOURCES

Barnsley, Michael. *Fractals Everywhere*. San Diego, CA: Academic, 1988.

Boles, Martha and Newman, Rochelle. *The Golden Relationship: Art, Math & Nature*. 4 Vols. Bradford, MA: Pythagorean Press.

Briggs, John. *Fractals: The Patterns of Chaos*. New York: Simon & Schuster, 1992.

Briggs, John, et al. *Turbulent Mirror*. New York: Harper, 1989.

Burger, Edward, et al. *The Heart of Mathematics*. Emeryville, CA: Key College Publishing, 2000.

Choate, Jonathan, et al. *Fractals, a Tool Kit of Dynamic Activities*. Emeryville, CA: Key Press, 1999.

Devaney, Robert. *The Mandelbrot and Julia Sets: A Tool Kit of Dynamic Activities*. Emeryville, CA: Key Press, 2000.

Gleik, James. *Chaos*. New York: Penguin, 1987.

Gulick, Denny. *Encounters with Chaos*. New York: McGraw-Hill, 1992.

Kappraff, Jay. *Connections: The Geometric Bridge between Art and Science*. New York: McGraw-Hill, 1990.

Mandelbrot, Benoit. *The Fractal Geometry of Nature*. San Francisco: Freeman, 1982.

Peitgen, Heinz-Otto, et al. *Fractals for the Classroom*. New York: Springer, 1992.

NOTES

1. Mandelbrot, p. 1.
2. Peitgen, p. 229.
3. Mandelbrot, p. 37.
4. Mandelbrot, p. 5.
5. A direct quote from *The Fractal Geometry of Nature*, p. 15, is "A fractal is by definition a set for which the Hausdorff Besicovich dimension strictly exceeds the topological dimension." Later on the same page, Mandelbrot calls the Hausdorff Besicovich dimension the "fractal dimension."
6. Mandelbrot, p. 4.
7. Douglas Adams. *Hitchhiker's Guide to the Galaxy*. New York: Pocket Books, 1979, p. 152.
8. From *On Poetry* (1733) by Jonathan Swift (1667–1745).
9. Peitgen, p. 212.
10. Peitgen, p. 17.
11. From *The Geometer's Sketchpad Learning Guide, Version 4*, Tour 8.
12. This program is adapted from Peitgen, et al., *Fractals for the Classroom, Strategic Activities*, Vol. 1, p. 49. You may have to make some changes for it to run on a different TI calculator. Also included is a program for the Casio calculator.
13. Peitgen, p. 70.
14. Livio, p. 219.
15. Mandelbrot, p. 23.
16. Leonardo goes to great length and minute detail in his descriptions of how to represent the deluge. See J. P. Richter, *The Notebooks of Leonardo Da Vinci*, Vol. 1, pp. 306–314.
17. J. P. Richter, Vol. 1, p. 254.
18. Mandelbrot, p. 23. The work to which he refers is *Voresungen uber die Theorie der Automorphen Functionen*. Leipzig, 1897.
19. Mandelbrot, p. 24.
20. Barnsley, *Fractals Everywhere*.
21. From "Characteristic Irregularity" by Freeman Dyson, *Science*, May 12, 1978, pp. 677–678.
22. From Boles and Newman, p. 200.

APPENDIX A

■

The Art—Math Tourist

This appendix lists the travel locations and historical sites mentioned in this book. It is not intended to be an exhaustive list of sites interesting to the math- and art-savvy tourist, and it does not include every artwork mentioned in the book. Consult a recent guidebook for more information on times, admission fees, and directions.

APPENDIX B

■

Biographical Profiles

●

This appendix lists the page numbers on which you will find the featured biographical profiles of historical figures in mathematics and in art and architecture that are featured in the text. Many other people are mentioned in the book, without biographical profile, and they may be found in the general index, along with additional references to the people listed here.

Historical Figure	Page
Pacioli, Luca (c. 1445–1519)	307
Palladio (1508–1580)	30
Plato (c. 427–347 B.C.)	21
Pythagoras (c. 580–c. 500 B.C.)	4
Vitruvius (70? –25 B.C.)	26

APPENDIX C

Listings of Constructions by Chapter

The following is a comprehensive listing of the constructions in the text, which can be completed by compass and straightedge, paper folding, or CAD program.

Chapter 4   **Page**

Chapter 8   **Page**

Chapter 9   **Page**

APPENDIX D

∎

Index to Geometric Signs and Symbols

●

The following is an index to geometric signs, symbols, and figures relevant in both mathematics and in art and architecture, which are referenced throughout the text. These are main listings only; for other occurrences see the general Index.

APPENDIX E

■

Number Symbolism and Famous Groupings in Art

●

"But thou hast arranged all things by measure and number and weight."

THE WISDOM OF SOLOMON, *BOOK XI, VERSE 20*

This appendix contains popular symbolism and famous groupings frequently associated with individual numbers. Number symbolism and groupings originate from a variety of places, times, and cultures, echoing the opening quote from the Old Testament Apocrypha (c. first century B.C.). As discussed in the text, symbolism and groupings are often represented in art. In addition to revisiting some of the ideas presented earlier in the text, this appendix also introduces a wealth of new and interesting information.

ONE

The number one, or *Unity*, was called the *monad* by the Pythagoreans. It was seen as the source of all numbers, good, desirable, essential, and indivisible. "Unity is the principle of all things and the most dominant of all that is: all things emanate from it and it emanates from nothing. It is indivisible and . . . it is immutable and never departs from its own nature through multiplication $(1 \times 1 = 1)$. Everything that is intelligible and not yet created exists in it. . . ."[1]

Vincent Hopper says that the first advance toward counting is with the use of words for *one* and for *many*, the differentiation of the self from the group. We still say "numero uno" to speak of ourselves.

One, the first, the beginning, is also usually identified with the Creator. "In the beginning God created. . . ." Therefore, in the beginning there was only one. One may also be identified with Christ, as in Figure E.1, a painting by Hildegarde von Bingen (1098–1179), who was a German nun, mystic, author, composer of music, and founder and abbess of the convent of Rupertsberg. Here Christ is shown within a set of concentric circles set within a rectangular frame. This "squaring" of the circle is appropriate for a figure believed to have united heaven and earth.

FIGURE E.1 ■ *Man in Sapphire Blue,* Hildegarde von Bingen

TWO

The number two was called the *dyad* by the Pythagoreans. It was the first feminine number and represented duality. Refer back to the Pythagorean Table of Opposites in Chapter 1, "Music of the Spheres," for more information about duality.

Some examples of the symbolism and groupings of two follow.

FIGURE E.2 ■ Janus

Janus: The two-faced Roman god Janus, shown in Figure E.2, saw both ways, inward and outward, and had the wisdom of both past and future. He was the god of beginnings and the guardian of doorways and gates. His image was often placed on a *term* (for *terminus*), a pillar marking a property boundary. Janus later became a symbol of the past and future. January was named for him because he saw the old year out and ushered in the new year. His attribute is a snake in the form of a circle, an ancient symbol of eternity.

Dualities in the Old Testament: The Old Testament has pairs of opposites, which are demonstrated in the familiar passage from Ecclesiastes 3:2–8. It reads "a time to be born, and a time to die; a time to plant, and a time to pluck up . . . a time for war, and a time for peace." Therefore, the number two has long been associated with the concept of opposites.

The Creation: The number two also signifies the possibility of creation and the bearing of offspring. Figure E.3 depicts Adam and Eve, whom, according to the Old Testament, were the first beings created.

Binary Digits: A modern instance of the generating power of a pair of opposites is found in the binary digits 0 and 1. These two digits can combine to form all the numbers, which can then be manipulated to perform the most complex computations via binary arithmetic. They also provide codes for letters of the alphabet and, therefore, can represent textual material. They are the basic building blocks of the digital computer and all the other digital devices of modern technology.

The Chinese Hexagrams: The two binary digits 0 and 1, which can combine in such amazing ways, have something in common with the two elements from

FIGURE E.3 ■ *Adam and Eve*, Dürer

the classic Chinese text *I Ching*. The Chinese divided nature into opposite parts, the masculine *yang* and the feminine *yin*. They believed that heaven and earth produced everything by the interaction of these opposites, just as offspring are the result of the interaction between a male and a female parent. The yin-yang symbol, which is shown in Figure E.4, is also called the *T'ai Chi* symbol, the *Great mondad*, and the *ovum mundi*. Note that the black field has a white spot and the white field has a black spot; this symbolically indicates that nothing is ever completely one or the other.

In the *I Ching*, the two elements are represented graphically by lines, broken and unbroken.

yin = broken line (b)
yang = unbroken line (u)

When objects are organized into groups of three, only eight combinations are possible, just as a three-digit binary number can take on only eight values, from zero to seven. The combinations, shown in Figure E.5, are called the *eight trigrams*. These eight trigrams are thought to represent the eight main forces of the universe and the consequences of the interactions between these forces.

	yin		yang
坤 (Kūn) Earth		乾 (Qián) Heaven	
巽 (Xùn) Wind		震 (Zhèn) Thunder	
離 (Lí) Flame		坎 (Kǎn) Water	
兌 (Dùi) Marsh		艮 (Gèn) Mountain	

FIGURE E.4 ■ The Yin-Yang Symbol

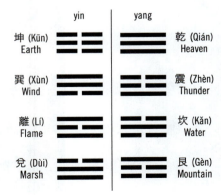

FIGURE E.5 ■ The Eight Trigrams

THREE

The number three was called the *triad* by the Pythagoreans. It was the first odd, masculine number.

In elementary number symbolism, three represents the triad of family: male, female, and child, as shown in Figure E.6. It also represents beginning, middle, and end, and birth, life, and death. In common usage, we say *good* of one, *better* of two, but *best* of three or more. Therefore, three also implies the superlative.

Further, a single occurrence of an event is usually of no statistical significance, and a second occurrence might just be a coincidence, but a third occurrence implies certainty. Such is the root of the Gypsy belief that if a dream comes three times, it is certain to be true.

Additional examples of the symbolism and groupings of three follow.

Reconciliation of Opposites: With three, the dualism associated with the number two was resolved. One way of dealing with dualities was to find a golden mean between two extremes or to find a middle path. Another way was to enclose the dualities within a *unifying principle*, such as is graphically done in the yin-yang symbol by enclosing opposites within the encompassing outer ring.

Three Fates: In Greek and Roman mythology, the three Fates, or daughters of the night, were three sisters who controlled destiny. Figure E.7 is a depiction of the Three Fates by Francesco Salviati. According to the myth, the length of someone's life was determined by the length of a single thread. The role of each sister varied, but Clotho usually held the spool, Lachesis pulled the thread, and Atropos snipped. They are mentioned in Plato's *Republic* (10:617).

Three Witches: The word *magic* comes from the Magi, a priestly caste in ancient Persia who supposedly possessed occult powers and knowledge of astrology, numerology, alchemy, demonology, and magic. By the first century A.D., the Magi were considered wise men and soothsayers, like the Biblical Magi who came from the East to worship the infant Jesus.

FIGURE E.6 ■ Statue of Family Group

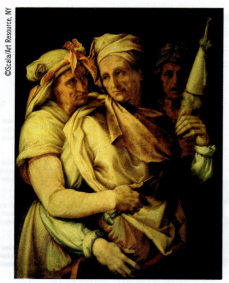

©Scala/Art Resource, NY

FIGURE E.7 ■ *Three Fates*, Salviati

FIGURE E.8 ■ *Three Witches,* Fuseli

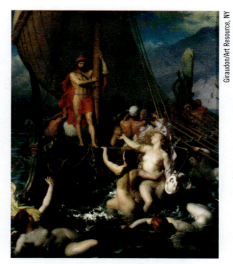

FIGURE E.9 ■ *The Sirens and Ulysses,* Leon Belly

Many numbers were considered magical to early man, but Hopper says that the most potent appear to have been 3, 4, 5, 7, and 9. Magic formulae were often repeated three times. To stop a hailstorm, one would throw three hailstones into a fire and repeat the Lord's Prayer three times; if the tempest was due to witchcraft, it would stop.

If the number three had strong magic, then three times three must have been awesome. The triple triad, or the power of three thrice over, appears in *Macbeth,* Act 1, Scene 3, with the three witches.

> *"The [three] weird sisters, hand in hand,*
>
> *Posters of the sea and land,*
>
> *Thus do go about, about:*
>
> *Thrice to thine and thrice to mine*
>
> *And thrice again, to make up nine.*
>
> *Peace! the charm's wound up."*

Figure E.8 is a depiction of the three witches by Fuseli.

Three Sirens: The three Sirens (Legeia, Leucosia, and Parthenope) shown in Figure E.9, are best known for their part in Homer's *Odyssey.* Their lovely singing lured sailors to their deaths by crashing their vessels into rocks in the Mediterranean Sea, but Odysseus (Ulysses in English) lashed his men to the masts and plugged their ears with wax.

Three Graces: A more appealing trio of women is the Three Graces, shown in Figure E.10. They are Aglaia, Euphrosyne, and Thalia, and they have been variously named Splendor, Mirth, Good Cheer, Beauty, Gentleness, Friendship, Desire, Fulfillment, Abundance, and Chastity. They were handmaidens of Venus and are often shown with the nine muses and Apollo. The Three Graces were a very popular art motif, with involved iconography, although some art historians suggest that it might have been just an excuse to portray nude women. The outer women usually face the viewer, and the middle figure faces away.

FIGURE E.10 ■ *The Three Graces,* Raphael

The Judgement of Paris: A trio of women is also the subject of another favorite art motif, the *judgement of Paris*, shown in Figure E.11.[2]

As the story goes, the king and queen of Troy heard a prophecy that their son Paris would be the ruin of Troy, so they put him on Mount Ida, where he was brought up by shepherds. Years later the three goddesses Hera, Athena, and Aphrodite were at a wedding when someone tossed a golden apple into the crowd inscribed "To the fairest." All three goddesses wanted it and asked Zeus to decide who was the fairest amongst them, but he declined to answer. Instead, he sent them to Mount Ida for Paris to make the decision.

Hera and Athena offered Paris riches, fame, empire, and military glory, but Aphrodite bribed him with Helen, the "most beautiful woman on earth." He voted for Aphrodite, even though Helen was already married to the king of Sparta and Paris himself was in love with a nymph. Eventually, Paris had to abduct Helen, which started the Trojan War.

Trios of Deities: The number three became the most universal number of the deity. The primary gods of the Babylonian pantheon were three in number, Anu, Bel, and Ea, representing heaven, earth, and the abyss.[3]

©Scala/Art Resource, NY

FIGURE E.11 ■ *Judgement of Paris,* Rubens, 1638

FIGURE E.12 ■ Egyptian Deities

One of the best known celestial families came from Egypt; Osiris, Isis, and their son Horus are shown in Figure E.12. Sun worship was one of the most primitive forms of religion. The Egyptians divided the sun god into three: *Horus*, the rising sun, *Ra* or *Rê*, the midday sun, and *Osiris*, the old setting sun who later became ruler of the dead.

The White Goddess: Just as the three positions of the sun are represented by three deities, the moon is represented by a goddess as well. Robert Graves writes about "the White Goddess, of Birth, Love, and Death, visibly appearing as the New, Full, and Old Moon, and worshipped under countless titles. She was beautiful, generous, fickle, wise, implacable."

Abraham and the Three Angels: In the Old Testament, three figures appeared to Abraham. "And he lifted his eyes and looked, and three men stood by him: he ran to meet them . . . and bowed himself toward the ground." The three men are usually interpreted as angels, and in the Christian era they were a symbol of the Trinity. They revealed that a son would soon be born to Abraham's wife Sarah even though she and Abraham were "old and well stricken in age." Sarah eventually bore Isaac, whose birth was later interpreted as a prefiguration of the Annunciation.[4]

The Trinity: In the Christian era, the number two became associated with the duality of the godhead: God the father and Christ. However, according to Vincent Hopper, this duality was a problem because three was the traditional number of deity. He cites the early theologians Clement, Origen, and Hipolytus, who lived in Alexandria and were steeped in number symbolism; they wrote that the duality of the godhead was one weakness with Christianity. This weakness was overcome by the addition of the Holy Ghost, thereby creating the Trinity.

The Holy Trinity represents three persons in one god—the *Father*, *Son*, and *Holy Ghost* or *Holy Spirit*. It is probably the most well-known of all triads and a major theme in religious art. In a typical trinity, as shown in Figure E.13, the Father is shown as an old, bearded man placed above Christ, with the Holy Ghost as a hovering dove. The Father is the only one ever depicted wearing a triangular halo.

Early Christianity showed few depictions of the Father as a man because he is unseen and unknowable. Instead he may have been depicted as an eye or as a hand reaching down from a cloud. In addition, artists sometimes resorted to geometric representations of the Trinity, such as three interlocking circles or the equilateral triangle.

Three Theological Virtues: Another trio in Christianity, and a common art subject, are Faith, Hope, and Charity.

FIGURE E.13 ■ *Trinity*, Dürer

FIGURE E.14 ■ *Charity*, Bougereau

Faith is often depicted holding a cross or a chalice, the cup used at the last supper. In later art, Faith is also depicted holding an open book representing the Bible, or a candle representing the light of faith. She wears a helmet to protect against unbelievers. She is sometimes shown holding her hand to her breast and sometimes with her foot resting on a block, the foundation of faith; nearby is a baptismal font symbolizing initiation into the faith.

Hope is often shown with her eyes raised to heaven and her hands in prayer or reaching for a crown, symbolizing the hope of future glory. She may also be shown holding flowers, the hope of fruit to come. Depictions of Hope with an anchor are from St. Paul, who wrote, "We might have a strong consolation, who have fled for refuge to lay hold upon the hope set before us: Which [hope] we have as an anchor of the soul, both sure and stedfast. . . ."[5]

Charity, shown in Figure E.14, is considered the greatest of the three virtues due to Paul's letter to the Corinthians. "And now abideth faith, hope, and charity, these three; but the greatest of these [is] charity." She is sometimes shown performing the acts of mercy mentioned in the book of Matthew. "For I was hungry, and ye gave me meat: I was thirsty, and ye gave me drink: I was a stranger, and ye took me in: Naked, and ye clothed me: I was sick, and ye visited me: I was in prison, and ye came unto me." Charity has also been interpreted as the love of God, so she is sometimes shown holding a flame in her hand or in a vase, a flaming heart, or a candle. An attribute of Charity is the pelican, which feeds its young with its own blood. According to Fisher, by the sixteenth century, depictions of a nursing Virgin Mary had fallen out of favor with the church. It was perhaps too earthy, and so Charity was depicted instead.

FOUR

The number four was known to the Pythagoreans as the *tetrad*. It was the first square number and symbolized justice, steadfastness, and the square.

©Scala/Art Resource, NY

FIGURE E.15 ■ The Symbols of the Evangelists, Ceiling Mosaic in the Archbishop's Palace, Ravenna, Italy

Additional examples of the symbolism and groupings of four follow:

Four Ages of the World: Ovid writes of the Four Ages of the World.[6]

Gold, the first, free of fear and conflict, a paradise

Silver, the second, where man had to seek shelter

Bronze, the third, aggressive but not yet entirely evil

Iron, the fourth, with treachery, violence, greed, deceit, and war

The Four Gospels and Four Evangelists: The four gospels of the New Testament were written by the four evangelists: Matthew, Mark, Luke, and John. It is hard to find a cathedral that does not portray the four evangelists. They are also a common theme in early manuscripts.

As shown in Figure E.15, three of the four evangelists are often represented by animals: the ox for Luke, the eagle for John, and the lion for Mark. Carl Jung says that the origin of the animal representations is Ezekiel 1:5, which states ". . . of the midst there came the likeness of four living creatures . . . And . . . they had the likeness of a man . . . And every one had four faces . . . the face of a man, and the face of a lion, on the right side: . . . the face of an ox on the left side; . . . they four also had the face of an eagle."

The beasts are also found in the New Testament. "And . . . in the midst of the throne, and round about the throne, [were] four beasts full of eyes before and behind. And the first beast [was] like a lion, and the second beast like a calf, and the third beast had a face as a man, and the fourth beast [was] like a flying eagle."[7] Because they appear in the Apocalypse, they are sometimes called the apocalyptic beasts. Perhaps the most familiar apocalyptic beast is the *Lion of St. Mark*, which has become a common artistic motif, especially in Venice and the surrounding Veneto.

Four Cardinal Virtues: Plato, in his *Republic*, mentions four virtues: Temperance, Justice, Fortitude, and Prudence, which are called the *four cardinal virtues*. Like the three Ecclesiastical virtues from the last section, they were a popular art motif.

Temperance is usually shown with a bridle, sometimes with bit in mouth. She is sometimes shown diluting wine with water, signifying abstinence from drink. When she is depicted holding a pitcher and a torch, Temperance is indicating the dousing of the fires of lust. The presence of a clock symbolizes a well-regulated life.

Justice is usually shown with a sword indicating power, or *fasces* (a bundle of wooden rods used to beat lawbreakers, including an axe used for beheading, bound with a strap), the Roman symbol of authority, or scales and a blindfold, representing impartiality.

Fortitude (or Courage, Endurance, Strength) often appears in armor with a helmet, spear or sword, and shield.

Like Janus, Prudence (or Wisdom) can be shown with a second male face, suggesting that she can see both past and future. The male and female faces suggest a person who can bridge two worlds: material and spiritual, eternal and temporal, upper and lower. In Andrea Della Robbia's *Allegory of Prudence* (not shown here), a mirror further suggests that the wise can see themselves clearly. The snake may symbolize Jesus' advice to his disciples, "I send you forth as sheep among wolves: be wise as serpents and harmless as doves."[8]

The Four Horsemen of the Apocalypse: The book of Revelation, or the Apocalypse of John, is filled with more numerical references than any other book of the New Testament. Perhaps the most famous group of four in Revelation is the four horsemen, who are depicted in Figure E.16. "Loose the four angels which are bound in the great river Euphrates. And the four angels were loosed . . . to slay the third part of men."[9]

Four Realms of the Universe:[10] We have seen the urge to carve the world into four pieces: four directions, four continents, four elements, four winds, four rivers of earth and four of hell, four seasons, four divisions of the day, and four ages of the world. This urge was perpetuated in the medieval notion of the *four hierarchies of the universe*, which are as follows:

The Cosmic Mind: Supercelestial, incorruptible, stable, contains prototypes for everything below

The Cosmic Soul: Translunary, incorruptible, not stable, celestial, contains the nine spheres of heaven (empyrium, fixed stars, sun, moon, and the five known planets)

The Realm of Nature: Sublunary, terrestrial, corruptible; man, animals, plants

The Realm of Matter: Formless, lifeless; dirt, rocks

FIGURE E.16 ■ *Four Horsemen,* Dürer, 1498

The Four Worlds and Four Elements of the Cabala: The four hierarchies of the universe are similar to, and may have been derived from, the four planes of reality of the Cabala.

Atziluth: Archetypal word, realm of pure idea

Briah: Creative world, containing patterns

Yetzirah: Formative world

Assiah: Material world

The Cabala (or Kabbalah) was a strong influence in the Middle Ages. It was a kind of mysticism that arose among Jews at the start of common era as a reaction against the sober and austere teachings of Rabbinical Judaism. The name is from the Hebrew *kibble* or *quibbel*, meaning "to receive," indicating the Cabala as a received doctrine professing to teach the secret or mystical sense of the holy scriptures.

Developing alongside the Cabala was the science of Gematria, from the Greek γεωμετρια, which is the same word for *Geometry*. Gematria was the practice of assigning numerical values to the letters of the alphabet and thereby deriving new meanings and relationships from Biblical passages.

Additional groups of four that are discussed in this book are listed here, along with chapter references.

- **Ten Sets of Four Things:** See Chapter 1.
- **Four Subjects of the Quadrivium:** See Chapter 1.
- **Four Cardinal Directions:** See Chapter 4.
- **The Four Elements:** See Chapter 4.
- **The Four Continents:** See Chapter 4.
- **Four Rivers of the Earth:** See Chapter 4.
- **Four Rivers of Hell:** See Chapter 4.
- **Four Winds:** See Chapter 4.
- **Four Humours:** See Chapter 4.
- **Four Seasons:** See Chapter 11.
- **Four Times of Day:** See Chapter 11.

FIVE

The number five was called the *pentad* by the Pythagoreans. It was their masculine marriage number, uniting the first female number and the first male number by addition. It was also the number of regular solids or polyhedra and the number of Platonic solids. It was considered incorruptible because multiples of 5 end in 5.

Five is the number of fingers or toes, and the four limbs plus the head also add up to five; therefore, five is the symbol for the flesh. The five senses are portrayed in art as follows: a musical instrument for hearing, a mirror for sight, fruit for taste, and flowers or perfume for smell. Touch has no consistent representation in art.

Additional examples of groups of five follow.

The Pentateuch: The number five or *Penta* appears in the *Pentateuch*, the name of *The Old Law*, which is comprised of the first five books of the Bible. They are attributed to Moses and are also called the *Torah*.

FIGURE E.17 ■ *St. Francis*, Dürer

FIGURE E.18 ■ The Passion Flower

The Five Stigmata and the Cross: The cross is generally associated with the number four, but it can also be associated with the number five if the intersection is included. It may then symbolize the four directions and *here*. The message might be that you can go North, South, East, or West, but you really belong *here*.

This five-point interpretation of the cross is connected to the other main symbolism of five in Christianity. Five is the number of wounds, or *stigmata*, received by Jesus on the cross and the number of stigmata received by St. Francis on Mt. Averna, as depicted in Figure E.17. It is the subject of many paintings.

The passion flower in Figure E.18 symbolizes events in the last hours of the life of Jesus, the Passion of Christ. The corona represents the crown of thorns; the styles represent the nails used in the Crucifixion; the stamens represent the five stigmata; and the five sepals and five petals represent ten of the twelve apostles. Excluded are Judas, who betrayed Jesus, and Peter, who denied him.

The Franciscan habit is in the shape of a Latin cross, and Franciscan churches are also built in the shape of a Latin cross. Catherine of Siena is depicted with stigmata.

> For the symbolism of six-sided or six-cornered figures, see the section on Hexagons and Hexagrams in Chapter 5, "Polygons, Tilings, and Sacred Geometry."

SEVEN

The number seven was called the *heptad* by the Pythagoreans. It was considered the virgin number because seven alone has no factors, and seven is not a factor of any number within the decad. A circle cannot be divided into exactly seven parts by any known construction.

There are seven openings in the head and seven colors of the rainbow: red, orange, yellow, green, blue, indigo, and violet. There are seven stars in the Big Dipper, which is the most prominent of the so-called indestructible stars (the circumpolar).

Additional examples of groupings of seven follow.

The Seven Sisters: The constellation Taurus contains several hundred stars, including a cluster of seven stars called the Pleiades, that can be seen without a telescope. The *Seven Sisters* name is derived from the Seven Sisters of Greek mythology: Alcyone, Maia, Electra, Merope, Taygete, Celaeno, and Sterope, the daughters of Atlas and Pleione. Figure E.19 is Vedder's depiction of the Seven Sisters.

FIGURE E.19 ■ *The Pleiades,* Elihu Vedder

These stars have also appeared in the myths and literature of many cultures. Tennyson wrote of them, "Many a night I saw the Pleiades, rising thro' the mellow shade / Glitter like a swarm of fireflies tangled in a silver braid."[11] In ancient times, the spring rising of the Pleiades in the northern hemisphere signaled the start of the farming season, and their autumn setting marked the end of that season.

The Seven Days of the Week: Seven is one of the main astrological numbers. Because the four phases of the moon make a complete lunar cycle of 28 days, each phase is seven days, defining the week. Three days are named for the sun, moon, and Saturn, and the remaining four days are named for gods of Teutonic mythology: *Wotan,* the supreme deity, his wife *Frigg, Thor,* the god of thunder, and *Tiw,* the war god.

Seven Planets: Seven is also identified with the seven "planets" known in the Middle Ages: the sun, the moon, Mars, Mercury, Jupiter, Venus, and Saturn. The sun and moon, which are not planets in the modern sense, were grouped with the five bodies we now consider planets. With seven planets and seven days of the week, it was natural to associate the days with planets, and even to name some of the days after planets, as shown in Figure E.20. We will discuss each planet later in this appendix under the nine spheres of heaven.

FIGURE E.20 ■ Symbols for the Planets and Days of the Week

Seven Metals: Medieval alchemy identified seven metals known at the time. Alchemy is usually considered to be the forerunner of modern chemistry.

The "Canon's Yeoman's Tale" from Chaucer's *The Canterbury Tales* relates these metals to the planets.

> *"As for the seven bodies that I mention*
> *Here they all are, if they are worth attention:*
> *Gold for the sun and silver for the moon,*
> *Iron for mars and quicksilver in tune*
> *with Mercury, lead which prefigures Saturn*
> *And tin for Jupiter. Copper takes the pattern*
> *Of Venus if you please!"*[12]

Correspondences Between Groups of Seven: Just as you have seen with the number four, seven also has its correspondences. Each planet is identified with a specific day of the week and with a particular metal. The relationships of these groups to the heptagon and the heptagram were shown in Figure 5.57 in Chapter 5, "Polygons, Tilings, and Sacred Geometry." The following table summarizes those relationships.

Relationship of Planets and Metals to the Days of the Week

Day	Origin of Name	Planet	Metal
Sunday	*Sun day*	Sun	Gold
Monday	*Moon day*	Moon	Silver
Tuesday	*Tiw's day*	Mars	Iron
Wednesday	*Wotan's day*	Mercury	Mercury
Thursday	*Thor's day*	Jupiter	Tin
Friday	*Frigg's day*	Venus	Copper
Saturday	*Saturn's day*	Saturn	Lead

The Seven Days of Creation: The number seven appears in several interesting places in the Old Testament. The most famous groupings of seven are mentioned in the first few pages of Genesis, the seven days of creation. Figure E.21

©Scala/Art Resource, NY

FIGURE E.21 ■ *Creation of Adam,* Michelangelo

shows Michelangelo's depiction of the creation of man on the sixth day. Note the strong diagonal separating the earthly realm from the celestial.

The notion that the seventh day was for rest led to the adoption of the weekly Sabbath as a day of rest. However, other sacred times of the calendar were also based on seven, such as the seventh or sabbatical year. "And . . . then shall the land keep a Sabbath. . . . Six years shall you sow your field, and prune thy vineyard, and gather fruit. But the seventh year shall be a Sabbath of rest unto the land."[13]

There was also the *Jubilee year* in which leases were to expire, everyone was to return to his ancestral estate, and slaves were to be freed. "And you shall number seven Sabbaths of years unto thee, seven times seven years; Then shalt thou cause the trumpet of the jubilee to sound."[14]

The Seven-Branched Menorah: Early in the Old Testament, we also see the number seven as the number of days of Passover and the number of branches of the menorah.

> *"And thou shalt make a candlestick [of] pure gold . . .*
>
> *. . . six branches shall come out of the sides of it;*
>
> *three branches . . . out of the one side,*
>
> *. . . and three branches out of the other side*
>
> *. . . Three bowls made like unto almonds . . . in one branch*
>
> *. . . and three bowls made like almonds in the other branch.*
>
> *. . . And you shall make the seven lamps thereof . . ."*[15]

Robert Graves gives the menorah cosmic significance by comparing the seven flames to the seven planets.[16] He cites Zechariah 4:1–10, which states, "And the angel . . . came again, and waked me. . . . And he said to me, What seest thou? And I said, I have seen . . . a candlestick all of gold, with its bowl upon the top of it, and its seven lamps thereon; . . . [these are] the eyes of Jehovah, which run to and fro through the whole earth." Graves interprets the seven eyes to mean the seven planets.

The Flood: Other groups of seven that appear in the Old Testament are included in the great flood. Noah had seven days to prepare before the flood. He was commanded to take seven pairs of clean beasts and birds: "Of every clean beast thou shalt take to thee by sevens, the male and the female: and of beasts that are not clean by two, the male and the female."[17] *Clean* meant ritually pure, similar to what is meant by *kosher.*

Seven Liberal Arts: Throughout the Middle Ages, the seven arts represented the sum of human learning. The medieval course of study for a Bachelor of Arts consisted of the *trivium,* or grammar, rhetoric, and logic. The course of study for a Master of Arts consisted of the *quadrivium,* or arithmetic, music, geometry, and astronomy. The quadrivium had its origins with the Pythagoreans, who gave us the very word *mathematics.* Looking for more correspondences, we note that Dante identified the seven arts with the seven planetary circles, relating each planet to the appropriate study.[18]

The seven liberal arts were a major theme in art, and one depiction is shown in Figure E.22.

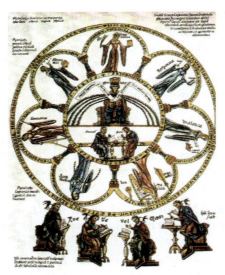

FIGURE E.22 ■ The Seven Liberal Arts, from Herrad of
Landsberg, *Hortus delicarum,* Twelfth Century

Allegories and Personifications: Each of the Seven Liberal Arts is personified
by a historical figure and represented by an allegorical figure, as shown in the
following table.

The Seven Liberal Arts Personified

Liberal Art	When personified by a historical figure is . . .	Allegories may show a female figure with . . .
Grammar	Priscian or Donatus	Writing instruments, rod for discipline
Rhetoric	Cicero	Scroll, book
Logic	Aristotle	Scorpion or snake, scales
Arithmetic	Pythagoras	Tablet with figures, abacus, fingers for counting
Music	Jubal	Musical instruments
Geometry	Euclid	Compass, square
Astronomy	Ptolemy	Astrolabe, celestial globe, armillary sphere

Interestingly, allegories to the liberal arts are usually portrayed by women,
such as the *Allegory to Geometry* in Figure E.23. Critchlow writes, "The beginning
of applied mathematics could be considered the achievement of the womenfolk
of any given community, discovering in the principles of weaving, knotting,
plaiting, braiding, and even counting and storing the foods for the winter
months. A particular example of this mathematical ability is found in the memo-
rizing of many highly complicated carpet patterns; in many communities in the
Islamic countries a woman's status is founded on the number of patterns she
has memorized . . . from threaded beads was presumably developed the abacus—
following from the lost language of knotted cords."[19]

A good source for the pictorial representation of allegories, vices, virtues,
character traits, and more is Cesare Ripa's *Iconologia* (1758–1760), containing
200 such depictions. See the Bibliography for further details.

FIGURE E.23 ■ *Allegory of Geometry*
Follower of Laurent de La Hyre (French), *Allegory of Geometry*, after 1649, oil on canvas, 40 × 62 (101.6 × 158.6 cm) Toledo
Museum of Art. Purchased with funds from the Libbey Endowment. Gift of Edward Drummond Libbey, 1964.124. © 2006

Seven Virtues: We have already discussed the three theological virtues: Faith, Hope, and Charity. To these we add the four cardinal virtues—Temperance, Justice, Fortitude, and Prudence—which are mentioned by Plato in his *Republic*. The three theological and four cardinal virtues combine to make one of the best-known heptads, the Seven Virtues.

Given the Seven Virtues and the Seven Liberal Arts, its not surprising that seven has been long been associated with the number of steps toward perfection, purification, or wisdom. Examples of this include the seven sages of Greece and the seven Hathors of Egypt, who were seven female genii who appeared at the birth of a child and foretold its future. From the Old Testament, we have, "Wisdom has built her house, she has hewn out her seven pillars."[20]

Seven Vices: To offset the Seven Virtues, we have the Seven Vices, or Seven Deadly Sins, as shown in Figure E.24. Gluttony, Lechery, Avarice, Luxury, Wrath, Envy, and Sloth.[21]

FIGURE E.24 ■ The Seven Deadly Sins

FIGURE E.25 ■ *The Woman Clothed in the Sun and the Seven-Headed Dragon,* Dürer

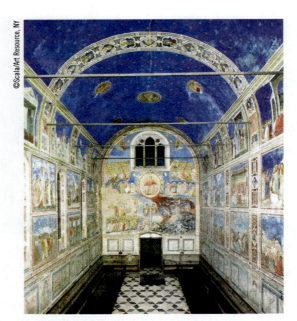

FIGURE E.26 ■ Scrovegni Chapel Frescoes, Giotto

The Seven Sins are sometimes depicted as the seven-headed dragon, such as in Figure E.25. It was a favorite subject of Dürer's. "And there appeared a great red dragon, having seven heads and ten horns, and seven crowns upon his heads. . . . And there was war in heaven: Michael and his angels fought against the dragon. . . . And the great dragon was cast out, that old serpent, called the Devil, and Satan."[22]

The seven-headed dragon has sometimes been equated with Rome, with its seven hills.

Vices Versus Virtues: With Seven Virtues and Seven Vices, we need to have a battle, of course, and vices and virtues are sometimes shown in direct opposition.[23] For example, Giotto's Scrovegni Chapel fresco, *The Last Judgement,* shown in Figure E.26, shows God with his right-hand palm up toward the saved, and along that wall are the Seven Virtues. His left hand is palm down toward the damned, and along that wall are the Seven Vices. Each vice is located directly opposite its corresponding virtue in the following table.

Virtue Versus Its Corresponding Vice

Virtue	Vice
Prudence	Foolishness
Fortitude	Inconstancy
Temperance	Ire
Justice	Injustice
Faith	Faithlessness
Hope	Desperation
Charity	Envy

Seven Ages of Man: The Seven Ages of Man is another common motif in art and literature. From Shakespeare we have the famous passage:

"All the world's a stage,

And all the men and women merely players:

They have their exits and their entrances;

And one man in his time plays many parts,

His act being seven ages." [24]

He goes on to describe each of the seven ages, from "the infant, mewling and puking in the nurse's arms" to "second childishness and mere oblivion, Sans teeth, sans eyes, sans taste, sans everything."

We also find reference to the three ages of man (childhood, adulthood, and old age), four ages of man corresponding to the four seasons,[25] and twelve ages, corresponding to the months. For each, the underlying theme is *vanitas*, the transience of human existence, the passage of youth and beauty, and the inevitability of death.

EIGHT

The number eight, the *ogdoad*, was the first feminine cube to the Pythagoreans, the cube of two.

Some popular groupings of eight follow.

Chanukah Candlestick: In contrast to the menorah of Exodus, which has seven candles, the Chanukah menorah has eight candles. It celebrates a miracle that occurred when the temple in Jerusalem was rededicated[26] after having been desecrated by Antiochus Epiphanes, after the war of the Maccabees (167–160 B.C.). Only a small amount of oil was found to light the menorah, but it lasted for eight days. A ninth candle, the Shamesh, is used to light the other eight, one night at a time, so most Chanukah candlesticks have nine candles. The candles are lit during the eight days of Chanukah, the festival of lights, which occurs near the winter solstice.

The Eight Winds: Earlier we mentioned the four winds; however, Vitruvius, in his *Ten Books of Architecture*, says there are eight, as shown in Figure E.27. "Some have held that there are only four winds: Solanus from due east, Auster from the south; Favonius from due west; Septentrio from the north. But more careful investigators tell us that there are eight."[27] Note that Vitruvius' names for the four winds are somewhat different from those given earlier in Chapter 4, "Ad Quadratum and the Sacred Cut." Vitruvius goes on to show how to lay out streets in order to avoid the disagreeable effects of the eight winds.

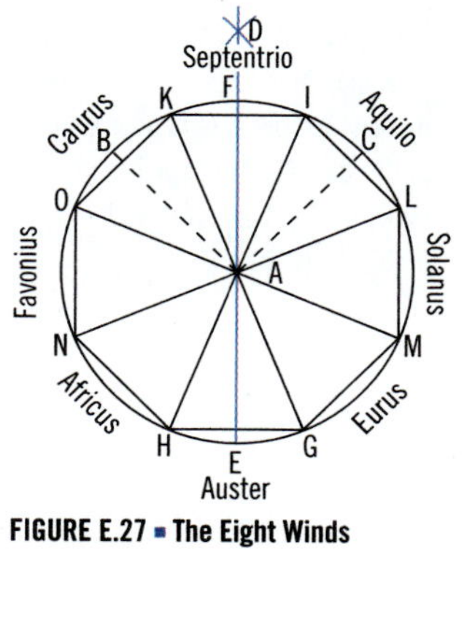

Other references to the symbolism of eight are as follow:

- **Eight as a Symbol of Baptism:** See Chapter 5.
- **The Eight Canonical Hours:** See Chapter 11.

FIGURE E.27 ■ The Eight Winds

NINE

The number nine, the *ennead*, was the first masculine square to the Pythagoreans. It was called *incorruptible* because it reproduces itself when multiplied by any number. For example, $9 \times 6 = 54$ and $5 + 4 = 9$. This rule applies when multiplying 9 by any number, however large. In the following section, you will

FIGURE E.28 ■ *La danza d Apollo con le Muse*, Baldassare Peruzzi

see that the number ten symbolizes *completeness*, so the number nine has come to represent *almost complete*. Troy was besieged for nine days and fell on the tenth. Odysseus wandered for nine years and arrived home on the tenth.

Some popular groupings of nine follow.

The Nine Muses: In Greek mythology, the nine muses were the nine daughters of Zeus and Mnemosyne, the goddess of memory. Refer to Figure E.28. The muses presided over the arts and sciences and were believed to inspire all artists, especially poets, philosophers, and musicians. Calliope was the muse of epic poetry, Clio of history, Euterpe of lyric poetry, Melpomene of tragedy, Terpsichore of choral songs and dance, Erato of love poetry, Polyhymnia of sacred poetry, Urania of astronomy, and Thalia of comedy. Note that Thalia the muse is not to be confused with Thalia, one of the three graces. The muses can be difficult to identify in art as their attributes are not always consistent from artist to artist.

The nine books of Herodotus' *History* are named for the nine muses—probably by someone other than the author.

> This material on the heavens is an extension of our discussion of the medieval spheres of heaven begun in Chapter 11, "The Sphere and Celestial Themes in Art and Architecture."

The Nine Medieval Spheres of Heaven: In medieval times, the universe was subdivided into four concentric realms or hierarchies, which were listed earlier under the number four. Realm 2, the Cosmic Soul, was further subdivided into the *nine spheres of the heavens*; the sun and moon, the five known planets, the fixed stars, and the primum mobile.

Nine Choirs of Angels: Each of the nine spheres of heaven had its own particular kind of angel, as shown in Figure E.29. An *angel*, from the Greek *angelos* or "messenger," is a messenger of the gods, such as Iris and Apollo. The angels' other functions included making announcements, protection, and punishment. There were supposedly nine distinct orders of angels, arranged in three ranks of three—a triple trinity.

Angels were assigned to particular spheres of heaven as follows: *angels* with the moon, *archangels* with Mercury, *princedoms* with Venus, *powers* with the sun, *virtues* with Mars, *dominations* with Jupiter, *thrones* with Saturn, *cherubim* with the fixed stars, and *seraphim* with the Empyrean. The source of this arrangement of angels is *The Celestial Hierarchy*, attributed to a sixth-century Neoplatonist called the *Pseudo-Dionysius*.

Dante's Nine Spheres of Heaven: In *Paradiso* Dante and Beatrice travel from Earth through the nine spheres of the medieval heavens. We will give some

FIGURE E.29 ■ *Nine Choirs of Angels*, Hildegarde of Bingen

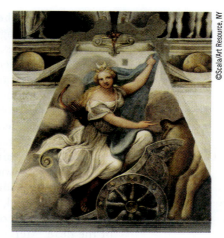

FIGURE E.30 ■ *Diana*, Correggio

information about each sphere here, identify any Greek god or angel with which it is associated, and point out any art motifs connected to it.

> ### Sphere 1: The Moon (*Paradiso*, Cantos II to IV)
>
> *"That orbéd maiden, With white fire laden,*
>
> *Whom mortals call the Moon"*
>
> SHELLY

The most common symbol for the moon is the crescent or lune, which we mentioned in Chapter 7, "Circular Designs in Architecture." Of course, the moon had associations long before Dante—for example, the moon goddess Selene who later became associated with the Greek goddess Artemis and the Roman goddesses Diana and Luna. As shown in Figure E.30, she usually wears a crescent in her hair.

Another association for the moon is, of course, night. For more on moon symbolism, read about Madonna standing on the moon in Chapter 11, "The Sphere and Celestial Themes in Art and Architecture."

> ### Sphere 2: Mercury (*Paradiso*, Cantos V to VI)
>
> *"The planet Mercury, whose place*
>
> *Is nearest to the sun in space*
>
> *Is my allotted sphere."*
>
> LONGFELLOW

FIGURE E.31 ■ *Mercury*, Giovanni da Bologna (Giambologna)

In mythology, Mercury (Greek: Hermes) was a messenger of the gods, and the planet may have been named for him because of its swift movement across the heavens. Hermes conducted the dead to the underworld and was a protector of travelers. Vertical stone pillars, or *herms*, were often set up at roadsides both as cult objects and for practical purposes such as milestones or boundary markers. We showed a herm in Figure 10.8 in Chapter 10, "The Solids."

In art, Mercury is often shown with winged feet or sandals and a winged hat, as in Figure E.31. He is sometimes shown with a *caduceus*, a wand about which snakes are entwined, which is now the symbol of the medical profession.

Mercury is the celestial sphere that has the most interesting rank of angels—the archangel. There are only three archangels: Raphael, Gabriel, and Michael. Gabriel is a messenger. His major role is archangel of the Annunciation, as shown earlier in Figure 4.34 by Fra Angelico, who did some of the countless portrayals of this theme. Raphael is the ideal guardian angel, protector of travelers and children. He appears in the Old Testament Apocrypha *Tobit* in which he is the traveling companion and protector of young Tobias. Michael is a protector of the Hebrews. He was adopted as a saint by the Church. He is often shown slaying a dragon or banishing Lucifer and the rebellious angels to hell. He wears armor, carries a sword, and looks very similar to St. George, the other slayer of dragons, with the exception that Michael has wings.

There are some interesting similarities between Mercury and the archangels. Gabriel, like Mercury, is a messenger of the gods. Raphael, like Mercury, is a protector of travelers. Michael, like Mercury, is involved with the souls of the dead. Mercury is often depicted in art as conducting the dead to the underworld to be weighed, and Michael is often shown actually doing the weighing. Further, sites dedicated to St. Michael are often found on hills where a temple to Mercury is known to have stood.

Sphere 3: Venus (*Paradiso*, Cantos VIII and IX)

"For a breeze of morning moves,
And the planet of love is on high . . ."

TENNYSON

Venus (Greek: Aphrodite) was the goddess of beauty, love, laughter, and reproduction. She was the patroness of prostitutes—sexy, cruel, and capricious. Note that the astrological symbol for Venus shown earlier in Figure E.20 is the same as the biology symbol for female.

From Plato comes the idea of two Venuses: *natural* Venus and *celestial* Venus.[28] Natural Venus was the daughter of Jupiter (Zeus) and Hera (Dione). In one version of the Greek myths, celestial Venus was born from the foam of the sea. Uranus had two of his sons thrown into the underworld. As revenge, their mother, Earth, had her son Saturn (Cronos) castrate Uranus. Drops of Uranus' blood landed on earth and gave birth to the *Three Furies*, but Uranus' private parts fell into the sea and Venus sprang from the foam. The name *Aphrodite* is perhaps derived from *aphros* meaning "foam."

Also from Plato comes the idea that each Venus symbolized a different kind of love, that of the soul and that of the body.[29] This notion of *sacred and profane love* was a common topic in the Renaissance, and Venus was one of the most common themes in art. As Kenneth Clark put it, "Since the earliest times the obsessive, unreasonable nature of physical desire has sought relief in images. . . ." As shown in Figures E.32 and E.33, celestial Venus is typically shown naked, signifying purity and innocence, while earthly Venus is richly dressed and bejeweled. Botticelli's *Birth of Venus*, shown earlier in Figures 4.29 and 4.30, shows the celestial Venus rising from the foam. This pose, with hands covering breasts and pubic area, is called *Venus pudica* or "Venus of modesty."

Again quoting Clark, "to give these images [of the obsessive, unreasonable nature of physical desire] a form by which Venus may cease to be vulgar and become celestial has been one of the recurring aims of European art."[30] Her attributes in art include the apple of discontent, which she won from Paris besting Hera and Athena, her son Cupid (Eros or Amor), and a scallop shell or dolphin, recalling the manner of her birth. She is sometimes attended by the Three Graces.[31]

FIGURE E.32 ■ *Aphrodite of Melos (Venus de Milo),* Second Century B.C.

FIGURE E.33 ■ *Sacred and Profane Love,* Titian, c. 1515

The planet may have been named for her because of its brightness and beauty. The planet Venus at dawn is the *morning star* and is called *Hesperus* or *Vesper*. When the *evening star* at dusk, Venus is called *Phosphorus* (*Lucifer* in Latin).

Sphere 4: The Sun (*Paradiso,* Cantos X to XIV)

"The sun, with all those planets revolving around it and dependent upon it,

can still ripen a bunch of grapes as if it had nothing else in the universe to do."

GALILEO

In Figure E.34, Dante and Beatrice reach the sun, which is depicted by Giovanni di Paolo as a golden wheel sending golden rays to the landscape below. The sun is located in the middle of the orbs, with three above and three below, like the heart in the middle of the body or a wise king in the middle of his kingdom. Recall from Chapter 6, "The Circle," that the circle was often used to symbolize the sun.

Earlier we mentioned that the Egyptians divided the sun god into three. Other sun gods include Helios ("sun" in Greek), who was a Greek sun god who drove his chariot daily across the sky. The *Colossus of Rhodes,* one of the seven wonders of the ancient world, was a bronze statue of Helios over 100 feet high. Apollo, depicted in Figure E.35, was often regarded as a sun god as well. His sister was Diana, the moon goddess, and the two shown together may symbolize universality or the reconciliation of opposites.

FIGURE E.34 ■ *Beatrice and Dante Approach the Sun,* Giovanni di Paolo, c. 1445

FIGURE E.35 ■ *Apollo,* Dürer

FIGURE E.36 ■ *Mars and Venus,* Botticelli, 1483

Sphere 5: Mars (*Paradiso* Cantos XV to XVII)

"As he glow'd like a ruddy shield on the Lion's breast."

TENNYSON

Mars (Ares in Greek), the god of war, was brutal and aggressive. Note that the astrological symbol for Mars, shown earlier in Figure E.20, is the same as the biology symbol for *male*. The month of March derives its name from Mars.

Mars was hated by everyone but Venus, who seemed to love everybody. While married to Vulcan, the deformed god of the forge, she had affairs with Mars, Mercury, Neptune, Dionysus, and the mortal Anchises with whom she bore Aeneas, hero of the *Aeneid*. Her power over men was enhanced by a magic girdle made for her by Vulcan.

A common art subject was Mars and Venus portrayed together, such as in Figure E.36. This depiction was taken as an allegory to valor and beauty and the combined qualities of warrior and lover. It was also an allegory to the triumph of love over strife and of peace over war.

Sphere 6: Jupiter (*Paradiso,* Cantos XVIII to XX)

"Turning I perceived

The whiteness round me of the temperate star

The sixth, whereinto I had been received."

PARADISO, *CANTO XVIII*

Jupiter or Jove (Zeus in Greek) was ruler of the gods and mortals. Like Venus, he had many lovers. He often seduced them disguised as a bull, swan, cloud, satyr, or shower of gold. Recall our earlier reference to Jupiter as the bull who carried off Europa, one of the four continents. His attributes are the thunderbolt and an eagle, which sometimes holds a thunderbolt.

Galileo's discovery of Jupiter's four moons (Io, Europa, Ganymede, and Callisto) was the first example of a celestial motion that did not have the Earth as its apparent center. These four moons are now known as the Galilean moons.

Sphere 7: Saturn (*Paradiso,* Cantos XXI and XXII)

"While Saturn whirls, his steadfast shade

Sleeps on his luminous ring."

TENNYSON

Remember Saturn? He castrated his father, Uranus, leading to the birth of Venus. He is portrayed in Figure E.37 as an old man with a sickle. The sickle or scythe, like the shears of Atropos (one of the three Fates) represents life cut short, the grim reaper eventually mowing down every living thing. The sickle

FIGURE E.37 ■ *Dante and Beatrice at Saturn,* Giovanni di Paolo, c. 1445   FIGURE E.38 ■ *Cronos Devouring Children,* Goya

may be the instrument he used to castrate his father or simply an agricultural implement for this god of agriculture.

According to prophesy, one of Saturn's children would usurp him; to keep that prophesy from coming true, he ate them. Pictures of Saturn devouring his children, such as the one shown in Figure E.38, may also symbolize "sharp-toothed time" devouring all that has been created.

Saturn (Cronos in Greek) came to represent Father Time, probably because of the similarity between *Chronos*, Greek for "time," and *Kronos*, Roman for "Saturn." Saturn's festival in December, the *Saturnalia*, may be the origin of Christmas. Saturday is named for Saturn. Melancholia was also referred to as a *saturnine temperament.*

Sphere 8: The Fixed Stars (*Paradiso,* Cantos XXII to XXVII)

"And verily in the heaven we have set mansions of the stars,

and we have beautified it for beholders."

THE KORAN

The *zodiac* is a belt around the heavens extending 9° on either side of the *ecliptic* (the plane of the earth's orbit and of the sun's apparent annual path). Most of the twelve constellations lying within the ecliptic represent animals, so the ancient Greeks called this zone *zodiakos kyklos,* or *circle of animals.* The signs of the zodiac are believed to have originated in Mesopotamia as early as 2000 B.C., and each month was eventually identified with a sign of the zodiac. See the following table for the signs of the zodiac and their corresponding months of the year.

The Greeks adopted the symbols from the Babylonians and passed them on to the other ancient civilizations. The Chinese also adopted the 12-fold division, but called the signs rat, ox, tiger, hare, dragon, serpent, horse, sheep, monkey, rooster, dog, and pig. Vitruvius wrote about the zodiac and other constellations in Book IX, Chapters III–VI.

We have not seen many angels since we left the archangels. Each of the nine spheres has its own type, but they have not often been depicted by artists. However, in this sphere and in the next, there are very distinct types. In the fixed stars, we have the *cherubim,* little baby-like creatures, perhaps copied from

For more representations of the zodiac, see the references to *the fixed stars* in Chapter 11.

Signs of the Zodiac

Zodiac Sign	Corresponding Dates	Zodiac Sign	Corresponding Dates	Zodiac Sign	Corresponding Dates
Aries (Ram) ♈	March 21–April 19	Leo (Lion) ♋	July 23–Aug. 22	Sagittarius (Archer) ♐	Nov. 22–Dec. 21
Taurus (Bull) ♉	April 20–May 20	Virgo (Virgin) ♍	Aug. 23–Sept. 22	Capricorn (Goat) ♑	Dec. 22–Jan. 19
Gemini (Twins) ♊	May 21–June 21	Libra (Balance) ♎	Sept. 23–Oct. 23	Aquarius (Water Bearer) ♒	Jan. 20–Feb. 18
Cancer (Crab) ♌	June 22–July 22	Scorpios (Scorpion) ♏	Oct. 24–Nov. 21	Pisces (Fish) ♓	Feb. 19–March 20

FIGURE E.39 ■ *Cherubim*, S. Marco Cupola

classical putti. Sometimes they are portrayed as winged heads only. Other times they are shown with three pairs of wings often in the shape of a cross, such as in Figure E.39. In the next higher sphere, we have the *seraphim* that look like the cherubim but are red.

Sphere 9: The Primum Mobile and the Empyrean (*Paradiso*, Cantos XXVIII to XXXIII)

Toward the end of their journey, Dante and Beatrice cross the primum mobile, or *first moved*, the sphere that dictates the motions of the other spheres. They finally reach the *Empyrean*, the highest heavenly sphere, which was supposedly composed by a kind of sublimated fire; it was the uppermost paradise, heaven, and the seat of God.

The image that dominates the closing cantos of *Paradiso* is the cosmic rose. The cosmic rose is depicted literally by Giovanni di Paolo as a flower with nine angels and the Trinity. Doré portrays the cosmic rose more like a Gothic rose window, as shown in Figure E.40.

FIGURE E.40 ■ *Cosmic Rose*, Gustave Doré

TEN

The number ten, which is the number of fingers or toes, is the base of the *decimal number system*. It is the number of *completeness* or *finality*. To the Pythagoreans, it was the *decad*, and the sum of the first four numbers ($1 + 2 + 3 + 4 = 10$). The number ten contains all the numbers, and after ten, the numbers merely repeat themselves.

Some common symbolism and groupings of ten follow.

The Sacred Tetraktys: See Chapter 1.

Ten Commandments: The Old Testament contains the Ten Commandments or Decalog.[32] "So Moses went down unto the people, and spake unto them. He gave the people the ten commandments that the Lord had given him."

Ten Spheres of the Sephiroth: The Cabala speaks of the Ten Spheres of the Sephiroth: (1) Crown, (2) Wisdom, (3) Intelligence, (4) Love, (5) Justice, (6) Beauty or Mercy, (7) Victory or Firmness, (8) Splendor, (9) Foundation, and (10) Kingdom. Because the sefiroth were considered archetypes for everything in the world of creation, it was believed by some that an understanding of their workings would illuminate the inner workings of the cosmos and of history.

The diagram in Figure E.41 shows the Ten Spheres of the Sephiroth, each with its associated name, number, and celestial body. The first four are associated with geometric figures (point, line, triangle, and plane), showing a Pythagorean influence that is also evident in their number, ten, which corresponds to the Pythagorean Tetraktys. The spheres are interconnected by 22 paths, representing the 22 letters of the Hebrew alphabet, the 22 Major Arcana of the Tarot, and by further correspondences too involved for us here.

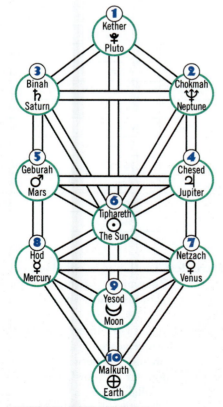

FIGURE E.41 ■ Cabalistic Tree of Life

FIGURE E.42 ■ *The Gods of Olympus,* Nineteenth-Century Engraving

TWELVE

The number twelve is one of the major numbers in astrology because twelve complete lunar cycles take approximately one year. Twelve, therefore, symbolizes a *complete cycle*. The number twelve is also used to divide the day, and also the night, into hours.

Some groupings of twelve follow.

Twelve Gods of Olympus: The table below gives the Greek and Roman names of the twelve gods and their realms or occupations. A depiction of the twelve gods is shown in Figure E.42.

The Twelve Gods and Their Realms

Number	Greek Name	Roman Name	Attribute
1	Zeus	Jupiter	Supreme ruler
2	Poseidon	Neptune	Ruler of the sea
3	Hades	Pluto	Ruler of the underworld
4	Hestia	Vesta	Virgin; hearth & home
5	Hera	Juno	Zeus' wife
6	Ares	Mars	War
7	Pallas Athena	Minerva	Virgin
8	Hephaestus	Vulcan	Fire, forge
9	Phoebus	Apollo	Beautiful
10	Artemis	Diana	Huntress, virgin
11	Hermes	Mercury	Messenger
12	Aphrodite	Venus	Love & beauty; mother of Aeneas

FIGURE E.43 ■ *Hercules and Antaeus,* Pollaiuolo

Twelve Labors of Hercules: Another very popular art subject is Hercules, or Heracles, who was the son of Zeus and an earthly woman. Zeus' wife Hera was enraged by his philandering and put two huge snakes in Hercules' cradle. He, of course, strangled them.

According to legend, Hercules later threw his own children into a fire. As punishment, he had to work twelve years for a certain king who made him do twelve dangerous tasks (some say nine), otherwise known as the *Twelve Labors of Hercules.* These tasks were a popular art motif, as was *Hercules resting from his labors.*[33] The Renaissance artist Pollaiuolo did a series of pieces on Hercules, one of which is shown in Figure E.43. Hercules is often presented as an allegory of good versus evil. He is often portrayed wearing a lion skin because the first of his labors was to kill the invincible Nemean lion.

Twelve Tribes of Israel: In the Old Testament, the twelve tribes of Israel were led by and named for the sons or grandsons of Jacob. They took possession of the Promised Land of Canaan after the death of Moses. "All these [are] the twelve tribes of Israel: and this [is it] that their father spake unto them, and blessed them; every one according to his blessing he blessed them."[34]

The twelve tribes of Israel are also mentioned in the New Testament; "And Jesus said unto them, Verily I say unto you, That ye which have followed me, in the regeneration when the Son of man shall sit in the throne of his glory, ye also shall sit upon twelve thrones, judging the twelve tribes of Israel."[35]

Twelve Apostles: In the New Testament, the main grouping of twelve is the twelve apostles or disciples. They are a popular art motif, especially for cathedrals, such as in Figure E.44. The twelve disciples are sometimes arranged in four groups of three. This arrangement also relates the numbers twelve and seven; both are composed of three and four, one by multiplication and one by addition. Four, seven, and twelve are also important astrological numbers.

Green Grow the Rushes, Ho: Also called The *Twelve Prophets* or *The Carol of Twelve Numbers,* this is a cumulative song with religious content.

"One is one and all alone, and ever more shall be so.

Two, the lily white boys, clothed all in green-ho.

Three, the rivals

Four, the gospel makers

Five for the symbols at your door

Six for the proud walkers

Seven for the seven stars in the sky

Eight for the April rainers

Nine for the nine bright shiners

Ten for the ten commandments

Eleven for the eleven that went up to heaven

Twelve for the twelve Apostles."[36]

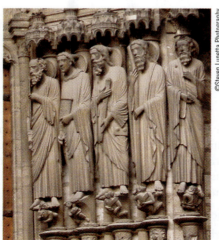

©Steven Lunetta Photography

FIGURE E.44 ■ Five of the Twelve Apostles on Chartres South Porch

Other groupings of twelve include the following:

- **Twelve Signs of the Zodiac:** See Dante's "Nine Spheres of Heaven" in this appendix.
- **Labors of the Twelve Months:** See Chapter 11.

THIRTEEN

Thirteen is the number of faithlessness and betrayal, no doubt stemming from the last supper where Jesus was accompanied by the twelve apostles and Judas.[37] Refer back to Leonardo's *Last Supper*, Figure 12.46.

The notion of the number thirteen as representing bad luck may have first been mentioned by Montaigne (1533–1592), the French writer who introduced the essay as a literary form. He wrote, "And it seemeth I may well be excused . . . if I had rather make a twelfth or fourteenth at a table, than a thirteenth . . .," which is a definite reference to the last supper.

In addition, a witch's coven usually has thirteen members, twelve plus a leader.

Echad Mi Yodea (Who Knows One): This is a traditional Passover song; it is cumulative like "Green Grow the Rushes, Ho." Here we give just the final verse. Bits of it may be heard in Visconti's *The Garden of the Finzi-Continis*, a movie based on the novel of the same name by Giorgio Bassani.

"Who knows thirteen?

I know thirteen.

>   *Thirteen are the temperaments of God;*
>
>   *Twelve are the tribes of Israel;*
>
>   *Eleven are the stars of the Joseph's dream;*
>
>   *Ten are the Commandments;*
>
>   *Nine are the months of the pregnant;*
>
>   *Eight are the days of the circumcision;*
>
>   *Seven are the days of the week;*
>
>   *Six are the books of the Mishnah;*
>
>   *Five are the books of the Torah;*
>
>   *Four are the Matriarchs;*
>
>   *Three are the patriarchs;*
>
>   *Two are the tablets of the covenant;*
>
>   *One is our God, in heaven and on earth."*

FORTY

Forty is the number of *trial* and *privation*. This symbolism possibly originated when the Babylonians observed the 40-day disappearance of the Pleides, which coincided with the rainy season, storms, floods, trial, and danger. The Pleides' return marked the start of the New Year Festival.

Other dire occurrences of forty include the following:

- 40 years in which the Hebrews wandered the desert
- 40 days and nights of the great flood
- 40 years of Philistine dominion over Israel
- 40 days of Moses on Sinai

FIGURE E.45 ■ *The Temptation of Christ on the Mountain,* Duccio, 1308–1311

- 40 days of Elijah's journey
- 40 days of mourning for Jacob
- 40 days and Nineveh shall be overthrown, said Jonah
- 40 days of Lent, the period of fasting, self-denial, and penitence traditionally observed by Christians in preparation for Easter
- 40-day period of isolation in a Venetian port, which survives in the word *quarantine,* after the Italian *quaranta giorni*

Forty Days in the Wilderness: After Jesus' baptism, he was in the wilderness for forty days, tempted by the devil, as depicted in Figure E.45. "And Jesus being full of the Holy Ghost returned from Jordan, and was led by the Spirit into the wilderness; Being forty days tempted of the devil. And in those days he did eat nothing: and when they were ended, he afterward hungered. And the devil said unto him, If thou be the Son of God, command this stone that it be made bread. And Jesus answered him, saying, It is written, That man shall not live by bread alone."[38]

This is probably a New Testament echo of the Old Testament's use of forty as a period of trial and privation; the forty days of Jesus' temptation recalls the forty days of Elijah's solitude and the forty days of the flood.[39] The forty days of lent commemorate this event.

666

The number 666 is perhaps the most provocative number in the Bible, and it is the mysterious Number of the Beast of Revelation. The Book of Revelation 13:18 says: "And here is wisdom. Let him that hath understanding count the number of the beast: for it is the number of a man; and his number is Six hundred threescore and six."

Some have interpreted the beast as the Antichrist, Nero, Mohammed, or the Pope. The theologian Peter Bengus wrote a 700-page book devoted mostly to the number 666, which he found to be equivalent to the name of Martin Luther. Luther replied with an analysis equating 666 to the duration of the Papal regime and rejoiced that it was so near its end. It seems apparent that by selecting the appropriate alphabet, word, translation, or system of interpretation, you can derive any meaning of the number that you want.

SOURCES

Carr-Gomm, Sarah. *Dictionary of Symbols in Western Art*. New York: Facts on File, 1995.
Clark, Kenneth. *The Nude*. New York: Doubleday, 1959.
Critchlow, Keith. *Time Stands Still*. New York: St. Martin's, 1982.
Fisher, Sally. *The Square Halo*. New York: Abrams, 1995.
Graves, Robert. *The White Goddess*. New York: Farrar, 1948.
Hopper, Vincent. *Medieval Number Symbolism*. New York: Columbia, 1938.
Panofsky, Erwin. *Studies in Iconology*. New York: Harper, 1939.

NOTES

1. Theon of Smyrna, *Mathematics Useful for Understanding Plato*, p. 66.
2. Carr-Gomm, p. 168; Fisher, p. 160.
3. Jacobus, et al., eds. *A New Standard Bible Dictionary*. New York: Funk and Wagnalls, 1926.
4. Genesis 18:1.
5. Hebrews 6:19.
6. Metamorphoses, I, 89–150, Carr-Gomm, p. 16.
7. Revelation 4:6.
8. Matthew 10:16.
9. Revelation 9:13.
10. Panofsky, *Studies in Iconology*, p. 132.
11. From "Locksley Hall" by Alfred Lord Tennyson (1809–1892). *English Literature*, p. 813.
12. Nevill Coghill translation, 1952. Penguin, p. 481.
13. Leviticus 25:1–4.
14. Leviticus 25: 8–11.
15. Exodus 25:31–37.
16. *Hebrew Myths*, p. 53.
17. Genesis 7:2–3.
18. Il Convito, Book ii.
19. Crichlow, p. 96.
20. Proverbs 9:1.
21. Panofsky, *Studies in Iconology*, p. 224.
22. Revelation 12:3.
23. Carr-Gomm, p. 226.
24. *As You Like It*, Act II, Scene vii.
25. According to Diogenes Laertius, the Pythagoreans divided the ages of life as a boy for twenty years, a young man for twenty years, a middle-aged man for twenty years, and an old man for twenty years. See Guthrie p. 143.
26. Graves, p. 469–470.
27. Vitruvius, Book 1, Chapter VI, para. 4.
28. *Symposium*, p. 169.
29. Panofsky, p. 141, Clark, p. 109.
30. Clark, p. 109.
31. Clark, p. 109.
32. Exodus 19:25.
33. Carr-Gomm, p. 111.
34. Genesis 49:28.
35. Matthew 19:28.
36. Margaret Boni, ed. *Fireside Book of Folk Songs*. New York: Simon and Schuster, 1947.
37. Hopper, p. 130.
38. Luke 4:1–4.
39. Hopper, p. 71.

APPENDIX F

Summary of Facts and Formulas

Appendix F presents every formula that appears in this text. Statements of mathematical facts, while not true "formulas," are also included. They are presented in sequence here, and may not always appear in the same order as in the text. The formula number is given in the first column, and the name of the formula, if applicable, is in the second column. The formula or statement itself is in the third column, and the page(s) on which it appears is in the fourth column. Note that many equations have quantities that are not defined here, but are given on the text page on which the formula appears.

Ratio and Proportion

No.	Name	Formula	Page
1		The product of means equals the product of extremes.	13
2		The extremes may be interchanged.	13
3		The means may be interchanged.	13
4		The means may be interchanged with the extremes.	13
5	Quadratic Formula	The solution to the quadratic equation $ax^2 + bx + c = 0$ is $$x = \frac{-b \pm \sqrt{b^2 - 4ac}}{2a}.$$	41
6	Golden Ratio	a. The golden ratio divides the whole so that the smaller part is to the larger as the larger is to the whole.	40
		b. $\Phi = \dfrac{1 + \sqrt{5}}{2} \cong 1.61803\ldots$	42
		c. $\Phi = \dfrac{1 - \sqrt{5}}{2} \cong -0.61803$	42
7	Reciprocal of the Golden Ratio	a. $\dfrac{1}{\Phi} = \dfrac{2}{1 + \sqrt{5}} \cong 0.61803$	42
		b. $\dfrac{1}{\Phi} = \Phi - 1$	

No.	Name	Formula	Page
8	Square of the Golden Ratio	$\Phi^2 = \Phi + 1$	42
9	Arithmetic Mean	Arithmetic Mean $b = \dfrac{a + c}{2}$	17
10	Geometric Mean or Mean Proportional	Geometric Mean $b = \pm\sqrt{ac}$	11, 18
11	Harmonic Mean	Harmonic Mean $b = \dfrac{2ac}{a + c}$	18

Sequences and Series

No.	Name	Formula	Page
12	Arithmetic Progression	Recursion Formula $$a_n = a_{n-1} + d,$$ where d = common difference	17
(13)	Arithmetic Progression	General Term $$a_n = a + (n - 1)d$$	17
14	Geometric Progression	Recursion Formula $$a_n = ra_{n-1},$$ where r = common ratio	18
(15)	Geometric Progression	General Term $$a_n = ra_{n-1}$$	18
16	Golden Sequence	$1 \quad \Phi \quad \Phi^2 \quad \Phi^3 \quad \Phi^4 \quad \Phi^5 \ldots$	44
17	Fibonacci Sequence	$1 \quad 1 \quad 2 \quad 3 \quad 5 \quad 8 \quad 13 \quad 21 \ldots$	47
18	Recursion Formula for the Fibonacci Sequence	$F_n = F_{n-1} + F_{n-2}$	47
19	Ratio of Two Consecutive Terms of a Fibonacci Sequence	$\Phi = \lim\limits_{n \to \infty} \dfrac{F_n}{F_{n-1}}$	47
20	Binet's Formula	a. $F_n = \dfrac{1}{\sqrt{5}}\left[\left(\dfrac{1 + \sqrt{5}}{2}\right)^n - \left(\dfrac{1 - \sqrt{5}}{2}\right)^n\right]$ b. $F_n = \dfrac{1}{\sqrt{5}}\left[\Phi^n - \left(-\dfrac{1}{\Phi}\right)^n\right]$	48

Straight Lines and Angles

No.	Name	Formula	Page
21	Units of Angular Measure	1 revolution $= 360° = 2\pi$ rad $1° = 60$ minutes 1 minute $= 60$ seconds	64, 173

No.	Name	Formula	Page
22	Opposite and Adjacent Angles	For two intersecting lines: a. Opposite angles are equal to each other. b. Adjacent angles are supplementary.	64
23	Parallels Cut by a Transversal	If two parallel lines are cut by a transversal: a. Corresponding angles are equal. b. Alternate interior angles are equal.	65
24	Parallels Cut by Two Transversals	When a number of parallel lines are cut by two transversals, the ratios of corresponding segments of the transversals are equal.	65

Any Polygon

No.	Name	Formula	Page
25	Sum of the Interior Angles of a Polygon	$$\text{Sum} = 180°(n - 2),$$ where n is the number of sides in the polygon.	133
26	Sum of the Exterior Angles of a Polygon	$$\text{Sum of exterior angles} = 360°$$	134

Any Triangle

No.	Name	Formula	Page
27	Altitude and Base	The *altitude* of a triangle is the perpendicular distance from a vertex to the opposite side, which is called the *base*.	69
28	Median	A *median* of a triangle is a line from a vertex to the midpoint of the opposite side.	70
29	Area of a Triangle	The area of a triangle equals one-half the product of the base b and the altitude h. $$A = \frac{bh}{2}$$	69
30	Hero's Formula	Area of triangle $= \sqrt{s(s - a)(s - b)(s - c)},$ where s is half the perimeter. $$s = \frac{a + b + c}{2}$$	70
31	Sum of the Interior Angles of Any Triangle	The sum of the three interior angles A, B, and C of any triangle is 180°. $$A + B + C = 180°$$	68
32	Exterior Angle of a Triangle	An exterior angle equals the sum of the two opposite interior angles. $$\theta = A + B$$	69

No.	Name	Formula	Page
33	Centers of a Triangle	a. The centroid is the point of intersection of the medians. b. The orthocenter is the point of intersection of the altitudes. c. The incenter is the point of intersection of the angle bisectors. d. The circumcenter is the point of intersection of the perpendicular bisectors of the sides.	71
34	Line Parallel to a Side	If a straight line is drawn parallel to one side of a triangle, it will cut the other sides proportionally. Conversely, if two sides of a triangle are subdivided proportionally, the line joining the subdividing points will be parallel to the third side of the triangle.	83
35	Napoleon's Theorem	a. If equilateral triangles are drawn externally or internally on the sides of any triangle, their centers form an equilateral triangle.	84
		b. The inner Napoleon triangle has the same center as the external Napoleon triangle.	85
36	Morley's Theorem	The lines of trisection of the vertices of a triangle intersect to form an equilateral triangle.	85
37	Pappus's Theorem	If points A, C, and E lie on one straight line and B, D, and F lie on another line, then the points of intersection of AB with EF, BC with DE, and CF with AD lie on a third straight line.	85
38	Fermat Point	Some triangles have a point such that lines drawn from it to each vertex form three 120° angles.	91

Right Triangles

No.	Name	Formula	Page
39	Pythagorean Theorem	The square of the hypotenuse of a right triangle is equal to the sum of the squares of the two legs. $$a^2 + b^2 = c^2$$	75
40	Trigonometric Ratios	a. Sine $$\sin \theta = \frac{\text{opposite side}}{\text{hypotenuse}}$$ b. Cosine $$\cos \theta = \frac{\text{adjacent side}}{\text{hypotenuse}}$$ c. Tangent $$\tan \theta = \frac{\text{opposite side}}{\text{adjacent side}}$$	88

No.	Name	Formula	Page
41	Altitude Drawn to the Hypotenuse	In a right triangle, the altitude drawn to the hypotenuse forms two right triangles that are similar to each other and to the original triangle.	86
42	Pythagorean Triples	a. Attributed to Plato: For any m, $$(2m)^2 + (m^2 - 1)^2 = (m^2 + 1)^2.$$ b. Rule of Pythagoras: m is an odd integer. $$m^2 + \left[\frac{m^2 - 1}{2}\right]^2 = \left[\frac{m^2 + 1}{2}\right]^2$$ c. Rule of Plato: m is an even integer. $$m^2 + \left[\frac{m^2}{4} - 1\right]^2 = \left[\frac{m^2}{4} + 1\right]^2$$ d. Rule of Masères: m and n are any integers, and $m > n$. $$(2mn)^2 + (m^2 - n^2)^2 = (m^2 + n^2)^2$$	90

Pairs of Triangles

No.	Name	Formula	Page
43	Congruent Triangles	Two triangles are congruent if: a. Two angles and a side of one are equal to two angles and a side of the other (ASA), (AAS). b. Two sides and the included angle of one are equal, respectively, to two sides and the included angle of the other (SAS). c. Three sides of one are equal to the three sides of the other (SSS).	78
44	Similar Triangles	a. Two triangles are similar if the angles of one triangle equal the angles of the other.	81
		b. If two triangles are similar, their corresponding sides are in proportion. The ratio between corresponding sides is called the scale factor.	81
		c. Areas of similar triangles are proportional to the square of the scale factor.	82
45	Triangles in Perspective	a. Two triangles are said to be in perspective from a point if the three lines joining pairs of corresponding points meet in a single point. b. *Desargues' Theorem:* If two triangles are in perspective from a point, then the three points of intersection of corresponding sides, extended, lie on a straight line.	85

Quadrilaterals

No.	Name	Formula	Page
46	Perimeter and Area of a Square	a. The perimeter of a square is the sum of the four sides. $$P = 4s$$ b. The area of a square is the square of one side. $$A = s^2$$	94
47	Perimeter and Area of a Rectangle	a. The perimeter is the sum of the four sides. b. The area is the product of the width and the length.	98
48	Parallelogram	a. Opposite angles of a parallelogram are equal. b. The diagonals of a parallelogram bisect each other. c. The area equals the product of the perpendicular distance between two opposite sides and the length of one of those sides.	101
49	Rhombus	a. The diagonals of a rhombus bisect each other at right angles. b. The diagonals of a rhombus bisect the interior angles of the rhombus.	101
50	Trapezoid	The area of a trapezoid equals the altitude times the average of the bases.	102

The Circle

No.	Name	Formula	Page
51	Circle Definitions	A *circle* is a plane curve consisting of all points at a given distance (called the *radius*) from a fixed point (called the *center*). The *diameter* is twice the radius. The diameter cuts the circle into two *semicircles*.	168
52	Pi (π)	$$\pi = \frac{Circumference}{Diameter} = \frac{C}{d} \cong 3.1416$$ π is the ratio of the circumference of any circle to its diameter.	168
53	Circumference of a Circle	a. The *circumference* of a circle is its total length. This term is also used to mean the circle itself, rather than its *interior*. b. The circumference of a circle equals π times its diameter d. $$C = \pi d = 2\pi r$$	168

No.	Name	Formula	Page
54	Area of a Circle	The area of a circle (that is, its interior) equals π times the square of the radius. $$A = \pi r^2$$	170
55	Angles in a Circle	a. A central angle is one whose vertex is at the center of the circle. b. An inscribed angle is one whose vertex is on the circle.	169
56	Angle Inscribed in a Semicircle	Any angle inscribed in a semicircle is a right angle.	169
57	Arc	An *arc* is a portion of the circle between two points on the circle. A smaller arc is called a *minor arc*, and a longer is called a *major arc*.	169
58	Sector	A *sector* is a plane region bounded by two radii and one of the arcs intercepted by those radii.	169
59	Area of a Sector	$$A_s = \frac{r^2\theta}{2}$$	174
60	An Inscribed and a Central Angle Subtending the Same Arc	If an inscribed angle ϕ and a central angle θ subtend the same arc, the central angle is twice the inscribed angle. $$\theta = 2\phi$$	160
61	The Radian	a. One *radian* (rad) is the central angle subtended by an arc equal in length to the radius of the circle. b. Radian Measure: The *radian measure* of a central angle θ is the ratio of the arc s it subtends to the radius r of the circle. $$\theta = \frac{s}{r}$$	172
62	Arc Length	The length of an arc of a circle equals the radius times the central angle (in radians) that it subtends. $$s = r\theta$$	173
63	Tangent, Secant, and Chord	a. A line that touches a circle at just one point is called a *tangent* to the circle. b. A line that cuts across a circle at two points is called a *secant* to the circle. c. A *chord* is the portion of a secant that joins two points on the circle.	175
64	Perpendicular Bisector of a Chord	The perpendicular bisector of a chord passes through the center of the circle.	175
65	Tangent and Radius	A tangent to a circle is perpendicular to the radius drawn through the point of contact.	175

No.	Name	Formula	Page
66	Two Tangents Drawn to a Circle	Two tangents drawn to a circle from a point outside the circle make equal angles with a line drawn from the circle's center to the external point. The distances from the external point to each point of tangency are equal.	176
67	A Tangent and a Secant Drawn to a Circle (Euclid, III, 36)	For a tangent and a secant drawn to a circle from a point outside the circle, the product of the portions of the secant equals the square of the tangent. Therefore, $$(OP)(OQ) = (OT)^2.$$	176
68	Intersecting Chords	If two chords in a circle intersect, the product of the parts of one chord equal the product of the parts of the other chord. Here, $$ab = cd.$$	176
69	Intersecting Chord and Diameter	If the diameter of a circle bisects a chord, it is perpendicular to it. Conversely, if the diameter of a circle is perpendicular to a chord, it bisects the chord.	177
70	Angles Subtended by a Chord (Euclid III, 21)	A chord subtends equal angles from any point on the circle that are on the same side of the chord. Here, $$\theta = \phi.$$	177
71	Segment of a Circle	A *segment of a circle* is a plane region bounded by a chord of a circle and the arc cut off by that chord. As with arcs and sectors, there is a *major* (larger) segment and a *minor* (smaller) segment. (We say "segment *of a circle*" to avoid confusion with *line* segment.)	177
72	Area of a Segment of a Circle	$$\text{Area} = \frac{1}{2} r^2(\theta - \sin\theta),$$ where θ is in radians.	191
73	Nonintersecting Circles	Nonintersecting circles have four common tangents.	196
74	Tangent Circles	Two tangent circles have three common tangents. The line of centers passes through the point of contact and is perpendicular to the common tangent at that point.	196
75	Circles Intersecting at Two Points	Two intersecting circles have two common tangents. The line of centers is the perpendicular bisector of the common chord. The converse is also true, that the perpendicular bisector of the common chord passes through the center of each circle.	196
76	Perpendicular Bisector of a Chord	The perpendicular bisector of a chord passes through the center of the circle.	196

No.	Name	Formula	Page
77	Incenter of a Triangle	The incenter of a circle inscribed in a triangle is at the intersection of the bisectors of the angles of the triangle.	197
78	Radius of the Incircle of a Triangle	The radius of a circle inscribed in a triangle equals twice the area of the triangle divided by the perimeter of the triangle. $$r = \frac{2(\text{Area})}{P}$$	198
79	Circumcenter of a Polygon	If a polygon has a circumcircle, its center (the circumcenter) is at the intersection of the perpendicular bisectors of the sides of the polygon.	198
80	Intersecting Chord and Diameter	If the diameter of a circle bisects a chord, it is perpendicular to that chord; or, if the diameter is perpendicular to a chord, it bisects that chord.	117
81	Radius of a Circle Inscribed in a Triangle	$$r = \frac{\sqrt{s(s-a)(s-b)(s-c)}}{s},$$ where s is half the perimeter of the triangle $$s = \tfrac{1}{2}(a+b+c)$$	225
82	Radius of a Circle Circumscribed About a Triangle	$$r = \frac{abc}{4\sqrt{s(s-a)(s-b)(s-c)}},$$ where s is half the perimeter of the triangle.	225
83	Euclid III, 22	The sum of opposite angles of any quadrilateral inscribed in a circle is equal to two right angles.	201
84	Simpson Line	The feet of the perpendiculars from the sides of a triangle to a point on the circumcircle lie on a straight line, called the *Simpson* line.	201
85	Nine-Point Circle	A nine-point circle is one that passes through a. the three midpoints of the sides of a triangle, b. the three points where each altitude touches its base, and c. the three midpoints of the line segments joining each vertex to the orthocenter.	225
86	Ptolemy's Theorem	If a quadrilateral is inscribed in a circle, the sum of the products of two opposite sides is equal to the product of the diagonals.	201

Analytic Geometry

No.	Name	Formula	Page
(87)	Equation of a Circle, Center at Origin	a. $x^2 + y^2 = r^2$ b. $y = \pm\sqrt{r^2 - x^2}$	259

No.	Name	Formula	Page
88	Definition of an Ellipse	An ellipse is the set of all points in a plane such that the sum of the distances from each point on the ellipse to two fixed points (called the *foci*) is constant.	249
89	Parts of an Ellipse	The ellipse has two axes of symmetry, the *major axis* and the *minor axis*. They intersect at the *center* of the ellipse. A *vertex* is a point where the ellipse crosses the major axis. Half the lengths of the axes are called the *semimajor* and *semiminor* axes. Each *focus* lies on the major axis. They are equidistant from the center.	249
90	Distance from Center to Focus of an Ellipse	$c = \sqrt{a^2 - b^2}$	250
91	Ellipse with Center at the Origin	a. $\dfrac{x^2}{a^2} + \dfrac{y^2}{b^2} = 1$ b. $y = \pm b\sqrt{1 - \dfrac{x^2}{a^2}}$	256
(92)	Definition of the Conics	The conics may be defined as the locus of a point whose distance from a fixed point (the *focus*) is ε times its distance from a fixed line (the *directrix*). The type of conic, ellipse, parabola, and so on, is determined by the value of ε (called the *eccentricity*).	275

Coordinate Systems

No.	Name	Formula	Page
93	Conversion of Coordinates	a. Polar to Rectangular $$x = r \cos \theta$$ $$y = r \sin \theta$$ b. Rectangular to Polar $$r = \sqrt{x^2 + y^2}$$ $$\theta = \arctan \frac{y}{x}$$	276

Similar Plane or Solid Figures

No.	Name	Formula	Page
94	Dimensions of Plane or Solid Similar Figures	Corresponding dimensions of plane or solid similar figures are in proportion. The ratio of their lengths is the scale factor.	134, 293
95	Areas of Plane or Solid Similar Figures	Corresponding areas of similar figures are in proportion, with the constant of proportionality equal to the square of the scale factor.	134, 293

No.	Name	Formula	Page
96	Volumes of Similar Solids	Corresponding volumes of similar solids are in proportion, with the constant of proportionality equal to the cube of the scale factor.	294

Transformations

No.	Name	Formula	Page
97	Isometries	*Translation:* A shift in position of a given distance *Rotation:* A turning through an angle *Reflection:* A mirror image about some mirror line To these isometries, we add: *Glide reflection:* A combination of translation and reflection	80
98	Similarity Transformations	The similarity transformations are translation, rotation, reflection, glide reflection, and scaling.	82

The Spiral

No.	Name	Formula	Page
99	Polar Equation of an Archimedean Spiral	$r = k\theta$	261
100	Polar Equation of an Equiangular Spiral	$r = ke^{a\theta}$, where $e \approx 2.71828$, and k and a are constants.	263
101	Polar Equation of a Hyperbolic Spiral	$r = a/\theta$	264
102	Polar Equation of a Parabolic (or Fermat's) Spiral	$r = a\sqrt{\theta}$	264
103	Polar Equation of a Lituus	$r = \dfrac{\sqrt{a}}{\theta}$	264

Solids

No.	Name	Formula	Page
104	Cube (side = a)	a. Volume = a^3 b. Surface area = $6a^2$	283
105	Rectangular Parallelepiped (dimensions *lwh*)	a. Volume = lwh b. Surface area = $2(lw + hw + lh)$	283
106	Any Cylinder or Prism	Volume = (area of base) (altitude)	283, 290
107	Right Cylinder or Prism	Lateral area = (perimeter of base) (altitude)	283, 290
108	Any Cone or Pyramid	Volume = $\frac{1}{3}$(area of base) (altitude)	283, 290

No.	Name	Formula	Page
109	Right Circular Cone or Regular Pyramid	Lateral area $= \frac{1}{2}$(slant height) (circumference of base)	283, 290
110	Frustum of any Cone or Pyramid	$\text{Volume} = \frac{h}{3}(A_1 + A_2 + \sqrt{A_1 A_2})$	283, 290
111	Frustum of Right Circular Cone or Pyramid	Lateral area $= \frac{1}{2}$(slant height) $\times$ (sum of base circumferences)	283, 290
112	Theorems of Pappus	a. The surface area of a solid of revolution generated by a plane figure C is equal to the perimeter of C times the distance L traveled by the center of C. b. The volume of the solid of revolution is equal to the area of C times L.	291

Sphere

No.	Name	Formula	Page
113	Definition of a Sphere	A sphere is the set of points in space at a given distance from a fixed point.	320
114	Surface Area	$\text{Area} = 4\pi r^2$	320
115	Volume	$\text{Volume} = \left(\frac{4}{3}\right)\pi r^3$	320
116	Surface Area of a Spherical Zone	The surface area of a spherical zone (of either one or two bases) is equal to the circumference of a great circle of that sphere times the altitude of the zone.	321
117	Surface Area of a Spherical Zone of One Base	The surface area of a spherical zone of one base is equal to that of a circle whose radius is the chord of the generating arc.	360
118	Volume of a Spherical Segment	$V = \frac{\pi h}{6}(3a^2 + 3b^2 + h^2)$ For a segment of one base, either a or b is zero.	321
119	Volume of a Spherical Sector	a. The volume of a spherical sector is one-third the product of the area of its zone and the radius of the sphere. b. The volume of a spherical sector of altitude h in a sphere of radius r is $V = \frac{2}{3}\pi r^2 h.$	321

Perspective

No.	Name	Formula	Page
120		Straight lines on the object appear as straight lines in the picture plane.	366
121		Vertical lines on the object appear as vertical lines in the picture plane.	366

No.	Name	Formula	Page
122		Horizontal lines on the object that are parallel to the picture plane appear as horizontal lines in the picture plane.	366
123		Receding parallel lines on the object will meet at a vanishing point somewhere in the picture plane.	367
124		Receding horizontal parallel lines on the object will meet at a vanishing point on the horizon line in the picture plane	368
125		Receding orthogonals on the object will meet at the principal vanishing point VP on the horizon line in the picture plane.	368
126		The vanishing point for parallel lines on the object is at the same angle from the eye as the parallels make with the picture plane.	369
127		The vanishing point V' for the diagonal of a square whose sides are parallel and/or perpendicular to the picture plane is at a distance from the principal vanishing point VP equal to the viewing distance D.	369

Fractals

No.	Name	Formula	Page
128	Euclidean Dimension E	Points are zero dimensional. Lines are one dimensional. Surfaces are two dimensional. Solids are three dimensional.	394
129	Topological Dimension D_T	a. A set has topological dimension 0 if every point has an arbitrarily small neighborhood whose boundary does not intersect the set. b. A set has topological dimension D_T if each point in the set has arbitrarily small neighborhoods whose boundaries meet the set in a set of dimension D_{T-1}.	395
130	Fractal Dimension D	$$D = \frac{\log n}{\log m}$$	397
131	Definition of a Fractal	A fractal is a set for which the fractal dimension exceeds the topological dimension. $$D > D_T$$	398

APPENDIX G

Bibliography

Aaboe, Asger. *Episodes from the Early History of Mathematics.* New York: Random House, 1964.

Abas, Syed. *Symmetries of Islamic Geometrical Patterns.* Singapore: World Scientific, 1995.

Abbott, Edwin. *Flatland: A Romance in Many Dimensions.* New York: Dover, 1992. First published in 1884.

Achen, Sven. *Symbols Around Us.* New York: Van Nostrand, 1978.

Ackerman. *Distance Points: Essays in Theory of Renaissance Art & Architecture.* Cambridge, MA: MIT Press, 1991.

Adams, Henry. *Mont-Saint-Michel and Chartres.* Boston, MA: Houghton, 1904.

Adzema, Robert, et al. *The Great Sundial Cutout Book.* New York: Hawthorn, 1978.

Alberti, Leon Battista. *On Painting.* A translation of *Della pittura.* New Haven, CT: Yale University Press, 1956.

Alberti, Leon Battista. *The Ten Books of Architecture.* New York: Dover, 1986. A 1775 edition of the work written in 1452.

Argüelles, José and Miriam. *Mandala.* Boston, MA: Shambhala, 1985.

Arnheim, Rudolph. *Art and Visual Perception: A Psychology of the Creative Eye.* Berkeley, CA: University of California Press, 1966.

Arnheim, Rudolph. *The Power of the Center.* Berkeley, CA: University of California Press, 1988.

Artmann, Benno. "The Cloisters of Hauterive." *Nexus: Architecture and Mathematics.* Kim Williams, ed. Fucecchio (Firenze): Edizione dell'Erba, 1996.

Atalay, Bulent. *Math and the Mona Lisa.* Washington, D.C.: Smithsonian, 2004.

Bairati, Eleonaora. *Piero della Francesca.* New York: Crescent, 1991.

Barnsley, Michael. *Fractals Everywhere.* San Diego, CA: Academic, 1988.

Barratt, Krome. *Logic and Design.* New York: Design Books, 1980.

Bayley, Harold. *The Lost Language of Symbolism,* 2 Vols. Philadelphia, PA: Lippincott, 1913.

Baxandall, Michael. *Painting and Experience in Fifteenth Century Italy.* Oxford: Clarendon Press, 1972.

Bell, Daniel Orth. "New Identifications in Raphael's School of Athens." *Art Bulletin,* Dec. 1995, p. 639.

Bentley, W. A. *Snow Crystals.* New York: Dover, 1962.

Boles, Martha, and Rochelle Newman. *The Golden Relationship: Art, Math & Nature.* 4 Vols. Bradford, MA: Pythagorean Press.

Boni, Margaret, ed. *Fireside Book of Folk Songs*. New York: Simon and Schuster, 1947.

Bord, Janet. *Mazes and Labyrinths of the World*. New York: Dutton, 1975.

Bouleau, Charles. *The Painter's Secret Geometry*. New York: Harcourt, 1963.

Brenni, Paolo, et al. *Orologi e Strumenti della Collezione Beltrame*. Florence: Instituto e Museo di Storia della Scienza, 1996.

Briggs, John. *Fractals: The Patterns of Chaos*. New York: Simon & Schuster, 1992.

Briggs, John, et al. *Turbulent Mirror*. New York: Harper, 1989.

Brunés, Tons. *The Secrets of Ancient Geometry—and Its Use*. Copenhagen: Rhodos, 1967.

Bühlmann, Josef. *Classical and Renaissance Architecture*. New York: Helburn, 1916. English translation of Bühlmann's *Die Architektur des Klassischen Altertums und der Renaissance*. Eszlingen: Neff, 1913–19.

Burckhardt, Jacob. *The Altarpiece in Renaissance Italy* (1898), ed. and trans. P. Humfrey. Cambridge, UK: Cambridge University Press, 1988.

Burckhardt, Jacob. *The Civilization of the Renaissance in Italy*. New York: Modern, 1954. First published in 1860.

Burckhardt, Titus. *Chartres and the Birth of the Cathedral*. Ipswitch, UK: Golgonooza, 1995.

Burckhardt, Titus. *Mirror of the Intellect: Essays on Traditional Science & Sacred Art*. Albany, NY: SUNY Press, 1987.

Burger, Edward, et al. *The Heart of Mathematics*. Emeryville, CA: Key Press, 2000.

Burnham, Jack. *The Structure of Art*. New York: Braziller, 1971.

Busch, Harald, and Bernd Lohse, eds. *Renaissance Sculpture*. New York: Macmillan, 1964.

Butler, Christopher. *Number Symbolism*. London: Routledge, 1970.

Calter, Paul. "Façade Measurement by Trigonometry." In *Geometry at Work*, C. Gorini, ed. Mathematical Association of America, 2000.

Calter, Paul. "How to Construct a Logarithmic Rosette (Without Even Knowing It)." *Nexus Network Journal*, April 2007.

Calter, Paul. "Mathematics of the St. Louis Arch." Presentation at the AAAS Annual Meeting, 2006. Printed in the *Nexus Network Journal*, 2006.

Calter, Paul. "Measuring Up to Michelangelo: A Methodology." Proceedings of NEXUS 2000, Ferrara, Italy, 2000.

Calter, Paul. *Raphael's School of Athens and the Art of Perspective*. Catalogue entry for Dartmouth College Hood Museum of Art Show, Visual Proof: The Experience of Mathematics in Art, 1999.

Calter, Paul. "Sun Disk, Moon Disk." In *Geometry at Work*, C. Gorini, ed. Mathematical Association of America, 2000.

Calter, Paul. *Technical Mathematics*, 5th ed. New York: Wiley, 2007.

Calter, Paul. "A Universal Method for Building Measurements by Theodolite." Proceedings of NEXUS '96, Fucecchio, Italy, 1996.

Calvino, Italo. *Six Memos for the Next Millennium*. Cambridge, MA: Harvard University Press, 1998.

Campbell, Joseph, with Bill Moyers. *The Power of Myth*. New York: Doubleday, 1988.

Canaday, John. *Mainstreams of Modern Art*. New York: Holt, 1959.

Canaday, John. *Masterpieces by Michelangelo*. New York: Crown, 1979.

Carr-Gomm, Sarah. *Dictionary of Symbols in Western Art*. New York: Facts on File, 1995.

Carter, David. *Dynamic Symmetry*. Exhibition Catalog, 1961.

Chanon, Steven. *The Geometer's Sketchpad Learning Guide*. Emeryville, CA: Key Press, 2000.

Chanon, Steven, ed. *101 Ideas for The Geometer's Sketchpad*. Emeryville, CA: Key Press, 2000.

Chiarini, Armando e Adriano. *Le Meridiana della Basilica di S. Petronio in Bologna*. Bologna, Italy: La Grafica Emiliana, 1992.

Chitham, Robert. *The Classical Orders of Architecture*. New York: Rizzoli, 1985.

Choate, Jonathan, et al. *Fractals: A Tool Kit of Dynamic Activities*. Emeryville, CA: Key Press, 1999.

Christiansen, Keith. *Italian Paintings.* New York: Hugh Lauter Levin Associates; New York: distributed by Macmillan, 1992.

Clark, Kenneth. *Civilization.* New York: Harper, 1969.

Clark, Kenneth. *Leonardo da Vinci.* Baltimore, MD: Penguin, 1939.

Clark, Kenneth. *The Nude.* New York: Doubleday, 1959.

Clark, Kenneth. *The Romantic Rebellion.* New York: Harper, 1972.

Cole, Alison. *Perspective.* London: Kindersley, 1992.

Cole, Rex. *Perspective for Artists.* New York: Dover, 1976. First published in 1921.

Conway, William Martin. *The Writings of Albrecht Dürer.* New York: Philosophical Library, 1958.

Cook, Roger. *The Tree of Life.* New York: Thames and Hudson, 1974.

Cook, Theodore. *The Curves of Life.* New York: Dover, 1979. First published in 1914.

Cornford, Francis. *Plato's Cosmology.* New York: Harcourt, 1937.

Cowen, Painton. *Rose Windows.* London: Thames and Hudson, 1979.

Coxeter, H. S. M. *Introduction to Geometry*, 2nd ed. New York: Wiley, 1989.

Coxeter, H. S. M., et al. *Geometry Revisited.* Washington, D.C.: Mathematical Association of America, 1967.

Critchlow, Keith. *Order in Space: A Design Source Book.* London: Thames & Hudson, 1969.

Critchlow, Keith. *Time Stands Still.* New York: St. Martin's, 1982.

Dabrowski, Magdalena. *Contrasts of Form: Geometric Abstract Art 1910–1980.* New York: Museum of Modern Art, 1985.

Dante. *The Divine Comedy.* Trans. H. F. Cary, III. Gustave Doré. London: Cassell, 1885.

Demus, Otto. *The Mosaic Decoration of San Marco, Venice.* Chicago: University of Chicago Press, 1988.

Devaney, Robert. *The Mandelbrot and Julia Sets: A Tool Kit of Dynamic Activities.* Emeryville, CA: Key Press, 2000.

De Vogel, C. J. *Pythagoras and Early Pythagoreanism.* Assen, the Netherlands: Van Gorcum, 1966.

Dixon, Laurinda S. "Giovanni di Paolo's Cosmology." *Art Bulletin*, December 1985, pp. 604–613.

Doczi, György. *The Power of Limits: Proportional Harmonies in Nature, Art, and Architecture.* Boston, MA: Shambhala, 1981.

Dunning, William V. *Changing Images of Pictorial Space.* Syracuse, NY: Syracuse University Press, 1991.

Dürer, Albrecht. *The Complete Engravings, Etchings and Drypoints*, Walter Strauss, ed. New York: Dover, 1972.

Dürer, Albrecht. *The Complete Woodcuts*, Willi Kurth, ed. New York: Dover, 1963.

Dürer, Albrecht. *Drawings*, Heinrich Wolfflin, ed. New York: Dover, 1970.

Dürer, Albrecht. *The Human Figure*, Walter Strauss, ed. New York: Dover, 1972. Originally printed c. 1528.

Dürer, Albrecht. *The Painter's Manual.* Walter Strauss, trans. New York: Abaris, 1977. Originally printed in 1525.

Eco, Umberto. *Art and Beauty in the Middle Ages.* New Haven, CT: Yale University Press, 1986.

Edgerton, Samuel. *The Heritage of Giotto's Geometry.* Ithaca, NY: Cornell University Press, 1991.

Edgerton, Samuel. *The Renaissance Rediscovery of Linear Perspective.* New York: Basic Books, 1975.

Edwards, Edward. *Pattern and Design with Dynamic Symmetry.* New York: Dover, 1967. Reprint of *Dynamarhythmic Design*, 1932.

Ekwall, Åke. "Violins and Volutes: Visual Parallels Between Music and Architecture." *Nexus Network Journal*, Autumn 2001.

El-Said, Issam, et al. *Geometric Concepts in Islamic Art.* Palo Alto, CA: Seymour, 1976.

Emmer, Michele, ed. *The Visual Mind: Art and Mathematics.* Cambridge, MA: MIT Press, 1993.

Escher, M. C. *Escher on Escher: Exploring the Infinite.* New York: Abrams, 1989.

Escher, M. C. *The Graphic Works of M. C. Escher.* New York: Ballantine, 1960.

Euclid. *The Thirteen Books of the Elements.* New York: Dover, 1956.

Eves, Howard. *An Introduction to the History of Mathematics.* New York: Holt, 1953.

Ferguson, George. *Signs & Symbols in Christian Art.* London: Oxford University Press, 1954.

Fibonacci, Leonardo. *Fibonacci's Liber Abaci, a Translation into Modern English of Leonardo Pisano's Book of Calculation.* Lawrence Sigler, trans. New York: Springer, 2002.

Fichten, John. *The Construction of Gothic Cathedrals.* Oxford, UK: Oxford University Press, 1961.

Fisher, Sally. *The Square Halo.* New York: Abrams, 1995.

Forseth, Kevin. *Graphics for Architecture.* New York: Van Nostrand, 1980.

Fox, Matthew. *Illuminations of Hildegard of Bingen.* Santa Fe, NM: Bear, 1985.

Frazer, James. *The Golden Bough.* New York: Collier, 1922.

Frings, Marcus. "The Golden Section in Architectural Theory." *Nexus Network Journal:* http://www.nexusjournal.com/Frings.html.

Furnari, Michele. *Formal Design in Renaissance Architecture.* New York: Rizzoli, 1995.

Gayley, Charles. *Classical Myths in English Literature.* Boston, MA: Ginn, 1893.

Ghyka, Matila. *The Geometry of Art and Life.* New York: Dover, 1977.

Gimpel, Jean. *The Cathedral Builders.* New York: Harper, 1961.

Gleik, James. *Chaos.* New York: Penguin, 1987.

Golding, John. *Cubism: A History and an Analysis 1907–1914.* Cambridge, MA: Belknap/Harvard, 1988.

Gombrich, E. H. *Art and Illusion.* New York: Pantheon, 1960.

Gombrich, E. H. *New Light on Old Masters.* Chicago: University of Chicago Press, 1986.

Gombrich, E. H. *The Story of Art*, 16th ed. Upper Saddle River, NJ: Prentice Hall, 1995.

Gorini, Catherine. *The Facts on File Geometry Handbook.* New York: Facts on File, 2003.

Grabow, Stephen. "Frozen Music: The Bridge Between Art and Science," article in *Companion to Contemporary Architectural Thought,* Farmer et al., eds. New York: Routledge, 1993, pp. 438–443.

Graves, Robert. *Hebrew Myths.* New York: Doubleday, 1963.

Graves, Robert. *The White Goddess.* New York: Farrar, 1948.

Grünbaum, Branko, et al. *Tilings and Patterns: An Introduction.* New York: Freeman, 1989.

Gulick, Denny. *Encounters with Chaos.* New York: McGraw-Hill, 1992.

Guthrie, Kenneth S. *The Pythagorean Sourcebook and Library.* Grand Rapids, MI: Phanes, 1987.

Hale, John R. *Encyclopedia of the Italian Renaissance.* New York: Thames and Hudson, 1981.

Hale, Jonathon. *The Old Way of Seeing.* Boston, MA: Houghton Mifflin, 1994.

Hall, James. *Dictionary of Subjects and Symbols in Art.* New York: Harper, 1974.

Hambidge, Jay. *Dynamic Symmetry: The Greek Vase.* New Haven, CT: Yale University Press, 1920.

Hambidge, Jay. *The Elements of Dynamic Symmetry.* New York: Dover, 1967.

Hambidge, Jay. *The Parthenon and Other Greek Temples: Their Dynamic Symmetry.* New Haven, CT: Yale University Press, 1924.

Hamilton, Edith. *Mythology.* New York: Mentor, 1942.

Hanks, Kurt. *Rapid Viz: A New Method for the Rapid Visualization of Ideas.* San Francisco, CA: Kaufmann, 1980.

Hargittai, István, ed. *Fivefold Symmetry.* New York: World Scientific, 1991.

Hargittai, István, and C. A. Pickover, eds. *Spiral Symmetry.* New York: World Scientific, 1991.

Hargittai, István. *Symmetry Through the Eyes of a Chemist.* New York: Plenum, 1995.
Hargittai, István, ed. *Symmetry 2.* New York: Pergammon, 1989.
Hargittai, István. *Symmetry: A Unifying Concept.* Bolinas, CA: Shelter, 1994.
Harris, Cyril. *Illustrated Dictionary of Historic Architecture.* New York: Dover, 1977.
Hartt, Frederic. *Italian Renaissance Art.* New York: Abrams, 1994.
Heath, Sir Thomas. *A History of Greek Mathematics.* New York: Dover, 1981. First published in 1921.
Heilbron, J. L. *The Sun in the Church, Cathedrals as Solar Observatories.* Cambridge, MA: Harvard University Press, 1999.
Heroditus. Trans. H. Carey. London: Bell, 1901.
Hersey, George L. *Architecture and Geometry in the Age of the Baroque.* Chicago: University of Chicago Press, 2000.
Hersey, John. *The Lost Meaning of Classical Architecture.* Cambridge, MA: MIT Press, 1988.
Hersey, John. *Possible Palladian Villas.* Cambridge, MA: MIT Press, 1922.
Hersey, John. *Pythagorean Palaces.* Ithaca, NY: Cornell University Press, 1976.
Hessemer, F. M. *Historic Designs and Patterns in Color from Arabic and Italian Sources.* New York: Dover, 1992. First published in 1842.
Herz-Fischler, Roger. *A Mathematical History of Division in Extreme and Mean Ratio.* Ontario, Canada: Laurier, 1987.
Hofstadter, Douglas. *Gödel, Escher, Bach: An Eternal Golden Braid.* New York: Vintage, 1979.
Hopper, Vincent. *Medieval Number Symbolism.* New York: Columbia University Press, 1938.
Humfrey, Peter. *The Altarpiece in Renaissance Venice.* New Haven, CT: Yale University Press, 1993. Reviewed by Alexander Nagal in *Art Bulletin*, March 1995, p. 139.
Huntley, H. E. *The Divine Proportion.* New York: Dover, 1970.

Ivins, William. *Art & Geometry: A Study in Space Intuitions.* Cambridge, MA: Harvard University Press, 1946.

Jacobs, et al. *A New Standard Bible Dictionary.* New York: Funk and Wagnalls, 1926.
Jameson, Anna. *Sacred and Legendary Art.* Boston, MA: Houghton, 1895.
Janson, H. W. *History of Art*, 5th ed. Revised by Anthony F. Janson. New York: Harry N. Abrams, Inc., 1995.
Jewish Museum (New York, NY), *Treasures of the Jewish Museum.* New York: Universe, 1986.
Jones, Lesley, ed., *Teaching Mathematics and Art.* Cheltenham, UK: Stanley Thornes (Publishers), 1991.
Jung, Carl G., et al. *Man and His Symbols.* New York: Dell, 1964.

Kappraff, Jay. *Connections: The Geometric Bridge Between Art and Science.* New York: McGraw-Hill, 1990.
Kappraff, Jay. "Musical Proportions at the Basis of Systems of Architectural Proportion Both Ancient and Modern." In *Nexus: Architecture and Mathematics*, Kim Williams, ed. Fucecchio, Florence: Edizioni dell'Erba, 1996, pp. 115–133.
Katzenellenbogen, Adolf. *The Sculptural Programs of Chartres Cathedral.* Baltimore, MD: Hopkins, 1959.
Keller, Sharon. *The Jews: A Treasury of Art and Literature.* New York: Levin Associates, 1992.
Kemp, Martin. *Leonardo on Painting.* New Haven, CT: Yale University Press, 1989.
Kemp, Martin. *The Science of Art.* New Haven, CT: Yale University Press, 1990.
Kepes, Gyorgy, ed. *The New Landscape in Art and Science.* Chicago: Theobald, 1953.
Kepes, Gyorgy, ed. *Structure in Art and Science.* New York: Braziller, 1965.
Kinsey, L. C., et al. *Symmetry, Shape, and Space.* Emeryville, CA: Key College Publishing, 2002.
Kitzinger, Ernst. *The Art of Byzantium and the Medieval West.* Bloomington, IN: Indiana University Press, 1976.

Kline, Morris. *Mathematics in Western Culture*. New York: Oxford University Press, 1953.

Koch, Rudolf. *The Book of Signs*. New York: Dover, 1955. Originally published in 1930.

Kubovy, Michael. *The Psychology of Perspective and Renaissance Art*. Cambridge, UK: Cambridge University Press, 1986.

Lawlor, Robert. *Sacred Geometry*. New York: Thames & Hudson, 1982.

Le Corbusier. *The Modulor*. Cambridge, MA: Harvard University Press, 1966.

Legge, James, trans. *I Ching: Book of Changes*. New York: Causeway, 1973.

Lehmann, Karl. "The Dome of Heaven." *Art Bulletin*, Vol. XXVII, 1945, pp. 1–27.

Lehner, Ernst. *Symbols, Signs & Signets*. New York: Dover, 1950.

Leonardo da Vinci. *Treatise on Painting* (Codex Urbinas Latinus 1270). Trans. A. Philip MacMahon, Princeton, NJ, 1956.

Levey, Michael. *Early Renaissance*. New York: Penguin, 1967.

Levin, Hugh, Associates. *Jewish Calendar*, 1997.

Linn, Charles. *The Golden Mean*. New York: Doubleday, 1974.

Lionass, Francois. *Time*. New York: Orion, 1959.

Lippard, Lucy. *Overlay*. New York: Pantheon, 1983.

Lippincott, Kristen. "Giovanni di Paolo's 'Creation of the World.'" *Burlington Magazine*, 1990, pp. 460–468.

Livio, Mario. *The Golden Ratio*. New York: Broadway Books, 2002.

Loomis, E. S. *The Pythagorean Proposition*, 2nd ed. Ann Arbor, MI: Edwards, 1940.

Lowrie, Walter. *Art in the Early Church*. New York: Norton, 1947.

Mackworth-Praed, Ben. *The Book of Kells*. London: Studio, 1993.

MacGillarvy, Caroline. *Symmetry Aspects of M. C. Escher's Periodic Drawings*. Utrecht: Uitgeversmaatschappij NV, 1965.

Madoff, Henry. "Vestiges and Ruins: Ethics and Geometric Art in the Twentieth Century." *Arts Magazine*, Vol. 61, 1986.

Male, Emile. *The Gothic Image*. New York: Harper, 1913.

Mandel, Gabriele. *How to Recognize Islamic Art*. New York: Penguin, 1979.

Mandelbrot, Benoit. *The Fractal Geometry of Nature*. San Francisco: Freeman, 1982.

Manetti, Antonio. *The Life of Brunelleschi*. Annotated by Howard Saalman. University Park, PA: Penn State University Press, 1970. Originally printed c. 1489.

Mannering, Douglas. *The Art of Leonardo Da Vinci*. New York: Excalibur, 1981.

March, Lionel. *Architectonics of Humanism*. Chichester, UK: Academy, 1998.

Markowsky, George. "Misconceptions about the Golden Ratio." *College Mathematics Journal*, January 1992.

McCurdy, Edward. *Leonardo da Vinci's Notebooks*. New York: Empire, 1923.

Metropolitan Museum of Art. *Treasures of Early Irish Art*. New York: Metropolitan Museum of Art, 1977.

Morrison, Stanley. *Pacioli's Classic Roman Alphabet*. New York: Dover, 1994. First published in 1933.

Nagel, Alexander. Review of *The Altarpiece in Renaissance Venice*, by Peter Humfrey. New Haven, CT: Yale University Press, 1993. In *Art Bulletin*, March 1995, p. 139.

Newman, James R., ed. *The World of Mathematics*. New York: Simon and Schuster, 1956.

Norling, Ernest. *Perspective Made Easy*. New York: Macmillan, 1939.

Olson, Alton T. *Mathematics Through Paper Folding*. Reston, VA: National Council of Teachers of Mathematics, 1975.

Ostrow, Steven. "Cigoli's Immacolata and Galileo's Moon." *Art Bulletin*, June 1996, pp. 218–235.

Pacholczyk, Josef. "Music and Astronomy in the Muslim World." *Leonardo*, Vol. 29, No. 2, pp. 145–150, 1996.

Palladio, Andrea. *The Four Books of Architecture*. New York: Dover, 1965.

Panofsky, Erwin. *Gothic Architecture and Scholasticism*. Cleveland, OH: Meridian Books, 1951.

Panofsky, Erwin. *The Life and Art of Albrecht Dürer*. Princeton, NJ: Princeton University Press, 1955.

Panofsky, Erwin. *Meaning in the Visual Arts*. Chicago: The University of Chicago Press, 1939.

Panofsky, Erwin. *Perspective as Symbolic Form*. New York: Zone Books, 1991.

Panofsky, Erwin. *Studies in Iconology*. New York: Harper, 1939.

Papadopoulos, Athanase. "Mathematics and Music Theory: From Pythagoras to Rameau." *The Mathematical Intelligencer*, Winter 2002, pp. 65–73.

Partridge, Loren. "The Room of Maps at Caprarola, 1573–75." *Art Bulletin*, September 1995, pp. 413–444.

Pederson, Mark A. "The Geometry of Piero della Francesca." *The Mathematical Intelligencer*, Vol. 19, No. 3, Summer 1997.

Pedoe, Dan. *Geometry and the Visual Arts*. New York: Dover, 1976.

Peitgen, Heinz-Otto, et al. *Fractals for the Classroom*. New York: Springer, 1992.

Pevsner, Nikolaus. *An Outline of European Architecture*. Baltimore, MD: Penguin, 1972.

Philip, J. A. *Pythagoras and Early Pythagoreanism*. Toronto: University of Toronto, 1966.

Pickover, Clifford A. "Mathematics and Beauty: A Sampling of Spirals and 'Strange' Spirals in Science, Nature, and Art." *Leonardo*, Vol. 21, No. 2, pp. 173–181, 1988.

Plato. *Republic*. Trans. by Francis Cornford. New York: Oxford, 1945.

Plato. *Timaeus*. Ed. and trans. by John Warrington. London: Dent, 1965. Original c. 360 B.C.

Plato. *Timaeus and Critias*. Trans. by Desmond Lee. London: Penguin, 1965. Original c. 360 B.C.

Pope-Hennessy, John. *Paradiso: The Illuminations of Dante's Divine Comedy by Giovanni di Paolo*. New York: Random House, 1993.

Pope-Hennessy, John. *The Study and Criticism of Italian Sculpture*. Princeton, NJ: Princeton University Press, 1980.

Pozzo, Andrea. *Perspective in Architecture and Painting*. New York: Dover, 1989. Originally printed c. 1707.

Prusinkiewicz, Przemyslaw, et al. *The Algorithmic Beauty of Plants*. New York: Springer, 1990.

Pseudo Dionysius. *Complete Works*. New York: Paulist, 1987.

Rehmel, Judy. *The Quilt I.D. Book: 4,000 Illustrated and Indexed Patterns*. New York: Prentice-Hall, 1986.

Reti, Ladislao, ed. *The Unknown Leonardo*. New York: McGraw-Hill, 1974.

Richter, Irma. *Rhythmic Forms in Art*. London: John Lane, 1932.

Richter, Jean Paul. *The Notebooks of Leonardo da Vinci*, Two Volumes. New York: Dover, 1970.

Ripa, Cesare. *Baroque and Rococo Pictorial Imagery*, Edward Maser, ed. New York: Dover, 1971. A republished version of Ripa's *Iconologia*, 1758–1760.

Robins, Gay, et al. *The Rhind Mathematical Papyrus*. New York: Dover, 1987.

Rosin, Paul L. "Rosettes and Other Arrangements of Circles." *Nexus Network Journal*, Autumn 2001.

Rosenbusch, Robert. "The Pantheon as an Image of the Universe." *Nexus Network Journal*, Vol. 6, No. 1 (Spring 2004).

Rotzler, Willy. *Constructive Concepts: A History of Constructive Art from Cubism to the Present*. New York: Rizzoli, 1989.

Rowe, Colin. *The Mathematics of the Ideal Villa and Other Essays*. Cambridge, MA: MIT Press, 1976.

Rucker, Rudolph. *Geometry, Relativity and the Fourth Dimension*. New York: Dover, 1977.

Runion, Garth E. *The Golden Section*. Palo Alto, CA: Seymour, 1990.

Saad-Cook, Janet. "Natural Phenomena, Earth, Sky, and Connections to Astronomy." *Leonardo*, Vol. 21, No. 2, 1988, pp. 123–134.

Sanders, Cathi. *Perspective Drawing with The Geometer's Sketchpad.* Emeryville, CA: Key Press, 1994.

Schattschneider, Doris. *Visions of Symmetry: Notebooks, Periodic Drawings, and Related Works of M.C. Escher.* New York: W. H. Freeman, 1990.

Schwaller de Lubicz, R. A. *The Egyptian Miracle.* Rochester, VT: Inner Traditions, 1985.

Serlio, Sebastiano. *The Five Books of Architecture.* New York: Dover, 1982. A reprint of the edition of 1611.

Sharp, John. "Spirals and the Golden Section." *Nexus Network Journal*, Vol. 4, No. 1 (Winter 2002).

Shearer, Rhonda. "Chaos Theory and Fractal Geometry." *Leonardo*, Vol. 25, No. 2, 1992, p. 143.

Shelby, Lon R, ed. *Gothic Design Techniques.* Carbondale, IL: Southern Illinois University Press, 1977.

Shlain, Leonard. *Art and Physics: Parallel Visions in Space, Time, and Light.* New York: Morrow, 1991.

Sill, Gertrude. *A Handbook of Symbols in Christian Art.* New York: Collier, 1975.

Smith, Baldwin. *The Dome.* Princeton, NJ: Princeton University Press, 1950.

Sobel, Dava. *Galileo's Daughter.* New York: Walker, 1999.

Sobel, Dava. *Longitude: The True Story of a Lone Genius Who Solved the Greatest Scientific Problem of His Time.* New York: Walker, 1995.

Stone, Irving. *The Agony and the Ecstasy.* New York: Doubleday, 1961.

Thom, Alexander. *Megalithic Sites in Britain.* Oxford, UK: Clarendon, 1967.

Thompson, Darcy. *On Growth and Form.* New York: Dover, 1992. First published in 1942.

Tillyard, E. M. W. *The Elizabethan World Picture.* New York: Vintage.

Tompkins, Peter. *Secrets of the Great Pyramid.* New York: Harper & Row, 1971.

Turner, Gerard. *Antique Scientific Instruments.* Dorset, UK: Blandford, 1980.

Vasari, Giorgio. *The Lives of the Artists.* Oxford, UK: Oxford University Press, 1991. Originally printed in 1550.

Venters, Diana, et al. *Mathematical Quilts.* Emeryville, CA: Key Press, 1999.

Vitruvius. *The Ten Books on Architecture.* New York: Dover, 1960.

Vredman de Vries, Jan. *Perspective.* New York: Dover, 1968. Originally printed c. 1604.

Ward, Roger. *Durer to Matisse*, Exhibition Catalog. Kansas City, MO: Nelson, 1996.

Wasserman, James. *Art and Symbols of the Occult.* Vermont: Destiny, 1993.

Watts, Carol Martin. "The Square and the Roman House: Architecture and Decoration at Pompeii and Herculaneum." In *Nexus: Architecture and Mathematics*, Kim Williams, ed. (Fucecchio, Florence: Edizioni dell'Erba, 1996), pp. 167–181.

Watts, Donald J., and Carol Martin Watts. "The Role of Monuments in the Geometrical Ordering of the Roman Master Plan of Gerasa." *Journal of the Society of Architectural Historians*, LI, No. 3, September 1992, pp. 306–314.

Watts, Donald J., and Carol Martin Watts. "A Roman Apartment Complex." In *Scientific American*, Vol. 255, No. 6 (December 1986), pp. 132–139.

Waugh, Albert E. *Sundials: Their Theory and Construction.* New York: Dover, 1973.

Weisstein, E., ed. *CRC Concise Encyclopedia of Mathematics*, 2nd ed. Boca Raton, FL: CRC, 2003.

Wenninger, Magnus J. *Polyhedron Models for the Classroom.* Reston, VA: National Council of Teachers of Mathematics, 1966.

Wilkins, Peter. "The Pantheon as a Globe-Shaped Conception." *Nexus Network Journal*, Vol. 6, No. 1 (Spring 2004).

Williams, Kim. *Italian Pavements, Patterns in Space.* Houston, TX: Anchorage, 1997.

Williams, Kim. "Michelangelo's Medici Chapel: The Cube, the Square, and the Root-2 Rectangle." *Leonardo*, Vol. 30, No. 2, 1997, pp. 105–112.

Williams, Kim, ed. *Nexus: Architecture and Mathematics*. Fucecchio: Edizioni dell'Erba, 1996.

Williams, Kim, ed. *Nexus II: Architecture and Mathematics*. Fucecchio: Edizioni dell'Erba, 1998.

Williams, Kim, ed. *Nexus III: Architecture and Mathematics*. Pisa: Pacini, 2000.

Williams, Kim. "Spirals and the Rosette in Architectural Ornament." *Nexus Network Journal*, Vol. 1 (1999), pp. 129–138.

Wills, Herbert, III. *Leonardo's Dessert, No Pi*. Reston, VA: National Council of Teachers of Mathematics, 1985.

Wittkower, Rudolf. *Architectural Principles in the Age of Humanism*. New York: Random House, 1965.

Wolfe, Tom. *The Painted Word*. New York: Farrar, Straus, 1975.

APPENDIX H

Selected Answers

Chapter 1

RATIO AND PROPORTION ▪ 13

2. **a.** 3/2   **b.** 8/3   **c.** 9/2   **d.** 1.39
3. **a.** ±10   **b.** ±8   **c.** ±30   **d.** ±15
4. **a.** Shorter = 54.6, longer = 145.5
   b. Shorter = 85.6, longer = 299.4
   c. Shorter = 114, longer = 171
   d. Shorter = 24.3, longer = 32.5
5. $93.21
6. Five masons
7. $11,398

SEQUENCES, SERIES, AND MEANS ▪ 20

2. **a.** 6   **b.** −8
3. **a.** 3   **b.** 2.50
4. **a.** 15   **b.** 389
5. **a.** ±10   **b.** ±6.22
6. **a.** 1.82   **b.** 5.43
7. Arithmetic mean = 15, geometric mean = 14.1, harmonic mean = 13.3

Chapter 2

EXERCISES AND PROJECTS ▪ 57

2. 376
12. Two hundred diameters = 146.7 m. The wheel circumference is 2.3050 m, so 100 revolutions is 230.5 m.

Chapter 3

ANGLES BETWEEN LINES ▪ 66

2. **a.** 62.8°   **b.** 64.6°   **c.** 37.7°
3. $A = D = G = 32°$, $F = 180 − 32 = 148°$, $B = C = E = F = 148°$
4. 50.5 ft
5. $A = 55°$, $B = 145°$
6. 15.3 ft

TRIANGLES ▪ 72

2. **a.** 792   **b.** 4.63   **c.** 44,500
3. **a.** 795   **b.** 11.4   **c.** 248,000
4. A = 170,568 sq. ft, cost = $4606
5. A = 367.5 sq. yd, cost = $1286.25
6. 43.4 acres
7. $B = 79.6°$, $C = 68.2°$, $A = 32.2°$
8. 2.62 m
19. The four centers are at the same point.
20. The centroid, circumcenter, and orthocenter are collinear.
22. The orthocenter of a right triangle is at the right-angled vertex.

RIGHT TRIANGLES ▪ 77

2. **a.** $c = 542$ **b.** $b = 3.59$ **c.** $a = 36.6$
3. **a.** $x = 9.96$ **b.** $x = 6.72$ **c.** $x = 4.57$
4. 16.5 ft
5. 35.4 ft
6. 6.05 m
7. $x = 156.9$ ft, $y = 180.8$ ft, $z = 247.0$ ft

8. $x = 36.0$ ft
9. $x = 15.5$ ft

EXERCISES AND PROJECTS ■ 84

45. 57.7 cm

Chapter 4

THE SQUARE ■ 97

2. a. Perimeter = 14 in., area = 12.25 sq. in.
   b. Perimeter = 309.2 cm, area = 5975 sq. cm
3. a. Side = 7.46 cm, perimeter = 29.8 cm
   b. Side = 13.6 ft, perimeter = 54.4 ft
4. Common ratio = 2

THE RECTANGLE ■ 100

2. a. Perimeter = 28 units,
      area = 45 square units
   b. Perimeter = 23.1 units,
      area = 29.7 square units
   c. Perimeter = 2260 units,
      area = 245,241 square units
3. 1024 tiles
4. $11,886
5. $2605.50
6. 24,893 sq. in.
10. a. 3.23   **b.** 3.59

OTHER QUADRILATERALS ■ 102

2. a. Perimeter = 7.40 units,
      area = 2.66 square units
   b. Perimeter = 96.9 units,
      area = 470.5 square units
   c. Perimeter = 964 units,
      area = 51,798 square units

Chapter 5

POLYGONS ■ 134

2. a. 98.9°   **b.** 142.4°
3. a. 99.7°   **b.** 54.6°
4. a. $x = 10.26$ in., $y = 12.45$ in.   **b.** 84.6 sq. in.
5. a. $x = 43.4$ cm, $y = 57.7$ cm   **b.** 4740 sq. cm
6. 71.7°
7. 75.5°
8. 11.0 ft
9. 1100 sq. ft
10. 66.0 sq. ft
12. Subdivide the perimeter into nine equal lengths
   and connect the ends of each to the center.

TILINGS WITH POLYGONS ■ 159

2. Regular polygons only; congruent polygons
   only; must be edge-to-edge tilings; only one
   kind of regular polygon
3. The first three restrictions listed in Exercise 2
   apply.
4. Edge-to-edge tilings only; mark the tiles so that
   two tiles can touch in only one way; no spaces
   or overlaps
5. Equilateral triangles, squares, regular hexagons
9. a.

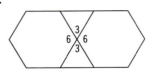

b.

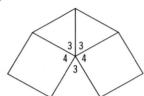

c.

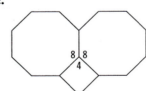

d.

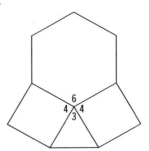

EXERCISES AND PROJECTS ■ 160

3. $1, \sqrt{2} + 1, (\sqrt{2} + 1)^2, (\sqrt{2} + 1)^3 \ldots$
35. $OA = \Phi$

Chapter 6

GEOMETRY OF THE CIRCLE ■ 171

3. 36.6 in.
4. 209 cm
5. 22.1 m
6. 61.1 in.
7. 6165 sq. in.
8. 2445 mm^2
9. 5.72 sq. ft
10. 4.51 cm
11. 96.6 in.
12. 1.13 m
13. 20.8 ft
14. 26.1 sq. ft
15. 313 sq. in.
16. 17.8 ft
17. a. 516 cm   **b.** 17,928 sq. cm
18. 10.2 m^2
19. Circ. = 15.2 m, area = 18.5 sq. m
20. 5.29
21. 74.8 ft

RADIANS ■ 174

2. 0.824 rad
3. 64.3°
4. 162.8
5. 40.9
6. 33.7
7. 1127
8. It has no units of measure.
9. Pi (π), specific gravity, trigonometric ratios, gear ratios, etc.
10. a. 1.06 rad   **b.** 2.88 rad   **c.** 0.0974 rad
11. a. 112°   **b.** 334°   **c.** 50.6°
12. 18.5
13. 2921
14. 6.73 m

LINES INTERSECTING A CIRCLE ■ 177

2. 9.39
3. 138
4. 6.45 ft
5. 0.82 m
6. AB = 3.53 m, AD = 3.53 m, AC = 3.67 m
7. 3.52 ft

EXERCISES AND PROJECTS ■ 188

25. 1.62 m
27. 1.48 m^2
28. 21.6 sq. cm

29. 1127 sq. cm
30. 245 sq. in.
31. 63.4 sq. ft
33. a. r = 2.28 m   **b.** s = 3.58 m
    c. Total area = 11.85 m^2
34. 1.48 m^2

Chapter 7

MORE CIRCLE GEOMETRY ■ 199

3. 118 cm
4. 0.846 ft
5. 1.85 ft
6. 2.64 m^2

EXERCISES AND PROJECTS ■ 219

49. 63.8 cm, 34.8 cm, 19.2 cm
52. a. 0.2887, 0.0962   **b.** 0.183

Chapter 8

THE GEOMETRY OF SQUARING THE CIRCLE ■ 233

4. b. 3.14142
5. 3.14154

EXERCISES AND PROJECTS ■ 241

20. Percent accuracy = 1.83%

Chapter 9

THE ELLIPSE ■ 252

4. 12.65 cm
8. 5.29 ft

GRAPHING THE ELLIPSE ■ 257

2.

3. $A(-2, 2.4)$; $B(-0.9, 1.75)$; $C(1.25, 1.9)$;
    $D(2, 1)$; $E(-1.8, -0.7)$; $F(-1.4, -1.4)$;
    $G(1.4, -0.6)$; $H(2.5, -1.9)$

4.

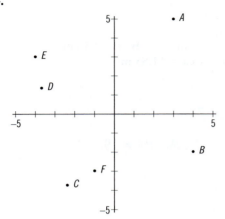

7.

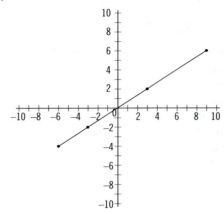

5.

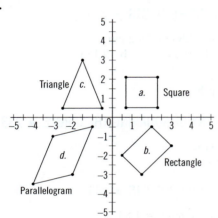

8.

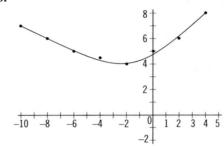

6.

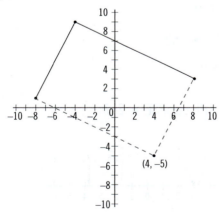

9.

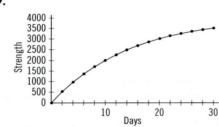

10. a.

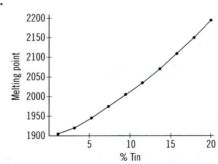

b. 2090° F   **c.** 19% tin

11. a. At $x = 2.5, y = 8.5$

x	y
−3	−8
−2	−5
−1	−2
0	1
1	4
2	7
3	10

d. At $x = 2.34, y = -1.48$

x	y
−3	−5
−2	0
−1	3
0	4
1	3
2	0
3	−5

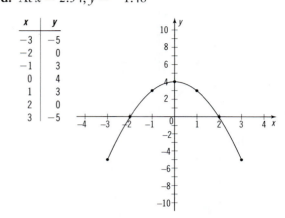

b. At $x = 1.75, y = 1.5$

x	y
−3	−8
−2	−6
−1	−4
0	−2
1	0
2	2
3	4

12. a. $\dfrac{x^2}{20.3} + \dfrac{y^2}{5.29} = 1$ **c.** $y = 2.17$

14. 33.1 ft, 43.3 ft, 48.4 ft, 50 ft

15. 9.72 ft

16. 479 ft

17. a. 5.27 ft   **b.** 8 ft

18. 10.2 ft

19. 7.79 ft

20. 12 ft

21. 200 ft

22. 7.19 ft, 28.8 ft, 64.7 ft

THE SPIRAL ■ 264

2.

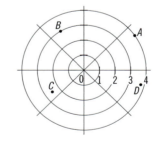

c. At $x = 1.33, y = -1.23$

x	y
−3	6
−2	1
−1	−2
0	−3
1	−2
2	1
3	6

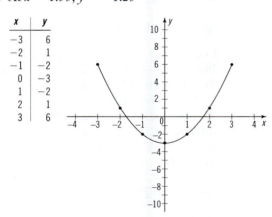

3. a.

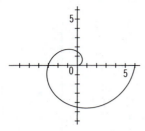

b.

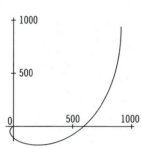

c.

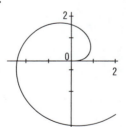

d.

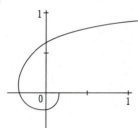

e.

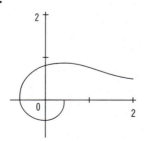

Chapter 11

THE SPHERE ■ 322

2. **a.** $V = 1,725,000,000$ cubic in.,
$SA = 6,956,000$ sq. in.
   b. $V = 15.6$ cubic m, $SA = 30.2$ sq. m
3. $r = 1.91$ cm, $V = 29.3$ cubic cm
4. $r = 4.796$ ft, $SA = 289$ sq. ft
5. 818 cubic in.
13. 308 sq. in.
14. 851 cubic cm
15. 1,095,000 cubic ft
16. 1,515,000 cubic ft

THE TERRESTRIAL SPHERE ■ 332

4. 3788 mi
5. 1161 mi

TIME ■ 342

2. 2:40 P.M.
3. 2:56 P.M.
4. 1:12 P.M.

Chapter 13

DIMENSION ■ 399

3. **a.** 1   **b.** 1
4. **a.** 1.585   **b.** 1.8928
   c. 1.262   **d.** 2.3219

SELF-SIMILARITY ■ 401

2. 2
3. **b.** 6   **c.** 9, 36
4. **b.** 8   **c.** 9, 64
5.

Stage	0	1	2	3	4
Magnification Factor	1	2	4	8	16
Number of Triangles	1	3	9	27	81

6.

Stage	0	1	2	3	4
Magnification Factor	1	3	9	27	81
Number of Squares	1	8	64	512	4096

Index

Page numbers in *italics* refer to artwork and architecture illustrations.

Photo Credits

8.21 ©Kavaler/Art Resource, NY; **8.24** ©The J. Paul Getty Museum, Los Angeles; **8.25** ©Steven Lunetta Photography; **8.26, 8.27** ©Erich Lessing/Art Resource, NY

Chapter 9

9.1 Réunion des Musées Nationaux/Art Resource, NY; **9.10** ©Steven Lunetta Photography; **9.11** ©Jo Ann Snover; **9.55** Image from *Illustrated Dictionary of Historic Architecture*, edited by Cyril M. Harris. Used with permission from Dover Publications.; **9.59a** ©George Bain, *Celtic Art* (Dover Publications, Inc.); **9.59b, 9.59d** ©Steven Lunetta Photography; **9.59e** ©Jastow, 2006. Gift of F. Doistau, 1919.; **9.64c** ©Steven Lunetta Photography; **9.64d** ©Aaron Kohri/Stockphoto.com; **9.66** From Sebastiano Serlio, *The Five Books of Architecture*; **9.67** From Leon Battista Alberti, *The Ten Books of Architecture*; **9.68** From Andrea Palladio, *The Four Books of Architecture*; **9.70** ©Steven Lunetta Photography

Chapter 10

10.1 ©Erich Lessing/Art Resource, NY; **10.8** ©Cyril M. Harris, *Illustrated Dictionary of Historic Architecture*. Reprinted by permission of Dover Publications, Inc.; **10.40** ©Erich Lessing/Art Resource, NY; **10.47** ©Scala/Art Resource, NY; **10.49** ©Landschaftsver band Rheinlan; **10.50** ©Cameraphoto/Art Resource, NY; **10.51** ©Giraudon/Art Resource, NY; **10.53** ©Steven Lunetta Photography; **10.54** ©2005 The M. C. Escher Company—Holland. All rights reserved. www.mcescher.com; **10.55** ©CNAC/MNAM/Dist. Réunion des Musées Nationaux/Art Resource, NY and Artist Rights Society, NY; **10.56** ©Francis G. Mayer/Corbis; ©Salvador Dali, Gala-Salvador Dali Foundation/Artists Rights Society (ARS), New York; **10.57** ©Tibor Bognar/Corbis; **10.59** Courtesy of Fondazione Lucio Saffaro, Bologna; **10.60** ©2005 The M. C. Escher Company—Holland. All rights reserved. www.mcescher.com

Chapter 11

11.1 The Metropolitan Museum of Art, Robert Lehman Collection, 1975 (1975.1.31). Image © The Metropolitan Museum of Art; **11. 8** Image from *Garner's Art Through the Ages*, 9/e, De la Croix, Tansey, Kirkpatrick, eds. Copyright © 1991, 1986, 1980, 1975, 1970, 1959. Copyright 1948, 1936, 1926 by Harcourt Brace Jovanovich, Inc. Copyright renewed 1954 by Louise Gardner. Copyright renewed 1964, 1976, 1987 by Harcourt Brace Jovanovich, Inc. Reprinted by permission.; **11.11** ©Adam Korzekwa; **11.12** ©Murat Taner/zefa/Corbis; **11.13** ©Scala/Art Resource, NY; **11.14** ©Steven Lunetta Photography; **11.15** ©Brian A. Vikander/Corbis; **11.16** ©Galen Rowell/Corbis; **11.18** ©Tate Gallery, London/Art Resource, NY; **11.20** ©2007 Greaves & Thomas; www.globemakers.com; **11.32** ©Lawrence Manning/Corbis; **11.36** ©Steven Lunetta Photography; **11.40** ©Ufficio Turstico Comune di Parma; **11.43** ©Erich Lessing/Art Resource, NY; **11.45** ©Réunion des Musées Nationaux/Art Resource, NY; **11.46** Image scanned from *The Complete Engravings, Etchings & Drypoints of Albrecht Dürer*, Walter Strauss, ed. New York: Dover 1972, Plate 7, page 17.; **11.47** The Samuel Courtauld Trust, Courtauld Institute of Art Gallery, London; **11.48** ©Cameraphoto/Art Resource, NY; **11.49, 11.51** ©Erich Lessing/Art Resource, NY; **11.52** ©Scala/Art Resource, NY; **11.53** ©Erich Lessing/Art Resource, NY; **11.54** ©Erich Lessing/Art Resource, NY;

11.55 ©Robert Holmes/CORBIS; **11.57, 11.58, 11.59** ©2005 The M. C. Escher Company—Holland. All rights reserved; www.mcescher.com; **11.60** ©Robert Morris via ARS-NY; **11.63** ©Image courtesy of Robert de Andrade Martins; **11.65** ©Tate Gallery, London/Art Resource, NY; **11.67** © The British Library Board. All rights reserved; **11.69** ©Fotgraphia Vasari; **11.73** Courtesy of NASA; **11.75** ©National Gallery Collection, by kind permission of the Trustees of the National Gallery, London/CORBIS; **11.76** ©Digital image, The Museum of Modern Art/Licensed by Scala/Art Resource, NY

Chapter 12

12.10 ©The J. Paul Getty Museum, Los Angeles; **12.23** ©Werner Forman/CORBIS; **12.25** ©Scala/Art Resource, NY; **12.26** ©Steven Lunetta Photography; **12.28** ©Erich Lessing/Art Resource, NY; **12.34, 12.35** ©Scala/Art Resource, NY; **12.37** ©Erich Lessing/Art Resource, NY; **12.38** ©Scala/Art Resource, NY; **12.40, 12.41** ©Alinari/Art Resource, NY; **12.42, 12.43, 12.44** ©Scala/Art Resource, NY; **12.45** ©Alinari/Art Resource, NY; **12.46** ©Réunion des Musées Nationaux/Art Resource, NY; **12.50** ©Burstein Collection/CORBIS; **12.51** ©The Museum of Modern Art/Licensed by Scala/Art Resource, NY; **12.52** Digital Image ©2005 The M. C. Escher Company—Holland; www.mcescher.com; **12.53, 12.54** ©2005 The M. C. Escher Company—Holland; www.mcescher.com

Chapter 13

13.1 ©Erich Lessing/Art Resource, NY; **13.2** ©2007. Used Courtesy of Benoit Mandelbrot; **13.15** ©2007, Steven Lunetta Photography; **13.23** ©Richard Dion Wilson/iStockphoto.com; **13.32** David E. Joyce, ©1994; **13.49** ©Christie's Images/CORBIS; **13.50, 13.51, 13.52** M. C. Escher's "Study Design Pulpit Ravello Cathedral" ©2007 The M. C. Escher Company—Holland. All rights reserved; www.mcescher.com; **13.53** ©2006 Dr. Susanne Krömker; **13.54** ©Tom Verhoeff. Sculpture designed by Koos Verhoeff and produced in bronze by Anton Bakker and Kevin Gallup. Used with permission.; **13.56** ©2006 B. B. Mandelbrot, from *The Fractal Geometry of Nature*, by B. B. Mandelbrot

Appendix E

E.1 ©Erich Lessing/Art Resource, NY; **E.3** Victoria & Albert Museum, London/Art Resource, NY; **E.7** ©Scala/Art Resource, NY; **E.8** ©Erich Lessing/Art Resource, NY; **E.9** Giraudon/Art Resource, NY; **E.11** ©Scala/Art Resource, NY; **E.13** Reproduced from *Symbols, Signs, and Signets* by Ernst Lehner, published by Dover Publications.; **E.15** ©Scala/Art Resource, NY; **E.18** ©Steven Lunetta Photography; **E.21** ©Scala/Art Resource, NY; **E.23** Follower of Laurent de La Hyre (French), *Allegory of Geometry*, after 1649, oil on canvas, 40×62 (101.6 $\times$ 158.6 cm) Toledo Museum of Art. Purchased with funds from the Libbey Endowment. Gift of Edward Drummond Libbey, 1964.124. © 2006; **E.26, E.28** ©Scala/Art Resource, NY; **E.29** ©Erich Lessing/Art Resource, NY; **E.30** ©Scala/Art Resource, NY; **E.32** ©Steven Lunetta Photography; **E.33** ©Scala/Art Resource, NY; **E.34** ©HIP/Art Resource, NY; **E.38** "Saturno Devorando a uno de sus Hijos" by Goya. ©Museo del Prado. Used with permission.; **E.39** ©Scala/Art Resource, NY; **E.44** ©Steven Lunetta Photography; **E.45** ©Francis G. Mayer/Corbis